Watershed Restoration: Principles and Practices

Funding for the publication of this book
was provided by the

National Fish and Wildlife Foundation

Plum Creek Foundation

Trout Unlimited

**U.S. Department of the Interior
Bureau of Land Management
Bureau of Reclamation**

**U.S. Department of Agriculture
Forest Service**

Watershed Restoration: Principles and Practices

Edited by

Jack E. Williams
U.S. Department of the Interior
Bureau of Land Management
Christopher A. Wood
U.S. Department of Agriculture
Forest Service
Michael P. Dombeck
U.S. Department of Agriculture
Forest Service

Bethesda, Maryland
1997

Suggested citation formats follow.

Entire book

Williams, J. E., C. A. Wood, and M. P. Dombeck, editors. 1997. Watershed restoration: principles and practices. American Fisheries Society, Bethesda, Maryland.

Chapter within the book

Preister, K., and J. A. Kent. 1997. Social ecology: a new pathway to watershed restoration. Pages 28–48 *in* J. E. Williams, C. A. Wood, and M. P. Dombeck, editors. Watershed restoration: principles and practices. American Fisheries Society, Bethesda, Maryland.

Cover artwork by Monte Dolack from an original painting of Montana's Blackfoot River commissioned by The Nature Conservancy.

Library of Congress Catalog Card Number: 97-61098

ISBN: 1-888569-04-2 (hard cover)

ISBN: 1-888569-05-0 (soft cover)

Printed in the United States of America on recycled, acid-free paper.

American Fisheries Society
5410 Grosvenor Lane, Suite 110
Bethesda, Maryland 20814-2199, USA

Contents

v

PART 3: KEY PRACTICES

PART 4: CASE STUDIES

PART 5: A VISION FOR THE FUTURE

Contributors

Gary Aitken (Chapter 23): Big Blackfoot Chapter, Trout Unlimited, 386 Klein-schmidten Flat Road, Ovando, Montana 59854, USA.

Paul L. Angermeier (Chapter 4): U.S. Geological Survey, Biological Resources Division, Department of Fisheries and Wildlife Sciences, Virginia Polytechnic Institute and State University, Blacksburg, Virginia 24061, USA.

D. Albrey Arrington (Chapter 24): South Florida Water Management District, Post Office Box 24680, West Palm Beach, Florida 33416, USA.

Glenn Begue (Chapter 24): U.S. Army Corps of Engineers, Post Office Box 4970, Jacksonville, Florida 32232, USA.

David L. Bowling, Jr. (Chapter 17): Tennessee Valley Authority, Aquatic Biology Laboratory, Highway 441, Norris, Tennessee 37828, USA.

John Cairns, Jr. (Chapter 28): Department of Biology, Virginia Polytechnic Institute and State University, Blacksburg, Virginia 24061, USA.

Terry S. Chilcoat (Chapter 17): Tennessee Valley Authority, Aquatic Biology Laboratory, Highway 441, Norris, Tennessee 37828, USA.

Joe P. Colletti (Chapter 19): Department of Forestry, Iowa State University, Ames, Iowa 50011, USA.

Jock Conyngham (Chapter 22): Trout Unlimited National, HC1, Box 1835, Starlight, Pennsylvania 18461, USA.

Janice P. Cox (Chapter 17): Tennessee Valley Authority, 1101 Market Street, Chattanooga, Tennessee 37402, USA.

Jim Cummins (Chapter 18): Interstate Commission on the Potomac River Basin, 6110 Executive Boulevard, Suite 300, Rockville, Maryland 20852, USA.

Michael P. Dombeck (Chapters 1, 2, and 25): U.S. Bureau of Land Management, 1849 C Street, N.W., Washington, D.C. 20240, USA. *Present address*: Office of the Chief, U.S. Forest Service, Post Office Box 96090, Washington, D.C. 20090, USA.

Bob Doppelt (Chapter 27): Pacific Rivers Council, Post Office Box 10798, Eugene, Oregon 97440, USA. *Present address*: Center for Watershed and Community Health, Post Office Box 10933, Eugene, Oregon 97440, USA.

Fred H. Everest (Chapter 20): U.S. Forest Service, Pacific Northwest Research Station, 2770 Sherwood Lane, Suite 2, Juneau, Alaska 99801, USA.

Harv Forsgren (Chapter 26): U.S. Forest Service, Fish, Wildlife and Rare Plants, Post Office Box 96090, Washington, D.C. 20090, USA.

Christopher A. Frissell (Chapter 7): Flathead Lake Biological Station, University of Montana, 311 Bio Station Lane, Polson, Montana 59860, USA.

Carol B. Griffin (Chapter 16): Henry's Fork Foundation, Post Office Box 852, Ashton, Idaho 83420, USA. *Present address*: Natural Resources Program, Grand Valley State College, Allendale, Michigan 49401, USA.

Laura A. Gutzwiller (Chapter 21): U.S. Fish and Wildlife Service, 5100 E. Winnemucca Boulevard, Winnemucca, Nevada 89445, USA.

James R. Hagerman (Chapter 17): Tennessee Valley Authority, Aquatic Biology Laboratory, Highway 441, Norris, Tennessee 37828, USA.

Bruce E. Hansen (Chapter 20): U.S. Forest Service, Pacific Northwest Research Station, 3200 S.W. Jefferson Way, Corvallis, Oregon 97331, USA.

Paul A. Heikkila (Chapter 15): Oregon State University Extension Service, 290 N. Central, Coquille, Oregon 97423, USA.

Tracii L. Hickman (Chapter 20): U.S. Forest Service, Mt. Hood National Forest, 595 N.W. Industrial Way, Estacada, Oregon 97023, USA.

David B. Hohler (Chapter 20): U.S. Forest Service, Deschutes National Forest, 1645 Highway 20 East, Bend, Oregon 97701, USA.

William F. Hudson (Chapter 15): U.S. Bureau of Land Management, Coos Bay District Office, 1300 Airport Lane, North Bend, Oregon 97459, USA.

Tom Isenhart (Chapter 19): Department of Forestry, Iowa State University, Ames, Iowa 50011, USA.

James A. Kent (Chapter 3): James Kent Associates, AABC Building 20, Suite E, Aspen, Colorado 81611, USA.

Jeffrey L. Kershner (Chapter 8): U.S. Forest Service, Department of Fisheries and Wildlife, Utah State University, Logan, Utah 84322, USA.

Kenneth D. Kimball (Chapter 12): Appalachian Mountain Club, Post Office Box 298, Gorham, New Hampshire 03581, USA.

Joseph M. McGurrin (Chapters 22 and 26): Trout Unlimited, 1500 Wilson Boulevard, Suite 310, Arlington, Virginia 22209, USA.

Randy M. McNatt (Chapter 21): U.S. Bureau of Land Management, 850 Harvard Way, Reno, Nevada 89502, USA.

David A. Nolte (Chapter 13): Trout Unlimited, 6322 N.W. Atkinson Avenue, Redmond, Oregon 97756, USA.

Edwin Philip Pister (Chapter 2): Desert Fishes Council, Post Office Box 337, Bishop, California 93615, USA.

Kevin Preister (Chapter 3): The Rogue Institute for Ecology and Economy, 762 A Street, Ashland, Oregon 97520, USA. *Present address*: Social Ecology Associates, 163 Beacon, Post Office Box 3493, Ashland, Oregon 97520, USA.

Roy D. Price (Chapter 21): U.S. Bureau of Land Management, Elko District Office, Post Office Box 831, Elko, Nevada 89803, USA.

Gordon H. Reeves (Chapter 20): U.S. Forest Service, Pacific Northwest Research Station, 3200 S.W. Jefferson Way, Corvallis, Oregon 97331, USA.

Richard C. Schultz (Chapter 19): Department of Forestry, Iowa State University, Ames, Iowa 50011, USA.

James R. Sedell (Chapter 20): U.S. Forest Service, Pacific Northwest Research Station, 3200 S.W. Jefferson Way, Corvallis, Oregon 97331, USA.

David Shepp (Chapter 18): Metropolitan Washington Council of Governments, 777 N. Capitol, Suite 300, Washington, D.C. 20002, USA.

Daniel Shively (Chapter 20): U.S. Forest Service, Mt. Hood National Forest, 595 N.W. Industrial Way, Estacada, Oregon 97023, USA.

Jack Ward Thomas (Chapter 9): U.S. Forest Service, Post Office Box 96090, Washington, D.C. 20090, USA. *Present address*: School of Forestry, University of Montana, Missoula, Montana 59812, USA.

Whitney Tilt (Chapter 10): National Fish and Wildlife Foundation, 1120 Connecticut Avenue, N.W., Suite 900, Washington, D.C. 20036, USA.

Louis A. Toth (Chapter 24): South Florida Water Management District, Post Office Box 24680, West Palm Beach, Florida 33416, USA.

William M. Turner (Chapter 11): Missouri Department of Conservation, 1014 Thompson Boulevard, Sedalia, Missouri 65301, USA.

Christopher D. Ungate (Chapter 17): Tennessee Valley Authority, 400 West Summit Hill Drive, Knoxville, Tennessee 37902, USA.

Robert W. Van Kirk (Chapter 16): Henry's Fork Foundation, Post Office Box 852, Ashton, Idaho 83420, USA.

Cindy A. Williams (Chapter 10): U.S. Forest Service, Fisheries, Wildlife and Range, 1720 Peachtree Road, N.W., Atlanta, Georgia 30367, USA. *Present address*: U.S. Fish and Wildlife Service, National Education Training Center, Route 3, Box 49, Kearneysville, West Virginia 25430, USA.

Gary G. Williams (Chapter 17): Tennessee Valley Authority, Aquatic Biology Laboratory, Highway 441, Norris, Tennessee 37828, USA.

Jack E. Williams (Chapters 1 and 25): U.S. Bureau of Land Management, 1387 S. Vinnell Way, Boise, Idaho 83709, USA.

Robert C. Wissmar (Chapter 5): School of Fisheries, University of Washington, Seattle, Washington 98195, USA.

Christopher A. Wood (Chapters 1, 9, and 25): U.S. Bureau of Land Management, 1849 C Street, N.W., Washington, D.C. 20240, USA. *Present address*: Office of the Chief, U.S. Forest Service, Post Office Box 96090, Washington, D.C. 20090, USA.

Robert R. Ziemer (Chapter 6): U.S. Forest Service, Pacific Southwest Research Station, 1700 Bayview Drive, Arcata, California 95521, USA.

Seth Zuckerman (Chapter 14): Mattole Restoration Council, Post Office Box 160, Petrolia, California 95558, USA.

Foreword

Like every reader, I approached this book from my own personal and professional experience which, most recently, meant unleashing a flood of biblical proportions in the Grand Canyon. The flood began on 18 March 1996 as I stood on a catwalk in front of the Glen Canyon Dam, turning a valve that opened the jet tubes, allowing water to surge through the dam and down into the Colorado River at 45,000 cubic feet per second. I watched in wonder as the river surged, cascading a fountain of mist hundreds of feet in the air and flooding the entire length of the canyon.

Our purpose in flooding the canyon was to restore the beaches and fish-spawning areas that had been eaten away by ruptured, artificial river flows that had fluctuated erratically from day to day and week to week in response to the electrical power demand from cities as distant as Phoenix and Salt Lake City.

The Glen Canyon Dam, one of two great keystone structures on the Colorado River, was built in the early 1960s. In that era—prior to federal environmental laws such as the Endangered Species Act and the National Environmental Policy Act—no one, not even the U.S. National Park Service, paused to consider, much less analyze, how those hydropower-driven fluctuations might affect the downstream habitat of one of the world's great national parks.

That lack of consideration was not exclusive to Glen Canyon Dam. Until very recently we have lived on and used the land and its forests and rivers as if the landscape were merely an assemblage of unrelated parts, each to be used, removed, or substituted without regard to the others.

Today we know better. With the insight of modern ecological science, we have come to understand that neither the Grand Canyon nor any other protected place is an island unto itself; that every part of the landscape is tributary to the whole; and that the operation of Glen Canyon Dam has consequences for the entire watershed.

Grasping this connection within watersheds is one thing, but making it a basis for policy is something else entirely. It involves taking the many frayed ethical and ecological strands that were torn loose by the dam and weaving them back together into one cohesive fabric.

Of course, the people who weave those ecological strands back together need not be the Secretary of the Interior acting at the Grand Canyon, nor the President of the United States acting to protect and restore the pristine watershed of the Yellowstone River. It need only be citizens, male or female, rich or poor, old or young, and of any race or creed in America who care deeply about the watershed in which they live. And those citizens, in turn, need only to understand both the natural and human laws, resources, and tools required to restore it.

This book is their blueprint.

As the authors show, watershed restoration has already begun in a nationwide movement that varies in scope and degree as much as the contours of the landscape itself. But despite the variations, in each case we find Americans who no longer view their land use decisions as separate or separable from those of others. Instead they look at the broader picture and see their actions and those of their neighbors as occurring within the same natural region, most often clearly defined by the geologic boundaries of their local watershed. A central theme of this book is that local watersheds embrace not just natural resources, but also social and economic cultures that are connected to and dependent on those watersheds.

The human confluence, the coming together of local citizens to resolve problems within their watersheds, is another critical theme. This book showcases restoration projects in which communities—often strangers—come together to restore their waters, and the process of restoration rebuilds the communities themselves. Most of these efforts have developed locally: all involve local landowners, farmers, and ranchers working in partnership with anglers, scientists, environmentalists, government agencies, and a host of local citizens to restore their lands. The process calls to mind the words of cultural anthropologist Margaret Mead: "Never doubt that a small group of thoughtful, committed citizens can change the world; indeed, it's the only thing that ever has."

Across the country people are awakening to the idea of using our laws, our voices, and our backs in the cause of watershed restoration. It is not enough to simply stop the decline—we must reverse it; not enough to preserve the isolated parts—we must reconnect entire landscapes and watersheds; and not enough to fence off the local greenway or trickling neighborhood stream—we must unite them with public lands, national forests, state and national parks, and the broad ocean-bound rivers.

Another core theme is the forging of sound scientific knowledge with the energy of local communities and watershed-based coalitions. I know the restoration of the Grand Canyon would never even have begun were it not for the careful and steady application of science, the many extensive comments of stakeholders, and the emerging and consensus-building catalyst of geographic information systems. The combination of sound science and community-based partnerships is a potent force for environmental restoration. This book brings these two forces together by providing science-based chapters on topics such as biological integrity, watershed analysis, and monitoring with chapters on social ecology and building local coalitions to foster restoration.

Watersheds, for all their complexity, perform three very basic and important functions. They catch, store, and safely release water over time. Every one of us lives in a watershed. All our actions, whether grazing livestock on a western range or applying herbicide to a neighborhood lawn, affect a watershed's ability to perform these basic functions.

The outcomes of all human activity within a watershed are accumulated by runoff across the landscape and eventually funnel into our rivers. Thus, individual human activities coalesce to determine the quality of downstream resources. This makes us ask: Into what stream does the storm sewer at the end of the road drain, and then into what larger river? How do the many miles of roads in a watershed disrupt surface runoff and affect stream flow? The quality and productivity of our rivers truly reflect the health of their watersheds.

Unfortunately, many of our nation's watersheds and their rivers are degraded. In my home state of Arizona, 31 of our 33 native fishes are threatened with extinction or soon will be at risk. Across the country, our vital riparian resources—those ribbons of green along our streams and rivers that are critically important to fish and wildlife, and for maintaining water quality—have been eroded, trampled, and abused. If we do not change our ways, we will leave a legacy of lost resources: eroded topsoil, rangelands dominated by exotic weeds, dry gullies where streams should be, and long lists of endangered or extinct species. The time is at hand to begin the restoration of our natural heritage, to begin to live within the limits of the land.

Despite the magnitude of the task before us, we must not become frustrated. The movement toward sustainability takes time. Persistence and determination will help to broaden our horizons and diversify our allies. Charles Wilkinson of the University of Colorado has said that "it should not be so hard to mesh the needs of the lands and waters and the people . . . in the last analysis, they are the same."

We all want healthy watersheds, open space, productive soils, and a good diversity of plants and animals. As we begin to restore our rivers and riparian areas, we must look at the broader picture. We must understand what is happening upstream of us and be aware of how our actions affect those downstream. From Oregon's Crooked River to New York's Beaverkill have come stories of dedication and persistence, of combining our best available scientific knowledge with the hard work of local citizens to successfully restore our landscape. And, the efforts of these local coalitions are working: they are making strong contributions to a better future for those of us here today and for generations that will follow.

This book is a call to action. We must work together to restore the integrity of our landscape and of our native fishes, riparian areas, forests, and streams. The health of our watersheds depends upon developing a proper land ethic, upon conducting ourselves in a manner consistent with the principles of that ethic, and upon correcting damage from past actions. The challenge is not an easy one, but one that is worthy of our best fight. Within this remarkable book are all the tools needed to accomplish one of the most important tasks facing our society: the healing of the land for the use and benefit of present and future generations.

Bruce E. Babbitt
Secretary of the Interior
Washington, D.C.

Preface

Society is poised to restore the health of our rivers and watersheds. The authors and editors of this book see an emerging culture and ethic that will embrace this restoration. The timing could not be more critical, for despite strong environmental regulation on many fronts, human demand on our natural resources continues to grow. Whether the issue is the decline of salmon stocks (local populations) in an Oregon river, or the nonpoint-source pollution of a midwestern stream, or the loss of wetlands in Florida's Everglades, we are increasingly reminded of the need to live within the limits of the land.

A restoration ethic must be nurtured if future generations are to enjoy the ecological essentials, such as clean water, productive soils, and diverse fish and wildlife populations. These sustain us daily and give us pleasure and sustenance, but we largely take them for granted. The chapters in this volume can facilitate the transition from a culture bent on resource extraction to one that embraces a sustainable resource ethic. The transition will not be easy, but it is of enormous consequence and worthy of our best effort.

The impetus for this volume largely comes from the continuing decline in the status of our aquatic resources and the ineffectiveness of many past efforts to reverse this disturbing trend. The scientific community has realized that we must focus our actions at a broader geographic scale to correct many environmental problems. At the same time, concerned citizens, as demonstrated by the case studies herein, increasingly demand that all parties work together to begin healing our lands and waters.

We must restore the health of watersheds to recover the endangered fishes within lakes and rivers, to secure clean drinking water, and to sustain productive crops and forage—in short, to ensure that society continues to derive the many benefits of a healthy environment. Therefore, the approach advocated in this volume is not one of engineered fixes at specific sites, but of broadly applying sound restoration principles at watershed scale through community and agency coalitions.

The need is scientifically clear for an approach to environmental problems that is ecologically sound, that views the entire watershed and not just portions of it, and that is long term. Many of the same problems that stimulated scientists to call for a new approach to conservation—problems such as degraded water quality and declining fish populations—have triggered local communities into action. Local problems are being solved through grassroots community stewardship. Across the nation, coalitions of people having diverse interests and perspectives are forming to address common problems.

The names of these coalitions reflect a widespread geography, but a common cause: The Blackfoot Challenge, Henry's Fork Coalition, Trout Creek Mountain Working Group, Mattole Restoration Council, Crooked River Ecosystem Education Council, and dozens more. The involvement of local citizens and communities in restoration is valuable, because it not only brings more resources to combat the problem of watershed degradation, but also forges a stronger connection between people and the land they live on. This invokes an attitude of caring and compassion for both land and community.

Applying sound principles of ecological and social science is critical for the long-term success of watershed coalitions. Equally important is our ability to learn from the experience of this initial wave of grassroots restoration. However, many restoration projects lack long-term monitoring data and evaluation, and this hinders our effort to identify truly successful projects and to determine which ones should serve as models. Adaptive management is the key to resolving these problems: we must establish clear goals, monitor our actions, evaluate their effectiveness, and adjust our methods to achieve restoration goals.

The complexity of watersheds and human communities means that no single ecological or social template can ensure successful restoration of all watersheds. Nonetheless, in reading the manuscripts that constitute this volume, we are struck by the abundance of sound scientific knowledge and community support available to assist practitioners in broad-scale restoration. The case studies are a mere sampling of the many worthwhile restorations underway across the country. More case studies, monitoring, and evaluations are badly needed. Yet, society cannot afford to await the perfect restoration model. Such delay only ensures continuing degradation and a diminished standard of living for future generations.

Although diverse interests within watersheds have been adversaries in the past, an encouraging number are recognizing the value of working together to combat the common problems of natural resource depletion that diminish our lives and those of our children and grandchildren. By documenting some of this knowledge and experience, we hope that society will begin to realize the benefits of restoration and the consequences of inaction. Prosperous communities depend on healthy watersheds, and healthy watersheds promote strong, sustainable economies. Such basic insights are a clarion call to action for us all.

This book's objective is to provide information necessary for initiating watershed restoration and for improving its success. To accomplish this, we present a philosophical and scientific framework for restoration, 13 watershed-scale case studies as examples and learning experiences, and critiques and comparisons of the case studies. The case studies, from varied U.S. geographic regions, provide a broad cross section of watershed-scale restoration projects in progress. The watersheds discussed are shown on the map on page *xix*. Our intent is to learn from these experiences, to foster a more scientifically rigorous approach to restoration, and to garner greater community support.

We have organized this book in five parts. Following a brief introduction that describes the need for watershed-scale restoration, Part 1 describes the principles of ethics, social science, and ecology and the adaptive management needed for successful restoration. In Part 2, the critical process of building partnerships and coalitions is described. Parts 3 and 4 build on the theme of diverse interests working together by presenting restoration case studies from around the country. The case studies in Part 3 focus on certain key restoration practices that are broadly applicable to other watersheds. The case studies in Part 4 describe

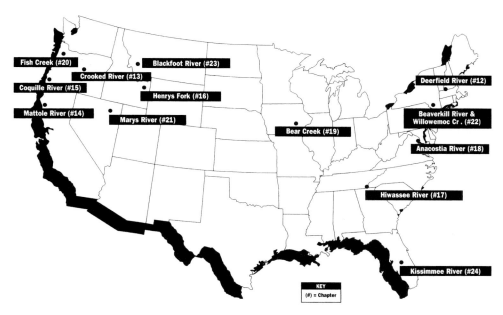

Locations of case study watersheds discussed in Chapters 12–24.

broad-scale restoration. Part 5 describes the key elements of successful restoration and presents a vision of our future based on the emerging restoration culture.

The authors of this collection represent state and federal agency scientists and managers, academicians, social scientists, conservationists, and anglers. They bring distinctive and invaluable perspectives to watershed restoration. We gave them considerable latitude in developing and writing their chapters, but we endeavored to make each chapter understandable to a broad readership. The diversity of chapter styles mirrors the diverse cross-disciplinary skills, backgrounds, and perspectives needed to restore the health of watersheds in which we live and work.

Jack E. Williams
Christopher A. Wood
Michael P. Dombeck

Acknowledgments

Writing and editing a book with some four dozen authors is both rewarding and challenging. We thank all the authors for their contributions. Many of them also kindly reviewed contributions from their colleagues.

In addition, we thank the following for their reviews of individual chapters: Alan Barta, Peter Bisson, Dan Bottom, Jan Brown, Noel Burkhead, Baird Callicott, Paul Calvert, Chris Campbell, Osborne Casey, James Clayton, Glenn Clemmer, Allen Coopereider, Walter Courtenay, Jr., Jim Decker, Joe Dillard, Don Duff, Julie Elfving, Martha Hahn, Eric Hammerling, Helen Hankins, Dave Heller, David Hohler, Robert House, Philip Hulbert, James Johnson, Russ Kraph, Russ LaFayette, Andy Martin, Pam McClelland, Cal McCluskey, Jack McIntyre, Gary Meffe, Robert Naiman, Willa Nehlsen, Charles Olchowski, Alan Olson, Kerry Overton, Frank Panek, Don Peters, Charles Pregler, Jill Silvey, Joel Snodgrass, Rick Swanson, Russ Thurow, Ron Wiley, and Cindy Deacon Williams.

Meggan Laxalt helped prepare several figures and the location maps that accompany the case studies. James Workman facilitated discussions with Interior Secretary Babbitt. The staff at Shenandoah National Park kindly provided facilities for the editors during a work session at Camp Hoover. We acknowledge the editorial contributions of Fred Schroyer and Eva Silverfine, and finally, we would like to thank the professional staff at the American Fisheries Society, particularly Beth Staehle, Robert Kendall, and Paul Brouha, for help in bringing this volume to completion.

Some specific acknowledgments for individual chapters are presented at the end of those chapters.

The acid test of our understanding is not whether we can take ecosystems to bits on pieces of paper, however scientifically, but whether we can put them together in practice and make them work.
—A. D. Bradshaw, 1983

CHAPTER 1

UNDERSTANDING WATERSHED-SCALE RESTORATION

Jack E. Williams, Christopher A. Wood, and Michael P. Dombeck

A quarter of a century ago, Professor Noel Hynes of the University of Waterloo in Ontario was among the first to describe how links between the soil and vegetation in a watershed combine with local climate to produce the physical structure and biological productivity of streams (Hynes 1975). Since Hynes's synthesis, the synergy between a river and its watershed has been described in various ways. At first, the river-watershed relation was viewed along the longitudinal gradient of a river from its headwaters to the ocean (called the river continuum concept). More recently, the connectivity of rivers has been viewed from upstream to downstream, from upslope to downslope, and from surface flows to subsurface (hyporheic) flows (Minshall et al. 1985; Naiman 1992a).

Our understanding of the dynamic nature of riverine ecosystems has evolved, so our ability to manage them should evolve as well. Successful management of aquatic and riparian ecosystems is predicated upon sound management of their watersheds. Whether we are dealing with a small headwater stream and its modest catchment (first order), or a large river system and its extensive basin (fourth or fifth order), society can derive the full benefit of healthy watersheds only if we understand how these systems function and how our activities are disrupting them.

Virtually all watersheds, except some smaller headwater catchments, have been modified and degraded by human development. A consequence has been a loss of products and functions provided by healthy watersheds. Under products we include consistent high-quality water and productive soils. Under functions we include the moderation of flood energy and drought and the maintenance of diverse plant and animal communities.

The present magnitude and geographic scope of degradation have made restoration a more important and more common component of watershed management. The increasing interest in restoration derives from a growing awareness that all of the goods, services, and values that society derives from the land ultimately depend on healthy, properly functioning watersheds. Legislation requiring rehabilitation of mine sites and other degraded areas is another motive in restoration. And specific restoration practices reflect the growing desire to use natural vegetation patterns and native species.

In this chapter, we define the concept of restoration and place it within the scale of the entire watershed. We then briefly describe why restoration at this scale is critical to reestablishment of healthy, functioning watersheds, including the many ecosystem services they provide and their complement of aquatic biological diversity.

Conducting successful restoration at the watershed scale is difficult. First, watersheds typically are disturbed by multiple factors. Tracking the cause of each disturbance and evaluating its cumulative effect is intricate and time-consuming. Second, addressing the scope and magnitude of watershed degradation requires substantial time, money, and energy, invested over years. Yet, as this book shows, the many community coalitions, foundations, local conservation society chapters, and landowner groups have formed to promote watershed restoration, thus demonstrating the keen interest in and need for restoration.

DEFINING RESTORATION

The terms restoration, reclamation, rehabilitation, and habitat creation often are misunderstood and misapplied. Reclamation and rehabilitation generally are more local and site-specific. Habitat creation refers to establishment of new habitat without regard to historic conditions.

But the term restoration, as used in this book, refers to a more holistic process than the others. *Restoration is reestablishment of the structure and function of an ecosystem, including its natural diversity* (Cairns 1988a; National Research Council [NRC] 1992a). In this definition, *structure* is an ecosystem's native species diversity. *Function* is an ecosystem's productivity (growth of plant biomass as the basis for food webs) and functions of hydrology, trophic structure, and transport. Conceptually, restoration may be viewed as reversing the decline of ecosystem health and returning a degraded ecosystem toward its historic structure and function (Figure 1.1).

A restored ecosystem is much more than just an assemblage of its components of soil, water, air, and biota. Rather, a restored ecosystem displays the interactions among all these components—interactions that are critical to healthy

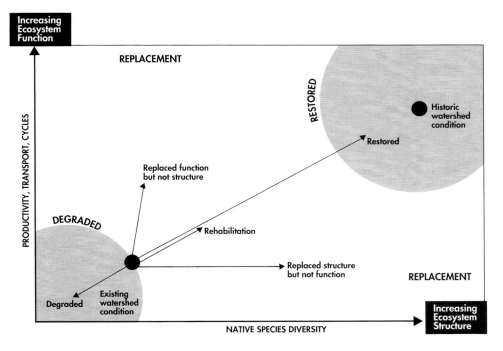

FIGURE 1.1.—A primary goal in restoration is to redirect the trajectory of a degraded ecosystem. This redirection applies both to its structure (species complexity) and function (biomass and nutrients). Trajectories away from its original structure and function result in further degradation, partial rehabilitation, or novel conditions. (Modified from Bradshaw 1984.)

ecosystem functioning. As pointed out by the NRC (1992a), merely recreating the form of an ecosystem without its functions, or creating the functions in some artificial configuration, does not constitute restoration.

Rehabilitation, reclamation, habitat creation, or mitigation can be achieved through manipulation of site-specific or isolated elements of ecosystems, but restoration is a more complex process. Successful restoration means that ecosystem structure (diversity) and function (e.g., plant productivity and hydrology) are recreated or repaired and that natural ecosystem processes can operate unimpeded. Our distinctions among these terms may seem subtle, but they are fundamental to defining and achieving successful restoration of healthy, productive watersheds.

People and their communities typically are integral components of the watersheds we seek to restore. A complete return of all conditions to entirely natural predisturbance levels, therefore, is impractical and unrealistic. Attempting to do so may jeopardize the broad community support necessary to restore watershed health. Many natural resource managers tend to disregard the human element of ecosystems, but doing so may confound our effort to institutionalize restoration within communities and agencies (Lee 1992). Thus, restoring the

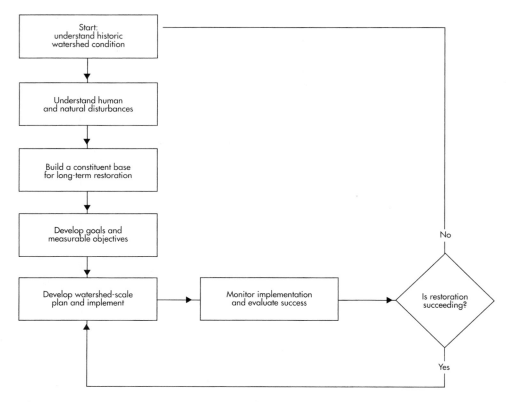

FIGURE 1.2.—Conceptual framework for watershed-scale restoration.

structural and functional components of an ecosystem rather than attempting to recreate pristine prehuman conditions provides a realistic alternative. Such restoration can reconcile the biodiversity goals of natural resource managers with the reality of human presence.

Ideally, watershed restoration is a rigorous, long-term, comprehensive, and adaptive process. Resource managers and scientists need to educate communities about the importance of healthy watersheds. They need to standardize a more scientific approach to restoration. They need to maximize learning from existing case studies. And they need to apply our new understanding through adaptive management (Walters 1986; Naiman et al. 1995).

Many important restoration research questions remain, and no detailed models of restored ecosystems exist (Naiman et al. 1995). Although no single procedure is appropriate for all watersheds, a conceptual framework for conducting watershed-scale restoration is outlined in Figure 1.2. The elements shown are developed and described in subsequent chapters.

Historically, the management of rivers and watersheds has applied simple solutions to complex problems. Often, surprisingly little consideration has been given to ecosystem structure and function, to the historical conditions of watersheds, or to an understanding of how human and natural disturbances shape each system. All of these factors are fundamental to achieving comprehensive and successful watershed-specific restoration. As an example, during the 1950s and 1960s log jams and woody debris were removed from Pacific coastal streams to improve fish passage. Subsequently, stream and watershed managers recognized the critical value of wood in streams as food for insects and as structure in creating pools. Thus, large quantities of wood have been reintroduced into coastal rivers (Maser and Sedell 1994).

As illustrated by their Flathead River basin experience in Montana, Stanford and Ward (1992) described how interactive and cumulative effects become seemingly untraceable and intractable without studies that produce specific knowledge of large and ecologically complex watersheds. Lacking an understanding of ecosystem complexity, management actions often produce results significantly different from those expected. Also, it is far easier to manipulate biological communities by introducing exotic species or by hatchery supplementation than it is to educate the public and to conduct long-term restoration that may alter how people use and value the land. Such manipulation, however, is less productive and sometimes destructive in the long run, making the case for careful study.

STRUCTURE AND FUNCTION OF WATERSHEDS

Watershed-scale restoration should begin with an understanding of watershed structure and function and of how human activities affect and shape watershed health. Riverine ecosystems are dynamic, exhibiting great variation from year to year and multiple states of equilibrium (Naiman et al. 1995). These systems are subject to periodic natural disturbances, such as flooding, drought, and wildfire, that alter habitat structure and may establish a new equilibrium within the watershed. Human disturbance also alters watersheds in complex and often synergistic ways. Therefore, restoration practitioners must distinguish between natural variation and human-induced modification that may reduce diversity and lower resiliency to disturbance.

Riverine ecosystems and their watersheds are multidimensional. Ward (1989) described four dimensions of riverine systems and a landscape connection for each: longitudinal (upstream to downstream), lateral (floodplains to uplands), vertical (subsurface to riparian), and temporal (because the other three dimensions are dynamic over time). These dimensions and landscape connections are summarized in Table 1.1. In many watersheds, human disturbance has severed or modified these connections. For example, rivers have been walled off from their floodplains by levees and channelization. Subsurface river flows have been reduced and isolated from surface flows.

This spatial and temporal connectivity within watersheds must be restored to maintain healthy and productive river systems (Naiman et al. 1992). Restoration practitioners must recognize the multidimensional nature of these systems, the repair of which may require reconnecting isolated patches to broader landscapes.

TABLE 1.1.—The multidimensional nature of riverine ecosystems and their watersheds. Our understanding of these connections is essential to successful restoration.

Dimension	Landscape connection
Longitudinal	Upstream to downstream
Lateral	Rivers to floodplains; floodplains to uplands
Vertical	Subsurface to riparian (streambank area)
Temporal (the other three dimensions are dynamic through time)	Altered landscapes over months, years, and centuries

Restoration also may include management decisions that affect local or regional economies. This reinforces the importance of communicating the long-term benefits of restoration to local communities. For example, much of the social and ecological disruption caused by the 1993 flooding in the Missouri–Mississippi watershed might have been avoided had levees been breached to allow flood pulses to reconnect with floodplains.

Riparian (streambank) areas are the critical interface between upland and aquatic habitats. As such, riparian zones are among the most dynamic and important components of watersheds (Gregory et al. 1991). Their linear nature increases their importance in linking landscapes within watersheds. The structure (species mix) of riparian areas is diverse, reflecting a high level of habitat diversity caused by disturbances of both aquatic systems (floods and drought) and terrestrial systems (fire, disease, and insect damage). This structural complexity and diversity of microhabitats leads to far greater species diversity in riparian areas than in associated upland habitats (Gregory et al. 1991).

Healthy riparian areas dissipate flood energy, moderate drought, store surface waters, recharge groundwater supplies, moderate water temperatures by providing shade, and reduce erosion. Riparian areas also provide large-sized woody debris, which is critical in creating structural diversity and habitat complexity in many stream systems (Maser and Sedell 1994). In combination with streams and rivers, riparian areas also are transportation corridors for fish and wildlife, energy, minerals, and sediment. Thus, the importance of riparian areas far exceeds the relatively minor proportion of the land they occupy in most watersheds.

The structure and function of riverine ecosystems are determined largely by their interaction with adjacent riparian communities. For example, the quantity and size of pools in a stream typically depend on the volume of large-sized woody debris, boulders, and other materials that enter streams from riparian areas. The quantity of stream sediment often is determined by riparian vegetation, which filters pulses of sediment that come from upslope. Because of this critical relationship among upslope, riparian, and aquatic systems, ecosystem-based management often gives riparian areas special emphasis and protection to maintain or restore their important functions (FEMAT 1993; Williams and Williams 1997). Recognizing the importance of riparian and bottomland habitats to the ecological functions of streams and rivers, and because these habitats are degrading increasingly, the NRC (1992a) recommended that their restoration be made a national priority.

Because riverine ecosystems are multidimensional and interactive, effective restoration should be conducted at a scale sufficient to incorporate all their significant components (NRC 1992a). A logical scale is that of an entire watershed. Past management, however, often has focused only on isolated fragments of watersheds or on certain species. Restoring high variability and diverse structure to riverine and riparian ecosystems is incongruent with past management approaches that were directed at producing some homogeneous or generic desired future condition across broad landscapes (Doppelt et al. 1993). Instead, watershed management should focus on identifying and restoring key ecological processes that historically provided the structure and function of watersheds (Holling 1992; Risser 1995). A focus on ecosystem processes requires a broader perspective of the entire watershed.

STATUS OF AQUATIC RESOURCES

Aquatic resources have suffered under increasing human population pressure. As proclaimed by Naiman et al. (1995), "Over the past fifty to two hundred years the fresh waters of the United States have undergone the most significant transformation they have experienced in nearly ten thousand years." Nearly one-third of North American freshwater fish species qualify for threatened, endangered, or some other sensitive status (Williams et al. 1989). Joint surveys of the U.S. Environmental Protection Agency and the U.S. Fish and Wildlife Service found that 81% of stream fish communities are adversely affected by environmental degradation (Judy et al. 1984). Hundreds of West Coast stocks (local populations) of salmon and steelhead are in decline and at least 106 stocks already have been wiped out (Nehlsen et al. 1991).

Many freshwater invertebrate groups also appear to be in general decline, as exemplified by a recent review of freshwater mussels of the United States and Canada. The report found 72% to qualify for classification as "endangered," "threatened," or "special concern" (Williams et al. 1993). In all cases, habitat loss is the primary culprit. In general, the risk of extinction for aquatic animals is much greater than for terrestrial species in the United States and Canada, as shown in Table 1.2.

These declines of freshwater species mirror the declining quality of our streams and rivers, which in turn reflects the health of their watersheds. Data for 1990 from the U.S. Bureau of the Census indicated that 85% of the nation's inland water surface area is manipulated by fixed weirs (barriers), dams, and other structures (NRC 1992a). Many of these structures have isolated portions of the landscape and fragmented aquatic habitats.

Although significant progress has been made in controlling point sources of water pollution (such as sewage or factory effluent), nonpoint source water pollution is a persistent problem in many watersheds. Nonpoint source pollutants include runoff from agriculture, municipalities, timber harvesting, mining,

TABLE 1.2.—Extinction risk of selected U.S. and Canadian aquatic and terrestrial animal groups. Data from American Fisheries Society for fishes (Williams et al. 1989) and freshwater mussels (Williams et al. 1993); all others from The Nature Conservancy (Master 1990).

Percentage of species rare or extinct	Total species	Animal group
72	297	Freshwater mussels
65	313	Crayfishes
28	790	Fishes
28	226	Amphibians
14	301	Reptiles
11	762	Birds
13	443	Mammals

and livestock grazing. Nonpoint source pollution now accounts for more than half of U.S. water quality impairment (USEPA 1987).

Wetland losses now exceed half the wetland acreage that existed in the conterminous United States in 1700. This loss continues and averages 287,000 acres annually (Dahl 1990). Riparian habitats, which are critical to proper functioning of aquatic systems, also continue to decline as a result of agricultural encroachment, erosion, channelization, flood control, and bank revetment (riprap) projects. A nationwide rivers inventory concluded that only 2% of waterways in the conterminous United States remained of sufficient quality to justify consideration as national wild or scenic rivers (Benke 1990).

The cumulative impact of human disturbance also affects the basic productivity (biomass growth) of the watershed. For example, soil erosion can reduce the long-term productivity of watersheds in several ways. Erosion reduces the soil's *A* horizon (uppermost soil layer where organic materials accumulate). Erosion also reduces the soil seed bank and soil porosity, which helps water and air to circulate in the soil. The net result is reduction of the soil's vegetative growth capability (Swanson et al. 1989). Soil erosion reduces plant growth primarily by loss of nutrients and loss of water-holding capacity (Clayton and Kennedy 1985; Swanson et al. 1989).

Soil erosion is not limited to streambanks, but includes slow downslope soil creep, debris slides, gully and rill erosion, overland flow erosion, slumps, and surface erosion. Debris slides, slope failures, and other mass movements of soil result in a slope scour with reduced soil production potential. In forested environments, such slope failures typically result from poorly constructed roads, too many roads, and removal of supporting root structures by logging (Powers et al. 1990). In steep areas, soil erosion can strip away the entire organic horizon, exposing infertile soil or bedrock.

Many factors influence erosion rates and few studies of undisturbed sites have quantified erosion that is accelerated above natural rates. One study in Oregon's Cascade Mountains found sediment yields in a clear-cut watershed to have accelerated more than an order of magnitude two years after logging. Soil erosion rates continued three times greater than unlogged watersheds 10 years after logging (Swanson et al. 1989). Regardless of the erosion rate, once depleted, restoration of soil productivity may require many decades or even centuries.

In many watersheds, human disturbance has eliminated or greatly restricted the natural connectivity of the landscape. Remaining patches of intact habitat continue to be degraded gradually by nonpoint source pollution, soil erosion, and other factors. As the health of our habitats has declined, so has the natural diversity of fishes, mussels, mollusks, other aquatic species, and the terrestrial community.

Rather than addressing the root causes of habitat degradation and fragmentation, agencies often have manipulated species richness in aquatic communities through introduction of native species and hatchery supplementation or replacement of declining native populations. Such relatively simple solutions to complex problems often are detrimental and lead to further loss of biological diversity (Stanford and Ward 1992). In a review of nonnative introductions, Moyle et al. (1986) documented the unforeseen problems they often cause by increased predation, competition, introduction of new parasites and diseases, and destabilized fish communities. They termed these unforeseen problems the "Frankenstein effect" for biological diversity. Past management failures indicate that comprehensive restoration is needed to reverse the decline in biological diversity, productivity, and health of aquatic ecosystems.

PAST EXPERIENCE WITH RESTORATION

Historically, conservation has focused on protection of resources. Although protection will continue as a priority, restoration is increasingly significant to management agencies. Although our restoration experience is limited and critique of it is scant, it is important to review past work and its results.

Probably the most extensive experience has been gained in recreating and restoring wetland habitats, although results have been mixed. Success has been achieved where natural biological diversity is not too depleted and habitats are not greatly damaged (Zedler 1994). Recreation of highly degraded salt marsh habitat in San Diego Bay, California, is perhaps the best monitored restoration so far. According to Zedler and Langis (1991), recreated salt marshes resembled reference marshes in their rate of nitrogen fixation, although they failed in reestablishing 10 other measures of ecosystem structure and function. As a consequence, a primary objective of the restoration failed—reestablishment of endangered light-footed clapper rails.

Like wetlands managers, fisheries biologists often employ mechanical habitat improvement and rehabilitation techniques. Modifying instream physical features to increase or enhance habitat for selected fishes has been a common practice in the United States (Platts and Rinne 1985; Rinne and Turner 1991; Hunt 1993). Modifications have included installing logs, rock gabions (wire baskets filled with stones to stabilize streambanks), rock riprap (large rocks used to stabilize streambanks), concrete blocks, and rock boulder structures, often in ways that were intended to increase the quantity of pools or to prevent erosion.

However, Frissell and Nawa (1992) found a high incidence of physical damage to instream structures in western Oregon and Washington streams and a high incidence of functional failure in those that were not physically damaged. They concluded that instream structures were inappropriate and counterproductive in streams that have high sediment loads, high peak flows, or highly erodible banks. Loss of pool habitat and high erosion rates are symptomatic of degraded ecosystem processes that operate at scales larger than most restoration efforts.

Many instream rehabilitation projects have not realized their anticipated benefits because the primary cause of degradation—poor land use management—was not adequately recognized or treated (Rinne and Turner 1991; Doppelt et al. 1993). Engineered stream rehabilitation that attempts to fix the channel to a particular form is incompatible with the dynamic rivers it seeks to restore. What does work is to reestablish ecosystem structure and function. This is especially critical for successful restoration of larger streams (fourth order and higher), which have greater importance because of their higher productivity and persistence during periods of drought or flood (Frissell and Nawa 1992). Although site-specific rehabilitation can slow the effects of stream degradation, it generally should be employed cautiously and in the broader context of a watershed-scale restoration.

Restoration should be designed for each watershed, based on basinwide analyses of historical conditions, trends, and causal factors. Watershed analysis (Ziemer 1997, this volume) provides one basis for developing restoration plans and priorities. Further, restoration should be designed in the larger context of major river basins and regions. Unfortunately, few examples exist of restoration on such broad, basin-scale knowledge.

One good example is a restoration strategy developed for the Skagit River basin in Washington State by Beechie et al. (1994). They determined historical loss of various habitat types from the headwaters to the estuary. Although they focused on coho salmon production, their data illustrate the relative loss of each habitat type, its contribution to coho smolt production (smolt are juvenile salmon migrating seaward), and those habitat types where restoration should prove most cost-effective.

Successful stream restoration is likely to begin in headwaters and should use existing ecological processes. Restoration should not rely on artificial in-channel structures to repair damage that ultimately is caused by poor land use in riparian and upland areas (Frissell and Nawa 1992; Doppelt et al. 1993). Past restoration attempts have concentrated on isolated sites and structural improvements, and thus have not operated at the necessary watershed scale.

In summary, common causes of failure in restoration, reclamation, rehabilitation, and habitat creation include

1. failure to understand the ecological history of area,
2. failure to look at proper scale (i.e., watershed scale),
3. failure to treat root causes of degradation, instead of symptoms,
4. failure to work with local communities and to solicit their support for project goals,
5. failure to integrate ecological principles,

6. failure to develop proper goals,
7. failure to institutionalize commitments within local communities and agencies, and
8. failure to monitor and adapt management accordingly.

Problems of past restoration can be attributed to planning and implementing projects that do not meet the fundamental characteristics and definition of restoration—*a holistic process aimed at reestablishing ecosystem structure and function*. Prior experience indicates a need for broader, watershed-scale restoration.

BENEFITS OF WATERSHED-SCALE RESTORATION

To review, watershed-scale restoration is a comprehensive, long-term process to reestablish the predisturbance structure and function of river basins. Restoration at this scale contrasts strongly with more local and often site-specific reclamation and rehabilitation. Whereas rehabilitation often is limited to specific sites or species, restoring whole watersheds benefits entire aquatic communities and general biological diversity. Restoration treats the primary causes of degradation rather than its symptoms by identifying why an ecosystem is dysfunctional and providing what it needs to heal itself.

For example, if pools have been lost because of diminished structural diversity in the channel, management action that reduces woody debris in riparian zones should be stopped. Additional wood may need to be stockpiled into riparian areas so that it can be captured during high streamflows and deposited naturally.

Restoration plans also should identify and correct processes that cause habitat loss and ecosystem dysfunction. For example, where accelerated streambank erosion is occurring, restoration should uncover its causes, such as poor road construction, and address them rather than simply lining streambanks with rock to improve stability. Recognizing the watershed-scale environment and the effects of disturbance are the first steps in restoration (Sear 1994).

Without a watershed-scale perspective, the risk of undesirable side effects from treatment increases. This is so because the inherent complexity of riverine ecosystems greatly increases the chance for detrimental effects to cascade downstream from treated sites. With a greater understanding of structure and function at the watershed scale, the consequences of restoration become more predictable.

Restoration of ecosystem structure and function should reestablish healthy, functioning watersheds and riverine systems. The benefits of doing so are great, because healthy streams and rivers and their riparian areas provide essential benefits that often are unrecognized and underappreciated. For example, most people value salmon populations for their importance to sport and commercial fisheries, but few appreciate how postspawning salmon carcasses sustain plant productivity in headwater streams. In these small streams, decomposed salmon carcasses provide up to approximately 45% of the carbon and nitrogen in resident trout, aquatic invertebrates, and riparian vegetation (Bilby et al. 1996).

Let us briefly consider some benefits of healthy watersheds. First, healthy watersheds provide high biotic integrity, meaning habitats that support adaptive animal and plant communities and that reflect natural evolutionary and biogeographic processes (Angermeier and Karr 1994). Benefits should accrue to all native species. Another aspect of such biotic integrity is that streams and rivers that exhibit flow similar to historical conditions will be more resistant to establishment of exotic species (Minckley and Meffe 1987; Deacon 1988).

Second, healthy ecosystems are resilient and recover rapidly from natural and human disturbance. Costanza (1992) defined a healthy ecosystem as one that is stable and sustainable, in that it maintains its organization and autonomy over time and is resilient to stress. High biological diversity and habitat complexity provide much of the resistance and resilience exhibited by healthy watersheds.

Third, healthy watersheds exhibit a high degree of connectivity from headwaters to downstream reaches, from streams to floodplains, and from subsurface to surface. Floods can spread onto floodplains, where their energy is dissipated and silt from floodwaters increase soil productivity. High connectivity also enables fish and wildlife populations to move freely throughout the watershed, which increases their viability and facilitates transfer of nutrients from rich downstream reaches to less-productive headwaters.

Fourth, healthy watersheds, with their robust riverine and riparian ecosystems, provide numerous ecosystem services. These include (1) high-quality and dependable water supplies; (2) moderation of the effects of climate change, flooding, and drought; (3) recharge of stream systems and groundwater aquifers; (4) maintenance of diverse and productive riparian plant communities that trap silt and buffer the high energy of floods; and (5) maintenance of healthy riparian areas that moderate stream temperature by shade and buffer sediment pulses from adjacent hillslopes. The diversity of native fish populations, which generally indicates the health of stream systems, will increase in response to ecosystem recovery, provided that recolonization sources are available.

Fifth, healthy watersheds also maintain long-term soil productivity. Rates of erosion may vary naturally from one first-order stream catchment to the next, but overall soil and nutrient loss in the watershed will not exceed soil formation rates. Sufficient vegetative cover, leaf litter, and large woody debris remain to allow gentle percolation of rainwater into soils and groundwater without excessive runoff and corresponding accelerated rates of erosion. Mineral and energy cycles continue without loss of efficiency.

Some of these benefits of healthy watersheds are summarized in Table 1.3.

This is not to say that all ecological repairs must be conducted at the watershed scale, or that a thorough understanding of ecological systems must precede restoration. As pointed out by John Cairns, Jr. " . . . some ecological destruction is so gross that even the most primitive and simple-minded measures will produce a condition that is ecologically superior to the damaged condition" (Cairns 1994). Some species may be at such great risk of extinction that action must be taken immediately. In some watersheds, site-specific or species-specific rehabilitation may be necessary to maintain some rare ecosystem component that

TABLE 1.3.—Some benefits of healthy, properly functioning watersheds.

Healthy characteristic	Benefit
High biotic integrity	Maintains habitat for native plants and animals
Resilient to natural and human disturbances	Vegetation diversity modifies wildfire effects
	Streams maintain perennial flow during moderate drought
High connectivity across landscape	Flood flows have access to floodplains
	Riverine habitats are not fragmented by dams
Ecosystem services are provided	Hillslope vegetation lets moderate rainfall percolate into soils rather than eroding them
	Watersheds produce good-quality drinking water
High long-term productivity	Soils and soil nutrients are not chronically removed by wind or water

otherwise would disappear before broad-scale restoration could become effective.

As described in the 13 case studies in this volume, many motivated and dedicated coalitions are conducting restoration in watersheds across the United States. Our hope is that by learning from these examples and by sharing and applying the ecological principles described in this book, we will foster greater support for restoration, improve our ability to restore damaged ecosystems, and, in the words of A. D. Bradshaw, make them work again.

ACKNOWLEDGMENTS

The editors acknowledge Cindy Deacon Williams (Pacific Rivers Council) and Robert J. Naiman (University of Washington) for their insightful reviews of earlier versions of this Chapter and Meggan Laxalt for her kind assistance in preparing the figures.

PART I

Principles

A consensus is growing that more restoration at the scale of entire watersheds is needed. But successful restoration at this large scale is challenging, not just because of large geographic areas, but because watersheds are complex. Rivers and riparian habitats are linear and connect portions of the landscape into functional watersheds, so correspondingly, restoration must be integrated across broad temporal and spatial scales.

What principles should guide watershed restoration? Where is restoration most effective? What skills and tools are needed? How do we enlist broad support within local communities? The seven chapters presented in this part provide a sound scientific foundation to answer these questions and to guide restoration. Here is a brief overview of the chapters.

Chapter 2 "Ethical Principles" provides the philosophical concepts and values that should guide natural resource management. Before restoration planning begins, we should, in the words of Aldo Leopold (1949), understand our role as "plain member and citizen of the land community."

With the ethical framework established, Chapter 3 "Social Ecology: A New Pathway to Watershed Restoration" describes the processes and structure of local human communities. People and how they interact within communities are the most pervasive force within most watersheds. Consequently, our ability to work effectively with local communities is a key to successful restoration.

Chapter 4 "Conceptual Roles of Biological Integrity and Diversity" distinguishes between biological integrity and species diversity. The chapter also provides a rationale for incorporating concepts of biological integrity into project goals. It describes the index of biotic integrity and other tools to measure watershed health and to determine the success of restoration.

Understanding past habitat condition is critical to understanding present habitat capability and to envisioning future watershed conditions. Thus, Chapters 5 and 6 present issues of temporal and spatial scale. Chapter 5 "Historical Perspectives" reviews the principles of historical context as they apply to restoration. Chapter 6 "Temporal and Spatial Scales" relates how distant activities in a watershed can directly affect downstream restoration. This chapter also describes how to conduct a watershed analysis, which is a tool for understanding ecological processes, problems, and solutions at the scale of entire watersheds.

Chapter 7 "Ecological Principles" overviews these principles as they apply to watershed restoration. This chapter also contains a strategy for prioritizing restoration activities within watersheds and helps determine where in a watershed restoration should begin and where it will be most cost-effective.

No restoration is complete without an analysis of what was accomplished and the results. Chapter 8 "Monitoring and Adaptive Management" describes how to establish effective monitoring plans that provide accountability and facilitate midcourse correction as restoration proceeds. In an era of limited funds, we must be able to demonstrate the effectiveness of restoration. This is the essence of monitoring.

Aldo Leopold (1949) remarked that "conservation is paved with good intentions which prove to be futile, or even dangerous, because they are devoid of critical understanding . . . " It is vital that we proceed with the task of watershed restoration, but it is equally vital that we bring the best scientific understanding to our efforts.

We end, I think, at what might be called the standard paradox
of the twentieth century: our tools are better than we are, and
grow better faster than we do. They suffice to crack the atom,
to command the tides. But they do not suffice for the oldest task
in human history: to live on a piece of land without spoiling it.
—Aldo Leopold, 1938 (in Flader and Callicott 1991)

CHAPTER 2

ETHICAL PRINCIPLES

Edwin Philip Pister

It became apparent to me some years ago that technology seemed to be creating far more resource management problems than it was solving. Philosophical issues—values—were almost totally ignored, both by agency administrators and the educational institutions that produced them. No doubt much of this problem stemmed from the fact that most biologists were taught by professors who had neither read a philosophy paper nor showed any real interest in doing so. Only in their later years, when biologists begin to mature and think more of leaving meaningful legacies than of achieving agency or academic advancement and prestige, does philosophical depth normally emerge.

Such depth was, of course, the hallmark of environmentalist Aldo Leopold (1887–1948), whose book *A Sand County Almanac* has become a philosophical Bible of conservation and preservation. I was fortunate to have studied wildlife conservation at Berkeley under A. Starker Leopold, Aldo's eldest son, beginning in 1949, the year following his father's untimely death. Starker was a chip off the old block, so wildlife values quickly became a major part of my education. Even so, I did not recognize the significance of this fortuitous association for more than a decade (Pister 1987).

In 1979 the new journal *Environmental Ethics* appeared, self-described as "An interdisciplinary journal dedicated to the philosophical aspects of environmental problems." Fascinated, I contacted the journal's editor Eugene C. Hargrove, then at the University of New Mexico, to learn about this promising field and how it might help me solve increasingly complex environmental management problems. I was not disappointed, for this refreshing perspective clarified and quantified many uncertainties in my work (Pister 1985, 1987). It became immediately apparent that most environmental philosophers know far more about biology than most biologists know about philosophy! I was fascinated as I learned more about the history and development of environmental ethics (Hargrove 1989; Nash 1989).

During my last decade with the California Department of Fish and Game, I increasingly examined philosophical issues related to my work, not simply because they were intriguing, but because they offered new insight toward solving problems. The need to incorporate values into research and management decisions is growing more evident throughout the profession.

In recent years I have found myself concerned with diverse habitats, from Devil's Hole in Death Valley National Park (described by the U.S. National Park Service as the world's most restricted habitat for a vertebrate species) to Lake Superior (perhaps the world's largest freshwater habitat). Yet their common problems of habitat integrity and species value are virtually identical.

Let us consider these two habitats. In northern latitudes, warming periods accompanied glacial recession toward the end of the Pleistocene epoch (12,000–15,000 years before the present). This warming diminished the rains and mountain snows that had created pluvial lakes (lakes that exist only when the climate is very wet). These lakes covered much of the Intermountain West. As the water supply dwindled and the lakes shrank, fishes in desert areas began evolving to accommodate the vastly restricted habitats. An example is Devil's Hole near Death Valley (about the size of a backyard swimming pool) and its endemic Devils Hole pupfish. The pupfish was separated from its progenitors by a declining water table, which isolated them more than 44,000 years ago (Winograd and Szabo 1991).

The other habitat includes specific areas of Lake Superior, and here the fish are lake trout. These trout depend on their spawning reefs as much as the Devils Hole pupfish depends on the few square feet of limestone shelf that provide its sole spawning and feeding area on Earth. From the evolutionary standpoint, specialized stocks (local populations) of several fishes within the Great Lakes are as unique as the Devils Hole pupfish.

Given this background, how do we place values upon things like pupfish and lake trout, and habitat integrity within the Great Lakes and tiny desert springs?

COMMON GROUND

To illustrate the often confusing concepts of ethics and values, here is a parable adapted from Bryan Norton (1987). Imagine a family living within the Mississippi River floodplain during the summer of 1993. Rains continue, and the family soon must leave for higher ground. Everything left behind will be lost. Only their most essential possessions are to be loaded into the family station wagon.

As the vehicle groans under the load, there is still room for one small addition. They can take the old family Bible, dating to the eighteenth century with all births and deaths recorded in the handwriting of the patriarchs and matriarchs of past generations, or they can take a small box of videotapes of the children's favorite television programs. The parents poll the family. The younger children want the videotapes, oblivious to the irreplaceable family history and values to be found within the Bible. A teenager suggests an easy way out by phoning a nearby pawn shop, with current cash value to be the determining factor. However, the parents exert their authority and take the Bible, secure in the knowledge that with added maturity, the children will understand the relative values involved and the wisdom of their decision.

This metaphor bears wide application in the natural world, especially regarding the way in which we undervalue irreplaceable species and ecosystems for material gain. About halfway into my career, two incidents demonstrated this concept to me. In 1975, I and other concerned biologists held all of the remaining individuals of a genus (a poolfish named *Empetrichthys*) in a small horse trough while preparing a temporary refuge pond on the Desert National Wildlife Refuge north of Las Vegas, Nevada.

In the other, in 1969, I held all that remained of the Owens pupfish in two buckets, one in either hand, while several colleagues, unaware of my plight, conducted creel censuses on a nearby rainbow trout fishery (Pister 1985, 1993). I remember trying to explain my unexpected absence from the census to an official from my agency who had a very different set of values, and I clearly recall his rebuke: "Maybe someday you will get your priorities straight!" Aldo Leopold's admonition comes to mind: "One of the penalties of an ecological education is that one lives alone in a world of wounds" (Leopold 1953).

EVOLUTION OF NATURAL VALUES

Although Native Americans had long held Nature in reverence (Vanderweth 1971; P. S. Wilson 1992), the European frontier mentality pervaded North America and remained firmly entrenched (and still does in places) until the nineteenth century. At that time, poet Ralph Waldo Emerson and writer-naturalist Henry David Thoreau suggested that uses other than utilitarian might be made of nature (Nash 1989; Callicott 1991). Naturalist John Muir (1894, 1901) picked up this theme, invoking the philosophies of Emerson and Thoreau in a campaign for appreciation and preservation of wilderness. (Value issues are detailed in Rolston 1981, 1986, 1988, 1994.)

In 1889 Gifford Pinchot graduated from Yale University with a deep interest in forest practices. Because Yale lacked a forestry program at that time, he traveled to Europe for a formal introduction to the profession. As the first chief of the embryonic U.S. Forest Service in 1905, Pinchot (1947) defined conservation as "the greatest good of the greatest number for the longest time." This concept, now termed the *resource conservation ethic*, fit nicely into the highly utilitarian values of the new U.S. Forest Service. Even today, it underlies and directs the basic conservation policy of most resource management agencies. However,

TABLE 2.1.—Contrasting the two primary ethical concepts as they relate to habitat restoration and rehabilitation.

Resource conservation ethic (Pinchot)	Evolutionary ecological land ethic (Leopold)
Maximize short-term gain	Maximize long-term perspective
Use engineered approaches to rehabilitate systems	Use natural ecosystem processes to promote restoration
Work to improve natural system in the short term	Work within constraints of the natural system
Focus on increasing specific resource outputs	Focus on restoring health of entire watershed

despite their different basic thinking, both conservationist Pinchot and preservationist Muir believed that only people possess intrinsic value, and that nature has only instrumental value. Both regarded only human interests as legitimate (Callicott 1991).

Aldo Leopold followed Pinchot from Yale in 1909. By then, Yale had a forestry department, funded by the wealthy Pinchot family. Leopold brought conservation into an ecological perspective and clarified the related ethical implications. Although Leopold began his career as a strong supporter of Pinchot's resource conservation ethic, his training and intuition as an ecologist caused him to break away and formulate his now-famous *evolutionary ecological land ethic*, which designated humans as part of the Earth ecosystem and not separate from it. Environmental impacts that affected other organisms, then, ultimately would affect us too (Callicott 1989a).

Leopold's land ethic placed *Homo sapiens* as a "plain member and citizen of the land community" and strongly implied intrinsic value for all creatures by granting "respect for his fellow members and also respect for the community as such" (Leopold 1949). Table 2.1 contrasts Pinchot's resource conservation ethic and Leopold's land ethic.

This same thinking was reflected in biologist Rachel Carson's 1962 classic *Silent Spring*. By exposing publicly the blatant misuse and insidious impact of pesticides (primarily DDT) upon nontarget organisms, she almost single-handedly launched the American environmental movement.

Based upon the land ethic, Callicott (1991) derived five commandments from Leopold's text.

1. Thou shalt not extirpate (exterminate) species or render them extinct;
2. Thou shalt exercise great caution in introducing exotic species into local ecosystems;
3. Thou shalt exercise great caution in extracting energy from the soil and releasing it into the biota;
4. Thou shalt exercise great caution in damming and polluting watercourses; and
5. Thou shalt be especially solicitous of predatory birds and mammals.

Each of these items is directly applicable to watershed restoration.

In *Why Should We Care About Rare Species?* A. S. Gunn aptly wrote:

I believe that human beings have a duty to maintain the fragile stability of what is left, and to endeavour, where possible, to recreate approximations to natural communities. In practical terms this means the attempt to provide suitable habitat for a wide range of mutually compatible plants and animals, and positive action to remedy the effects of past destructions—for example, the U.S. Fish and Wildlife Service's efforts to reintroduce peregrine falcons to their former haunts. (Gunn 1980)

LEOPOLD'S A–B DICHOTOMY

Early on, Leopold detected a philosophical rift within the conservation community (Leopold 1949; Pister 1987, 1992). Accordingly, he presented in *A Sand County Almanac* his concept of the A–B dichotomy: "Conservationists are notorious for their dissensions. . . . In each field one group (A) regards the land as soil, and its function as commodity-production; another group (B) regards the land as a biota, and its function as something broader" (Leopold 1949).

Both groups, A's and B's, long have been apparent in the management of fish and wildlife resources. Group A's remain strong supporters of maximizing commercial and sportfishing yields, whereas B's are surely interested in this goal, but would prefer to attain it by using native species, and most assuredly not by any means that might jeopardize the native fish fauna. Philosopher Hugh Nibley (1978) observed that " . . . we have taught our children by precept and example that every living thing exists to be converted into cash, and that whatever would not yield a return should be quickly exterminated to make way for creatures that do."

WORKS OF ART

During an expedition to the East Indies in the early 1860s, English naturalist Alfred Russel Wallace was deeply impressed by the area's natural history, ranging from indigenous human populations to insects and plants. Wallace wished to impress the British (and world) scientific communities with the enormous potential value of what he had seen. Thus, he wrote convincingly of the need to add these species to the collections of Europe's national museums (and, by inference, to preserve them in their natural habitats). He characterized " . . . every species of animal and plant now living as the individual letters which go to make up one of the volumes of our earth's history . . . " Wallace made an eloquent plea to collect and preserve these organisms and concluded with a prophetic statement that might well have been printed in this morning's edition of *USA Today*:

If this is not done, future ages will certainly look back upon us as a people so immersed in the pursuit of wealth as to be blind to higher considerations. They will charge us with having culpably allowed the destruction of some of those records of Creation which we had it in our power to preserve; and while professing to regard every living thing as the direct handiwork and best evidence of a Creator, yet, with a strange inconsistency, seeing many of them perish irrecoverably from the face of the earth, uncared for and unknown. (Wallace 1863)

Preoccupation with economics and political expediency often subordinates our marvel at indigenous populations and life-forms and the habitats they occupy. We normally fail to consider our own transient and insignificant role in the overall evolutionary process and assume an attitude characterized by David Ehrenfeld as "the arrogance of humanism" (Ehrenfeld 1978).

To place this in a more understandable time perspective, Milbrath (1989) used the height of New York City's Empire State Building as an analogy. Including the television tower, the building stands at about 1,473 feet. Allowing this to represent geological time since Earth's beginning (4.6 billion years), the time since Columbus arrived in America (500 years) would be equivalent to the thickness of one sheet of paper. The 100 years of industrialization that have caused much of our current environmental dilemma would be equivalent to one-fifth the thickness of that one sheet!

Yet the evolutionary marvel represented by all species and their habitats is almost entirely overlooked by a society "immersed in the pursuit of wealth" and other materialistic goals (Ludwig et al. 1993). We can only hope to hold out until more mature values concerning species and ecosystems are accepted by the industrialized nations and made part of the legal structure (Moyle and Moyle 1995).

When compared to the absolute masterpieces represented by North America's watersheds and their floral and faunal complexes, the paintings of Rembrandt and da Vinci pale into insignificance. Yet collectors pay millions to own a single piece of their artwork. A logical question follows: How much are we willing to pay to restore our natural studios (watersheds) and their artworks (terrestrial and aquatic life-forms)?

A related point is that Earth's life-forms and complexes should be considered as much for their intrinsic value as for their instrumental value. Instrumental values are subject to change with the whim of societal and economic interests. But intrinsic values retain a stability that can only increase as our perception heightens with the passage of time.

NEW TRENDS

Unfortunately, society assumes that the status quo has always been and thus will continue indefinitely. For instance, commercial and sport fishing for lake trout in the Great Lakes were thought to be permanent enterprises by their practitioners. However, the trout were taken to the point of no return in perhaps 100 years (abetted and hastened by sea lampreys and pollution). In this case, the resource was simply overtaxed.

But just as the status quo can shift because of a resource change, so can it shift because of our attitude—a value change. Today, societal values are showing definite signs of change. For example, California fishing license sales declined

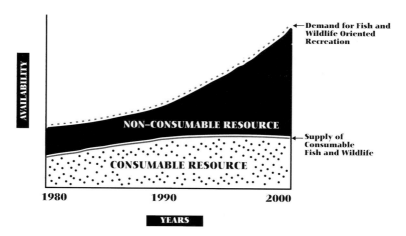

FIGURE 2.1.—Probable trends of supply and demand for North American fish and wildlife resources. (Adapted from Pister 1991b.)

sharply during the 1980s. In 1980 about 1 in 10 Californians bought a fishing license, whereas in 1989 only 1 in 20 did so (Pister 1992).

Analysts attributed this decline to various factors, including demographic changes that have resulted in a new diversity of recreational interests. The traditional recreations of hunting and fishing now face increased competition from other sports. Analysts also attribute the decline in licenses to a diminishing rural population, which traditionally has been the staunchest supporter of hunting and fishing. Hunting license sales dropped by a similar figure during the 1980s.

Values and priorities in other areas of North America probably will change too, likely in the direction led by California. We can state with some certainty that changes in interests and values will be greater between the years 2000 and 2100 than they were between 1900 and 2000. As North Americans increase in sophistication, outdoor interests seem likely to fall more into what Aldo Leopold (1949) defined as wildlife research.

This concept is illustrated by Figure 2.1, which presents hypothetical supply and demand curves for fish and wildlife resources as we move into the next century. It underscores the need to restore watersheds and their related biodiversity. Components of the total biota, existing in secure habitats, will then essentially constitute a resource from which we may fulfill the needs of an unquestionably more sophisticated and perceptive society (Pister 1976, 1991a, 1991b).

This phenomenon is also reflected in television programming. Whereas a decade ago most television productions about fish and wildlife concerned some aspect of consumptive harvest, more recent productions focus almost entirely on animal behavior and ecology. Referring to the earlier metaphor of the family Bible and videotapes, it appears that some North Americans are now experiencing less difficulty in making this distinction in values. Some are even venturing into the controversial area of granting rights to species and ecosystems (Pister 1995).

THE UPSHOT: OUR ETHICAL OBLIGATION

The reader may still wonder why I have laid out the preceding array of ethical history and application. The reason is simple: I want to make it clear that, individually and collectively, we carry an ethical obligation to do three things.

1. *Restore the physical components of watersheds wherever possible.* "Wherever possible" means that candidate watersheds cannot be so degraded that restoration is unfeasible. Restorable watersheds must be carefully selected so they will become the success stories necessary to gain public acceptance and support. Such programs carry with them great philosophical and practical significance. (The subject of restoration is covered in depth by MacMahon and Jordan [1994] and their associated essayists.)
2. *Restore within selected watersheds only native flora and fauna and their habitats.* We should be especially critical of new programs that suggest otherwise (Callicott 1991).
3. *Develop programs that communicate to the public the intrinsic value of all life-forms,* and our obligation to protect and preserve them (Ehrenfeld 1976; Norton 1983; Regan 1983; Callicott 1989b).

Concepts such as intrinsic value are new to our profession and are by no means universally accepted by decision makers (Norton 1987). We cannot in all cases expect watershed restoration to reflect obvious and immediate instrumental value.

We must emphasize those intrinsic values that thoughtful scientists and managers should associate with *all* life-forms (Ehrenfeld 1976). For instance, stocks of certain lake trout within Lake Superior add an intrinsic value that likely warrants their consideration under the Endangered Species Act of 1973, especially as technological refinements allow more precise means of stock identification (Krueger et al. 1989). Other native fishes within the ecosystem or watershed add to this value.

Adherence to the three objectives listed above appears to align with the mission of the U.S. Geological Survey, Biological Resources Division "to gather, analyze, and disseminate the information necessary for wise stewardship of our Nation's natural resources, and to foster an understanding of our biological systems and the benefits they provide to society" (NRC 1993a).

Although restoration and rehabilitation are commendable and a major step in the right direction, we must acknowledge that they seldom can fully replace the original or pristine state of an ecosystem, for various reasons. If you were given a choice of hiking into a "restored" forest planted 50 years ago, or into an "original" old-growth forest, which would you choose? How does one restore an old-growth forest?

If we assume that damaged habitat can be restored fully, we open the door to exploitive interests to conduct activities with the promise that, when they are finished, they will "make the affected area as good as new" (Rolston 1994). Habitats that require restoration have experienced interruption of their historical genesis, and their value thereafter is always something less than a pristine equivalent (Elliot 1982; Katz 1992). However, this should never discourage us from doing our very best. Rolston (1994) discusses these concepts in depth.

ENVIRONMENTAL TRADEOFFS

Through the years, many tradeoffs were made in which watershed integrity was sacrificed to achieve economic gain. An outstanding example is provided by the Great Lakes lake trout fishery. Less than a century ago, commercial and sport fisheries of the Great Lakes flourished, accompanied and supported by a marvelous complex of noneconomic species. Perhaps inadvertently, tradeoffs were made, primarily between industrial and agricultural development on the one side and water quality on the other. These tradeoffs attended and precipitated dramatic declines in native fish populations, especially lake trout.

Of the entire Great Lakes watershed and faunal complex— one of the world's great treasures— only portions of Lake Superior now retain but a semblance of their pristine condition. Native species were sacrificed on the altar of industrial pollution and excessive harvest, and complicated by an invasion of exotic and introduced species through canals dug for the shipping industry. When we view related Pacific salmonid introductions, we know that only the passage of time will allow us to judge their long-term wisdom, but recent indications already instill doubt.

Several parallels are provided by the current decline and extinction of salmon stocks in the Pacific Northwest. Here, fisheries of enormous economic and social value were traded for cheap hydroelectric power and irrigation water, and for shoddy logging and land use practices. The situation was aggravated and complicated further by massive introductions of hatchery-reared salmon. Relatively speaking, problems in the Great Lakes may prove easier to solve, despite their enormity (Hartig and Thomas 1988). It will likely prove simpler, both physically and politically, to clean up and rehabilitate the Great Lakes than to remove the dams that are the major cause for salmon decline in the Columbia River basin.

MAINTAINING BIOLOGICAL INTEGRITY IN A CHANGING WORLD

As we move into a new millennium, fisheries professionals should heed the gentle breezes that inevitably precede winds of change. A value shift from traditional programs to watershed restoration should be welcomed by top administrators, painful though the change might be. I fully recognize that rebels or rebellious expression seldom are welcomed by those who must adjust budgets and explain innovative programs to an ever-more critical and inflexible public. We may, however, find a ray of hope from German physicist Max Planck:

A new truth does not triumph by convincing its opponents and making them see the light, but rather because its opponents eventually die, and a new generation grows up that is familiar with it. (*Scientific Autobiography and Other Papers*, 1950, in Platt 1992)

The need for a value shift in conservation is eloquently and firmly stated by Indiana University's Lynton K. Caldwell:

The environmental crisis is an outward manifestation of a crisis of mind and spirit. There could be no greater misconception of its meaning than to believe it to be concerned only with endangered wildlife, human-made ugliness, and pollution. These are part of it, but more importantly, *the crisis is concerned with the kind of creatures we are and what we must become in order to survive* [emphasis added]. (in Miller 1988)

Survival, of course, ultimately includes such paramount endeavors as watershed restoration.

Each of us, after all, has only one opportunity to leave a meaningful legacy to future generations. I hope that the legacy from this watershed restoration volume might lie somewhere within the thinking of H. Rolston, one of the nation's leading contemporary environmental philosophers:

A species is what it is inseparably from the environmental niche into which it fits. Although a creative response within it, the species has the form of the niche. . . . The species stands off the world; at the same time it interacts with its environment, functions in the ecosystem, and is supported and shaped by it. . . . Integrity in the species fits into integrity in the ecosystem. . . . It is not preservation of *species* that we wish but the preservation of *species in the system*. It is not merely *what* they are but *where* they are that we must value correctly. (Rolston 1988)

Rolston's thinking blends well with a renowned corollary to the land ethic. The following excerpt from *A Sand County Almanac* appears tailor-made for the problems inherent in watershed restoration:

The "key-log" which must be moved to release the evolutionary process for an ethic is simply this: quit thinking about decent land-use [or water-use] as solely an economic problem. Examine each question in terms of what is ethically and aesthetically right, as well as what is economically expedient. *A thing is right when it tends to preserve the integrity, stability, and beauty of the biotic community. It is wrong when it tends otherwise* [emphasis added]. (Leopold 1949)

It of course goes without saying that economic feasibility limits the tether of what can or cannot be done for land. It always has and it always will. The fallacy the economic determinists have tied around our collective neck, and which we now need to cast off, is the belief that economics determines *all* land use. This is simply not true. An innumerable host of actions and attitudes, comprising perhaps the bulk of all land relations, is determined by the land-users' tastes and predilections, rather than by his purse. The bulk of all land relations hinges on investments of time, forethought, skill, and faith rather than on investments of cash. *As a land-user thinketh, so is he* [emphasis added].

What I have presented up to this point constitutes, in effect, what I feel to be an ethically acceptable road map indicating the major routes leading from Point A in watershed restoration (where I believe we are now) to Point B (our destination). The off-ramps, traffic lights, and service stations must be identified by those more familiar with watershed restoration than I. It is gratifying that great talent and commitment exists among those parties concerned with the future of watersheds. But irrespective of how it might be reached, we must keep firmly in mind the values inherent in reaching Point B, values of much greater significance to future generations than to us.

Some of what I have written may appear, on the surface, to be excessively idealistic, but the principles are valid and supportable both philosophically and biologically. I have learned that ethically sound programs inevitably are biologically sound and enduring. A set of values and ideals provides a strong basis for formulating more pragmatic means of achieving them, and these same values and ideals apply universally to watershed restoration. I have drawn heavily on the thinking of Aldo Leopold because *A Sand County Almanac* served as my primary guidebook during 40 years as a practicing fisheries biologist, spending my entire career in the trenches of environmental dilemmas. Leopold's philosophies have never failed me, and they are powerfully applicable to the subject at hand.

Is our goal attainable? A statement attributed to the German philosopher Goethe perhaps says it best: "Every man has only enough strength to complete those assignments of which he is fully convinced of their importance." Our primary challenge is to convince the electorate that our values are correct and worth pursuing, and that within the concept of worldwide watershed restoration may ultimately lie the key to the future of humankind.

ACKNOWLEDGMENTS

The author expresses gratitude to Baird Callicott, Michael Dombeck, Charles Krueger, Gary Meffe, Holmes Rolston, Jack Williams, and Chris Wood for their suggestions and criticism of this and an earlier draft.

Why don't we take all this money we are spending on regulation and pay farmers to have fish in the river? We could do it.
—Applegate resident's remark, 1995 town meeting with Oregon Department of Fish and Wildlife

CHAPTER 3

SOCIAL ECOLOGY: A NEW PATHWAY TO WATERSHED RESTORATION

Kevin Preister and James A. Kent

A story from the South Pacific Islands about a protected species, the green sea turtle, helps explain social ecology concepts and their application. This turtle has become endangered because local opportunists sell its shell on the black market. The turtle population is dwindling to extinction. The federal agency that protects the turtle sends agents onto the island to trap and prosecute offenders. When they arrest one person, another opportunist emerges and the illegal killing resumes. Too few agents are present, so this enforcement model is not stopping the killing of the sea turtle.

Restoring the sea turtle requires an expanded approach that involves both the agency and the islanders. Local residents need to become involved in protecting the turtle, supplementing the government's enforcement program. And the agency must look beyond its narrow mandate and recognize that recovery of this species will depend on something larger: the *cultural restoration* of island residents.

In the traditional culture of these island residents, the sea turtle is sacred. Today, however, the culture has changed, weakened by modern pressures. If the residents revitalize their culture, they will ensure that the sea turtle is protected as a sacred species. In this case, species protection can be accomplished through cultural restoration because the culture gives people a reason to once again protect the species. Thus, the mission of the protection agency is accomplished, an endangered species is saved, and an endangered culture is brought back to life. The most difficult part of this scenario is for the agency to recognize that their goal of wildlife restoration can be accomplished only through direct interaction with the culture as part of a broader concept of restoration.

Stories like this need not be taken from exotic locations. We have in the United States a recent history of environmental legislation designed to institutionalize restoration of ecosystems or watersheds. This legislation is wide-ranging. It includes the Comprehensive Environmental Response, Compensation, and Liability Act of 1980 (commonly called Superfund), in which the U.S. Environmental Protection Agency (USEPA) regulates environmental cleanup. It also includes the national restructuring of the U.S. Forest Service and the U.S. Bureau of Land Management (USBLM) under the philosophy of ecosystem management. And it encompasses state watershed health programs in Maryland, Oregon, and elsewhere.

Our central thesis is that, in many instances, *successful watershed restoration depends on cultural restoration*, meaning the good will, stewardship values, and participation of citizens. Hence, restoration programs and policies must reflect local watershed knowledge, create an integration between community and scientific concerns, and develop incentives that favor stewardship behavior. In short, we must *work through the culture* to succeed, not manipulate the people.

This chapter explains the conceptual development of social ecology, particularly a model of productive harmony, and the elements of bio-social ecosystems. We use two case studies to show how productive harmony can be implemented through an issue management approach. A new generation of cooperation in ecosystem restoration is described using the Applegate Partnership of Oregon's Applegate Valley. We conclude with an examination of the civic culture as it relates to watershed restoration.

THE DEVELOPMENT OF SOCIAL ECOLOGY

Social ecology is most frequently associated with the writings of Murray Bookchin (1990). He cited the importance of geographic place, local control (through the concept of the municipality), empowerment of citizens, and the meshing of social and environmental goals. Underlying his work is a philosophy of anarchism that calls for the elimination of societal hierarchies (class, race, gender) as a means of creating ecologically sound living.

Although we retain some key elements of Bookchin's thinking about geographic place and empowerment of citizens, our conception of social ecology could better be considered an *applied* social ecology. Applied social ecology seeks first to understand the relation between physical and social environments in an area, and then to act upon that understanding to create adaptive change.

A social ecosystem is a culturally defined geographic area within which people manage their lives and resources. In this perspective, we recognize that humans, like wildlife, are naturally organized into geographic territories. Humans

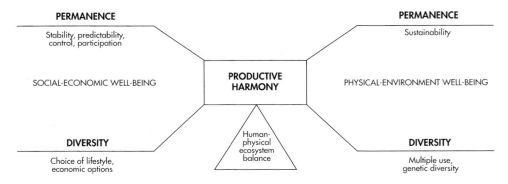

FIGURE 3.1.—The Bio-social ecosystem model of productive harmony recognizes the social and physical intent of the National Environmental Policy Act.

are embedded within informal systems, comprised of horizontal networks and natural gathering places, that can be systematically observed, measured, mapped, and mobilized. It is through cultural boundaries that the social ecosystem sustains and rejuvenates itself. (These concepts are summarized in several articles: Greiwe 1980; FUND Pacific Associates 1981; Preister and Kent 1984; Kent 1992.)

Productive Harmony

The National Environmental Policy Act of 1969 describes the related concepts of physical and social environmental health (Kent et al. 1994). Section 102 of the act, which presents the requirements for conducting an environmental assessment or for preparing an environmental impact statement, has received the most attention over the years.

But Section 101 has offered the most guidance in defining the social ecosystem assessment process. Section 101(a) introduces the concept of productive harmony, stating that environmental decisions and actions shall be made in ways that "create and maintain conditions under which man and nature can exist in *productive harmony*, and fulfill the social, economic, and other requirements of present and future generations of Americans" (emphasis added).

Productive harmony is a healthy, balanced state of an environment in which both social resources and physical resources have high levels of permanence and diversity, enabling their sustainability. Figure 3.1 shows the balance of the productive harmony concept.

To define our terms, *social resources* are the people in a culturally defined geographic area, including their "survival networks" of friends, families, and associates, and their self-described boundaries around their living patterns and activities (called cultural descriptors). The authors have developed and used seven cultural descriptors to conduct community assessment and management over the past 25 years (Greiwe 1980; Kent 1992).

Physical resources include all the natural and biological attributes of a given geographic area, except the people. Such resources may be renewable (timber, wildlife, water, and solar energy) or nonrenewable (minerals and fossil fuels).

Where social resources are concerned, *permanence* includes people's sense of stability coupled with their ability to participate in, predict, and control events affecting their lives and the lives of their children, neighbors, and kin. Where physical resources are concerned, permanence means that the presence of both renewable and nonrenewable resources will continue well into the future.

Diversity of social resources is the range of options people have in a human geographic unit for social, cultural, and economic activities: with whom to associate (networks), where to live (settlement), how to earn a living (work), how to get and give help (support services), and where and how to have fun (recreation). Diversity of physical resources is the variety and variability of natural resources that are interdependent in a systematic way (habitat continuum) such that each affects the viability of all other components (see also OTA 1987).

From a resource management perspective, productive harmony is achieved when actions affecting the entire environment are judged to increase permanence and diversity in the long run. This end can be achieved when diversity and permanence are recovered, enhanced, and sustained through integrating both the social and physical arenas (D. Baharav, J. A. Kent, and E. Baharav, paper presented at National Symposium on Issues and Technology in the Management of Impacted Wildlife, 1991). We call this integration a bio-social ecosystem.

The Bio-Social Ecosystem

A *bio-social ecosystem* recognizes the social ecosystem and the physical ecosystem as equal partners. To the extent that federal land managers still drive environmental decisions strictly from a physical ecosystem perspective, current conflicts of federal land use will not be alleviated. For example, the national ecological mapping project undertaken in 1993 by the U.S. Department of the Interior was mainly concerned with mapping the physical ecosystem. It did not recognize and therefore did not incorporate the social ecosystem.

Conflicts that still surround the northern spotted owl, grazing on public lands, and numerous other biological issues on federal, state, and private land are evidence of failure to recognize the social (human) ecosystem as critical to productive harmony and long-term sustainability. The human element is still treated as a by-product of the management of the physical resource (Kent et al. 1994).

Human geographic boundaries, a key element in identifying the social ecosystem, rarely coincide with traditional administrative–political boundaries of counties, states, or national forests. As we come to view geographic regions as diversified biological and socioeconomic systems, it is becoming harder to constrain them within unnaturally imposed boundaries (Garreau 1982). For example, northern California and southern Oregon, unified in a common history and geography, have identified themselves as the "State of Jefferson" for at least three generations. This area actually seceded from the United States just prior to World War II in reaction to the unresponsiveness of the respective state capitals. Thus, the artificial administrative boundary of a state line clashed powerfully with the real human geographic boundary. (This interesting story won a Pulitzer prize for the *San Francisco Chronicle* reporter who covered it [Olson 1987].)

One undesirable consequence of using administrative rather than cultural boundaries is that the impact zone of decisions is not recognized, thereby leading over time to the fragmentation of either the social or the biological system. For example, in Colorado, the county land use code (also known as the 1041 Hazard Review) requires biologists to provide information on wildlife within the boundaries of a proposed project. However, the major impact zone may lie a short distance outside the project boundary (Baharav 1991). Once informal boundaries are identified and their influence is recognized, public land managers can be more successful in recognizing trends, be more proactive in addressing public concerns, and manage change in a cultural context.

Today, in ecosystem management, the U.S. Forest Service and USBLM are redrawing their management boundaries to coincide more with watersheds, natural ecosystems, and geologic provinces. From mapping informal cultural systems over the years, we have found that the human geographic units often correspond at a certain scale to these ecological units. Differences occur when technological intrusions or economic changes affect boundaries, such as a new highway, the loss of a sawmill, or other developments (J. A. Kent and A. K. Quinkert, paper presented at the 24th annual conference of the Urban and Regional Information Systems Association, 1986).

Implied in Figure 3.1 is a new attention to the social ecosystem by resource scientists and public land use managers. The social aspect is essential in today's environmental management, because of the nature of residents. Residents are keen observers of change in their environment and know it best (they have local knowledge). Residents are strong stewards and are engaged in various steward-ship activities (local values). Residents have well-defined concerns about their social and physical environments (local issues) which, if acted upon in their terms, can generate a sense of empowerment and engagement in natural resource management and restoration (Preister and Kent 1984).

IMPLEMENTING PRODUCTIVE HARMONY: THE ISSUE MANAGEMENT PROCESS

Issue management is the process of implementing productive harmony, which is the desired balance of social and physical resources (Figure 3.2). Issue management allows an organization to identify the three stages of issues shown in the figure—Stage 1, emerging issue; Stage 2, existing issue; and Stage 3, disruptive issue. Effective issue management responds to emerging concerns so that the goals of both the organization and the community are met.

A public issue is defined as a statement made by an individual that can be acted upon. If citizen language is too general ("I'm against growth" or "You can't trust government") we call it a theme that may reflect a community value and may reflect possible issues which further communication can reveal ("I'm against growth *because* there is not enough park space for my kids to play," or "You

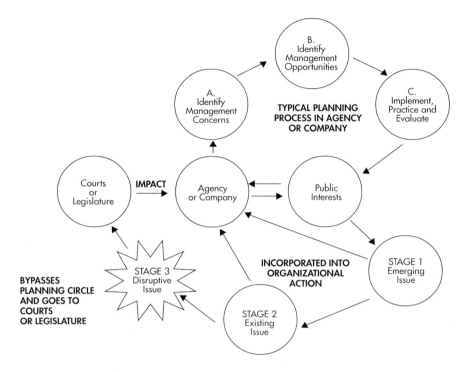

FIGURE 3.2.—The process of issue management.

can't trust [a particular agency] *because* it understated the true board footage offered in the timber sale'').

In the figure, the three circles at the top show the typical planning process of a company or agency—(a) identify concerns, (b) identify management opportunities, and (c) implement management practices and evaluate. The figure indicates that if public issues are identified in the emerging stage, or even in the existing stage, they have the greatest chance to be incorporated into organizational planning. Thus, emerging issues can be resolved using the least time and resources.

Untreated, an emerging issue escalates to an existing issue, and eventually to a disruptive issue. By definition, disruptive issues are beyond the ability of local management to resolve and are deferred to higher levels for court decisions or legislation. An organization confronting a disruptive issue is a reactive organization, bombarded with changes it cannot understand or manage. At this point, organizational staff typically project negative characteristics onto issueholders or their advocates, calling them a ''vocal minority,'' the ''radical few,'' or ''trouble-makers,'' and communication suffers further.

Organizations work best that can place into cultural context the information they receive from the social environment, and that can respond in an effective and timely fashion. The broadest public involvement is generated when emerging issues become the basis for management and decision making. As noted, unresolved emerging issues typically escalate to become existing issues and then disruptive issues. Then, they often are appropriated by formal bodies like environmental or industry groups, which use them to bolster support for their own political, economic, or ideological agenda, to be imposed on the local physical–social habitat.

During the past generation, legal and political developments permitted the authors and others to apply the model shown in Figure 3.2 in various settings. Two stories are related here to illustrate issue management in action in a bio-social context—the Beaver Creek ski development and the Smuggler mine site cleanup, both in Colorado.

The Beaver Creek Ski Development

One of the earliest applications of the bio-social ecosystem approach to resource management evolved from impact management of a proposed ski development at Beaver Creek, Colorado, by Vail Associates. This development, which began in 1971, was engineered to save the historic Hispanic mining community of Minturn, local wildlife, and family ranches. The National Environmental Policy Act had just been passed, and the proposed development provided a chance to test the new law.

The U.S. Forest Service used issue management to address off-site effects beyond the White River National Forest boundary. By incorporating the human habitat into the early stages of Beaver Creek decision making, potential acrimony was avoided during preparation of an environmental assessment. The U.S. Forest Service, much to its surprise, learned that its forest boundary was merely administrative. They drew new boundaries to combine the social and physical environment in a single bio-social ecosystem.

This early recognition and care in dealing with the community saved both the Beaver Creek project and the town of Minturn. Issue identification and resolution were so complete that the Beaver Creek Ski Area was approved in 1976 at the environmental assessment level. It never rose to the environmental impact statement level, which is required if "significant" effects are discovered during the environmental assessment.

The main employer in this small mining town, New Jersey Zinc Mine, closed in 1977, and the miners were laid off. This left them vulnerable to being driven from their valley by an intruding economic force with which they were culturally unfamiliar: the proposed skiing and resort–catering industry.

This emerging issue was handled during the environmental assessment by cooperative issue identification among the Minturn community, the U.S. Forest Service, and Vail Associates. Construction permit conditions were written (1) to enhance Minturn's economic diversification, leading to its long-term stability, and (2) to protect the environment.

At Minturn, 200 mining families and their community habitat were retained through a developer-funded small business ownership program to help the miners become business owners. Today more than 30 Hispanic-owned businesses employ more than 400 people throughout this area. A Manpower Conversion program was developed to move people into new skilled jobs, such as ski instructor and lift supervisor. A Life Cycle Mitigation program encouraged Hispanics in high school to attend college for hotel, restaurant, and ski area jobs that included management positions.

Thus, using the bio-social ecosystem approach and issue management, both the physical integrity of the community and the habitat for surrounding wildlife and ranchlands were preserved. The U.S. Forest Service, with help from the Minturn community and Vail Associates, secured US$5.9 million from the Land and Water Conservation Fund to purchase 3,000 acres of private land surrounding Minturn. This created a "green belt" buffer where residential and commercial development otherwise would have occurred. These sociophysical environmental enhancement programs were estimated in 1993 to be worth more than $1.2 billion as a cash credit to society. (For a complete discussion of this process, see Larsh 1995.)

This case study is refreshing compared to the many stories of resource development where people lose jobs, leave the community, and open land and natural ecosystems are destroyed. Because management recognized the value of the bio-social ecosystem approach and made creative use of issue management to address boundary concerns, the productive harmony model worked, and no personal or financial destruction occurred in Minturn.

The Smuggler Mine Site Cleanup

In 1989, the USEPA was given responsibility of cleaning up a Superfund site, the Smuggler mining site in Aspen, Colorado. People here live atop lead-bearing mine tailings, so cleanup threatened massive disruption. In human geographic terms, the mining site is a neighborhood unit within the larger Aspen community, physically separated from the larger community by the Roaring Fork River, with access limited to two main entrances and a foot-and-bike path to the community. This isolation created a strong sense of identity among working residents, who help one another solve everyday problems. They communicate through word-of-mouth and in gathering places like their kitchens and the community center. To outsiders like the USEPA, this community looked like a disparate group of people who were powerless to deal with strong outside forces. This turned out to be a completely inaccurate assessment.

The Superfund law is a litigation law. Once an area is placed on the National Priority List, the people in that area are in litigation. In such a situation, the USEPA relies on their regulatory and legal authority to impose solutions on the people at Superfund sites. It was announced that each individual homeowner was personally liable (a "Personal Responsible Party") and therefore responsible for the $16 million cleanup bill. This took the citizens by surprise, which evolved into fear, and then into anger. When they tried to talk to the USEPA representatives, the agency had nothing to say about the specific points of the cleanup, the worst of which was to remove a 24-inch depth of dirt from their lawns and replace it with clean fill. Trees would be lost and homes disrupted. The USEPA failed to recognize that people who have such a strong sense of place and who understand their geographic boundary can become formidable opponents of a cleanup.

The community mobilized to fight the cleanup. They bridged out of their neighborhood boundary and involved influential people from the greater community of Aspen, and eventually from the region and the nation. During the next two years, they launched a battle that drove the USEPA back from its original plans and changed the way Superfund is applied in mine cleanup sites nationwide. The bio-social ecosystem approach was used to help the USEPA understand what was happening and how they could relate in a more culturally sensitive manner to the residents. The learning process for the USEPA was to realize that their administrative and legal world, although right by law, had to fit into the cultural world of the residents. The USEPA learned that they had to communicate through the local system of networks and gathering places and not just through formal meetings and newsletters.

Once the USEPA understood how to recognize and communicate with the informal networks within the Smuggler area, they better understood the values of those who opposed the cleanup. Through the participation of local people, who demanded and obtained the appointment of nationally recognized lead experts to evaluate and monitor the site, the design of the cleanup was changed to accommodate the values and ideas of the residents. For instance, instead of stripping the entire site (over 400 homes), the residents suggested "hot spot" testing and cleanup along with a health monitoring program. The USEPA finally agreed to this alternative (USEPA 1993).

In this case, the agency initially did not understand the importance of the human geographic boundary that people saw as their protection. As long as the agency was outside the boundary and intruding, they were seen as the enemy. Once that boundary was understood and the agency stepped inside to work within the cultural context of the community to resolve their issues, solutions were forthcoming. The cleanup is now proceeding, creating productive harmony between the physical resource base and the social ecosystem of the area (Preister 1989).

A NEW COOPERATION: THE APPLEGATE PARTNERSHIP

The 1990 decision by the U.S. Fish and Wildlife Service to list the northern spotted owl as a threatened species created significant economic change in the Pacific Northwest, particularly in southern Oregon. Public timber sales were halted, perhaps one third of timber-related employment was lost during this period (for various reasons), the politics of forest management became even more contentious and polarized, and logging on private land escalated sharply.

Other social and demographic changes were occurring as well. Both the retirement population and the in-migration of displaced urbanites were increasing. The economic shift from basic production (farming, ranching, and fishing) to the trade and service sectors deepened and widened. Tourism and recreation became more important. Suburbanization of the countryside became more widespread with higher-income, better-educated residents owning larger homes.

The local economic base continued to shrink and the new norm became commuting to urban jobs by rural residents.

In this environment lies the Applegate River watershed. It is the focus of the following sections.

The Applegate Partnership

The Applegate River watershed in southwestern Oregon contains nearly one-half million acres. It is a biologically diverse area that escaped the last glaciation and served as the gene pool for the repopulation of flora and fauna in the region (Wallace 1983). Its rural nature and high percentage of public land meant that it was hit particularly hard by the changing politics of forest management resulting from loss of old-growth forests, automation of the timber industry, and listing of the northern spotted owl. Against the backdrop of social, demographic, and economic change, the Applegate Partnership was formed in 1992 as a response to the political gridlock affecting timber production and environmental protection. It was begun by residents, environmentalists, and timber industry representatives. The partnership later included personnel from the U.S. Forest Service and the U.S. Bureau of Land Management.

The Applegate Partnership is community based. It encourages and facilitates the use of natural resource principles that promote ecosystem health and diversity. Its major objective is to link community health and forest health. The partnership's early success in promoting collaborative planning among diverse interests prompted visits by Interior Secretary Bruce Babbitt and other officials, and it was cited in President Bill Clinton's Northwest Forest Plan as a model of community-based forestry.

The early history of the partnership was invigorating because participants were able to break through stereotypes and misconceptions about the "other side." Outside observers variously characterized the Applegate Partnership as the pawns of industry or of environmental groups. Locally, it was perceived to be an environmental group, and long-time residents showed little interest in participating. Over time, however, the partnership created a diverse base of participation that includes farmers, ranchers, miners, loggers, timber lobbying associations, and environmental groups. Its strength has been based on

1. a nonideological, practical approach to issue resolution;
2. remarkable individuals who have developed trust in each other despite enormous differences in outlook and despite regular and serious conflicts;
3. frequent contacts, most often through formal weekly meetings and informal contact in gathering places; and
4. a commitment to outreach and to broadening the base of its community contacts.

The partnership called for both physical and social ecosystem assessments, the ultimate aim being to integrate the two over time so that community and forest health would be synonymous within newly drawn human geographic boundaries. Reports documented the "state of health" in the physical and social

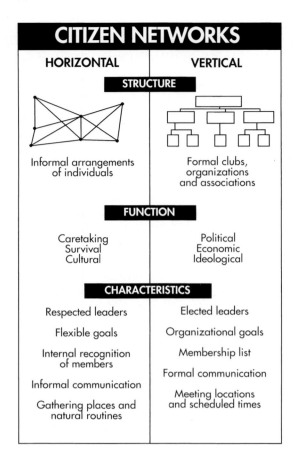

FIGURE 3.3.—Two forms of human organization.

environment (Applegate River Watershed Council 1994; USBLM et al. 1994; LaLande 1995). Preister (1994) performed a community assessment using the methods described in this chapter. An economic assessment followed (Reid and Russell 1995).

Horizontal Systems: Key to Bio-Social Ecosystem Management

Humans organize themselves vertically and horizontally (Figure 3.3). Vertical organization is hierarchical and formal, with established membership, leadership, chains of command, and formal communication and gatherings. The pervasiveness of vertical organization developed during the industrial era, and was born of society's need for centralization to meet consumptive demands. The function of this formal system is to maintain political, economic, and ideological control.

As the industrial era has shifted to the information age, the formal system has grown increasingly ineffective because of its inability to decentralize and link to the more informal horizontal or grassroots system. The informal system is characterized by social networks that support individuals in predictable ways, flexible goals, word-of-mouth communication through daily routines and gathering places, and respected leadership. The function of the informal system is survival, maintaining culture, and caretaking.

This formal–informal distinction guided our work in the Applegate watershed so that a balanced assessment was accomplished. There was surprising consensus at the informal community level about forest management (Preister 1994). Despite all the press rhetoric about differences, and the expectations of natural resource agencies regarding differences, residents reported common agreement about what constituted "balanced" forestry practice

1. need for management,
2. no clear-cuts or use of selective cuts,
3. harvest dead and dying trees,
4. use innovative ground methods of harvest,
5. keep canopy intact,
6. protect riparian (riverbank) areas,
7. avoid southern and western slopes,
8. plant sooner and link success to future sales,
9. use drought-tolerant species,
10. protect species diversity,
11. hold everyone accountable for their actions, and
12. create value-added economic activity.

Two observations can be made about the politics of timber harvest in light of this list, using Figure 3.3 as a guide. First, controversy and competition occur among regional and national leadership of the environmental community, industry, and land management agencies. The formal organizations fight over political, economic, or ideological concerns, whereas at the informal level of community in the valley, the high level of agreement about "balanced" forest practices was born pragmatically from issues of caretaking, survival, and cultural values (Preister 1994).

Statements from Thomas Freeson, a third-generation farmer and logger in the Applegate Valley, illustrate the dichotomy between formal and informal social settings. At the first public hearing regarding water and fish in the valley, his public statement was, "The hell with the fish. My family's more important!" This was a remarkable and telling statement because, in informal discussions with author Preister in Freeson's kitchen, Freeson had expressed pride in his front-yard sequoia, the rookeries down by the river, and all he had done to maintain his habitat. In everyday life, Freeson practices ecological values; he is an informal leader who has been influential in his community regarding restoration; and in late 1995, he joined the board of the Applegate Partnership. But the public setting, with its atmosphere of politics and regulation, brought out his worst.

The second observation about the politics of timber harvest is that the polarized nature of the debate, which took place in formal meetings, missed local knowledge and cultural aspects of forest management. For example, it was discovered that the logging culture in the Applegate Valley had strong values regarding selective cutting. In keeping with the crop metaphor that loggers use to describe the appropriateness of timber harvest, seed trees were left and clear-cuts were a breach of values. The following was stated by a well-known Applegate logger:

> This damn agency doesn't know what it's doing. We used to log selectively in this area 30 years ago and you couldn't tell we were in there. We left the trees under 16 inches and then came back in four or five years and got them. We used to plant, if we thought that it was getting too thin. We had controlled burns too. Then some fellow comes down from Portland and he starts telling us we aren't doing it right and makes us start clear-cutting.

It is true that comments like this are self-serving. But it is also true that, with the emergence of clear-cutting policies in the 1970s, large logging firms responded to the opportunity. The point is that values of the local culture were diminished by these changes. Selective cuts kept a harmonious scale of operation functioning, such as the sawmill, small-scale logger suppliers, and other support services. This scale, ecologically, is sustainable and can go on indefinitely. Clear-cutting changed the scale and local predictability, shifting power to the formal, large operators and unbalancing productive harmony in the bio-social ecosystem.

Local Stewardship Practices

When environmentalists castigate the agricultural community in public for mistakes made in the past, and when they publicly call for regulation of current problems, they ignore the history and value of stewardship shown by farmers, and the debate becomes polarized. The farmers and ranchers we met have a strong stewardship ethic whereby they protect the land in return for the productivity it provides. For example, one local stewardship practice involved farmers along the Applegate River. They used to have an informal management system for minimum streamflows, fishing requirements, and swimming holes. When the Applegate Dam and Lake were constructed in 1980 (Figure 3.4), management shifted to federal and state agencies, and local participation and control were lost. In addition, as farmers recognized the destructiveness of irrigation ditches to fish populations over time, they worked out cost-sharing procedures with the state, installing fish screens and shifting to sprinkler irrigation.

Our point is that this stewardship ethic must be rewarded and used to create cultural alignment between old and new practices. The current debate over pesticides, discussed over morning coffee in the gathering places, will result in behavioral shifts as well. If professionals address this concern at this emerging stage and work within informal networks, it will shorten the time required to implement changes.

The lack of incentive for management to talk with citizens and the polarization of public debate have caused the loss of local knowledge and practices. For example, on agency tours of proposed timber sale areas, author Preister has seen

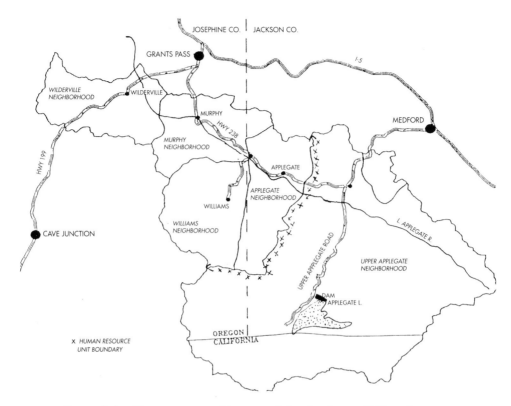

FIGURE 3.4.—Human geographic map of the Applegate Valley, Oregon.

the wildlife concerns of residents dismissed because they were not grounded in the science of the agency. The dismissal could be observed in the body language of agency personnel, who physically turned away from the speakers; in the failure of agency people to take notes about the concerns for possible follow-up; and in disparaging remarks made after the residents had gone home. Here, local knowledge and interest in the forest could have been used to further educate citizens, and the perceptual worlds of the agency and the residents could have been brought closer together, as happened in the Smuggler case.

Through the Applegate Partnership, we have been able to use the community assessment results to build participatory watershed restoration. Part of the effort has been to help the federal agencies adopt methods of geography-based citizen contact and follow-up. Traditionally, contact between residents and agencies often was limited to formal meetings and the use of outdated mailing lists. But under the partnership, when the USBLM proposed a modest timber sale involving "thinning from below" (taking younger age-classes, and leaving the older), the USBLM staff met informally with people who identified their issues and reasons and even brought neighbors to participate in the discussion.

In the past, agency people had no training in "closing the loop" with citizens.

That is, citizens were never recontacted regarding whether or how their concerns were addressed, nor were they included in the development of opportunities. But when the agency released the environmental assessment, the concerns of local residents were highlighted as the basis of its analysis. When the environmental assessment was presented at gathering places to inform residents, the citizens felt involved and remained in the discussion and planning. This process has remained stable because it respected the permanence of the local culture. People were not taken by surprise, nor was the integrity of their social ecosystem compromised.

This early learning paid off in the USBLM's thinning and fire hazard reduction program on lower Thompson Creek in the central Applegate Valley. This program was designed to implement ecosystem-based management and is among the first to incorporate ecological, economic, and local community objectives. A solid, positive, and enthusiastic partnership was created between residents and the agency over the course of action. Earlier attempts using traditional methods of contact had failed—a mass mailing attracted only nine people to a meeting. On the other hand, by using the social ties of Applegate Partnership members, 30 to 50 people regularly began attending the numerous field tours. Casual conversation worked best to access the strong informal networks in the area. The level of trust grew so high that a local citizen represented the agency on a tour for residents from a nearby subwatershed.

Through this informal process, citizens informed the agency about the multiple uses of the back roads, the areas that residents considered most in need of treatment, and the timing and sequencing of these activities. In turn, residents became educated about the science of fire hazard assessment and treatment. The suspicion that timbering was the hidden objective of this thinning project was laid to rest, and people developed trust in the agency's true motive of fire reduction. This trust eliminated the timbering issue, allowing the project to proceed.

A map of the human neighborhoods of the Applegate watershed has guided our work along several fronts (Figure 3.4). Two "social scales" are depicted. The five *neighborhood resource units* are predictors of very small-scale social ties and issue distribution. The two larger *human resource units*, one in Josephine County and the other in Jackson County, are an aggregation of neighborhood areas that share a common history, economic base, and values.

Traditionally, the boundary line between the two human resource units has roughly corresponded to county boundaries, but the line has shifted eastward to the Thompson Creek drainage. This happened because of a growing perception that Jackson County is too "urban" (containing the city of Medford) and because the area between the county line and the Thompson Creek drainage has a greater affinity with the "rural" identity of Josephine County. A vote by residents of this area in 1994 to join the Josephine County school district rather than the more urban Medford district in Jackson County is evidence of this rural identity.

The neighborhood map, drawn to reflect the physical and social ecosystem boundaries, guides project planning, public participation, and issue management. It has been used specifically to organize forestry workers for reviewing potential sales of small-diameter trees, to guide project review in other programs, to prioritize sites and treatment approaches for the fire hazard plan, and to guide education outreach. Once the human geographic boundary is drawn, resources can be managed to increase permanence and diversity within this bio-social ecosystem.

In addition to working with federal agencies, the partnership has now become the Applegate River Watershed Council, under the auspices of the state of Oregon's Watershed Health Program. The council's purpose is to develop and implement restoration projects in the watershed. The broadened contact, appreciation for local concerns, and heightened awareness of the Applegate Partnership created trust and goodwill regarding the program.

Initially, we heard language like this in the Applegate Valley: "This is another government program"; "I don't need government help to do things I've been doing all along"; "This is another ploy by the state to get control of our water." Such language characterized and fractured programs in other areas of the state, where successes have been much less dramatic. In contrast, the cooperative approach to watershed health in the Applegate Valley has led to several accomplishments

1. grading, graveling, and developing sediment catch basins on the Rush Creek Road on the Little Applegate River,
2. building a new headgate with a solar-powered fish screen on the Kubli Ditch near Rush and completing the Laurel Hill ditch headgate and fish screen,
3. installing fish structures and rearing habitat along the Little Applegate River,
4. developing habitat for the northwestern pond turtle along the Little Applegate River,
5. planting 100,000 hardwoods and conifers on more than 200 private properties throughout the watershed,
6. working with the U.S. Forest Service and the ranching community to develop an experimental project on the Applegate grazing allotments, and
7. producing the bimonthly *Applegator* newsletter, which is distributed to all 12,000 residents of the Applegate watershed.

These accomplishments are easy to list in retrospect, but the following story reveals the process one must go through to become grounded in the local culture.

Nobody Talked to Johnnie Ray

Johnnie Ray (J.R.) Fisher is the "last and the youngest" of the old-timers in the valley. His family owned one of the area's last sawmills and J.R. operates a feedlot on his property. He is valued by old-timers as a hard worker—as one of the few "custom" farmers, he leases property from others for growing alfalfa and other hay crops. He is derided by newcomers and environmentalists because of the smell of his feedlot and the degradation of the stream caused by having the feedlot so near. Agency people generally fell into this latter critical category, and for years they drove by his property secretly wishing it would disappear.

Nobody, however, had talked to Johnnie Ray. Finally, author Preister and a U.S. Forest Service employee went to make his acquaintance within his natural habitat. We stood in his muddy field in our street shoes, listening to the idle of J.R.'s tractor while he considered our opening greeting. Eventually, after we began to squirm in the silence, he said, "You know, it's all about communication." This statement opened the door for a dialogue based on acceptance and respect. We informed him about the Applegate Partnership and its goals. He described the challenges of his agricultural operation. We learned from this conversation that he was passionate about soils, one of his key interests. Within two months, a soils specialist from the U.S. Forest Service and Preister visited his home for about three hours in the evening, talking soils and eventually getting around to the feedlot. He smiled and said, "I know what my reputation is out there. Times change and so do I."

We learned that J.R. had intimate knowledge of his resources but needed a personal connection to create a pathway for becoming involved in resource restoration. From these initial contacts in his home, J.R. applied for a state watershed health program grant to move his feedlot back from the stream, to fence cattle out of the stream, and to plant several hundred trees. The restoration project is currently underway and has been chronicled in the local newspaper.

Next Steps

When selective burning was eliminated in the Applegate Valley 80 years ago, the consequence for the physical environment was the creation of densely overstocked stands of small-diameter trees. The drought cycle of the last 12 years, just coming to a close, favored the proliferation of insect pests. This has further decimated tree stands, leaving large areas of dead and dying trees. Representing the loss of permanence and diversity, productive harmony has been put directly at risk because of the lower land productivity and the risk of catastrophic fire. Fire hazard is probably the number-one environmental concern in the valley, with residents pointing out that they have addressed thinning and insect problems on their land, only to look over the fence at public land to see that the same had not been done.

As called for in the Northwest Forest Plan (USFS and USBLM 1994), the agencies are proceeding with a fuels (fuelwood) management program. The professionals bring their scientific and technical information to this program, such as identifying the highest risk areas with respect to vegetation, slope, and aspect (which direction a slope faces); sensitive wildlife habitat; air quality standards; wind patterns; and so on.

The community residents have their own standards for fire hazard, because they live there and are threatened by potential wildfires. Residents have identified their own fire-hazard priority sites: areas at lower elevations, areas next to homes, and areas that are insect-affected. High-elevation sites and previously unroaded areas are ranked low in priority. Residents are directly influencing which areas are selected, the nature and timing of treatment, and long-term elements of the program.

Agreement is being reached through the horizontal, informal communication network to focus on pragmatic, environmentally and geographically sensitive approaches. By working directly with the stakeholders, rather than large groups representing political, economic, or ideological interests, specific management programs are being designed that fit the local ecosystem requirements.

These and other coming projects illustrate the importance of maintaining permanence and diversity in the bio-social ecosystem. The efforts are clearly directed toward forest health (maintaining biodiversity and promoting long-term stability) and are oriented to include community health as an equal partner (diversity of economic options, current stewardship interests, and participation in decision making).

THE CIVIC CULTURE MODEL

We have stressed the importance of working within a cultural context to achieve successful watershed restoration. Our stories illustrate the importance of informal networks, word-of-mouth communication, gathering places, local knowledge, mutual respect, sensitivity to emerging issues, and the human geographic boundaries within which people bond to their land and their community. These features are the basis of a social–ecological understanding of human habitat and its relationship with the physical environment. We maintain that social ecology must be fully integrated with physical resource management to achieve successful watershed restoration. Cultural restoration is often the key to ecological restoration, and is embodied in the concept of productive harmony in the National Environmental Policy Act.

One of Alexis de Tocqueville's key observations of the United States in the 1830s was the impressive breadth and depth of our civic culture—numerous civic associations, for example, dealing with everything from religion to commerce. Today, however, in a 1995 *Journal of Democracy* article titled "Bowling Alone," Robert Putnam points out that "By almost every measure, Americans' direct engagement in politics and government has fallen steadily and sharply over the last generation" (Putnam 1995). His facetious measure for this change is bowling: he observes that more Americans than ever before are bowling, but that they are less involved in leagues and are "bowling alone." Putnam's observations concerning participation in politics and government relate to the collapse of citizens' trust in formal structures. However, he does not account for the engagement people have in directly influencing their specific geographic setting.

People or organizations that wish to restore watershed health must understand and use the invisible but powerful informal horizontal networks that local communities employ for day-to-day survival, cultural maintenance, and caretaking. These informal networks represent a new civic order for the information age and must be identified and engaged during watershed planning and restoration.

Figure 3.5 illustrates a model of civic culture with three levels. At the highest level is law, involving cultural recovery from disruptive issues. The intermediate

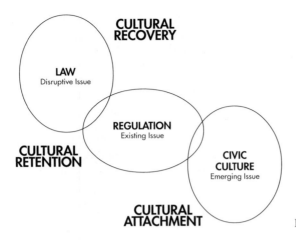

FIGURE 3.5.—A model of civic culture.

level is cultural retention, requiring regulation of existing issues. At the grassroots level is cultural attachment, where the civic culture deals with emerging issues.

Disruptive issues exceed the capabilities of local resolution, and thus must be resolved by law, using the outside political and judicial entities of society. Disruptive issues like civil rights, Love Canal, and the spotted owl, have resulted in national legislation for resolution. The legislation creates regulations, the intermediate level, with which agencies attempt to resolve existing issues.

Civic culture, on the other hand, is the setting in which emerging issues are best resolved. When you hear language like, "This is the way we do things around here," and, "It used to be that a handshake was all you would need," you are hearing civic culture in action. Civic culture is the traditional means of issue resolution at the local level. Our civic culture goes into action when you have a problem with your child's teacher, or the bus stop is inconvenient, or a street-widening happens without notification, or a timber sale is announced in your watershed. That is, you have people with whom you can discuss problems and mobilize local resources to resolve the issue. A civic culture is a true meshing of local interests and professional leadership if established leadership is attuned to local issues, indigenous methods of problem-solving, and can respond in a cultural context.

One objective of current watershed restoration is to recognize and support the civic culture. This culture is a birthright, as Tocqueville pointed out so long ago, and policies can support it, as Putnam (1995) argued. We want residents to be able to say, "Yes, we called the U.S. Forest Service yesterday and they said they'd come out this week and see about getting this area treated by next month." This is customer-oriented service, focused on professionals empowering citizens to learn and function in the best interest of their own watershed. This is done by responding immediately when people have an issue.

In the old model of resource management, these emerging and existing issues were not resolved, and they festered in the community until the agency sought support for a new program or project. Then agencies were blasted, not on the merits of the project, but because of the backlog of frustration and unresolved problems. We term this dynamic "issue loading." Issue loading often prevents agency staff from accomplishing the objectives of new projects. They end up fighting old battles over problems they no longer can resolve.

If we work within the civic culture to answer the question, "What can citizens do for themselves?" then regulation and law become less necessary. In truth, the U.S. Forest Service operated effectively for decades because it knew how to operate within the civic culture at the local level. Innumerable formal and informal agreements with constituents gave all players in U.S. Forest Service decisions a measure of predictability and confidence in the future.

But as society changed and as the agency became more centralized and commodity-driven, its link to the civic culture was disrupted, and reliance on regulation and law became the norm. Power thereby moved from informal cultural systems to formal economic and ideological groups. Today, land management agencies are shifting away from this legalistic position. Indeed, the Northwest Forest Plan calls for social and ecological experimentation in 10 regional pilot areas known as *adaptive management areas* (USFS and USBLM 1994). The current struggle is what to shift to, so as not to repeat the same cycle. We suggest that the key is recognizing humans as part of the ecosystem. The focus then shifts from "public involvement in watershed management" to a real understanding of the existing close relationship of humans to the land.

The advantage of staying in the civic culture is that it builds on the existing social and physical assets of a geographic area, and avoids the pitfalls of an outside regulatory approach to problem solving. In highly disruptive situations, laws are relied upon to deal with the disruption because the civic culture was not used. As an example, for all the money spent on formal approaches to drug problems—more law enforcement, more jails, the "three strikes and you're out" policy—the only lasting solution takes place when citizens physically and socially reclaim their streets and neighborhoods. The same is true for watershed restoration.

PUTTING SOCIAL ECOSYSTEMS TO WORK

The traditional regulatory approach is turned on its head through social ecology. Instead of asking, "How can we get those South Sea Islanders to stop killing sea turtles, or farmers to stop polluting watersheds by overfertilizing, or logging companies from clear-cutting?" we should be asking, "How can we use the social ecosystem to encourage South Sea Islanders to sustain the turtle population, farmers to maintain fish habitat, and logging companies to selectively cut?"

The National Marine Fisheries Service is considering listing as threatened numerous individual stocks (local populations) of coho salmon and steelhead (rainbow trout) in the Pacific Northwest. If listing occurs, farmers and others expect heavy human and financial cost. Also anticipated is rancorous public discord from political, economic, or ideological interests. The state of Oregon is planning to avoid additional listings by relying on more volunteer measures and by working with regulatory agencies.

The development of site-specific knowledge (Applegate River watershed, Rogue River basin, and others) will improve the accountability of each watershed. As local residents put it, they want a "report card" (that is, a social accounting) that will show progress toward a goal and give credit for the many thousands of stewardship activities by local residents—tree planting, erosion control, streambank stabilization, irrigation ditch rehabilitation, and water conservation (Loucks and Preister 1995). The authors see the emergence of such social accounting as indispensable to measuring the benefits of bio-social management and watershed restoration.

With the proper bio-social ecosystem boundaries drawn around the watershed, problems can be addressed at the optimum scale. An issue that cannot be resolved at the level of the local bio-social unit can be addressed to the next level for resolution. This process allows citizens and agency personnel to track the issue and to mobilize resources to resolve it. The social ecosystem must become visible if the civic culture is to be mobilized to participate in bio-social ecosystem restoration.

ACKNOWLEDGMENTS

This is an expanded version of a paper presented by the authors at the 1995 Annual Meeting of the Society for Ecological Restoration, Seattle, 15–17 September. The authors wish to thank reviewers Jack Williams, Martha Hahn, Russ Kraph, and Charles Pregler for their care and thoughtfulness. Any errors or misjudgments, of course, remain ours.

*Every country has three forms of wealth: material, cultural,
and biological. The first two we understand well because they
are the substance of our everyday lives. The essence of the
biological diversity problem is that biological wealth is taken
much less seriously. This is a major strategic error, one that
will be increasingly regretted as time passes.*
—*E. O. Wilson, 1992*

CHAPTER 4

CONCEPTUAL ROLES OF BIOLOGICAL INTEGRITY AND DIVERSITY

Paul L. Angermeier

Ecosystems are valuable to society for many reasons. Some value stems from our consumptive uses, such as water supply, agricultural production, or hunting. But a major share of the value—an underestimated share—stems from benefits that are not consumed directly, such as maintenance of life-supporting conditions (including clean air and water), aesthetics, and most recreation. Much of the perceived value of ecosystems is linked directly to a broad array of biological elements, including genes, species, communities of organisms, and landscapes. These elements can be grouped into three categories: taxonomic, genetic, and ecological (Table 4.1). They have evolved through and are maintained by a broad array of natural processes, including mutation, speciation, predation, competition, and disturbance. Human uses of ecosystems can cause direct loss of biotic elements through overconsumption or indirect loss through alteration of key processes.

As the human population continues to grow exponentially, so does its consumption of natural resources and its impact on regional landscapes and the global environment. Human impact eventually overwhelms ecosystem limits,

TABLE 4.1.—Biological diversity encompasses three independent hierarchies—taxonomic, genetic, and ecological. Organization of each hierarchy is nested, with elements at each level composed of multiple elements from the level beneath it. Thus, levels shown here are in diminishing order of size and complexity. Any element from any level (e.g., a species, a gene complex, or a landscape) may be an appropriate target of conservation or restoration. However, any comprehensive view of biodiversity must integrate all levels from all three hierarchies. (Adapted from Angermeier and Karr 1994.)

Taxonomic elements	Genetic elements	Ecological elements
Biota—all living organisms	Genome—all genetic information carried by a species	Biosphere—all portions of Earth that support life
Kingdom (Animalia)		Biome (temperate rain forest)
Division or phylum (Chordata, backboned organisms)	Chromosome set—a complete complement of unique chromosomes	Landscape (mountains with mixed evergreen forest)
Class (Osteichthyes, bony fishes)	Chromosome—a linear cell body that carries genes	Ecosystem or community (stream)
Order (Salmoniformes)	Gene complex—a sequence of genes linked on a chromosome	Guild—group of populations requiring similar food (insectivorous fish)
Family (Salmonidae, trouts)		
Genus (*Oncorhynchus*, some trouts and salmons)	Gene—a unit of inheritance, usually associated with a particular position along a chromosome	Population (rainbow trout)
Species (*mykiss*, rainbow trout)		
	Allele—an alternate form of a gene	

resulting in diminished or lost ecosystem value. In other words, natural systems lose their ability to provide goods and services. Historically, society's response to such degradation was to find another ecosystem to exploit. Now, however, Earth's finiteness is abundantly clear and our human impact is essentially inescapable. To sustain the value of ecosystems, we must become proficient at minimizing or reversing the many effects of human impact on biological elements and ultimately on the ecological and evolutionary processes that maintain life.

A crucial assumption in this endeavor is that ecosystems are resilient. That is, if the ecological conditions responsible for the original value of an ecosystem are reinstated, that value should accrue again. This is the basic tenet of *ecological restoration*.

Ecological restoration is increasingly common (Gore 1985; Cairns 1988b; Cooke et al. 1993), although the goals of its practitioners are highly variable. This reflects the lack of a unifying framework to guide restoration. Two concepts that should contribute to such a framework include biological diversity and biological integrity (Angermeier and Karr 1994), to be defined shortly. Both are central to current developments in natural resource policy. For example, protection of biological integrity is a primary goal of federal water quality legislation in the United States (see Water Quality Act Amendments of 1972 and subsequent reauthorizations), and estimated losses of biological diversity are major concerns of biologists worldwide (Wilson 1985; Ehrlich and Wilson 1991). Establishing appropriate roles for diversity and integrity concepts in the rationale and practice of restoring ecosystems should help make ecological restoration a more unified, mature discipline.

This chapter addresses five main topics: (1) the meanings of biological diversity and integrity and their relation to ecological restoration, (2) goals of restoration and the main constraints to achieving it, (3) key ecological processes and linkages requisite for restoration, (4) typical restoration programs and strategies, and (5) the use of biological monitoring to evaluate restoration success. The chapter concludes with 10 general principles for a strategy that has biological integrity as its primary goal for ecological restoration.

DEFINING DIVERSITY, INTEGRITY, AND RESTORATION

The terms biological diversity, biological integrity, and ecological restoration each may convey multiple meanings, so it is important to define them explicitly before proceeding.

Biological Diversity

Biological diversity (biodiversity) is "the variety and variability among living organisms and the ecological complexes in which they occur"; moreover, biological diversity "encompasses different ecosystems, species, [and] genes" (OTA 1987). These multiple levels are a fundamental feature of biodiversity (Reid and Miller 1989; Noss 1990), but they make the concept difficult to apply. Herein, I follow Reid and Miller (1989) in referring to the biological units at each level as "elements" (i.e., each ecosystem, species, or gene is a biotic element).

Biological diversity can be viewed in three distinct hierarchies of elements—taxonomic, genetic, and ecological. Organization within each hierarchy is nested; that is, elements at a given level are composed of elements from the level just beneath it (Table 4.1). For example, a class comprises multiple orders, a chromosome comprises multiple gene complexes, and a landscape comprises multiple communities. An element at any level of any hierarchy could be an appropriate focus for conservation or restoration, although such efforts typically focus on lower levels (such as species or populations).

Biological diversity often is misinterpreted and the term often is used inaccurately. The most common misconception is that biological diversity is synonymous with the much simpler concept of species diversity (Angermeier and Karr 1994). A second misconception is that biological diversity can be compared objectively between places or times. In fact, no calculus that would allow meaningful comparison through space or time has been developed to integrate diversity among different levels of a biotic hierarchy (such as population, community, or biome) or among different hierarchies (Angermeier and Karr 1994).

Our inability to compare biological diversity among systems is illustrated when we ask the simple question, "Is biological diversity greater on land or in the sea?" The oceans support more phyla (primary divisions of the animal kingdom) than the continents but there are more terrestrial species than marine species (Ray and Grassle 1991). Thus, neither the land nor the sea is unequivocally more diverse biologically.

Biological diversity is typically viewed as desirable to restore. However, unless restoration goals include maximizing diversity (which they generally do not), the concept of biological diversity offers little guidance on precisely how diverse a restored system should be or which biotic elements should contribute to that diversity. Too much diversity could be just as undesirable as too little.

Biological Integrity

The most influential definition of *biological integrity* is "the capability of supporting and maintaining a balanced, integrated, adaptive community of organisms having a species composition, diversity, and functional organization comparable to that of natural habitat of the region" (Karr and Dudley 1981). A more general definition, "a system's ability to generate and maintain adaptive biotic elements through natural evolutionary processes," developed by Angermeier and Karr (1994), applies to all levels of the three biotic hierarchies discussed above (Table 4.1).

Three features of the biological integrity concept are especially applicable to ecological restoration.

1. Integrity is a function of both the biological elements present and the natural processes maintaining them, whereas diversity is a function of the elements only. Thus, diversity is a component of integrity.
2. Integrity must be viewed in an evolutionary context. The elements and processes indicative of integrity are different in regions that have different environmental conditions or biogeographical histories.
3. Integrity can be examined at multiple levels in the ecological hierarchy (e.g., population, community, landscape), thereby implying a wide range of appropriate scales over space and time.

Integrity refers more to a system's capacity and resilience than to its particular state. Because community composition at any given time may reflect the influence of initial conditions, random events, or neighboring systems (Drake 1991; Pickett et al. 1992), a system with integrity may exhibit multiple sequential states. Adopting integrity as a management goal does not imply maximizing any particular process rate (such as production) or compositional attribute (such as diversity); rather, it implies maximizing similarity to previously evolved ranges of states and process rates.

Five classes of factors organize ecological systems and provide an operational framework for assessing biological integrity: physicochemical conditions, trophic base (source and packaging of energy), habitat structure, temporal variation, and biotic interactions (Table 4.2). As is evident from the table, each class contains many specific factors that can influence ecological elements and processes.

Virtually all such factors can be affected by human activity. Human impact on ecosystems typically stems from changes in multiple factors from more than one class, as well as complex interactions among factors. Consequently, restoring

TABLE 4.2.—Five classes of factors organize ecological systems and provide a framework for assessing biological integrity. In each class, some factors are marked (A) to indicate their special applicability to aquatic systems or (T) to terrestrial systems. (Adapted from Angermeier and Karr 1994.)

Physicochemical conditions		
Temperature	Nutrients	Oxygen (A)
pH	Salinity	Contaminants
Insolation (sunlight)	Precipitation (T)	
Trophic base (the food supply)		
Energy source	Standing stock (biomass)	Energy transfer efficiency
Productivity	Nutritional content of food	Complexity of trophic web
Food particle size	Spatial distribution of food	(connected food chains)
Habitat structure		
Spatial complexity	Vegetation height (T)	Water depth (A)
Cover and refugia	Vegetation form (T)	Current velocity (A)
Topography (T)	Basin and channel form (A)	
Soil composition (T)	Streambed substrate (e.g., clay, gravel, bedrock) (A)	
Temporal variation		
Seasonal	Fire	Weather (T)
Annual	Amplitude	Flow regime (A)
Climate change	Predictability	
Biotic interactions		
Competition	Herbivory (consumption of living plants)	Coevolution
Parasitism	Mutualism (mutually beneficial relations	
Predation	among organisms)	

biological integrity must be based on a broad, holistic perspective that recognizes myriad potential constraints.

Both diversity and integrity can be examined over a wide range of organizational levels, including population, community, and landscape. However, the evolutionary and ecological processes that generate and maintain biotic elements are explicit components of integrity but not of diversity. For example, information on rates and intensity of competition or physical disturbance is not required to assess diversity within a community, but it is required to assess community integrity. Diversity is a useful (and commonly used) indicator of integrity, but the two are not necessarily positively correlated. As an example, when naturally species-poor systems are transformed into species-rich systems through human intervention, biological integrity is reduced.

The distinction is not always clear between natural biological variation that indicates integrity and human-induced variation that diminishes integrity. However, physiological and evolutionary limits of biota frequently are exceeded by rapid or severe environmental changes induced by humans, especially in industrial societies (Pickett et al. 1992). The dramatic shifts in distribution and abundance of biota typically caused by such changes clearly represent losses of biological integrity. Although humans need not be categorically excluded when characterizing a system's integrity, a reasonable rule of thumb is that preindustrial conditions provide a benchmark against which subsequent conditions can be compared.

Ecological Restoration

Ecological restoration involves replacing lost or damaged biological elements (such as populations and species) and reestablishing ecological processes at historical rates (such as dispersal and succession). Perhaps the most important step in restoration is deciding which preexisting conditions to emulate—that is, which conditions to select as restoration benchmarks. A particular ecosystem can assume a wide range of states, and each state may support a distinctive suite of elements and process rates.

Some potential states reflect natural (historical) environmental variation, whereas others are induced only by humans. For example, most natural landscapes encompass a complex mosaic of vegetation types where age distribution and species composition reflect the local climate and prevailing soil characteristics and disturbance regime. (Disturbance regime refers to the timing, duration, and intensity of disruptions such as floods or wildfires.) In contrast, highly managed landscapes (e.g., agricultural) typically support fewer vegetation types and species, forming a simpler mosaic than would occur naturally.

Evolutionary context is essential to defining objective restoration goals. If evolutionary history is not the primary basis for selecting a restoration benchmark, myriad contrived anthropogenic states become equally plausible management goals. Pursuit of such goals can be justified by societal values like aesthetics or economics, but it does not constitute ecological restoration. For example, Cairns (1988b) erroneously implied that ecological restoration includes replacement of damaged ecosystem components with alternative components if the alternatives enhance the "social, economic, *or* ecological value" of the damaged system (emphasis added). By this definition, replacing the decimated native salmonids of Lake Michigan with Pacific salmonids to provide a sport fishery could be considered a restorative action, even though biological integrity is diminished.

Ecological flexibility may be appropriate for "reclaiming" severely degraded ecosystems, but it should not be central to restoration strategies. For example, abandoned mined lands are often converted to rangelands that are very different in productivity and species composition from the original landscape, but they are preferred for socioeconomic reasons (Gillis 1991). Similarly, ecological engineers often "create" ecosystems such as wetlands to perform certain ecological services like amelioration of pollution (Mitsch 1992). These manipulations can mitigate the loss of specific ecosystem values, but they are too narrowly conceived to qualify as ecological restoration. Ecological restoration should include only those actions that deliberately enhance biological integrity. Ecosystem manipulations based on goals other than integrity are more appropriately considered to be enhancement or mitigation.

Historical maps, accounts, and data often reveal ecological conditions that now are rare or unrecognized, but that help establish restoration goals. For one example, accounts from the 1800s show that North American streams and rivers contained vastly more woody debris and habitat complexity than is typically observed today (Harmon et al. 1986; Naiman et al. 1988). In other cases, historical hydrologic data provide a basis for restoring flow regimes (discharge rates, as well as daily, seasonal, and annual variations in them) in the Missouri River (Hesse and Mestl 1993), Florida's Kissimmee River (Toth et al. 1993), and in the Everglades (Fennema et al. 1994). When interpreted cautiously, present conditions in wilderness areas may also provide benchmark information for ecological restoration (Flebbe and Dolloff 1995).

The complex and dynamic nature of ecosystems probably precludes their complete restoration. By definition, a system that has been restored (i.e., manipulated) cannot be completely natural; for some, such reconstructions can never be as valued as the original pristine system (Katz 1992). Although natural systems may not be completely restorable, what often *can* be restored is a system's ability to generate and maintain biological elements through natural evolutionary processes (i.e., its integrity).

Adopting integrity as the goal for restoration does not imply that an ecosystem must be returned to its pristine condition (i.e., no exploitation). But it does imply maintenance of key elements and processes that would enable the system to move toward pristine conditions if the human impact were reduced. In theory, restoration involves putting back all natural elements and reestablishing processes at rates appropriate for maintaining the elements. In practice, restorationists rarely expect to fully restore ecosystems because too many societal constraints exist (see following section). A reasonable alternative to complete restoration is rehabilitation, which focuses on attainable replacement of elements and reestablishment of processes (Cooke and Jordan 1995).

RESTORATION GOALS AND CONSTRAINTS

Management goals for ecosystems (e.g., restoration or exploitation) are not selected by society scientifically, but are based on prevailing values. Scientists and managers rarely are charged with choosing large-scale management goals. Instead, their role may include (1) describing past, present, or future ecosystem states, (2) developing prescriptions for guiding ecosystems toward preferred states, or (3) articulating the costs and benefits of maintaining ecosystems in selected states. The integration of physical, biological, and socioeconomic expertise needed to restore an ecosystem makes restoration a truly multidisciplinary enterprise.

When restoration is the management goal, the level of restoration selected is likely to be a balance between the credibility of scientific information and the perceived socioeconomic cost of restoration. Because historical conditions typically are poorly documented, pristine characteristics of biota and ecological processes become increasingly speculative and scientifically indefensible. The short-term socioeconomic cost of restoration probably increases continuously as the ecosystem approaches pristine conditions. Criteria for balancing scientific defensibility and society's willingness to pay should be developed case by case.

Important constraints on ecological restoration are informational, intellectual, technological, biological, institutional, and socioeconomic. An informational constraint might be that critical data on historical conditions or processes simply are unobtainable. Intellectual constraints might include a narrow focus on particular biotic elements or human impact, or failure to properly synthesize available information into a workable restoration strategy. Technologically, some restoration objectives may be infeasible, such as removal of toxic sediments or exotic species. Other objectives may be biologically infeasible, such as replacing missing species or gene complexes.

Institutional constraints on watershed-scale restoration can stem from involvement by multiple institutions or stakeholders who have conflicting objectives. Resolving watershed-level issues requires a degree of coordination rarely achieved in human society (Naiman 1992b). For example, an interim strategy called PACFISH was developed to restore watersheds that support anadromous salmonids (salmon and trout that migrate from the sea to spawn in freshwater) in federally owned portions of Oregon, northern Washington, Idaho, and California (Williams and Williams 1997). The PACFISH strategy encompasses activities in 15 national forests, 7 districts of the U.S. Bureau of Land Management, 4 regions of the U.S. Forest Service, and 4 states.

The coordination required for PACFISH is substantial, yet it is minuscule compared to that expected in a privately owned system of similar size. The difficulty of coordinating multiple institutions ensures that few restoration strategies are applied at geographic scales large enough or time scales long enough to meet watershed needs.

Perhaps the most powerful and common constraints on restoration are socioeconomic. Local stakeholders may be unwilling to sacrifice immediate management benefits (e.g., flood control and timber harvest) to provide future public benefits. Such preferences typically reflect short-term economics, which may encourage unsustainable use of an ecosystem. Unfortunately, because restoration cost typically increases dramatically as ecosystems deteriorate (Milton et al. 1994), stakeholder support for restoration may become increasingly unlikely.

KEY ECOLOGICAL PROCESSES AND LINKAGES

Successful restoration requires recognition and reestablishment of key processes and linkages. Each level of ecological organization (e.g., population, community, landscape) is associated with a suite of key processes that regulate the distribution and abundance of biotic elements (Table 4.3). The long-term success of restoration is more likely if it focuses on reestablishing these key processes rather than on manipulating those elements affected by human impact. For example, local populations occasionally become extinct under natural conditions but eventually are reestablished by individuals that disperse from other populations. Artificial barriers such as dams or pollution can impair recolonization. Restoration would be made more effective by removing dispersal

TABLE 4.3.—Four levels of ecological organization each encompass representative processes and potential indicators of biological integrity. The biological integrity of a particular ecosystem should be assessed using indicators from multiple levels.

Level of ecological organization	Representative process	Potential indicator
Individual	Dispersal Growth Reproduction	Vagility (mobility of an organism) Biomass or volume Fertility
Population	Abundance fluctuation Colonization or extinction Evolution	Age or size structure of the population Persistence Gene frequency
Community	Competitive exclusion Disturbance Energy flow Nutrient cycling	Number of species Life history traits Number of trophic (feeding) levels Organic matter distribution
Landscape	Disturbance Succession (change in communities over time) Sediment transfer	Fragmentation and connectivity Number of community types Fluvial geomorphology (stream-carved topography)

barriers than by reestablishing populations through culture and release of affected species.

Physical processes are fundamental to ecosystems because they produce the habitat template that allows only a subset of all possible populations to persist (Poff and Ward 1990). In aquatic systems, all important physical processes are intimately linked to the flow regime, which is a primary determinant of habitat features such as water depth, current velocity, bottom type, and channel shape (Rosgen 1994), as well as the delivery of nutrients, sediment, and wood (Naiman et al. 1992). Because the flow regime also regulates wet and dry cycles and delivery of organic matter, it exerts considerable control over the rate of biomass production (Junk et al. 1989).

Patterns of fish community composition (which species are present and their ecological attributes) are closely linked to the flow regime (Schlosser 1985; Poff and Allan 1995). For example, in Florida's Kissimmee River, partial reestablishment of the natural flow regime quickly improved habitat quality and the status of biotic communities (Toth et al. 1993, 1997, this volume). Restoring or mimicking the natural flow regime is an essential component of riverine conservation throughout the Northern Hemisphere (Dynesius and Nilsson 1994).

In terrestrial systems, key determinants of physical processes include weather, fire, and geomorphic (land-shaping) events, such as landslides. These processes affect delivery of sediment and organic matter into water bodies, and they influence water runoff and absorption rates, which are important determinants of flow regime. Infrequent events such as landslides may be locally catastrophic for aquatic systems but essential for maintaining high regional productivity (Reeves et al. 1995). Natural disturbances throughout the Pacific Northwest reset local succession (physical and biological), thereby maintaining a dynamic mosaic of habitat types. However, large-scale conversion of old-growth forest to other land uses has increased the frequency and extent of disturbances, which has dramatically reduced the availability of high-quality habitat for anadromous salmonids (Reeves et al. 1995).

Many biotic processes interact to determine the distribution and abundance of biota, but few can be manipulated directly for restoration. In aquatic restoration, the process most often targeted is colonization by dispersing organisms. Colonization is essential to regional persistence of populations, especially those requiring widely separated habitats during their life cycle (e.g., anadromous fish) or those that periodically experience local extinction (Schlosser and Angermeier 1995). Much is unknown regarding ecological factors that determine dispersal rates and distances, but recent work in streams indicates that colonization success is related to high-quality habitat (Sedell et al. 1990), flow rates, and predation pressure (Schlosser 1995).

In terrestrial restoration, the key biotic processes include colonization and succession, especially by plants. Successful colonization by many plants is complicated by the need for symbiotic species such as nitrogen-fixing bacteria and animals that disperse seeds and pollinate flowers (Milton et al. 1994). For both aquatic and terrestrial species, enhancing dispersal and colonization generally can be done only indirectly, by maintaining appropriate habitat, dispersal corridors, and nearby refugia (localities where populations of interest persist).

To sustain appropriate interactions between physical and biotic processes, key linkages among ecosystem components must be maintained. Stanford and Ward (1992) offer a useful framework for depicting three important linkages in aquatic systems: longitudinal (upstream to downstream), lateral (floodplains to uplands), and vertical (subsurface to riparian). To varying degrees, each of these linkages regulates the persistence of populations by providing contact with refugia and corridors for dispersal.

However, these linkages are easily uncoupled by human activities such as release of toxins, construction of dams, alteration of stream channels or riparian areas (streambanks), and water extraction. The linear nature of drainage networks makes them especially vulnerable to fragmentation (Dynesius and Nilsson 1994), which ultimately may lead to widespread extirpation of species and loss of high-quality habitat (Frissell 1993; Angermeier 1995).

Much recent work has demonstrated clearly the fundamental importance of lateral linkages for maintaining aquatic productivity and habitat quality (Junk et al. 1989; Gregory et al. 1991), and the particular importance of riparian zones (Naiman et al. 1992; Osborne et al. 1993). Because lateral linkages extend throughout watersheds, restoration of upland communities is an important component of aquatic restoration. Watershed managers only recently have begun to shift focus from maintenance of particular physicochemical conditions in water bodies to more holistic protection of key ecological processes and linkages (Gore and Bryant 1988; Bisson et al. 1992).

RESTORATION STRATEGIES

Aquatic restoration involves manipulation of five components: water quality, habitat structure, terrestrial vegetation (riparian or upland), flow regime, and biological populations. These components are listed in order by decreasing frequency of manipulation. Of these five components, usually only one or two are manipulated (Gore 1985; NRC 1992a; Cooke et al. 1993; Gore and Shields 1995). Below, I briefly discuss the most common activities of past restoration to illustrate how restoration is constrained in different ecosystems.

Water quality is by far the most common focus of restoration, especially in lakes and estuaries. Quality is improved by reducing nutrient, sediment, and toxin loadings.

Restoration of habitat structure is becoming increasingly common and sophisticated. Instream structures to enhance habitat (e.g., current deflectors and artificial cover) have been installed by fisheries managers for more than 60 years. But recent advances in knowledge of hydraulic processes (physical forces of flowing water) and their relation to channel shape are increasing the effectiveness of such structures (Rosgen 1994; Shields et al. 1995). Restoration of habitat structure in lakes usually involves managing the species composition and spatial distribution of aquatic macrophytes (plants other than algae) (Cooke et al. 1993).

A key step in restoring terrestrial vegetation (riparian or upland) is to establish ground cover, often through planting. Other restoration methods focus on regulating plant succession, which in riparian zones and floodplains is strongly influenced by flooding frequency.

Flow regimes typically are reestablished by manipulating water release from dams or other flow-control structures.

Direct manipulation of populations and species is relatively uncommon in restoration, except for particularly valuable or undesirable populations or where natural dispersal is severely impaired. Stocking is often used to establish populations that have key functions, such as wetland macrophytes, or to maintain densities of valuable populations, such as salmonids. An important objective for any restoration is to ensure that stocked populations are genetically adapted to the system being restored. Restoration also includes control (or eradication) of introduced populations (e.g., the sea lamprey in the Great Lakes) and livestock.

Common Shortcomings

Strategies for ecological restoration should share three features: (1) an explicit goal to enhance biological integrity, (2) a broad perspective that recognizes many potentially limiting factors (Table 4.2), and (3) a focus on ecosystem processes rather than on achieving a particular condition. Subsequent chapters in this volume present some of the best examples of restoration that share these features.

Unfortunately, many efforts called ''restoration'' in the literature lack one or more of these three features. For example, ecosystems often are manipulated primarily to enhance aesthetics or the production of a valued commodity (e.g., a fishery). Some prescriptions and short-term objectives that stem from such goals may coincidentally be consistent with restoration, especially in a severely degraded system, but long-term objectives generally diverge.

A common shortcoming is a narrow focus on a particular element, such as a fish population, or on a single human impact, such as nutrient enrichment. Habitat manipulations and stocking programs of fisheries biologists often are designed to enhance fishing opportunities rather than to enhance biological integrity (e.g., Nickelson et al. 1992a; Newbury and Gaboury 1993).

Narrowly focused restoration is likely to succeed only in ecosystems that have experienced limited human impact. For example, a narrow restoration strategy is unlikely to succeed in Chesapeake Bay watersheds. Despite their complex history of excessive nutrient and sediment loading, overharvest of fish and shellfish, habitat loss, and altered flow regimes, restoration of Bay watersheds has been focused primarily on reducing nutrient and sediment inputs. Even if these inputs approached historical levels, the ecosystem would not be restored because of the dramatic alteration of many key elements (anadromous fish, oyster beds, wetlands) and their attendant processes (predation, particulate filtering, nutrient cycling).

Another common shortcoming of restoration attempts is their focus on particular physical or biotic states (symptoms) rather than underlying processes (causes). Symptoms typically are treated with technology or engineering, which often fails because knowledge of ecosystem organization is inadequate or because the treatments are incompatible with key processes. Much stream restoration reconstructs or stabilizes channel shape without adequate attention to flow regime or to basic hydraulic processes (Beschta and Platts 1986; Brookes 1988). Consequently, many so-called channel improvements produce undesir-able effects (Rosgen 1994) or fail altogether (Frissell and Nawa 1992). In constructed wetlands, restoring species composition is relatively common, but ecological functions such as nutrient storage often are not restored, even after many years (Gibson et al. 1994).

A related problem is the treatment of areas that are too small or treatments that are not sustained long enough to be effective. For example, restoration of large rivers typically is focused on a few selected segments rather than on major portions of the entire system (Gore and Shields 1995). Fortunately, restoration that analyzes and treats human impact at the watershed level is becoming increasingly common (e.g., Megahan et al. 1992; Davis et al. 1994).

Technological Fixes

Due to our limited intellectual and technological capability, successful resto-ration usually has less to do with skillful manipulation of ecosystems than it does with staying out of nature's way. Most ecosystems are resilient and natural restoration will occur if we allow it. To the extent possible, restoration should promote and complement natural recovery rather than attempt to repair undesired conditions.

Gore (1985) described the ecological restoration of streams as primarily enhancing recovery through physicochemical improvements in habitat, which in turn encourage recolonization and persistence of adapted populations. In contrast, much of the technical literature emphasizes highly engineered strategies for restoration. These may be appropriate when the problem is very simple (a point source pollutant) or conditions are severely degraded (an absence of vegetation), but engineered solutions typically are too simplistic or narrowly focused to restore biological integrity. Even if physicochemical and biotic conditions could be perfectly reconstructed through engineering, an ecosystem would not be restored unless the processes that maintain those conditions (self-sustainability) also were reestablished.

Heavy reliance on technological fixes is dangerous for two reasons: they may cause further ecological damage, and they promote "techno-arrogance." When ecosystems are damaged by technological fixes, limited restoration resources are wasted. At worst, such damage may further reduce ecosystem value. For example, efforts to restore streams and watersheds near Mount St. Helens in Washington State after its 1980 eruption included removing woody debris from stream channels and planting exotic vegetation on slopes that had been covered with volcanic ash (tephra) (Franklin et al. 1988). These measures did not adequately account for the needs of native biota and may have impaired rather than enhanced natural recovery. Similarly, ecological damage incurred by engineered stream structures (Rosgen 1994) and their high failure rate (Frissell and Nawa 1992) suggest that they are poor approaches to restoration.

Engineered solutions promote techno-arrogance by sustaining the myth that technology can solve complex ecological problems. A widely cited example is from the Pacific Northwest, where hatcheries and other mitigative structures consistently have failed to avert the precipitous decline in salmon stocks (local populations). This decline stems from overharvest, wholesale loss of habitat, and changes in flow regime (Meffe 1992). Therefore, real restoration of these systems requires fundamental change in the ways water, land, and salmon are used and reestablishment of the hydrological and population processes under which the salmon evolved.

Simple technological fixes to complex ecological problems often allay immediate crises, but they worsen later ones (Holling et al. 1994). Meanwhile, biotic resources continue to deteriorate, and the true long-term cost of restoration is obscured. The inability of technology to emulate real ecosystems is most pronounced when we attempt to provide cost-effective substitutes for large-scale ecological services such as soil maintenance, flood control, and water purification (Ehrlich and Mooney 1983).

The shortcomings of the technological fix are illustrated clearly in the southwestern United States, where decades of livestock grazing have transformed vast areas of native grassland into less-productive and less-diverse shrubland (Whitford 1995). Ecologically, overgrazing compacts the soil, leading to increased runoff and erosion and decreased groundwater recharge, all of which are unfavorable to native perennial grasses. Cultural implications include greater flood damage, reduced livestock production, and fewer options for land use.

In the Southwest, two types of technological fix have failed to restore native grassland (Whitford 1995). First, an African grass was introduced to reduce erosion and to provide additional livestock fodder. The exotic grass, a poor-quality forage, transformed large areas of native shrubland and grassland into inferior grassland. Second, shrubland is widely treated with herbicides and surface disturbance (plowing and chaining), which enhance short-term herbaceous production but not the reestablishment of native grasses. The biotic shift from grassland to shrubland reflects fundamental change in water–soil–plant relations, mediated by livestock grazing (Whitford 1995). Restoration of southwestern native grassland is likely to require dramatic long-term reduction in grazing pressure and innovative technological strategies to reinstate the original environmental conditions.

MONITORING RESTORATION SUCCESS

Although often neglected, monitoring is essential to assessing restoration success and to determining additional needs (Kershner 1997, this volume). Effective monitoring depends on selection of appropriate indicators (ecological attributes such as diversity or abundance). These indicators should be coupled with key elements and processes (Table 4.3) and should be sensitive to major human impacts in the region.

Some physicochemical factors may be useful indicators, but restoration monitoring should focus primarily on the biotic conditions being restored. Biological monitoring is more effective than physicochemical monitoring because (1) biota are valued directly by society; (2) biota are sensitive to a wide range of human impact; (3) biota respond to physicochemical changes in complex, unpredictable ways; and (4) biota integrate the effects of various impacts and do so through time. (Recent overviews of potential indicators include Noss 1990, Karr 1991, and Angermeier and Karr 1994.)

Each type of community within a region (e.g., stream, wetland, forest) requires a distinctive suite of indicators to assess biological integrity. The ideal biological indicator is

1. easy to measure and interpret,
2. sensitive to human impact prior to severe ecological damage,
3. sensitive to a wide range of impact types and levels,
4. able to distinguish between natural variation and impact-induced variation,
5. applicable over multiple regions,
6. helpful in identifying the cause of an ecological problem, and
7. meaningful to the public.

No single indicator is likely to meet all these criteria, but ecologists familiar with the system being assessed generally can select a suite of indicators that collectively meet all criteria.

Over the past decade, protocols (standardized procedures) for incorporating information from suites of indicators have emerged as powerful tools for assessing the biological integrity of ecosystems (Karr et al. 1986; Karr 1987;

Fausch et al. 1990). Collectively, such suites of indicators are selected (1) to have broad sensitivity to human impact; (2) to represent multiple levels of ecological organization such as individual, population, community, or landscape (Table 4.3); and (3) to reflect key elements and processes.

Information from many indicators often is converted to an index that summarizes biotic status (e.g., from excellent to poor). Indices such as the index of biotic integrity (IBI) are useful for reducing complex biological information to a simple numerical form that can be communicated easily to nonexperts (Karr et al. 1986; Karr 1991). The IBI compares 12 aspects (metrics) of the fish community being assessed with conditions expected in the absence of human impact. The metrics reflect the composition of the community, population abundances, and the health of individuals. Numerical scores for each metric reflect the degree to which the community being assessed deviates from benchmark conditions (e.g., the restoration goal). The IBI is the sum of all metric scores.

Tools analogous to the IBI could be developed to monitor and assess restoration success in various ecosystems. The original IBI was developed to assess the integrity of small streams, but analogs have been developed for rivers, lakes, and estuaries (Angermeier and Karr 1994; Lyons et al. 1996). Both fish and macroinvertebrates (invertebrates large enough to be visible to the unaided eye, such as insects or mollusks) are commonly used for such indices.

Although similar protocols for upland communities have been slow to develop, recent work in wetland and riparian communities shows promise (Croonquist and Brooks 1991; Keddy et al. 1993). Development of reliable, ecologically based monitoring tools for riparian and upland communities would greatly enhance the ability of managers to assess landscape integrity. For example, independent assessments of all major community types and of an entire landscape (based on indicators from Table 4.3) could be synthesized to assess restoration success for a particular watershed.

CONCLUSIONS

Ecological restoration is a rapidly evolving discipline that for too long has been relegated to the sidelines of natural resource management. As human impact on ecosystems becomes increasingly pervasive, the ability of regional landscapes to provide natural values, including essential life support services, will require that restoration emerge as a key management activity. Effective restoration coupled with protection of habitats that remain relatively intact is critical to conserving major portions of Earth's biological diversity.

Biological integrity plays two important roles in ecological restoration: it helps to define restoration goals and to evaluate restoration success. Only activities that enhance a system's biological integrity should be called restoration. Restored ecosystems are dynamic and self-sustaining, with all key elements viable and all key processes intact. The benefits of ecological restoration for society stem from the sustainability of myriad ecosystem values.

Specific restoration prescriptions vary widely but should be based on best-available knowledge and innovative approaches. The following ten principles provide general guidance for restoration.

1. Identify relations between human impacts and key ecological processes. Be especially aware of large-scale, long-term processes, because they are the most difficult to manipulate.
2. Account for complex ecological limitations on damaged ecosystems. Consider potential limitations among all five classes of environmental factors (physicochemical conditions, trophic base, habitat structure, temporal variation, biotic interactions—Table 4.2).
3. Recognize the system's inherent limitations—climatic, zoogeographic, physiographic, etc. Restoration goals should reflect those limitations, which vary regionally.
4. Reconnect severed linkages—stream channel with floodplain, upstream with downstream, and so on. Minimize the fragmentation of watersheds.
5. Do not expect to find "smoking guns." The causes and cumulative effects of degradation are complex and pervasive. Consequently, an individual system's response to restoration may be complex and unpredictable.
6. Eliminate the causes of problems rather than just treating the symptoms. A long-term, large-scale perspective is crucial.
7. Be skeptical of engineered fixes. Enhancing natural recovery processes, or at least not obstructing them, is more likely to achieve restoration.
8. Measure progress in biotic terms. Focus on key biotic elements, such as rare communities or sensitive species that are the ultimate arbiters of restoration success.
9. Address and monitor multiple organizational levels (such as individual, population, community, and landscape) and multiple taxa (such as fish, birds, and plants) that have differing sensitivities to human impact.
10. Evaluate restoration success and use the findings to improve subsequent restoration. Recognize the appropriate time frame for evaluating each response variable.

Ongoing examples that embrace these principles and the goal of restoring biological integrity include restoration of Florida's Kissimmee River (Toth 1993; Toth et al. 1997) and the Everglades (Davis and Ogden 1994a; Davis et al. 1994). In both systems, the major impact on flow regime stemmed from large-scale simplification of the drainage network and adjacent landscapes. Natural vegetation dynamics in the Everglades also were upset by fire suppression.

Current restoration of aquatic components of these systems is partially focused on reestablishing natural flow regimes, which largely regulate water and habitat quality. More natural flow regimes reconnect the various aquatic habitats (such as a stream channel and its adjacent wetlands) in time frames that provide crucial functions for the biota (e.g., refugia during drought).

Restoration prescriptions also extend beyond flow regime to problems of nutrient enrichment (from agriculture) and structural change in the drainage network. For example, the old (natural) channel bends of the Kissimmee River are being reconnected, and plans are being developed to backfill the artificial straight channels. Restoration success is being assessed by monitoring the response of selected populations and communities of plants, invertebrates, fish, and birds. Restoration prescriptions are being implemented incrementally under a long-term plan. Biotic responses are being measured carefully and findings are used to adjust subsequent restoration steps.

Only time will tell if the Kissimmee River and the Everglades eventually can be restored. In any case, these efforts and others in this volume indicate that ecological restoration is a maturing discipline with much to offer, both to science and society.

ACKNOWLEDGMENTS

This chapter benefited from numerous helpful comments by C. A. Dolloff, J. Kershner, and two anonymous reviewers.

The environment, to be sure, is not infinitely malleable. It limits as well as creates human possibilities, but it simultaneously reflects the actions of human beings upon it.
—*Richard White, 1991*

CHAPTER 5

HISTORICAL PERSPECTIVES

Robert C. Wissmar

A common goal for restoring watershed ecosystems is to return them to "natural conditions." This means restoring their original conditions—as we perceive them—of structure (e.g., species diversity) and function (e.g., plant productivity, feeding, and migration) This "natural conditions" goal is derived from certain perspectives, correct or incorrect, of past ecosystems. They are believed to have existed as abundant natural habitats, to have had diverse species assemblages and abundant populations, and to have enjoyed self-regulating and sustainable processes that buffered nature's cycles. Although a "natural conditions" restoration goal is laudable, it must be viewed with caution because our expectations and the capacity of restored ecosystems to recover can easily be confounded by the complex interaction of natural processes and human activities (NRC 1992a).

This chapter describes how an environmental history (retrospective analysis) of changing watershed and river conditions can improve our understanding of aquatic ecosystems (e.g., streams) and riparian ecosystems (shore and stream-bank), and thus their restoration.

Historical analysis of watersheds commonly documents the effects of human action on historical and current ecosystem structures and functions. Human actions can include natural resource extraction, other land and water uses, management practices, and socioeconomic development (e.g., agriculture, timber harvest, and urbanization; Wissmar et al. 1994). Environmental responses that are documented include numerous physical and biological ecosystem processes that interact through time and space (Pivnicka 1992; Sear 1994).

Historical baseline data often are assumed to depict habitat conditions and biological diversity that prevailed prior to some change or alteration. Subsequent applications of historical baseline data can be simplistic and therefore misleading unless the retrospective analysis takes care to estimate the temporal–spatial variability of the baseline data (McIntosh et al. 1994).

Historical studies can be focused by asking specific questions about past events: their temporal scale (years, decades, or centuries), spatial scale (from whole regions or watersheds down to sediment particles), magnitude of change (from minor to catastrophic), and location. Restoration plans for aquatic and riparian ecosystems rely on accurate answers to these questions. Thus, historical studies of human actions and natural disturbances that alter ecosystems can facilitate restoration design. For example, studies that identify the sequence of timber harvesting and livestock grazing practices that have caused continuing soil erosion and stream sedimentation can help identify both the problems and the conditions to be restored. Historical records of human and environmental change also are a basis for predicting future conditions in terms of expected recovery times and for evaluating restoration success.

The central question of this chapter is: "How do we use historical information to develop restoration plans?" Step one is to determine the overall objective of the proposed restoration. Step two is to decide what historical information could be useful in developing a restoration strategy. This requires composing specific questions to be answered by the historical analysis. Once the restoration questions have been stated and data needs identified, historical analysis can begin.

In this chapter, several historical studies illustrate the types of retrospective information that might be useful in developing restoration plans. These examples include natural physical and biological processes and human modifications that change aquatic and riparian ecosystems. The case study summaries address several questions.

1. What are the spatial and temporal landscape scales of the restoration area?
2. How can an understanding of past climate changes be used in restoration planning?
3. How can an understanding of the history of natural disturbance regimes (e.g., floods and fire events) be used in restoration planning?
4. What kind of historical study can be used to develop restoration strategies that emphasize natural processes?
5. How can an understanding of the environmental history of human actions be used in restoration planning?
6. How can historical baselines for "natural conditions" of aquatic and riparian habitats be used to identify restoration opportunities?
7. How can historical records of rivers that have been modified by flood and flow regulation be used to identify restoration opportunities?
8. What are the basic steps in conducting a historical analysis?

TABLE 5.1.—Variability of temporal and spatial scales that operate within natural river systems. These scales are important in restoration. (Modified from Sear 1994.)

Geomorphic structure (landform or stream feature)	Geomorphic function	Spatial scale (square feet)	Temporal scale (years)
River network	Water and sediment transfer	10–100 million	Greater than 10,000
Valley floor	Water and sediment storage	100–1 million	1,000–10,000
Floodplain	Water and sediment storage	10–100,000	100–1,000
Riparian corridor flanking stream	Water and sediment storage	1–10,000	10–100
Channel meander (snakelike shape of stream)	Water and sediment transfer and storage	10–10,000	10–100
Pools and riffle habitats	Water and sediment transfer and storage	10–1,000	10–100
Deposited sediment (e.g., sand or gravel bars)	Sediment storage	1–1,000	1–100
Sediment particles	Erosion product	Less than 0.01	1–100

SPATIAL AND TEMPORAL LANDSCAPE SCALES

Studies of past environmental changes can be used to define scales for restoration, both temporal (time frame) and spatial (geographic). Knowing these scales helps determine the appropriate dimensions and timetable for restoring a system. For example, studying past disturbances can help define the baseline variability of spatial and temporal scales that operate within natural river systems (Table 5.1). The largest spatial and temporal scale is that of the watershed (measured in square miles) and the smallest scale is that for sediment particles (measured in fractions of a square inch and from 1 to 100 years; Sear 1994).

A watershed can be viewed as a hierarchical system containing progressively smaller structures nested within one another (Newbury and Gaboury 1993).

- Starting at the largest scale, the complete watershed contains smaller watersheds of tributary streams and segments of the primary river valley. The river valley segments can be described by valley width and by channel gradient (slope).
- Within the valley segments lie channel reaches (stretches of channel) that have various shapes, habitats, and substrates. Examples of channel shape are straight, meandering, and braided (in which the stream breaks into multiple fingers that intertwine like randomly braided hair). Examples of habitat are quiet pools and turbulent riffles (shallow areas of swift current with a choppy whitewater surface). Examples of substrate are fine sediments, gravels, and boulders.
- The next-smaller scale includes local flow conditions within a given channel reach.
- Within local flow conditions exist the smaller surface boundary layer conditions of individual organisms and habitats.
- Additional subsurface flow systems and habitats (Stanford and Ward 1988) can exist at varying scales, depending on stream deposits and bedrock formations.

Each scale within a watershed contains valuable information about how the ecosystem functions. An even more important concept is that all the scales of a drainage system function together to create and maintain habitats. Functions commonly provide the basis for restoration plans. An example of a function is the interaction between a channel's hydraulics and shape (geomorphology) and the habitat requirements of fish and riparian vegetation (NRC 1992a; Rosgen 1994).

CLIMATE CHANGE IN RESTORATION PLANNING

A major reason for studying historical climate change is to discover the temporal scales of long-term change in physiographic conditions and the biosphere. Such information can guide restoration plans in identifying, securing, and modifying portions of watersheds that could be buffers or that could benefit from predicted climate change.

A common research procedure is to develop chronologies of climatic change that are related to different disturbance regimes (e.g., frequency and magnitude of occurrence of flood and fire events). Such chronologies indicate how various disturbances operate over different time frames and spatial scales. For example, in the Pacific Northwest, Chatters et al. (1991) developed a chronology to evaluate how past disturbances relate to long-term climate and landscape changes in watersheds. Their purpose was to discern the potential impact of climate change on chinook salmon. This study (Chatters et al. 1991) and one by Neitzel et al. (1991) examined the Holocene epoch (approximately the past 10,000 years) for insight regarding past and future climate change in the Columbia River basin (Table 5.2).

An example of such change is the elevated-temperature interval that occurred 6,000 to 9,000 years before the present (during the middle Holocene geologic epoch). It provides a model of how future warming may affect river ecosystems. The interval appears to have averaged 2 to 5°F warmer and 33 to 38% drier than today. Timberlines in the Cascade and Rocky mountains were 450 to 900 feet higher then than today (Osborn and Luckman 1988). Reconstruction of prehistoric riverine conditions made from floodplain deposits containing fossil mussel shells, and moisture estimates made from pollen suggest that stream discharge was about 30% less than the average annual runoff. The warmer water, reduced discharge, and deposition of fine substrates (fine sediments) during the mid-Holocene epoch could have occurred in watersheds throughout the Columbia River basin. If such conditions recur in the near future, perennial streams could become intermittent and the warm waters could become uninhabitable by many fish.

The value of this historical reconstruction of mid-Holocene climate is in demonstrating how future climatic warming might influence watersheds. For example, an elevated-temperature interval and drier climates combined with today's depressed salmon populations could further reduce fish abundance. Such conditions might also increase evaporation and reduce forest and riparian habitats (Chatters et al. 1991; Neitzel et al. 1991; Firth and Fisher 1992). Climate-warming combined with excessive human water use could greatly damage fish stocks (local populations) that already occupy marginal environments. Studying sequences of climate change can also provide knowledge essential to predicting the long-term cumulative effects of human modification in both aquatic and terrestrial ecosystems.

NATURAL DISTURBANCE REGIMES IN RESTORATION PLANNING

Significant examples of natural disturbance in more recent times (0 to 6,000 years before present) include floods, avalanches, debris flows into stream channels, fire, wind, glacial activity, and tectonic events such as volcanic eruptions and earthquakes (Table 5.2). The frequencies of large wildfires and floods are of interest because they recur on decadal intervals (Wissmar and Swanson 1990; Covington et al. 1994). For example, wildfires in different types of forest stands in eastern Washington and Oregon show presettlement fire intervals ranging from 2 to 38 years for Ponderosa pine, 20 to 60 years for Douglas fir, 6 to 146 years for mixed conifers, and 25 to 500 years for lodgepole pine (Covington et al. 1994).

The frequency of large wildfires and floods also overlaps with the frequency of tectonic disturbances (earthquakes and volcanoes). Volcanic events in the Pacific Northwest include eruptions on a temporal scale of centuries. In the Cascade Mountains, Mounts Shasta, Hood, Baker, Lassen, and Rainier have shown volcanic activity within the past 60 to 200 years (Harris 1980) and Mount St. Helens experienced its major eruption in 1980 (Wissmar et al. 1982; Wissmar 1990).

Many of these events recur at intervals of decades to a few centuries. They can all reshape surface landforms and river courses and features. For example, small stream valleys can be reshaped by peak flows that occur annually, whereas larger

TABLE 5.2.—Chronology of disturbances and environmental responses in eastern Washington State during the Holocene and Pleistocene epochs. The Holocene epoch is the timespan from the present back approximately 10,000 years. The Pleistocene epoch spans approximately 10,000 to 2 million years before the present. (Modified from Wissmar et al. 1994; data from Chatters et al. 1991 and Neitzel et al. 1991.)

Epoch	Years before the present	Historic disturbance
Holocene	Less than 200 (modern)	Humans modify land and water resources
	Less than 1,000	Floods; earth debris flows; fires; eruption of Cascade Range volcanoes including Mount St. Helens
	4,000	Wind storms
	6,000	Elevated-temperature (hypsithermal) periods; Mt. Mazama eruption
	8,000	Elevated-temperature periods; wind storms; humans appear in region
	10,000	Elevated-temperature periods; floods, earth flows
	12,000	Glacier Peak eruption
Pleistocene	Greater than 10,000	Floods, tectonic uplift of land
	18,000–22,000	Glacial Lake Missoula floods much of eastern Washington State
	Around 50,000	Glacial ice dams breach; flooding

areas tend to be altered by high flows and channel adjustments that recur during major floods on decadal scales. These floods, along with the accompanying erosion and deposition of materials, contribute to the formation of riverine habitats (Benda et al. 1992). Knowledge of the history of different natural disturbance regimes contributes to our understanding of how natural ecosystems function over long periods. Such retrospective perspectives also provide a basis for assessing how and when human actions might interact with these processes.

RESTORATION STRATEGIES AND RETROSPECTIVE MODELS

Retrospective studies of riparian ecosystems can be used to develop restoration strategies. Several such studies point to the importance of natural disturbance frequency in influencing the stages of plant succession in riparian ecosystems (Wissmar and Swanson 1990; Gregory et al. 1991; Auble et al. 1994).

Figure 5.1 provides a conceptual model of how riparian forest patches might respond to different disturbance frequencies (higher frequency on top, lower frequency on bottom). In riparian forests along mountain streams, where disturbance can be chronic to frequent (e.g., beavers damming streams or flash flooding), the succession of forest patches may be interrupted by events that recur on a scale of years to several decades. Consequently, the successional stages of the forest community commonly remain young. During recovery, the stand initiation stage (initial development) and exclusion stage (emergence of specific plant species) tend to be continually reset by each disturbance.

In some locations, disturbances are less frequent, examples being large floods and wind storms that recur every 100 years. Such is the case in riparian areas of lowland rivers having broad floodplains, where the infrequent disturbances might give succession the opportunity to proceed beyond stand initiation and understory exclusion and persist in understory reinitiation and old-growth (Figure 5.1, bottom; Wissmar and Swanson 1990). Thus, the ability to predict these successional changes could help in setting long-term restoration objectives.

TABLE 5.2.— extended

Physiographic response	Climatic trend
Erosion and deposition Volcanic ash deposits; debris flows; erosion and deposition	Cool, moist
Erosion and deposition Stream flows about 30% less than today; volcanic ash deposition	Warm, dry
Stream flows about 30% less than today; erosion and deposition	Warm, dry
Stream flows about 30% less than today; erosion and deposition	Warm, dry
Volcanic ash deposition Erosion and deposition Catastrophic erosion and deposition Erosion and deposition	Warm, dry Ice Age begins

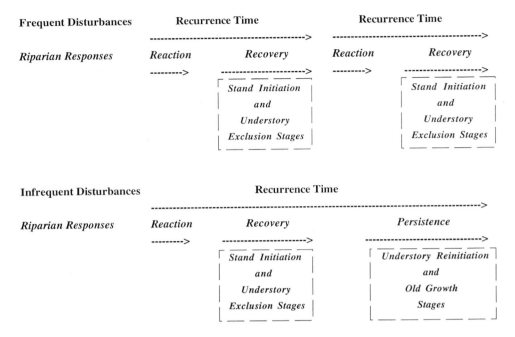

FIGURE 5.1.—How forest patches in riparian areas respond to frequent and infrequent disturbances. Stand initiation refers to the initial developmental stage; exclusion refers to the developmental stage for specific plant species.

Such objectives could be based on expected recovery times and desired future conditions within aquatic and riparian ecosystems.

A retrospective analysis of watersheds in Voyageurs National Park, Minnesota (1940–1981), demonstrates that simple mathematical models can predict how beaver activity can initiate significant change in stream and pond hydrology, nutrient cycling, and plant succession (Pastor and Johnston 1992). The study shows that beaver dams on streams changed their flow regimes (frequency and magnitude), which in turn altered the distribution and abundance of different riparian areas and wetlands. Temporal changes in four types of habitat were identified and mapped from aerial photos. The habitat changes were quantified as fractions of the total area, in four categories: (1) valley bottoms not currently in ponds, (2) flooded areas or open water, (3) seasonally flooded areas with emergent aquatic plants, and (4) moist meadows without standing water but with saturated soils most of the year.

The predicted changes in fractional areas of the four types of pond–wetland systems included three important patterns: (1) lowland areas of the watershed were occupied by ponds, (2) seasonally flooded areas were commonly a small proportion of the watershed, and (3) proportions of a watershed in the moist meadow class tended to increase and decrease in inverse relation to that in the flooded state.

The third pattern proved to be of greatest interest because flooded areas caused the major changes in plant succession. In several watersheds, Pastor and Johnston (1992) demonstrated a high inverse correlation of variations in flooded areas with moist meadows. They attributed these successional changes to variations in flooding regimes, which induced fluctuations in nutrient cycling and carbon accumulation over time.

These retrospective studies suggest that a restoration strategy for watersheds that contain connected stream–riparian–wetland habitats could use biological disturbances and treatments to induce flooding regimes and other vital ecosystem processes such as nutrient cycling and plant development. Examples include the reintroduction of beavers, muskrats, moose, and other animals whose activities can modify aquatic and riparian ecosystems. Objectives for recovery and future conditions could be evaluated by monitoring the developmental patterns in flooded and moist areas of ponds and wetlands as well as successional stages of riparian forests. The ultimate monitoring objective might be the development of mathematical methods for predicting transitional trends in succession of different vegetated areas in response to the restoration treatment (e.g., natural and physical).

ENVIRONMENTAL HISTORY OF HUMAN ACTION

Environmental histories of human action during modern times reveal how environmental conditions have changed since the arrival of Europeans in North America 500 years ago. In the western United States, most human influence includes land use and hydrologic modifications made over the past 150 years (Bisson et al. 1992; Wissmar et al. 1994). Such histories can illuminate the interaction between natural and human activity because the sequence of many natural disturbance regimes coincides with landscape changes induced by humans. Where retrospective studies provide baselines for comparing past and current conditions, they can be useful in identifying cumulative effects of human activity that may continue to alter environmental conditions during future decades. These studies can be very useful in designing restoration projects.

Several environmental histories of river drainage modification in eastern Washington and Oregon show that stream and riparian ecosystems have experienced long-term cumulative effects caused by numerous resource uses and management practices (Hanson 1987; McIntosh et al. 1994; Wissmar et al. 1994). In many areas, the order of importance (decreasing) of these factors is (1) livestock grazing, (2) timber harvesting practices, (3) agricultural practices, (4) road construction, and (5) flood events.

Table 5.3 shows an example chronology of resource use and change in fish and riparian habitats in north-central Washington State. The continuous occurrence of human impacts over the past 100 to 150 years (e.g., livestock grazing, sedimentation resulting from logging and agriculture, dam-induced water loss and temperature changes, irrigation diversions), combined with neglect, has caused considerable damage to watersheds, fish, and riparian habitats.

It is important to recognize that long before Europeans colonized North

TABLE 5.3.—Chronology of major land and water use practices influencing fish habitats in north-central Washington State from 1811 to about 1990. (Modified from Wissmar et al. 1994.)

Years	Resource use	Habitat change
1990s	Management modifies watersheds and stream channels	Watersheds and streams altered by large floods and wildfires
1980s	Water management in response to drought	Fish habitat capacity decreases as water recharge and storage decline
1950s	Truck logging; increased human population and recreational pressure	Habitat loss; erosion and deposition; water uses increase
1930s	Dam construction on Columbia River	Decline in fish habitat; agricultural and water uses increase
1920–1930s	Railroad logging, mining	Habitat loss; erosion and deposition
1916	Homestead Act	Increased agricultural and water uses
1915–1920s	Large fires damage watershed	Erosion and deposition
1905	Irrigation diversions	Decline in habitat volume; agricultural and water uses increase
1890s	Large floods alter watersheds and streams	Erosion and deposition
1880–1915	Increased railroad construction; settlement; livestock overgrazing	Habitat loss; erosion and deposition
1860s	Settlers harvest salmon from rivers; large cattle drives begin	Fish stocks depleted
1858–1890s	Mining	Habitat loss; erosion and deposition
1811–1847	Fur trade	Removal of beaver; aquatic ecosystems lose water-storage capacity

America, Native Americans also modified the landscape. Forest, shrubs, and grasslands of the Rocky Mountains, the Pacific Northwest, and Great Plains were burned for various purposes. At first, burning of lowland forests helped grasslands to spread and increased the size of deer herds. In more recent times (about 200 to 300 years before present), similar burning increased the number of horses. Burning also allowed growth of berries and other useful forage. Native Americans not only altered the land with fire but diverted water to grow plants in the Great Basin (e.g., Paiutes), the Southwest (e.g., Pueblos), and other lands bordering the Great Plains (White 1991).

RESTORATION BASELINES AND HISTORICAL CONDITIONS

An interesting retrospective study in the Pacific Northwest provides baselines for comparing past and current conditions. It also identifies habitat problems that might require restoration. This study is of watersheds in central and eastern Washington and eastern Oregon (McIntosh et al. 1994). The objective was to determine how fish habitats have responded to the cumulative effects of long-term land use. The study used 1930s–1940s habitat surveys performed by the U.S. Bureau of Fisheries (now the U.S. Fish and Wildlife Service) as historical benchmarks for resurveys to assess habitat change over the past half-century.

The resurveys were conducted during 1990–1992 by the U.S. Forest Service, Oregon State University, and the University of Washington. In general, the frequency of large, deep pools in central Washington watersheds increased regardless of management history (specifically, the Methow, Wenatchee, and Yakima rivers). The increase was twice as great in unmanaged watersheds. These trends were attributed to many of the study areas being within national forests, where about 65% of the land is in undeveloped areas (wilderness or roadless).

In contrast to Washington State, loss of pool habitat was greater in eastern Oregon where less land was allocated to wilderness and roadless areas. Some of the major losses in large pool habitats (33 to 66%) occurred in the managed watersheds in the Blue Mountains. These habitat losses coincided with greater intensities of land use and their long-term cumulative impact (mining, livestock grazing, and timber harvesting). Many effects that persist today were initiated by land use practices begun prior to the 1930s, with some extending back into the nineteenth century. Common ecosystem responses include elevated water temperature, inadequate streamflow, and increased fine sediment within stream channels of many managed watersheds. In Oregon's Grande Ronde River, ecological responses include not only loss of fish habitat but of fish populations. Of the three native stocks of anadromous fish (which spawn in freshwater but live in saltwater), one is extinct and another is listed as threatened under the Endangered Species Act.

Historical studies by McIntosh et al. (1994) and Wissmar et al. (1994) indicate that restoration of stream channel and riparian areas of many watersheds in eastern Washington and Oregon will require fundamental change in upland and riparian management. Restoration also will require long-term commitment to science-based ecosystem management. For example, the long-term objective of a restoration plan for the upper Grande Ronde River calls for protecting the best habitats as well as restoring degraded ones (Anderson et al. 1992). Short-term objectives include restoring spawning and rearing habitat for anadromous fish stocks, by providing refuge habitat for fish and wildlife by increasing streamflows and reconnecting habitats of stream channels with adjoining side channels and riparian areas.

The importance of reconnecting habitats in the upper Grande Ronde River and the possible time frames required are described by McIntosh et al. (1994). This study indicated that the best opportunity for reestablishing stream and riparian habitat connectivity exists in unconstrained stretches of floodplain in the wider valleys. Historically, the unconstrained stretches showed the highest frequency of pool habitat.

RESTORATION OPPORTUNITIES USING HISTORICAL RECORDS

Historical studies of rivers that have been modified by flood control and flow regulation have been used to identify restoration opportunities (Sear 1994; Dahm et al. 1995; Gore and Shields 1995; Johnson et al. 1995). An excellent example is the Cedar River, which flows into Lake Washington near Seattle. The Cedar River is a fifth-order stream—a large river having a hierarchy of four levels of

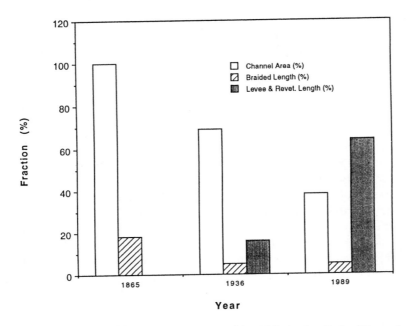

FIGURE 5.2.—Percentage of historical change (1865–1989) in the Cedar River channel in western Washington State. The lengths of braided channel, channel with levees, and channel with revetments extend from the river mouth upstream for 22 miles. Percentage is relative to the total channel length. (Modified from Wissmar and Beschta, in press and Perkins 1994.)

tributaries flowing into it. It also is an alluvial river, flowing through and continually reworking its own sediment as well as eroding streambanks.

Historical changes have occurred in the Cedar River's channel patterns and floodplains within eight reaches (stretches) of its lower 22 miles. The changes result from construction of levees, revetments (facing added to retain an embankment), and upstream dams. The chronology of these modifications was analyzed using aerial photographs that date from the 1930s and maps that date from 1865 (Perkins 1994).

Like most managed alluvial rivers, the Cedar River has responded to reduced sediment load by becoming narrower and less braided (it has fewer interwoven multiple channels; Schumm 1977). Since 1936, the placement of levees and revetments on the outer banks of most bends (the severest erosion points) has caused a 35% decrease in the river's width. The channel of the main stream currently occupies only 45% of its former active channel area (Figure 5.2). The decreased area and reduced channel complexity coincide with the loss of sediment storage within certain reaches, loss of river habitats, and isolation of the floodplain from the river.

Starting in 1904, headwater dams were constructed upstream of river mile 22 to improve flow regulation and water supply. This reduced the magnitude and frequency of sediment-transporting floods. Historical flow records indicate that under predam conditions the 2-year flood recurrence interval now approaches recurrence of once every 25 years. The combined influence of flow regulation by dams and channel constraint by levees and revetments have been the prime factors in diminishing active channel area (Perkins 1994).

This historical study points to the importance of flood events, changes in sediment transport, and flood management in determining the variability in connectivity of floodplains with channels in managed rivers, especially alluvial rivers. Supporting this are dramatic changes in sediment routing and storage in the lower Cedar River and in other rivers of Oregon and Washington following large floods in 1990, 1995, and 1996. The chronology of loss of floodplain riparian area and fish habitat and their reduced availability to fish and wildlife also have been documented in other retrospective studies of land use and water diversion (Wernecke 1936; King County 1993).

Restoration plans for the lower Cedar River watershed provide an excellent example of restorative measures designed with the aid of historical information. Knowledge from Perkins (1994) and retrospective analyses of flood frequencies and sediment rating curves by King County (1993) have identified major depositional and erosional reaches within the Cedar River. This has permitted designation of different restoration measures within specific stretches of the river.

King County has jurisdictional authority within the Cedar River catchment. The county recently enacted a Cedar River Basin Action Plan (King County 1996) that documents the current conditions and proposes solutions to problems of flooding, property damage, water quality and quantity, and declining populations of salmon and steelhead (rainbow trout). Two sets of objectives were identified. One includes resolving the hazardous flooding threat to about 130 homes by acquiring and removing them and restoring floodplain water and sediment storage through levee manipulation and removal. The other set of objectives includes protecting valuable aquatic habitats, restoring habitats that have the best chance for recovery, ensuring the long-term productivity of salmon populations, and maintaining the Cedar River's excellent water quality.

BASIC STEPS IN A HISTORICAL ANALYSIS

The case studies described above indicate that several basic steps can be taken in planning retrospective studies for restoration purposes:

1. state the objective of the restoration plan,
2. identify specific questions to be answered by the historical analysis,
3. determine what retrospective information is needed to address the questions,
4. plan the restoration (here, additional questions can focus on what types of retrospective information are needed to implement and evaluate the restoration).

Historical study design should recognize the value of pilot investigations, which can identify historical baseline data for comparing past and current conditions. Examples of baseline data types include stream channel hydraulics, geomorphic features of channels, and fish habitat conditions. The approach should be flexible to allow use of unexpected data sources that emerge during the study. Both expected and unexpected research findings can be useful in refining and improving subsequent phases of the study and in developing plans for restoration.

Important sources of historical information include old biological surveys, museum samples, field stations, university archives, diaries of amateur and professional naturalists, nonpeer-reviewed literature (agency reports, environmental impact statements), and scientific journals. Searches for historical information can be time-consuming and frustrating, especially if the central questions are poorly defined or unreasonable. This can be avoided by concisely stating the objectives, questions, and study design. Additional suggestions for conducting historical studies are in Nielsen (1995).

Once retrospective analysis begins, it is important to determine the validity of historical documents and potential flaws in past methods. For example, when possible, one should document the goals and procedures of past resource management, such as resource use, allocation, and inventory. A full understanding of historical information sources can be especially important if historical baselines are compared with contemporary conditions. In such cases, one should assure that the resurvey methods replicate historical methods.

Often, historical research generates questions about the timing of past land and water use impacts on riparian areas and streams and their legacy today. In such cases, the methods should specify data sources and procedures for identifying ecosystem change that results from modification by human action. Such studies could employ analysis of (1) tree rings in tree corings to determine the age of riparian forests, (2) a time series of aerial photos to assess stream channel change, and (3) the use of geographic information systems (GIS) technology to assess spatial change, such as the connectivity between river channels and floodplains.

CONCLUSIONS

This chapter has reviewed historical studies of aquatic and riparian ecosystems to gain insight into how past human actions and changing environments operate on different spatial scales and through time. Numerous environmental histories of watersheds are proving valuable because they demonstrate how natural and altered areas function and how they interact with riparian and salmon habitats (Sedell and Everest 1990; Lamberti et al. 1991; Smith 1993; Sear et al. 1994; Wissmar and Beer 1994; Reeves et al. 1995). Other studies help evaluate the cumulative effect of natural disturbances, human modifications, and their interactions (Bisson et al. 1992; Wissmar et al. 1994). Excellent examples of recent investigations that use historical environmental records to formulate restoration plans for watersheds and fisheries resources include FEMAT (1993), Beechie et al. (1994), Sear (1994), Dahm et al. (1995), and Lichatowich et al. (1995).

Awareness of the environmental history of human actions and changing ecosystems permits development of successful restoration strategies. Long-term perspectives of how terrestrial and aquatic ecosystems function over time and space can be used to predict conditions and to provide new options for designing, maintaining, and monitoring restoration (NRC 1992a; Frissell et al. 1993; Wissmar 1993).

Historical perspectives can also help change society's perception that today's stream and fish habitats are similar to historical ones. For example, younger people often believe that what they see now is how it has always been. This leads to accepting the status quo of environmental conditions and issues as norms, both now and for tomorrow. As a consequence, a lack of historical information perpetuates our unawareness of continually changing environmental conditions. Retrospective analysis and the increased appreciation of past conditions that it brings are central to improving environmental education.

ACKNOWLEDGMENTS

This research was supported by grants from the U.S. Forest Service, Pacific Northwest Research Station, Juneau, Alaska (grants PNW 93-0365, 93-3447, and 95-0706). The author appreciates review of an earlier version of this chapter by Peter A. Bisson.

I fear that, in recent years, too many ecologists have yielded to the temptation of finding a problem that can be studied on a conveniently small spatial and temporal scale, rather than striving first to identify the important problems, and then to ask what is the appropriate spatial scale on which to study them.
—Robert M. May, 1994

CHAPTER 6

TEMPORAL AND SPATIAL SCALES

Robert R. Ziemer

Human activities have degraded substantial portions of the nation's ecological resources, including physical and biological aquatic systems. The effects are continuing and cumulative, and few high-quality aquatic ecosystems remain in the United States. Concern about these diminishing resources has resulted in numerous restoration programs. Some are well conceived and address complex ecosystem interactions. However, most restoration begins with a broad ecosystem issue and quickly narrows because of jurisdictional politics, land ownership, user interest, funding, or time. Too often, this narrowed view leads to restoration that is well designed and well intentioned but irrelevant and ineffective. In some cases, expensive projects are conducted where they will have little effect. In other cases, a restoration project is completed only to be destroyed by the next moderate storm. In still other cases, restoration designed to benefit one component of the ecosystem severely damages other components.

A common thread through such failed restoration is that the plans consider only a particular site or problem and ignore the greater context of geography, time, and ecology. For example, restoration to address a dwindling run of anadromous salmonids (salmon or trout that live in salt water but migrate to spawn in freshwater) must not only discern the complex reasons why the run is dwindling, but how local projects might contribute to the solution. In some cases, a proposed local project may be ineffective because it covers too small an area or because of conditions outside the project area. Successful restoration is based on more than a thorough understanding of the problem. It also is based on understanding the interaction of the problem with other ecosystem components, both locally and beyond the project's boundaries.

RESTORATION AS PART OF A BROADER STRATEGY

In the Pacific Northwest, polarized views concerning use of public forest lands have produced lawsuits and counterlawsuits on widely varied issues. The resulting gridlock over federal forest management led President Clinton to convene the Forest Conference in Portland, Oregon, on 2 April 1993. Following the conference, the President formed the Forest Ecosystem Management Analysis Team (FEMAT). Their task was to identify management alternatives that attain the greatest economic and social contribution from the forests while conforming to the Endangered Species Act, National Forest Management Act, Federal Land Policy and Management Act, National Environmental Policy Act, and other laws and regulations. More than 600 scientists, technicians, and support personnel contributed to that effort.

An important part of the FEMAT (1993) report is the Aquatic Conservation Strategy. This strategy includes four main components.

1. *Riparian reserves*—portions of the watershed that govern the hydrologic, geomorphic, and ecological processes that directly affect streams, fish habitat, and riparian ecosystems (land areas flanking streams).
2. *Key watersheds*—a system of large watershed areas throughout the Pacific Northwest where genetic lines of fish can take refuge and survive despite hostile environmental changes elsewhere. These watersheds contain the best remaining habitat for at-risk fish species or they contain degraded habitat of high restoration potential.
3. *Watershed analysis*—an assessment that characterizes a watershed's human, aquatic, riparian, and terrestrial features, conditions, processes, and interactions.
4. *Watershed restoration*—a comprehensive, long-term program to restore watershed health, riparian ecosystems, and fish habitats.

Note that watershed restoration is only one of the four components of this strategy. Restoration should not be considered independently of other land management prescriptions.

MODELS OF RESPONSE TO DISTURBANCE

Most aquatic conservation strategies assume that salmonid habitat quality deteriorates as watershed disturbance increases. It is well documented that the best habitat for wild salmonids is the least disturbed, whereas greatly disturbed areas are where salmonids are most likely to have been wiped out. Less well documented are the rates and degrees of the relation between disturbance and habitat quality. We will examine two models of this relation, shown in Figure 6.1.

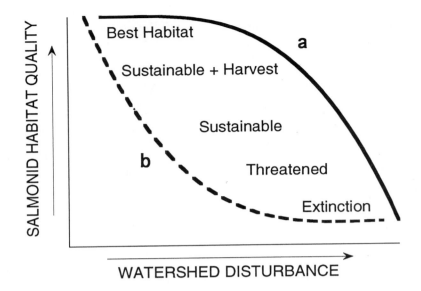

FIGURE 6.1.—Two conceptual models of the relation among watershed disturbance, salmonid habitat, and risk to fish stocks (local populations): (a) habitat quality is not degraded until substantial watershed disturbance is reached; (b) habitat quality is degraded most quickly during initial stages of disturbance.

Historically, watershed restoration has focused on improving the most severely degraded areas. Managers have assumed that commodity extraction can safely proceed where habitat is still good. That is, the least-disturbed areas have been assumed to be where future land-disturbing activities can proceed with the least concern. This strategy assumes a disturbance threshold beyond which degradation becomes significant (shown by the curve in Figure 6.1a). Before that level is reached, land use is assumed to cause watershed disturbance but without significant environmental cost. However, once that disturbance threshold is reached, habitat quality declines quickly with even a small additional disturbance.

According to this model, managers should become concerned about habitat degradation only when some threshold disturbance level is approached. A management and restoration strategy based on Figure 6.1a, then, would attempt to prevent the level of disturbance from reaching the threshold. This curve's shape also implies that, should the threshold level be exceeded, limited and focused restoration could quickly push watershed conditions back into the "good" habitat range.

A very different management approach is required if habitat quality degrades faster during the initial disturbance stages than during later stages, as shown by the curve in Figure 6.1b. In this case, protection becomes important from the first disturbance. This implies that any remaining good habitat is a valuable and fragile resource. Further, it implies that degraded habitat requires substantial work before recovery occurs. Highly disturbed areas require disproportionately more effort for an incremental habitat improvement than less-disturbed areas. Under this model (Figure 6.1b), for a given funding level it is better to focus on protecting or improving the best remaining habitat and allowing natural long-term processes to heal the most damaged areas.

Given the broad response range between the curves in Figures 6.1a and b, it is important to determine the correct response model when designing a restoration strategy.

Reeves et al. (1995) provided an excellent example of management strategies to maintain and restore freshwater habitat for anadromous salmonids in the Pacific Northwest. These strategies are based on designing a new disturbance regime around human activities. (Disturbance regime refers to the characteristics of a disturbance—timing, duration, and intensity.) The purpose is to create and maintain habitat conditions within and between watersheds that mimic conditions produced during natural cycles of disturbance and recovery.

Judging which model is correct for a given situation has important implications, not only for managing commodity outputs but for formulating an appropriate restoration strategy.

PRIORITIES FOR RESTORATION

Both models in Figure 6.1 assume some level of disturbance beyond which incremental restoration becomes ineffective. This might seem to dictate restoration priorities, but other realities include:

- A social imperative operates in our society to identify and concentrate restoration on the worst cases, regardless of their amenability to recovery. This is evident in legislation such as the Comprehensive Environmental Response, Compensation, and Liability Act (commonly called Superfund), which targets precisely those areas that often have the lowest probability of success in restoring the land to "satisfactory" condition.
- The public generally demands quick results, despite the fact that ecological recovery may require decades, if not centuries.
- Given limited financial and human resources, the incremental success ("bang for the buck") is usually greatest when a given expenditure is applied to preventing potential problems, rather than to fixing existing problems.
- A strategy of repairing many small problems before they become large problems is more effective than attempting to repair a single huge problem.

In the U.S. Bureau of Land Management's Proper Functioning Condition assessment for riparian areas, neither the best nor the worst areas receive the highest priority for restoration. The highest priority is assigned to those areas that are on a declining trend and are nearing some assumed threshold of loss of ecological function (Prichard et al. 1993).

Recalling the FEMAT four-component Aquatic Conservation Strategy (riparian reserves, key watersheds, watershed analysis, restoration), an oft-expressed approach is to "protect the best, restore the rest." The first two components, riparian reserves and key watersheds, fall under "protect the best." The fourth component, restoration, falls under "restore the rest." The third component, watershed analysis, bridges the "best" and "rest" by identifying the issues, displaying their linkages, and considering alternative solutions. For example, the watershed analysis might identify upslope restoration as the most effective way to protect the best remaining aquatic habitat by fixing potential problems at the source before channel disturbance actually occurs.

The FEMAT (1993) report identified eight guidelines to assist in developing restoration strategies or in choosing among potential projects. The strategy or project should

1. begin with a watershed analysis;
2. provide a broad range of benefits to riparian and aquatic ecosystems;
3. address the causes of degradation, rather than the symptoms;
4. have a well-defined project life span and understanding of expected benefits over time;
5. be self-sustaining once completed, requiring minimum maintenance or operation;
6. contribute to restoring historical composition, biodiversity, and disturbance regime;
7. link refugia and other isolated habitat units; and
8. integrate watershed protection, including adjustment or cessation of management practices that are responsible for degraded habitat.

WATERSHED ANALYSIS AND MULTIPLE VIEWPOINTS

The Regional Interagency Executive Committee (1995a) offered the following definition of watershed analysis:

Watershed analysis is a procedure used to characterize the human, aquatic, riparian, and terrestrial features, conditions, processes, and interactions within a watershed. It provides a systematic way to understand and organize ecosystem information. In so doing, watershed analysis enhances our ability to estimate direct, indirect, and cumulative effects of our management activities and guide the general type, location, and sequence of appropriate management activities within a watershed. As one of the principal analyses for implementing the Aquatic Conservation Strategy . . . it provides the watershed context for fishery protection, restoration, and enhancement efforts. Federal agencies are conducting watershed analyses to shift their focus from species and sites to the ecosystems that support them in order to understand the consequences of actions *before* implementation. Analysis teams identify and describe ecological processes of greatest concern, establish how well or poorly those processes are functioning, and determine the conditions under which management activities, including restoration, should and should not take place. Watershed analysis is not a decision making process. Rather it is a *stage-setting* process. The results of watershed analysis establish the context

ANALYSIS SCALES

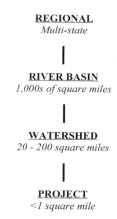

REGIONAL
Multi-state

RIVER BASIN
1,000s of square miles

WATERSHED
20 - 200 square miles

PROJECT
<1 square mile

FIGURE 6.2.—Hierarchy of four scales to establish the need for and context of restoration.

for subsequent decision making processes, including planning, project development, and regulatory compliance.

Watershed analysis originated from a recognition that planning directed at single issues by individual agencies does not work. For example, a single-issue management plan to harvest timber may meet the silvicultural and economic objectives of one landowner, but that plan may not adequately consider the effect of that activity on other owners, values, or activities within the watershed (such as spotted owls, fish, erosion, fire hazard, or restoration).

Similarly, a single-issue conservation plan protecting spotted owls also needs to incorporate plans for conserving fish, reducing fire hazard, controlling erosion, restoring and maintaining roads, and maintaining forest commodity production (timber, recreation, water, hunting, etc.). A timber program that requires substantial investment in restoration, and continued maintenance may not be cost-effective or ecologically effective, either in the short term or the long term.

For these reasons, the FEMAT report identified four analysis scales needed to establish the context of a plan (Figure 6.2): specific site prescription (less than 1 square mile), watershed (20 to 200 square miles), river basin (1,000s of square miles), and region (multiple river basins).

Watershed analysis simply identifies conflicting values and expectations and the social, biological, and physical processes that are important when viewed at the watershed scale (roughly an area of 20 to 200 square miles). This size of watershed is small enough to provide a useful level of precision, while being large enough to exhibit many of the interactions important to environmental issues. How to accomplish a watershed analysis has been described in procedural guides for private and public lands in Washington State (Washington Forest

TABLE 6.1.—Steps for conducting federal watershed analysis in the Pacific Northwest. (Regional Interagency Executive Committee 1995a.)

Step	Purpose	Information resources
1 Characterization	Identify dominant physical, biological, and human processes affecting watershed function; provide context to larger-scale processes	Existing maps, planning documents, literature
2 Key issues and questions	Focus analysis on key elements most relevant to managing the watershed	Existing basin plans, results of public meetings, interviews
3 Current conditions	Evaluate physical, biological, and human elements affecting key issues in the watershed	Existing reports, surveys, inventories, maps; narratives and anecdotal information
4 Changes that have occurred	Identify the rate and kinds of change occurring in watershed over time	Historical information, knowledge of basic ecosystem processes
5 Interpretation	Compare current and reference conditions to explain how and why key elements in watershed have changed over time	Evaluation of information obtained during previous steps
6 Recommendations	Determine efficacy of alternative practices to manage key elements and issues in watershed	Information from previous steps and objectives of the public and watershed managers

Practices Board 1993) and federal lands in the Pacific Northwest (Regional Interagency Executive Committee 1995a, 1995b).

The federal guide describes six steps for conducting watershed analysis in the Pacific Northwest: (1) characterize the watershed, (2) identify key issues and questions, (3) describe current conditions, (4) describe changes that have occurred, (5) interpret how and why they happened, and (6) recommend how the watershed's condition might change with alternative activities in the watershed. These steps are summarized in Table 6.1.

Those conducting a watershed analysis can become overly concerned with procedure and thereby distracted from the objective of the analysis. The principal objective of any watershed analysis is simply to expand the way we think about issues and their interactions. We need to consider the effect of multiple projects and activities of all landowners and managers within the watershed and river basin. We need to consider the overlapping, often contradictory objectives of individuals and multiple agencies at the private, local, state, and federal levels, including landowners, land managers, regulatory agencies, economic development agencies, and social agencies.

We need to involve the views of multiple disciplines when we evaluate an issue. In the past, "interdisciplinary" analysis typically has meant presenting a particular problem to a representative of each "appropriate" discipline, expecting each to comment from that discipline's perspective. But a better approach is for each discipline to be intimately represented as part of development, planning, and implementation from beginning through completion. These representatives should not be consultants but active team members with a stake in the outcome.

Often, nontraditional disciplines have been excluded because they were presumed to be noncontributing or threatened to confuse the issue by introducing irrelevant concerns. For example, a stream restoration team might traditionally include a fisheries biologist, hydrologist, geologist, and possibly a forester, but not a sociologist, economist, or terrestrial biologist. Serious consideration of these "irrelevant" concerns is precisely what is needed to avert failure of restoration projects designed with the tunnel vision of focused "action" groups.

SPATIAL SCALE

Just as it is a mistake to plan restoration in isolation, without knowledge of other projects in the vicinity, it is a mistake for watershed analysis to be concerned only with a single watershed. In part, watershed analysis is a scoping exercise to identify ecosystem processes and needs, including restoration, at the intermediate watershed scale, and to place these within the broader context of the larger river basin and regional settings.

The FEMAT strategy contains a hierarchy of four geographic scales: regional, river basin, watershed, and site (Figure 6.2).

1. *Regional*—the regional scale is used to evaluate how resources can be targeted to best influence values or concerns throughout a large multistate region. It is at this scale that an interconnected regional network of habitat protection might be established, based on regionwide habitat conditions or availability of refugia.
2. *River basin*—river basins within the region can be ranked by importance, based on opportunities and ability to contribute to meeting specific restoration objectives.
3. *Watershed*—within river basins targeted for restoration, individual watersheds can be further ranked by importance to identify the most effective placement of resources to accomplish restoration objectives.
4. *Site*—within the selected watersheds, individual sites can be identified and specific projects designed that will be most effective in accomplishing the objectives identified at the other three scales.

Using this hierarchy of scales, we can ask questions: What issues does the restoration attempt to correct? How large a program is necessary to significantly improve the situation? Which owners and agencies need to be involved? Where are the priorities of places that require restoration? What processes must be corrected to accomplish the objectives?

Traditionally, restoration has been tactical rather than strategic in nature. Much restoration has been small-scale and site specific, done for individual projects covering areas smaller than a few acres. Increasing concern exists about off-site problems that affect restoration and about the restoration's impact on other on-site and off-site values. Historically, off-site issues have been considered only in the immediate vicinity of the restoration, such as individual pools or lengths of streams that drain small upland watersheds. But it is becoming more apparent that to be successful, restoration must evaluate the effectiveness of a proposed project, not only within the context of small watersheds, but in the context of entire large river basins. For some restoration issues, such as restoring salmon runs, even an entire river basin is too small for establishing context. Consequently, a regional perspective is necessary, often covering multiple states.

It is at the larger scales that the efficacy of proposed restoration can be evaluated. For example, assume a problem of excessive sediment in a stream. The budget is sufficient to repair 20 culverts within a watershed to reduce the risk of failure and subsequent erosion of a stream crossing. But the watershed analysis suggests that 2,000 culverts have a comparable risk within the watershed. Further, 200,000 culverts exist within the river basin. One must question the efficacy of repairing just 20 culverts!

One also must ask whether the available resources could be better spent on an alternate program to reduce the sediment delivered to the stream. For example, for the same cost, one might reduce the diversion potential for the entire 2,000 culverts in the watershed by simple road engineering techniques, such as constructing dips in the road and grading the road surface to slope outward. This would prevent water from being diverted down the road in the event of culvert failure. Reducing the potential volume of erosion and sediment delivery to the stream network caused by diverted water from 2,000 culverts might be much more effective than preventing the failure of only 20 of the 2,000 culverts at risk.

In addition, the watershed analysis might reveal that, while culverts are being upgraded in one part of the watershed on one ownership, roads are being constructed in other portions of the watershed by other owners who are using the old inadequate design. In other words, the watershed analysis would suggest that this restoration is not accomplishing the overall objective of reducing culvert vulnerability or sediment input on a watershed scale. A river basin analysis, in turn, might reveal that restoration resources could be more effective in an entirely different watershed.

Identification of the appropriate scale is often ignored. Unfortunately, no single scale fits all issues. For example, the appropriate regional scale for restoration of anadromous fish extends throughout the Pacific Northwest and includes the land, the streams, and the ocean. In contrast, the entire range for a species of salamander may encompass only a small portion of a single river basin. Forcing any analysis to some standardized scale will be incorrect most of the time. The spatial scale must be appropriately tailored to the problem being considered.

TEMPORAL SCALE

Selecting an appropriate time scale upon which restoration is evaluated is as important as selecting the appropriate spatial scale. The time scale that is conventionally considered appropriate usually depends on the audience. A number of examples follow. Corporate decisions are often based on quarterly budget reports. Political decisions are driven by election cycles of 2, 4, or 6 years. A domestic water user might be concerned about changes in turbidity during a single storm. Changes in insect populations might be resolved at annual scales. Trends in anadromous fish populations might need a sequence of several cycles of 4 to 6 years. Silvicultural concerns traditionally operate in time frames of 50 to 100 years.

Geomorphic processes that determine the physical condition of streams operate at time scales from decades to several centuries. For example, coarse sediment introduced by placer mining into streams of the Sierra Nevada in California during the 1840s continues to enter the Sacramento River system 150 years later. In many cases, rare and unusual events like wildfire or flooding are the principal mechanisms that set the physical or ecological structure of an area for decades.

Consequently, planning of any restoration must consider the appropriate time scale upon which natural systems operate. If projects are designed using inappropriate time scales, at best the restoration will be ineffective, and at worst it may produce additional degradation of the very resource it was intended to repair.

CASE STUDY: REDWOOD CREEK

A specific example to illustrate some of these points is Redwood Creek in northwestern California. Redwood Creek drains a basin of 285 square miles and flows into the Pacific Ocean near Orick, California (Figure 6.3). In 1968, Redwood National Park was created to protect representative stands of old-growth coastal redwood. The park includes several groves that contain the world's tallest trees. They grow along the lower stretches of Redwood Creek, on natural terraces formed of stream alluvium.

In 1964, 1972, and 1975, flooding, bank erosion, and deposition of coarse sediment in the main channel of Redwood Creek damaged these unique alluvial groves. Accumulating evidence strongly suggested that timber harvesting and associated road building on private lands within the Redwood Creek watershed were partially responsible for the flood damage (Harden et al. 1978).

Nationwide concern over the threat to park resources and particularly the long-term safety of the Tall Trees groves culminated in the transfer of an additional 48,000 acres in the lower portion (northern end) of Redwood Creek basin from private ownership to an expanded Redwood National Park. Presently, the lower 40% of the watershed is within Redwood National and State Parks, whereas the upper 60% remains mostly in private ownership (Figure 6.3).

Because the land was purchased as a legal ''taking'' from private owners and given protection under national park status, this action implied that land management by the previous owners was abusive. Thus it became both technically and politically necessary to restore that abused land to protect the Tall Trees groves and other park values.

The accuracy of the abuse implication has aroused heated debate for several decades. Resolution is not simple because the geology of Redwood Creek is very unstable and numerous natural landslides throughout the basin were activated

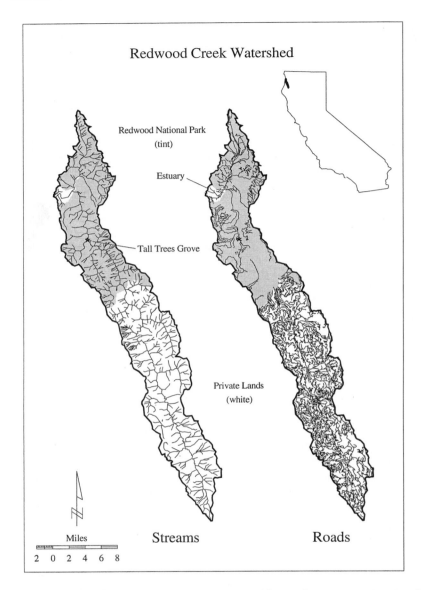

FIGURE 6.3.—Redwood Creek watershed. (Geographic information systems database provided by Redwood National Park.)

during the large floods in the 1960s and 1970s (Nolan et al. 1995). However, sufficient scientific evidence exists, both within the Redwood Creek watershed and elsewhere, to conclude that among land uses in mountainous terrain, roads are a primary cause of human-induced sedimentation (Kelsey et al. 1981; Hagans and Weaver 1987).

Extensive Road Network

The watershed's extensive road network (Figure 6.3) is the most important source of sediment delivery caused by humans to streams in the uplands. Roads modify the natural hillslope drainage and accelerate erosion. About one-third (400 miles) of the roads in the Redwood Creek watershed are on unstable bedrock, and about one-sixth (200 miles) are on soils that are particularly susceptible to landslides. In addition, common causes of accelerated erosion from the roads include unstable road fills, oversteepened road cuts, intercepted and rerouted surface and subsurface water, undersized and poorly placed culverts, and the diversion of streams at crossings.

A road survey in the Redwood Creek watershed revealed that about one-third of them either have been abandoned or are no longer maintained (V. Ozaki, Redwood National and State Parks, personal communication). Abandoned or unmaintained roads have been shown consistently to pose long-term problems. Such roads are increasingly likely to fail during large storms because road drainage features no longer function as designed and culverts deteriorate or become clogged with debris. This results either in failure of the road fill at the stream crossing or diversion of water from the stream channel and down the road to areas unaccustomed to increased water discharge.

The Redwood Creek watershed has experienced substantial land use over the past century. By the time the park was expanded in 1978, about 1,200 miles of roads (4.2 miles per square mile of watershed) and 5,400 miles of logging skid trails (18.9 miles per square mile) had been constructed within the watershed (Best 1984). In 1978, the road densities within and outside of the park were similar: about one-fourth (300 miles) of roads in the watershed were within the expanded park (D. L. Steensen and T. A. Spreiter, paper presented at the national meeting of the American Society for Surface Mining and Reclamation, 1992), and the remaining three-fourths (900 miles) were on private lands upstream of the park and the Tall Trees groves. Most of the road network present in 1978 was constructed before the introduction of modern forest practice regulations.

Further, because of the large size of the old-growth trees being tractor-logged, many skid trails are large enough to have a hydrologic effect similar to that of a road. However, these large skid trails were constructed without even the basic engineering design and drainage features that would be required for a road.

The federal legislation that expanded Redwood National Park included provisions not only to pay for the acquired land, timber, and other assets, but to establish programs for displaced workers and to initiate a major restoration of logged lands and roads within the park. Congress directed the restoration to focus on minimizing erosion from past land uses, reestablishing native vegetation, and protecting aquatic and riparian resources along park streams.

As of 1986, US$364 million had been paid to companies and individuals for their lands, timber, and other assets that were included in the expanded national park (Redwood National Park 1987). In addition, as of 1986, more than $100 million in economic development and employee assistance benefits had been paid to displaced forest product workers. As of 1991, Redwood National Park had expended about $10 million for watershed restoration. The total expenditure for adding 48,000 acres (about 26% of the Redwood Creek watershed) to the park is about $500 million.

Redwood National Park personnel are undisputed experts in road restoration. They have developed, tested, and applied road restoration techniques at a scale virtually unprecedented worldwide. Since the park was expanded in 1978, 134 miles of the 300 miles of road within park boundaries have been restored or obliterated. This work has removed about 1,300,000 cubic yards of material from stream crossings, landings, and unstable road benches. This volume approximately equals the long-term average annual sediment discharge near the mouth of Redwood Creek (A. T. Ringgold, Redwood National and State Parks, personal communication).

To evaluate the success of removing this volume of material, one must know (1) the delivery mechanism, (2) the timing, (3) the proportion of removed material that would have reached the channel without restoration, (4) the quantity of new material from erosion caused by the restoration itself, and (5) the proportion of treated and untreated areas having comparable risk in the basin. But by any measure, the U.S. National Park Service has been successful in restoring and obliterating a large portion of the unstable roads within the park.

Despite this effort, road mileage in the Redwood Creek watershed increased from 1,200 miles in 1978 to 1,266 miles in 1992 (D. W. Best, Redwood National and State Parks, personal communication). Between 1978 and 1992, 200 miles of road and 1,421 new stream crossings were constructed within the watershed above the park boundary for harvesting timber (Ozaki, personal communication). Even these figures underestimate the actual road construction in the basin during the 14-year period because they exclude roads that are exempt from a state Timber Harvest Plan, such as ranch roads, access roads for home construction, and other local access. These new roads more than offset the 134 miles of road restored or obliterated by the U.S. National Park Service.

Redwood National Park personnel recognize the need to deal with erosion sources outside the park boundaries. A watershed analysis of the Redwood Creek basin has been completed and a number of potential restoration opportunities have been identified. Unfortunately, restoration outside the park often is constrained by legislation and whether cooperation can be obtained from the upstream landowners. The past political battle that converted private land into national park property was heated and strikes at the heart of the private property rights debate. Any suggestion of additional measures by the federal government to control activities on the remaining private land in the watershed is politically risky.

Recently, however, the U.S. National Park Service and the U.S. Fish and Wildlife Service have made progress in implementing cooperative erosion control on private land in the upper Redwood Creek watershed. Funding for this cooperative restoration on private land is now roughly equivalent to that available for restoration within the park (Ringgold, personal communication).

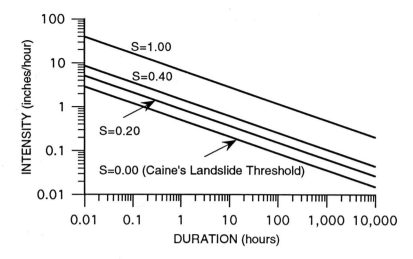

FIGURE 6.4.—Storm severity (S) as a function of the duration and intensity of rainfall. (Ziemer 1992).

Success or Failure?

Despite the unprecedented effort to acquire and restore land near the Tall Trees groves, there is yet no way to determine whether restoration is succeeding or failing. Such a determination can be made only following a major storm, because severe storms are a primary cause of erosion in steep forestland.

Caine (1980) presented a relation between rainfall intensity and duration that seems to describe a threshold for landslide occurrence worldwide (labeled curve in Figure 6.4). Rice et al. (1982) reasons that a good index of storm severity, and hence the probability of a landslide, might be how much larger a storm's intensity and duration are, relative to Caine's landslide threshold (Figure 6.4).

Amounts of erosion, sediment transport, and change in channel bed elevation depend on the timing and severity of storms (Figure 6.5). Most of the time the landscape and channel are recovering from the large rare events that produce nearly all of the erosion and channel changes. Consequently, before the success of restoration can be evaluated, the watershed must be subjected to a significant triggering event.

In a simulation, suppose a minimum storm severity of 0.25 were required to produce a triggering event to test a restoration program. That is, the size of the storm would be above Caine's (1980) landslide threshold (Figure 6.4). Only three storms (at years 44, 120, and 279) occurred during the 300-year simulation period (Figure 6.5). Between these three landslide-producing storms there were periods of 44, 76, and 159 years when there were no storms large enough to cause a landslide. Although Figure 6.5 is just a simulation, it demonstrates that only a few rare, large storms can be expected to test the effectiveness of restoration. A century or more may pass before the next large storm.

In the case of Redwood Creek, the 1964 storm produced massive hillslope

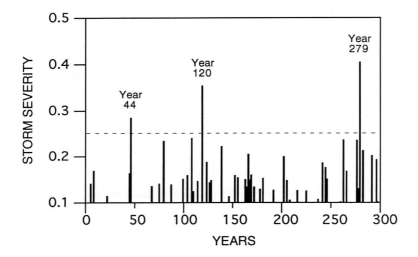

FIGURE 6.5.—Distribution of severe storms, based on measured rainfall duration and intensity in northwestern California during a single 300-year simulation (Ziemer 1992). Storm severity was calculated as a function of the duration and intensity of rainfall (Figure 6.4). The dotted line represents a storm severity of 0.25 (see text).

erosion and a large volume of coarse sediment was deposited in the channels of the upper and middle watershed. Landslides and other forms of erosion continue to deliver some coarse sediment to the channel each year. However, no significant storms have occurred for the past 20 years and sediment movement in Redwood Creek since the park was expanded and restoration began has been largely a redistribution of sediment deposited in 1964 (Madej 1984). The consequences of the next storm that equals or exceeds the 1964 storm in intensity will show the success or failure of the restoration program to protect the world's tallest trees.

CONCLUSIONS

The success of any restoration depends upon being able to identify a local concern, to objectively analyze the information, and then to design projects that effectively address concerns. This includes not only the local concern, but those at progressively larger spatial and temporal scales and complexity. It is at these larger scales that the efficacy of proposed local restoration can be evaluated. Each local restoration should be studied to determine whether the location, effort, and timing will produce a significant effect on larger-scale concerns. Without the larger-scale context, local restoration too often is of the wrong design and wrong size in the wrong location at the wrong time.

An excellent example of comprehensive restoration is the Aquatic Conservation Strategy in the FEMAT (1993) report. This strategy describes a process for analyzing problems across various scales (watershed analysis), identifies important portions of the landscape requiring protection (key watersheds and riparian reserves), and includes a component to improve degraded land (watershed restoration).

ACKNOWLEDGMENTS

The Redwood Creek example benefited from ideas and information from the very helpful personnel at Redwood National Park. The U.S. National Park Service kindly provided a review draft of the Redwood Creek watershed analysis. David Best and David Lamphear produced the digital geographic information systems files. Bret Harvey, David Hohler, Thomas Lisle, Mary Ann Madej, Alan Olson, Vicki Ozaki, Leslie Reid, and Jack Williams provided helpful reviews of early versions of this chapter.

*The Way is to preserve what you already have, not to seek
what you haven't got. If you seek what you haven't got, then
what you have is lost; if you go along with what you have,
then what you want will come.*
—*Wen Tzu, Han Dynasty, ca. 200* B.C.

CHAPTER 7

ECOLOGICAL PRINCIPLES

Christopher A. Frissell

What ecological assumptions and strategic considerations underlie restoration of aquatic ecosystems, particularly riverine systems? Managers sometimes pursue aquatic habitat restoration as a self-evident need, with little or no deliberation of the appropriateness of goals, objectives, and methods. In this sense, restoration is treated as a routine maintenance activity, rather than as the science it must be. In this chapter, I argue that restoration and habitat management have been hindered by inattention to ecological context and ignorance of the ecosystem processes involved in the degradation of aquatic resources. Neglect of ecological context has contributed to a long-standing, artificially imposed dichotomy between habitat restoration and habitat protection that has undermined conservation efforts (Doppelt et al. 1993).

After diagnosing the often-implicit ecological assumptions of past restoration, I suggest an alternative logic and conceptual framework for developing integrated and ecologically sustainable strategies for watershed restoration and protection, and for establishing new priorities in restoration tactics. This approach explicitly considers the spatial and temporal patterns in the physical and ecological function of watersheds, with special attention to the natural–historical template provided by existing patterns of aquatic biodiversity. Such an approach has been widely advocated by many scientists and nongovernmental organizations (e.g., Doppelt et al. 1993). Variations of it have been at least

ECOLOGICAL PRINCIPLES   **97**

tentatively implemented in recent federal and state restoration projects in the Pacific Northwest.

STRATEGY AS ECOLOGY; TACTICS AS TECHNOLOGY

In environmental planning and management, it is important to distinguish between strategy and tactics (Bella and Overton 1972). *Strategy* concerns the comprehensive, large-scale marshaling and allocation of resources, whereas *tactics* concern local, immediate, and short-term activities. It is critical that tactics be congruent with, and directed by, an overall strategy. It is also necessary that strategy be shaped by the limitations of tactical capabilities.

The scientific and technical literature on restoration is inordinately focused on tactics, with relatively little regard for strategic issues in conservation and environmental planning. Failure to address strategic issues in research, planning, and evaluation leads to wasted resources, misperceived and misrepresented success or failure of projects and plans, underestimated risk of cumulative or synergistic effects from multiple activities or projects, and ultimately to increased probability of irreversible, large-scale environmental crises (Bella and Overton 1972).

In aquatic habitat and fisheries restoration, both the technical literature and government policies are largely focused on techniques and tactics (Platts and Rinne 1985; Reeves et al. 1991a). Examples abound: Which is better, log structures or boulder structures? How do we get structures to remain in place? What designs and materials are optimal? How can cost per structure be minimized?

Relatively little attention has been paid to development of an ecologically sound, guiding strategy for restoration. Such a strategy must address broader questions.

- What processes are causing habitat loss?
- How can these processes be reversed?
- Are structures even feasible? Or are other kinds of treatment necessary?
- Should effort be concentrated in certain localities, or dispersed across the watershed?
- Which species will benefit from a given action, and will the benefits be long-term?
- What is the risk that unwanted side effects could accrue from a particular set of treatments?

Answers to these questions lie in ecological and geophysical analysis of watersheds, riverine habitat, and the conservation biology of aquatic and riparian (streambank) communities and populations.

In recent decades, fish habitat management has emphasized two related activities. The first is regulation of land use and other human activities to reduce their adverse, site-specific effect on fish. This activity has consisted primarily of leaving buffer strips of vegetation. The second activity is direct, structural modification of aquatic habitats to produce conditions that fish presumably require for survival.

A fundamental assumption underlying both activities has been that rivers consist of extensive "good" habitat with abundant fish populations and local patches of "poor" habitat that are sparsely populated or support relatively few species (Frissell and Bayles 1996). In this widely held view, local disturbance caused by clearing a farm, logging riparian trees, triggering a landslide, or channelizing a river segment are easily tolerated by an inherently forgiving and elastic ecosystem.

This view assumes that certain valued species should be the target of habitat management, and that we know fully what changes in natural systems will promote production of the target species. Parks and other protected areas are established to preserve scenic values and unique features that are seen as largely disconnected from the conservation and management of exploited natural resources like salmon or river mussels.

In this local-disturbance perspective, recovery of fish, wildlife, and plant populations from dispersed disturbances is presumed to begin instantaneously and to proceed independently, limited only by the rate of physical recovery of the disturbed site. Habitat "improvement" or rehabilitation is taken to involve little more than ameliorating the local site's physical condition or treating some factor that is assumed to be locally limiting. Such factors might be nutrient loads, spawning gravel quality, shade and water temperature, cover, or quiet water refuge—whatever is deemed locally necessary for the target species to survive and grow. Rehabilitation is intended merely to speed the rate of inherent recovery of local conditions, and other habitat management includes installation of artificial structures intended to replace specific natural habitat functions.

The crucial, underlying, but usually unstated assumptions of the local-disturbance view are as follows (Frissell and Bayles 1996; Frissell et al. 1997):

1. disturbances are isolated and independent in their effect, and the ecosystem as a whole remains intact;
2. recovery at each disturbed site proceeds relatively rapidly and independent of the site's context in the ecosystem;
3. a steady, unlimited supply of fish and other organisms is available from surrounding areas to colonize disturbed habitats as they recover or are artificially replaced; and
4. aquatic species and riverine habitats are largely homogeneous in distribution, so that habitat and biological populations are readily replaceable, generic techniques of habitat modification are widely applicable, and the risk of failure or unintended side effect is minimal.

These assumptions are deeply embedded in existing approaches to environmental regulation and planning (Doppelt et al. 1993). Given these assumptions, managers could endorse increased disturbance in riverine ecosystems by making projects contingent (1) on specific habitat improvement or (2) on rehabilitation techniques presumed to mitigate damage or perhaps even to improve or enhance the fishery resource beyond its natural condition. Such projects are routinely proposed and constructed as mitigation for timber sales, overgrazing, dams, and other disturbances (Brouha 1987).

For example, most current management plans for national forests assume—with little or no scientific support—that increased fish production from new artificial structure projects (partly funded by timber sale receipts) will more than compensate for the logging of critical ecosystems in sensitive watersheds of the Pacific Northwest during the next decade.

In another example, in arid ecosystems, federal agencies commonly install artificial structures in streams that have been damaged by decades of livestock grazing. However, researchers have repeatedly demonstrated that this cannot restore aquatic habitat when the primary process causing the damage—grazing—is not controlled (Platts and Nelson 1985; Platts and Rinne 1985; Elmore and Beschta 1987; Rinne and Turner 1991).

SHORTCOMINGS OF THE TACTICAL APPROACH

Past approaches to stream habitat restoration in the traditional paradigm have several distinguishing features (Doppelt et al. 1993). Identification and diagnosis of habitat problems tend to focus on finding sites or stream segments that are physically amenable to predetermined generic tactics. For example, much past and current fish restoration relies heavily on building fixed log weirs to create pools in streams. This approach generally identifies stream segments that have a gradient (rate of fall) and streambank structure suited to the installation of such devices, and that happen to be accessible to the heavy equipment needed.

Little consideration is given to whether the fish community, the physical and ecological capabilities of the site, and the watershed as a whole are suited to the habitat change that such structures are intended to induce. It is commonly assumed that all fish benefit equally from the sequence of plunge pools (basins scoured by vertical water movement) created by such devices, and that the construction of weir pools will compensate for the many complex and linked changes in the ecosystem that result from human disturbance.

Careful evaluation of such projects often indicates serious limitation or dysfunction. First, disturbance from grazing, logging, mining, and other activities can affect the entire stream network and the vast majority of streams are inaccessible to heavy equipment or are otherwise unsuited to structural modification (House et al. 1989). Second, failure rates appear to be high for all kinds of structures in severely damaged watersheds (Table 7.1; Figure 7.1) or stream reaches where disturbance is ongoing or where its effect is persistent (Platts and Nelson 1985; Frissell and Nawa 1992).

Third, many artificial structures have unintended and damaging physical side effects, whether they remain in place or fail. This eventually may promote undesirable ecological outcomes (Table 7.1). Examples include severe bank erosion and blockage of juvenile fish migration (Beschta and Platts 1986; Elmore

TABLE 7.1.—Examples of structural failure or unanticipated physical or biological outcomes reported for stream habitat manipulations in the western United States.

Type of project	Location	Years after installation	Type of failure or adverse response	Source
Log and boulder weirs and lateral deflectors	East Fork Kaweah River, California	18	75% lost or physically dysfunctional due to washout, endcutting, and sediment accumulation	Ehlers (1956)
Log and rock check dams	Sagehen Creek, California	12	57% undercut, laterally washed out, or rotted	Gard (1972)
Rock and log check dams, lateral deflectors, and cover structures in relocated channel	Tenmile Creek, Colorado	5	75% ineffective due to washout, sediment deposition, and instability of engineered channel	Babcock (1986)
Log weirs, deflectors, and cover structures	Southwestern Oregon coastal streams	1–5	27–100% dysfunctional due to bank erosion, washout, sediment accumulation, anchor failure, unstable watersheds	Frissell and Nawa (1992)
Wire gabions	Big Creek, Utah	8–9	Structures damaged by livestock trampling; accumulation of fine sediments from livestock-impacted reaches upstream	Platts and Nelson (1985)
Boulder wing deflectors	Nooning Creek, northern California	1	68% destroyed or nonfunctional, reduced condition factor and residence of steelhead parr (young trout or salmon) in treated reaches	Hamilton (1989)
Boulder deflectors	Hurdygurdy Creek, northern California	3	Loss of habitat and local breeding population of ranid frogs, increased predation on frog larvae	Fuller and Lind (1992)

and Beschta 1987; Frissell and Nawa 1992). Conditions in the riparian zone and the watershed as a whole are at least as important as structure design in determining whether structures will function appropriately.

Fourth and finally, numerous studies suggest that the effects of artificial structures are inconsistent, difficult to predict, and sometimes biologically adverse. Where an artificial structure achieves its physical objective, its effect on

Figure 7.1.—A highly sediment-charged, unstable reach of Euchre Creek, a coastal stream in southwestern Oregon. Log deflectors, weirs, and other artificial structures placed in Euchre Creek to rehabilitate salmon habitat have completely failed at rates exceeding 90%. (Frissell and Nawa 1992.)

native fish or other species may be negative. Examples include modification that inadvertently depletes habitat for a certain life stage (e.g., Everest et al. 1986b), or favors exotic species over native ones (Rinne and Turner 1991), or exposes target species (or sensitive nontarget taxa) to increased predation. In many watersheds or stream segments, target species are so widely depleted by extensive habitat degradation and other factors that few or no fish are available to colonize the artificially created habitats (e.g., Nickelson et al. 1992a).

Although superficially determined by tactical considerations, traditional aquatic restoration rests on implicit strategic assumptions about ecosystem behavior and management (Doppelt et al. 1993; Frissell et al. 1993). Agencies often set restoration priorities by identifying the most-degraded, most biotically impoverished sites and expending all resources to bring these areas "up to standard." Once the improvements have been made, managers assume that further habitat disturbance in the watershed can be allowed to proceed. In other words, attempted improvement of degraded, biologically impoverished habitat is explicitly or implicitly framed as mitigation, allowing the ongoing deterioration of pristine or productive habitat elsewhere in the ecosystem.

The result of this strategy is predictable: disturbance is maximally dispersed across the landscape and virtually all sites ultimately become degraded. The worst may be partially repaired, but sensitive species likely will have disappeared throughout the habitat network. No effort is made to identify and maintain particular habitat patches as refugia, geographic areas where recolonizing species survive, safe from the effect of human disturbance. Because of this, the most productive and diverse habitats endure ongoing disruption, whereas the most severely degraded and biologically impoverished—those least amenable to structural improvement, and most prone to project failure—receive all the restoration resources (Doppelt at al. 1993; Frissell and Bayles 1996).

From an ecological viewpoint, traditional landscape and aquatic habitat management is a blueprint for degrading natural watersheds and other relatively secure landscape refugia. This leads predictably to the cumulative decline and extermination of formerly abundant but sensitive species over large areas (Frissell and Bayles 1996). Instead of the ideal matrix of high-quality habitat with patches of disturbed habitat, traditional management has created a matrix of disturbed, degraded, and species-poor habitat that surrounds a few remnants of high-quality habitat having associated fragments of the watershed's historical biotic diversity (Frissell and Bayles 1996).

ECOSYSTEMIC VIEW OF HUMAN DISTURBANCES AND RECOVERY

The traditional management view has little room for the possibility that human disturbance can permeate or resonate over large scales, cumulatively changing the dynamics not only of stream segments over years, but of entire watersheds and landscapes over many decades or centuries. However, scientific studies reveal that logging, grazing, crop agriculture, urbanization, and channelization have environmental consequences that accrue across large areas and persist for decades or centuries (Leopold 1980; Coats and Miller 1981; Regier et al. 1989). These human disturbances involve permanent or persistent change in the entire watershed. The change is reflected directly or indirectly as change in the aquatic ecosystem. The changes are much different in magnitude, frequency, or quality compared to natural disturbance regimes that historically shaped the aquatic habitat mosaic.

These activities cause sustained alteration of habitat structure and biotic communities, often shifting them into persistent states that are different from the pre-alteration condition (Niemi et al. 1990; Yount and Niemi 1990). Recovery of fish populations and other aspects of stream community structure from such disturbances often requires several decades. A return to pre-existing conditions may never occur if the natural environmental disturbance regime remains altered (Niemi et al. 1990).

Disruptions that are relatively instantaneous and locally alter biotic communities without persistent change in the system's biophysical structure are called *pulse* disturbances. Examples are local chemical spills, small-to-moderate floods, and natural drought cycles (Niemi et al. 1990; Yount and Niemi 1990).

In relatively unaltered landscape mosaics, where human activity is temporary, infrequent, or isolated, the pulse model may describe the ecological effects. But in landscapes where human influence is widespread and frequent, the impact of land and water alteration more closely resembles the sustained disturbance model (Frissell and Bayles 1996).

Many long-term studies of land use effects on streams, rivers, and lakes have reported long lags between the disturbance event and physical–biological recovery of the aquatic ecosystem. The most detailed studies have detected at least some effects that persist more than 20 years after disturbance, despite an initial recovery trend. Such a pattern of arrested recovery strongly suggests that many land use disturbances produce permanent or long-term residual impacts. Thus, these constitute sustained disturbances that essentially preclude a return to any semblance of initial conditions, either physically or biologically (Yount and Niemi 1990; Ebersole et al. 1997).

The literature contains numerous good examples of the effect of sustained disturbance on streams in the Pacific Northwest. Logging followed by major storms dramatically increased the fine-sediment concentration in the bed of the South Fork Salmon River, Idaho, during the 1960s. Salmon populations declined precipitously. Following a logging moratorium and an initial decline in fine sediment, recovery has since become arrested, and the river retains residual elevated levels of fine sediment. Further, in the river's current state, the physical sensitivity of the system remains high, in that dramatic responses are seen wherever new disturbances deliver even small volumes of sediment to streams (Platts et al. 1989).

Hagans et al. (1986) summarized extensive research in the Redwood Creek watershed of northern California. They determined that the effects of land use-generated sediment, channel widening and aggradation, and channel stability loss would persist for at least several decades, and perhaps more than a century, even if no further disturbance occurred in the watershed following the 1964 flood. But further disturbance continues, particularly on private land and in highway corridors upstream of protected areas of Redwood National Park, potentially offsetting many benefits to the aquatic ecosystem that could result from the park's major watershed restoration.

In these examples, the effects of a major flood (a natural disturbance event) were compounded and qualitatively transformed by human disturbance. Human activity effectively curtailed many natural ecological processes that conferred resilience and resistance to the streams and their watersheds. The same 1964 flood had relatively limited, short-term, or perhaps even ecologically beneficial effects in nearby watersheds that had been less altered by humans.

Many long-term effects found to be significant in watershed studies were not anticipated from previous research. They involve causal mechanisms that are poorly understood. For example, logging in Carnation Creek, British Columbia, caused an unanticipated increase in winter stream temperature. This effect, along with change in spawning gravels, contributed to collapse of the chum salmon population (Holtby and Scrivener 1989). After more than a decade, the thermal regime of the stream had not returned to prelogging conditions, and scientists remained stymied by the mechanism driving these changes (Hartman and

Scrivener 1990). Fine sediment concentrations in the streambed also remained elevated (Scrivener and Brownlee 1989).

In another example, Hicks et al. (1991a) reassessed an earlier analysis which implied that the streamflow regime of an Oregon Cascades basin had recovered within a decade after logging. However, they found that, after the first 7 years, summertime low flows declined substantially below prelogging levels for at least the following 18 years of forest regrowth. In yet another instance, Schwartz (1991) reported that the population of cutthroat trout in coastal Oregon's Needle Branch remained depressed more than 20 years after it collapsed following logging of the basin. The mechanisms that caused this collapse and that inhibit population recovery remain unknown.

Although specific causal mechanisms and contingencies remain obscure, human landscape alteration does create an identifiable response in aquatic ecosystems. This is demonstrated in an important paper that has not received sufficient attention from restorationists. Regier et al. (1989) described a gradient of ecological degradation that incorporates predictable patterns of disorganization, fragmentation, and impairment of self-restorative processes in terrestrial and associated riverine ecosystems.

Further, this phenomenon is well illustrated by the work of McIntosh et al. (1994), who surveyed many rivers in the Columbia River basin. The density of large pools is a good indicator of many aspects of channel morphology and habitat complexity. These researchers found that large-pool density had diminished in abundance 30 to 85% during 40 years in virtually all streams except those protected in wilderness areas. The pattern of long-term habitat loss occurred across land ownership and stream types, regardless of dominant land use—logging, grazing, or crop agriculture.

SPATIAL REFUGIA AND THEIR ECOLOGICAL FUNCTIONS

Yount and Niemi (1990) and Niemi et al. (1990) reviewed case histories of aquatic system recovery following disturbance. They found considerable evidence that *spatial refugia*—undisturbed habitats providing a source of colonists to adjacent areas—was critical to enabling the recovery of disturbed systems. In stream systems where disturbance was spatially extensive and no accessible refugia remained, biological recovery was delayed or precluded.

Sedell et al. (1990) and Moyle and Sato (1991) discussed several kinds of riverine and hyporheic (subsurface) habitats that can serve as refugia. They provide examples of how these refugia may function in the recovery of populations from natural catastrophe and anthropogenic disturbance. Unfortunately, a major characteristic of logging, agriculture, channelizing, damming, and other land use disturbances is that they typically cause a long-term loss of habitat diversity that propagates downstream, diminishing the abundance and connectivity of actual or potential refuge habitats (Sedell et al. 1990).

One critical but increasingly rare kind of refuge is the ecologically intact drainage basin that is relatively unimpaired by human activity (Sedell et al. 1990; Moyle and Sato 1991). Natural headwater habitats can harbor many sensitive species and thus serve as efficient sources of colonists to downstream areas. They are the safest and most secure kind of refuge for many fishes, by virtue of large size, diversity of natural habitat, and relative impermeability to invasion by exotic species (Moyle and Sato 1991; Williams 1991).

Undisturbed watersheds also contribute to high-quality water and channel conditions in downstream areas. These areas may be important for additional native species (Williams 1991; Frissell et al. 1993). Whole-watershed refugia, by definition, lack sources of human disturbance that can propagate downstream from sensitive upslope and headwater areas (Moyle and Sato 1991; Frissell and Bayles 1996). Increasingly, headwater tributaries serve as de facto refugia for native salmonid fishes (salmon, trout, and chars). This is so because many populations of these coldwater fishes have been eliminated downstream and fragmented by logging, agriculture, channelization, urbanization, and introduction of nonnative species or hatchery fish (Howell and Buchanan 1992; Rieman and McIntyre 1993; Young 1995).

However, it is important to recognize that headwater streams rarely contain the full range of habitat necessary to sustain all native fish and other vertebrate taxa (Sheldon 1988). This is particularly obvious in the case of migratory species such as Pacific salmon or bull trout, and for species confined to lowland habitats such as the foothill yellow-legged frog in California (Drost and Fellers 1996). Moreover, headwater refugia, especially very small ones, may be relatively vulnerable to natural stress or catastrophe, such as drought, flash floods, and landslides. Consequently, it is unlikely that headwater refugia in isolation can sustain native populations of many species indefinitely.

For long-term persistence, many headwater-dependent species probably require redundant mosaics of headwater refugia, connected through downstream channel networks that allow free dispersal of individuals among them (Rieman and McIntyre 1993). In such a connected system of habitat patches, a local extinction can be reversed if a similar population of the species persists in a nearby refuge, providing a source of dispersers to recolonize as the disturbed habitat recovers.

Evidence is growing that many riverine organisms have maintained their historical distribution in the face of natural disturbances (e.g., floods, drought, landslides) through such a dynamic balance of local extinctions and recolonizations. A major, underappreciated ecological effect of human activity is disruption of such processes within and among watersheds (Rieman and McIntyre 1993; Frissell and Bayles 1996).

Other critical habitats or refugia function to sustain aquatic biodiversity in disturbed systems. These also require protection if restoration is to succeed (Moyle and Sato 1991; Doppelt et al. 1993; Frissell et al. 1993). Such refugia tend to occur in lower-elevation areas downstream that have a long history of human disturbance. In many cases, these are habitats that possess unusual resilience or resistance to disturbance. Although fragmented, at least some of their natural diversity and function remains relatively intact. Occasionally these habitats have escaped severe disturbance locally by chance or by virtue of unusual ownership or management history. Estuaries and forested floodplain segments of rivers often function as such downstream refugia. They provide islands of high-quality habitat for downstream-dependent species and life stages, as well as sources of colonists for upstream segments (Sedell et al. 1990).

Also important, and more widely distributed, are small-scale or "internal" refugia, such as the subsurface zone of a floodplain river segment, a floodplain wetland or pond, a cold water plume at a tributary mouth, or a groundwater-fed spring (Williams 1991). All these habitats offer at least some degree of physical shelter or buffering from pervasive, catchment-wide influences that cumulatively degrade mainstream habitats. The habitats may do so through drainage network structure or groundwater influences. Stream segments rich in such refugia may be critical to sustaining the current complement of species. Maintaining and propagating such habitats will become a necessary component of successful restoration (Sedell et al. 1990; Doppelt et al. 1993; Frissell et al. 1993; Ebersole et al. 1997).

Although many protected areas function as aquatic refugia, such as national parks and wilderness areas, only three federal refuges exist specifically to protect native fishes, all in the southwestern United States (Williams 1991). In the Pacific Northwest, the interagency Forest Ecosystem Management Assessment Team (FEMAT 1993) identified a network of watershed reserves on federal land to improve protection of key populations of imperiled anadromous fish (those that spawn in freshwater and live in saltwater).

An extensive inventory by the Oregon Chapter of the American Fisheries Society identified watershed refugia and "critical areas" across that state. This inventory was used by Henjum et al. (1994) as a basis for a regional conservation strategy. Areas identified in this survey include (1) intact whole watersheds that harbor broad aquatic diversity, (2) watersheds critical to downstream water quality, (3) stream segments that have relatively high natural habitat complexity and native species diversity, and (4) estuaries, lakes, and wetland complexes that are least disturbed in their region and least affected by introduced species (Li et al. 1995).

AN ECOLOGICALLY BASED STRATEGY

A more successful strategy for restoring riverine ecosystems might arise from a new synthesis of geomorphic knowledge (landforms and physical processes) and ecological knowledge about how watersheds, fish populations, and other riverine biota are organized. Such an ecologically based strategy can begin with new goals grounded in past management experience, together with knowledge of the biophysical dynamics of watershed and aquatic ecosystems. These goals might be

1. to maintain future recovery options by ensuring long-term maintenance of secure, well-distributed, diverse natural habitats and coadapted populations, and local examples of natural ecosystem processes;

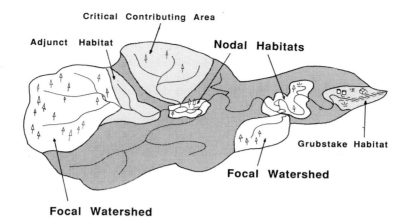

Critical Contributing Area

Adjunct Habitat

Nodal Habitats

Grubstake Habitat

Focal Watershed

Focal Watershed

FIGURE 7.2.—Simple conceptual diagram depicting spatial relations between functional habitat types identified in the "rapid biotic and ecosystem response" strategy (Doppelt et al. 1993; Frissell et al. 1993), as it might be expressed in a typical Pacific coastal drainage basin.

2. to secure existing populations of native aquatic species and maintain the critical refugia that support natural, historical ecosystem function;

3. to foster recovery pathways that offer the greatest probability of measurably improving the status and abundance of existing target populations, expanding from existing centers of native species diversity, and incrementally extending the influence of natural ecosystem processes.

These strategic goals stem from the following seven principles (Doppelt et al. 1993):

1. Instream habitat conditions and biota are largely determined by processes in the basin, riparian, and floodplain areas, and cannot be manipulated independent of this context (Warren 1979; Lisle 1981; Frissell et al. 1986; Platts et al. 1989; Gregory et al. 1991; Schlosser 1991).

2. Many disturbances propagate downstream from headwater sources and along a river network (Figure 7.2). Consequently, disturbances from multiple sources interact and can do so cumulatively and synergistically (Leopold 1980; Coats and Miller 1981). Therefore, protection of sensitive headwater areas is critical to maintenance and recovery of aquatic habitats throughout the river system (Hicks et al. 1991b; Moyle and Sato 1991).

3. Effective restoration treatment does not merely add structures or otherwise attempt to salvage the worst degraded or most visibly damaged areas. Instead, it changes the underlying processes that cause habitat deterioration (Platts and Nelson 1985; Beschta and Platts 1986; Elmore and Beschta 1987; Frissell and Nawa 1992). This requires comprehensive, rigorous research and monitoring to diagnose causal mechanisms of habitat loss and to evaluate the

effectiveness of various treatments (Everest et al. 1986b; Weaver et al. 1987; Reeves et al. 1991a; Stanford et al. 1996).

4. Riverine habitats are highly variable and patchy in space and time, even under natural conditions. Restoration must be directed not at producing homogeneous or generic conditions but at restoring the temporal regimes and spatial diversity of the natural habitat system (Frissell et al. 1986; Poff and Ward 1990; Sedell et al. 1990; Gregory et al. 1991; Schlosser 1991). This can be achieved by affecting the processes that determine these patterns (Bravard et al. 1986; Amoros et al. 1987).

5. Maintenance and restoration of a well-dispersed network of habitat refugia— including headwater watersheds, and relatively intact lower-river segments—is necessary to sustain current fish populations. It also is necessary to ensure persistent sources of colonists to seed habitats that become available following natural recovery or restoration (Moyle and Sato 1991). Most current natural production of anadromous salmonids is sustained by the relatively small proportion of habitat that remains relatively undisturbed. Restoration that secures and improves these habitats will have the greatest immediate effect on protecting and increasing fish populations.

6. The current distribution and life history patterns of native fish populations are largely governed by the nature and distribution of available habitat refugia in a basin, and sometimes by the history of nonnative species introductions. Thus, their distribution and life history determine the ability of fish populations to respond to habitat change (Niemi et al. 1990; Schlosser 1990). Restoration that first secures existing refugia, and then reestablishes similar and nearby habitat that requires little adjustment of life history patterns, is most likely to provide the habitat critical to existing fish populations. In contrast, restoration to reestablish long-lost habitat or to restore habitat at a great distance from present centers of productivity and diversity is likely to experience delayed biological response and slow colonization. In some cases, many years or decades could be required for the reevolution of life history types suited to such habitats (Warren and Liss 1980; Frissell et al. 1997).

7. Recovery of degraded and biotically impoverished watersheds requires a long time—many decades, and perhaps centuries (Hagans et al. 1986; Niemi et al. 1990; Yount and Niemi 1990). Restoration in such areas is likely to prove unsuccessful with unpredictable results in the near term (less than 20 years). Over the long term, if restoration of severely degraded catchments is accompanied and guided by careful monitoring, iterative planning, and experience from less-degraded ecosystems, it might yield significant benefits. Long-term success in highly degraded and biotically impoverished ecosystems will depend on preservation of biotic diversity and ecosystem functions in other areas suitable for natural colonization or reintroduction of organisms.

CLASSIFICATION OF RESTORATION UNITS

An example in which these principles were incorporated into a comprehensive watershed restoration strategy was presented as the "Rapid Biotic and

Ecosystem Response" approach in Doppelt et al. (1993) and Frissell et al. (1993). In conceptualizing and implementing this strategy, it is helpful to identify and classify critical areas within a large watershed by their functional significance as critical habitats for fish and other biota and their role in sustaining natural ecosystem processes. Based on functional significance and location in the basin, different parts of the watershed and stream network have different restoration requirements, and the urgency of intervention differs inherently. Watershed areas also vary in the potential cost-efficiency of projects, in terms of measurable improvement in the security and productivity of aquatic populations.

Doppelt et al. (1993) and Frissell et al. (1993) have described five classifications of habitat. These are detailed below and shown schematically in Figure 7.2. The five categories provide a general framework for planning, prioritization, and implementation of restoration. The terms used (focal, nodal, etc.) were selected for descriptive and heuristic value; they are not established in the scientific literature (in fact, no established terms exist). But these classifications have fostered understanding and dialogue at scientific workshops and in assessment and planning exercises. The terminology and descriptions are not as important as the concept that different kinds of habitats differ in value and urgency for restoration by virtue of their location, ecological processes, ecological setting, and history of previous alteration.

Some examples of habitat types in these functional categories are provided in Frissell et al. (1993). The classification scheme provides a working template for identifying specific areas where the need for and benefits of intervention are greatest or most urgent. However, further work is clearly needed to develop locally appropriate criteria and protocols for identifying and mapping habitat functional types.

1. Focal watersheds.—A focal watershed is defined as a drainage basin that remains hydrologically and biotically unimpaired by human landscape alteration or aquatic species introductions. These critical areas support a mosaic of high-quality, hydrologically intact habitats that sustain a diverse or unusually productive complement of native species. Typically, they are relatively undisturbed headwater drainages that foster spawning and rearing habitat for remnant populations of sensitive fishes and other organisms.

These areas have a very high priority for restoration for three reasons. First, many species benefit from protection and maintenance of the area. Second, the cost of protection or restoration is low relative to the biological benefits gained or secured (Moyle and Sato 1991). And third, the near-term success is very likely because the habitats already are inhabited by many of the populations or species assemblages of interest.

These refugia are only moderately well-connected to other portions of the stream network. I say moderately because colonists can move easily downstream to colonize other habitat, but organisms may have difficulty moving upstream into the focal watershed. On the other hand, the potential for large gain in production within these areas is low because habitat quality is already high relative to most lowland rivers and floodplains. Further, these headwater streams can encounter high frequencies of natural disturbance. Consequently, over a century or more, the likelihood of population persistence in such a refuge is not great. This problem can be minimized by establishing larger focal watersheds that encompass a network of tributary systems, increasing the internal redundancy of habitat within the refuge.

Larger focal watersheds (greater than 100 square miles) can be identified in certain relatively unmodified landscapes. However, due to widespread human alteration of most environments in the temperate zone (Frissell and Bayles 1996), most focal watersheds are likely to be small (less than 40 square miles). Most focal watersheds defined by these criteria are substantially smaller and more numerous than the majority of "key watersheds" identified in the Forest Ecosystem Management Analysis Team (FEMAT) process and other recent federal agency aquatic assessments in the West. Using this approach, a typical federal "key watershed" may consist of multiple focal watersheds and associated nodal habitat segments (see next), with intervening segments that have been more heavily impacted and depleted of sensitive species.

2. Nodal habitats.—These areas are spatially dissociated from refuge habitats, but serve critical life history functions for individual organisms that originate from populations and species in refuge habitats throughout the basin. For example, as juvenile salmonids grow, they move downstream to seasonally occupy habitats that are suitable for summer or winter rearing. Nodal habitats are often partly impaired by past riparian and upstream disturbance, but still retain vital components that are critical for certain life history stages. These components include forested alluvial valley flats, floodplain ponds, woody debris complexes, or groundwater-fed side channels.

Among this class of habitat are most coastal estuaries. Although partly degraded, they still serve vital functions for many anadromous and marine fish species. Protection of these habitats is necessary to maintain existing biotic diversity and production.

Like focal watersheds, nodal habitats also have a high restoration priority, for three reasons. First, they are critical to sustaining existing populations in the watershed. Second, they are highly connected and accessible to organisms moving upstream or downstream. And third, because they are the connecting links among life histories of numerous populations, these habitats have high potential productivity. Therefore, as in refuge habitats, restoration that secures or enhances habitat recovery is highly likely to be followed rapidly by positive biotic response.

3. Adjunct habitats.—These areas are directly adjacent to and typically downstream from focal watersheds or nodal habitats. However, they have been degraded by human or natural disturbances and do not presently support high diversity or abundant native species. Considerable restoration may be necessary to reestablish a natural biotic community.

Although restoration of these areas may have moderate cost, they are a high priority for, again, three reasons. First, watershed processes are controlled, or at least buffered, by the adjacent focal or nodal area, so that riparian and in-channel restoration stand a good chance of succeeding. Second, the adjacency to focal or nodal habitat ensures that appropriately adapted colonists are close at hand, so habitat recovery is likely to be closely followed by biotic recovery. And third, restoration of these areas can directly improve the productivity and viability of existing populations that are centered in focal or nodal areas.

4. Critical contributing areas.—These are portions of the watershed that do not directly support habitat for the species of interest, but are important sources of high-quality water and stable watershed conditions for downstream focal or nodal habitats. Thus, protection and restoration of these areas is necessary to secure the functional value of associated focal, nodal, and adjunct habitats. Examples include tributary basins contributing high quality cool water to downstream habitats, unstable slope areas that are directly adjacent to nodal habitat areas, and wetland complexes or alluvial aquifers that contribute to water quality and habitat integrity in adjacent critical habitats.

5. Grubstake habitats.—These habitats tend to occur in low-elevation, heavily disturbed portions of the drainage basin, generally associated with lowland floodplain rivers. They are called ''grubstake'' habitats because restoration may require extensive planning and experimental work, and in many cases the cost will be high. The potential payoff could be large, however, because these habitats were historically extremely productive for many species. Rapid biotic response is unlikely because eliciting change in these habitats may require a long time, and even then, sources of preadapted populations may be relatively distant.

Examples of possible grubstake habitats include estuarine marshes, floodplain wetland complexes, and the main stem habitats of lowland rivers. Almost universally, they are heavily disturbed and usually support only low levels of production and relict populations of native species. Historically, however, they were probably the most richly productive habitats for anadromous fish and many other organisms, and they remain the largest single reservoir for potential increase in production of highly valued species, such as chinook salmon.

Despite being disturbed, grubstake habitats, if restored, can be highly connected to the stream system and could be accessible to organisms from other parts of the basin. However, they are geographically distant from focal and nodal habitats that support high natural diversity, and they do not directly receive water quality and other ecosystem benefits that can be provided by headwater watershed refugia. Also, conditions in these habitats have been degraded or hostile for many decades. Consequently, the life history forms of fish and other organisms that used these habitats have been strongly selected against. Even after grubstake habitats are restored, it might take decades for populations to reevolve the life history patterns necessary to efficiently exploit these habitats.

The low-elevation, floodplain, and wetland areas that can contain grubstake habitats are geographically and ecologically complex, are directly affected by multiple ownership, and are sensitive to physical processes throughout the basin. Thus, the social and technical challenges of their restoration are great.

Priority for grubstake habitat restoration is moderate to low because the certainty of success is low, techniques are poorly understood, and the biological urgency of intervention is less than in other areas.

The primary incentives for beginning work in grubstake habitats are the need for experimentation to develop effective treatment and that the sooner work begins, the sooner results will be known. Once the long-term effects of restoration become evident, the potential could be great for production increase and fisheries improvement. It is essential that work in grubstake habitats be designed and implemented as experimental projects, including careful monitoring and feedback mechanisms to ensure that appropriate adjustments are made as restoration proceeds. (See Kershner 1997, this volume, for more information on monitoring and adaptive management.)

OPERATIONAL IMPLICATIONS

The ecological strategy described above for establishing restoration priorities is being implemented on a large scale in the Pacific Northwest and elsewhere. From preliminary tests, three important principles of watershed restoration priorities have emerged: address internal threats, maintain the greatest area of high-quality habitat and diverse aquatic biota, and create policies to facilitate restoring critical riverine habitats. We will look at each.

Address Internal Threats

Although Moyle and Sato (1991) focused on concerns about edge effects or external threats to the integrity of aquatic reserves, any watershed reserve network will also inherit internal threats—artifacts of previous human disturbance. It appears equally important to address this. In fact, a principal function of restoration should be to identify and defuse existing internal "time bombs" that set up the landscape for future damage.

Time lags on the order of years to decades exist between headwater slope disturbance and impacts to riverine habitat downstream (Coats and Miller 1981; Hagans et al. 1986; Frissell 1991). This means that many habitats that today function as refugia lie in watersheds where slopes and headwaters have been disturbed by people in recent decades, but have not yet experienced a major storm or similar catastrophic event that triggers the downstream cascade of impacts. An example is formerly roadless watersheds that have been partly penetrated by logging roads in the past 10 to 20 years. Once such a catastrophic triggering event occurs, the damage to the watershed and riverine ecosystem cannot be easily or rapidly reversed. Preventive treatment is therefore far more feasible, successful, and cost-effective than repair (Weaver et al. 1987; Reeves et al. 1991a).

Such watersheds have sometimes been referred to as "loaded guns" in which biota are fated to inexorable deterioration unless humans intervene in a focused way to remove potential sources of disturbance. Because roads are the single most persistent, preventable, and often the largest source of slope erosion and downstream disturbance (Figure 7.3), the most effective intervention that has

FIGURE 7.3.—Flow diversion and stream erosion caused by a plugged culvert at a road crossing of a tributary to Chesnimnus Creek, northeastern Oregon. Road systems are universally a principal and essentially permanent cause of stream habitat deterioration. Unless chronic and potential problem sites along road networks are correctly identified and treated, recovery of watershed and aquatic habitats usually will not occur.

been widely identified is obliteration or relocation of roads and reestablishment of natural drainage (Weaver et al. 1987). A similar example might be the removal of small populations of introduced species that could, if left unchecked, spread and displace native biota in the refuge watershed.

Maintain the Greatest Area of High-Quality Habitat

Restoration should be focused where minimal investment can maintain the greatest area of high-quality habitat and diverse aquatic biota. Few completely roadless, large watersheds remain in the Pacific Northwest, but those that continue relatively undisturbed are critical in sustaining sensitive native species and important ecosystem processes (Sedell et al. 1990; Moyle and Sato 1991; Williams 1991; McIntosh et al. 1994; Frissell and Bayles 1996). With few exceptions, even the least-disturbed basins have a road network and a history of logging or other human disturbance that greatly magnifies the risk of deteriorating riverine habitats in the watershed.

In other words, most currently functional riverine refugia are imperiled by past, recent, or proposed human disturbance. Focused restoration intervention on these sites, as well as land management changes, are necessary to secure key watershed refugia. A small investment in a watershed that retains much of its natural integrity can secure far more critical resources and can far better safeguard the future of sensitive fish species than a very expensive effort in a watershed that has already suffered severe, long-standing degradation.

Create Policies to Facilitate Restoring Critical Riverine Habitat (Nodal / Grubstake)

New policies are needed to facilitate restoration of low-elevation floodplains, wetlands, and other critical riverine habitats. Many such areas are in private ownership and a floodplain restoration policy will involve complex social and political dynamics.

One way to cost-effectively recover floodplains and wetlands could be to regulate land use following large floods (Sparks 1995). Such regulation should restrict and discourage channelization, revetment (building of retention structures), debris removal, diking and draining, reconstruction or reoccupation of floodplain roads and structures, and other activities that impede or reverse the positive natural effect of flooding in floodplain rivers. Particularly where the sediment load from upstream has stabilized at more natural levels, these regulations could allow lowland, floodplain rivers themselves to reclaim habitat and restore natural processes. This would eliminate large capital investment and the engineering risk of massive structural projects (Bravard et al. 1986; Amoros et al. 1987; Sparks et al. 1990; Stanford et al. 1996). Fiscal resources could be devoted largely to education, development of creative, relatively nonintrusive regulatory policies, and provision of financial incentives for floodplain and wetland disinvestment.

CONCLUSIONS

Past and present strategies for managing and restoring watersheds and riverine habitat are insufficient to ensure survival and recovery of anadromous salmonids and other native aquatic species. These strategies fail to address critical watershed and riverine ecosystem processes and fail to safeguard the future of fragmented and declining populations of fish and other riverine species. A major change in the philosophy and assumptions of habitat restoration is necessary.

A new approach to riverine habitat management is founded on principles of watershed function, ecosystem processes, and conservation biology. It calls for a network of remnant, diverse, and relatively unimpacted habitats as the focus for ecological restoration. Restoration priorities should begin by identifying and securing existing watershed refugia and downstream critical habitats that function as convergence nodes for existing populations and life histories of key species. The next priority is recovery of habitat in reaches adjacent to watershed refugia and nodal habitats. Finally, the long-term restoration of downstream, lowland habitats is necessary to reestablish historical levels of productivity and secure the future of native species.

In the long run, no system of fragmented refugia can alone sustain a viable and productive riverine ecosystem and thoroughly protect aquatic biodiversity. The long-term goal of sustainable management must be to restore landscape conditions so that functional habitat refugia become widely distributed and well dispersed in the river system. If we can successfully restore less-disruptive ecosystem processes and patterns basinwide, then what today are considered "critical habitats" could become less rare, more connected, and less individually indispensable to the overall function and persistence of fish populations and other aquatic species.

ACKNOWLEDGMENTS

I thank the many experts who volunteered their time and creative genius for the Pacific Rivers Council's technical workshops in 1992 and 1993. The constructive comments of D. Bayles, C. Dewberry, B. Doppelt, R. Franklin, S. Gregory, W. Liss, R. Nawa, W. Nehlsen, J. Sedell, R. Van Kirk, J. E. Williams, and others greatly improved drafts of this manuscript. Much of the material in this chapter was originally included in a report prepared for the Pacific Rivers Council in 1993.

. . . restoration ecology now faces two major conceptual challenges. One is to decide what we really mean by our goals when we pretend that we are restoring natural, self-sustaining communities—which we rarely, if ever, are. The second is to decide, given limited amounts of time and money, what we most urgently need to know in order to achieve our goals.
—Jared Diamond, 1987

CHAPTER 8

MONITORING AND ADAPTIVE MANAGEMENT

Jeffrey L. Kershner

By now the reader should be convinced that restoring degraded watersheds is important. Millions of dollars have been spent to reclaim our aquatic and riparian resources and millions more probably will be. Intuitively, we know that restoration can improve these resources, but clearly restoration dollars must be spent wisely.

Monitoring is the measure of success of any restoration. Well-designed monitoring should (1) indicate whether the restoration measures were designed and implemented properly, (2) determine whether the restoration met the objectives, and (3) give us new insights into ecosystem structure and function. Monitoring should help us reexamine our understanding of aquatic and riparian ecosystems and provide information needed to adapt the goals for restoring those systems. Significantly, as much or more is learned about systems by monitoring and reporting failure as is learned by reporting success.

If monitoring is so important, why is so little effective monitoring undertaken in proportion to the number of restoration projects? Probably foremost is the lack of funding for monitoring, an institutional problem that persists (Noss and Cooperrider 1994). Notable exceptions exist (e.g., Toth et al. 1997, this volume), but many resource management agencies are reluctant to commit funding for monitoring, particularly for the long-term monitoring of restoration.

Part of the problem is that much restoration implemented today may not yield significant benefit for years or even decades. Watershed restoration practitioners need to understand the trajectory of recovery and how to adapt our management to changing environmental conditions. This requires a long-term approach to funding by agencies, foundations, industry, tribal groups, and others. Unfortunately, this long-term perspective is currently missing from much restoration monitoring.

A second, more pervasive reason that monitoring gets short shrift is that restoration practitioners are intimidated by monitoring. They are intimidated because the job looks so large: Did the treatments work? How long did they work? Did they make a difference? Were the restoration objectives chosen correctly? How much money and time should be spent on monitoring?

Although increasing the funding for monitoring may not be possible, we can develop monitoring that is well-designed, that reflects realistically the available personnel and funding, and that answers questions about restoration measures.

The purposes of this chapter are (1) to identify the key components of watershed restoration monitoring and (2) to aid the monitoring practitioner in designing credible monitoring for watershed restoration, given varying levels of funding and personnel.

TYPES OF MONITORING

The restoration practitioner must understand what types of monitoring fit a particular project (MacDonald et al. 1991). Three types are particularly useful for restoration: *implementation monitoring*, *effectiveness monitoring*, and *validation monitoring*.

Implementation monitoring asks: Was the restoration implemented properly? This monitoring should be part of every restoration project and is normally performed during or shortly after restoration. During the project, implementation monitoring continually evaluates the project design to determine its appropriateness in the field. Midcourse corrections are often necessary because field conditions may make the original design unworkable. Implementation monitoring during the project can identify these problems early and suggest workable solutions. This is particularly important where restoration contracts identify specific restoration measures, materials to be used, and designs to be followed. (Nothing is more frustrating than to follow contract specifications in the field, only to realize that the project is doomed because the design was inappropriate!)

Implementation monitoring sets the stage for other types of monitoring by demonstrating that the restoration treatments were done correctly and followed the design. The practitioner can then concentrate on identifying and correcting design problems if failures occur.

Effectiveness monitoring asks: Was restoration effective in attaining the desired future condition and in meeting restoration objectives? Effectiveness monitoring is more complex than implementation monitoring and requires understanding of the physical, biological, and sometimes the social factors that influence aquatic ecosystems. This understanding is translated into quantifiable objectives or benchmarks that describe the function of healthy aquatic systems. The primary purpose of effectiveness monitoring is to measure whether objectives are met by restoration.

Trend monitoring is a less rigorous form of effectiveness monitoring; it often involves visual estimates or photographs of changing resource conditions over time.

Validation monitoring is more specialized and primarily has a research focus. Validation monitoring verifies the basic assumptions behind effectiveness monitoring. For example, until 20 years ago, large woody debris in streams was removed to facilitate fish movement (Sedell et al. 1988). More recent research has shown that woody debris is important in structuring stream communities in many areas of the country and is an important link between the physical and biological functions of streams (Maser and Sedell 1994). Validation monitoring is a research tool with which to examine the basic scientific understanding of how aquatic systems work. Effectiveness and validation monitoring are necessary steps to evaluate adaptive management prescriptions.

A RESTORATION MONITORING PROCEDURE

When multiple restoration activities are underway, the additional task of monitoring can appear overwhelming, so practitioners may neglect it. A common lament is that "it will take too much time or money," or "I don't have the statistical background," or "it's not a priority for my supervisor." Understanding the starting point and taking the first steps will alleviate this anxiety and get monitoring in motion. The following steps provide a template for sound restoration monitoring:

1. define participants;
2. establish clear goals and objectives;
3. design monitoring to detect change to (a) distinguish treatment effects from other variations, and (b) take replicate samples over space and time;
4. prioritize monitoring activities;
5. implement field prescriptions and techniques;
6. analyze data and report results; and
7. adapt goals and objectives to new information.

Further explanation of each step is provided next.

Define Participants

Watershed-scale restoration may involve myriad resource specialists, agency personnel, and nonagency partners. All should develop ownership in the monitoring. Interdisciplinary development of monitoring goals and objectives usually is best. For example, fisheries objectives may have hydrologic or geomorphic (landform) components that will determine the success of the restoration. Thus, hydrologists or fluvial geomorphologists (specialists in stream patterns and stream-related landforms) should actively participate in setting objectives, study design, analysis, and other appropriate phases of the project. Other specialists may be required to successfully complete restoration monitoring.

State and federal agencies, private landowners, tribal groups, and citizen groups often have ownership in watershed restoration projects and are interested in their success or failure. These groups can provide labor, financial support, and technical assistance. Their involvement can extend monitoring resources and allow expanded monitoring that otherwise would be impossible.

However, use of nontechnical personnel requires caution. Monitoring tasks that demand high technical expertise may require extensive training. It is best to question nontechnical participants to determine their expertise and desired level of participation before committing them to monitoring.

Establish Clear Goals and Objectives

Monitoring often fails because the overall mission, goals, and objectives are not clearly articulated. Establish the purpose of monitoring by developing clear goals and objectives (see Appendix Examples 1 and 2 at the end of the chapter). A successful monitoring plan has clear objectives to use as benchmarks for the analysis. These objectives define the project's purpose and determine the type and extent of restoration. Objectives should come from analyses of limiting factors for the species or community of interest and normally are determined during a full watershed analysis (Kershner, in press; Ziemer 1997, this volume).

Objectives are typically measurable and quantifiable and represent some desired future condition, within the constraints of resources. It is important to understand the objectives within the actual spatial and temporal scales that are operating on the subject landscape (Frissell et al. 1986; Wissmar 1997, this volume; Ziemer 1997). Thus, the spatial and time scales of interest need to be described specifically (Conquest et al. 1994). In doing so, the actual study design and sampling protocol can be more easily defined.

Objectives may represent standards and guidelines from broad planning or policy documents, or they may represent benchmark conditions from "healthy" aquatic systems. In all cases, these broader objectives must be modified for local conditions. Watershed-specific objectives should consider local disturbance processes and geomorphic conditions. For example, a watershed goal might be to reduce the total sediment input from roads. Thus, an effectiveness monitoring objective might be to quantify change in residual pool depths in low-gradient, unconfined channels that are sensitive to sediment inputs, or to quantify change in fine sediments in spawning areas.

In any case, one should select monitoring objectives that are the best indicators of change and measure them in the appropriate areas that are responsive to change. For example, it may be difficult to determine whether bank stabilization is reducing fine sediment if fine sediment inputs are measured in

TABLE 8.1.—Ten principles for conducting environmental field studies. (Green 1979.)

1. State concisely to someone else the research question. (The results will be as coherent and comprehensible as the initial conception of the problem.)
2. Take replicate (multiple) samples within each combination of time, location, and any other controlled variable. (Differences among sample groups can be demonstrated only by comparison to differences within the groups.)
3. Take an equal number of randomly allocated replicate samples for each combination of controlled variables. (Putting samples in "representative" or "typical" places is not random sampling.)
4. To test whether a condition has an effect, collect samples both where the condition is present and where the condition is absent, but all else is the same. (An effect can be demonstrated only by comparison with a control.)
5. Conduct preliminary sampling to provide a basis for evaluation of sampling design and statistical analysis options. (Those who skip this step because they do not have enough time usually end up losing time.)
6. Verify that the sampling device or method is sampling the population it is supposed to be sampling, and with equal and adequate efficiency over the entire range of sampling conditions to be encountered. (Variation in the efficiency of sampling from area to area will bias among area comparisons.)
7. If the area to be sampled has a large-scale environmental diversity, break the area into relatively homogenous subareas and allocate samples to each in proportion to the size of the subarea. If estimating abundance over the entire area, make the area allocation proportional to the number of organisms in the subarea (by adjusting the numbers of sample units).
8. Verify that the sample unit size is appropriate to the size, density, and spatial distribution of each organism being sampled. Then estimate how many replicate samples are required to obtain the desired precision.
9. Test the data to determine whether the error variation is homogenous, normally distributed, and independent of the mean. If it is not, as will be the case for most field data, then (a) appropriately transform the data, (b) use a distribution-free (nonparametric) procedure, (c) use an appropriate sequential sampling design, or (d) test against simulated H_o data.
10. Having chosen the best statistical method to test the hypothesis, stick with the result. (An unexpected or undesired result is not a valid reason for rejecting the method and hunting for a "better" one.)

high-gradient riffles. These areas typically transport sediment and may not be responsive to changing inputs of fine sediment.

Design Monitoring to Detect Change

Monitoring must detect change. While this in itself is a simple concept, it actually places an enormous burden on how monitoring is conducted and how the data are analyzed and interpreted. Often it is difficult to distinguish the effect or result of a particular activity among many interacting factors and great natural variability. It is useful to recall the ten principles described by Green (1979) for designing environmental field studies, presented in Table 8.1.

Implementation monitoring is relatively straightforward, and may involve simple yes–no answers. It may help to develop a checklist that identifies the project prescription and indicates whether each element was completed properly. This obvious step often is overlooked, but it is nearly impossible to conduct effectiveness monitoring unless it is clear that the project was implemented correctly.

If the project was implemented correctly, effectiveness monitoring is conducted to determine whether project objectives are being met. The following two general design principles are appropriate for effectiveness monitoring studies (Armour et al. 1983).

Distinguish treatment effects from other variations.—The study design must allow treatment effects (explained sources of variation) to be distinguished from all other sources of variation. This can be done by isolating "treatments" or restoration from other "untreated" sites. In this way, it can be determined whether restoration caused the change or whether similar changes would have occurred naturally without restoration.

It may be difficult to detect change from site-specific restoration in larger watersheds. An important factor is natural variability, which can make difficult the detection of change between restored and untreated areas. It is critical to minimize this variability by choosing areas of similar size, geology, morphology, stream discharge, and other characteristics. Ideally, they should vary only in the extent of restoration and should consist of one or more pairs of treated–untreated smaller subwatersheds within a larger watershed.

Realistically, practitioners may encounter other environmental factors that prevent the "ideal" situation. Although comparisons between partially and wholly treated smaller watersheds can be used, the contrast in restoration extent should be as great as possible, given the inherent difficulties in detecting change. Untreated areas can serve as the "control" or "reference," while restored areas can serve as the "treatment."

It is useful to collect pretreatment data in areas being considered for restoration (House 1996). Pretreatment data are often useful as a benchmark where suitable control areas may be unavailable. If possible, information should be collected on physical and biological characteristics of sites before treatment so that change resulting from restoration can be measured and documented. Pretreatment data may not be available over a long period, but whatever can be acquired is useful.

Take replicate samples over space and time.—Replication refers to the number of sites or samples needed over space and time. How many are needed depends on natural variation in the variables measured and the precision and accuracy desired. Two things are important in determining the sample quantity: (1) level of significance and (2) the probability of detecting a difference when one exists (statistical power).

Selection of the level of significance generally depends on the values and risks associated with the variable being measured (MacDonald et al. 1991). The level of significance must be appropriate for the project. For example, many field studies set a high level of significance of 0.05 to detect changes caused by management. (This level indicates that only 5% of the observed difference is due to chance.) Although the reliability of the answer might be quite high, more samples are generally required to reach this higher level of statistical significance. Picking a lower level may still provide reasonable certainty but may require fewer samples and be more cost-effective over the long term.

Another consideration when choosing sample size is being able to detect a difference when it exists, which is called statistical power (Sokal and Rohlf 1981; Peterman 1989). Similar to determining the level of significance, an increase in sample size will generally increase the statistical power of the test. When identifying the level of significance, it is important to remember that decreasing the level of significance will generally increase the statistical power. This may have advantages where funding and personnel limitations reduce how many samples can be obtained.

If the practitioner is interested in monitoring the direct, localized effects of a spatially limited activity, then the appropriate study design might use paired sites (see Appendix Example 1). This may involve selecting pairs of monitoring stations upstream and downstream or perhaps side by side.

If one is monitoring the combined effectiveness of varied restoration activities within a watershed, then the paired control–treatment approach should be used for a number of objectives and integrated throughout the watershed (see Monitoring Example 2). In these cases, an expanded study design would include comparisons between whole tributary watersheds or even entire watersheds.

Because multiple activities must be monitored in a watershed, single-activity monitoring cannot distinguish all the sources of change. For example, a decreasing ratio of stream channel width to depth could be the combined result of three different events: (1) decreased hillslope sediment delivery from a landslide stabilization project, (2) culvert and road fill removals in headwater areas, and (3) vegetation enhancement projects at the site to reduce bank erosion and resulting sedimentation. A combination of implementation monitoring and effectiveness monitoring will be needed to assess overall restoration effectiveness in the watershed. The watershed where these activities are taking place should be compared against a watershed or smaller subwatershed where similar management exists, but without similar restoration.

Because multiple activities are involved, multiple measurements are necessary on a variety of variables over an extended period. Perhaps the "best" variables are those that integrate a variety of processes and that serve as appropriate biological and physical monitoring surrogates (see Monitoring Example 2). Biological indices such as the index of biotic integrity (Karr 1991, 1993; Angermeier 1997, this volume) or aquatic invertebrate metrics (Plafkin et al. 1989) may be particularly suited to monitoring when restoration emphasizes the recovery of aquatic communities.

The data and analysis results are greatly influenced by spatial and temporal factors and by observer error. For example, changes in flow and channel morphology (shape) along the stream affect many variables commonly used for monitoring. Our inability to consistently identify features can be a very large source of error. Standardizing and using quantitative methods helps to improve repeatability in monitoring.

Prioritize Monitoring Activities

A well-designed restoration plan should identify all of the key elements to be monitored, but should prioritize them according to importance and availability of resources. The plan should estimate the time, money, personnel, and equipment required to implement each plan element. This information is essential to the monitoring practitioner and to supervisory personnel when planning a work force and budget.

Rarely can practitioners do all of the restoration monitoring they would like. During times of decreasing personnel and budgets, some monitoring elements may be deferred or dropped. An effective strategy is to prioritize program elements so that at least some level of monitoring can be accomplished in any year. For example, implementation monitoring of a riparian (streambank) planting project may be done annually for the first few years of the program, and deferred to every 3 years if it appears that vegetation has been successfully reestablished. This may make money and personnel available for other aspects of the monitoring plan.

Another approach is to share personnel and funds among administrative units. Multiple agencies and groups may be involved in the monitoring of large watersheds. It is often useful to combine resources and develop monitoring teams where similar types of monitoring activities are being implemented. Teams can be dedicated to limited work assignments, and once training in field techniques and data collection has been accomplished, these teams are often more efficient at collecting information. Typically, this improves the reliability and precision of data and should save time and money. The only down side is that it may be difficult to establish priorities for these teams in a multiple group or agency setting.

Implement Field Prescriptions and Techniques

Monitoring analyses can be only as good as the data gathered. Because different personnel often perform the same monitoring in different years, it is important to establish consistent field protocols to collect field measurements in the same way, year after year. This step is missing in many monitoring programs. It may be useful to develop a set of field protocols for each phase of the monitoring plan. A simple checklist, like the one mentioned above for implementation monitoring, can help ensure that each critical element of a restoration is evaluated. Trend monitoring often requires that photos be taken in successive years to evaluate changing conditions. A narrative outlining the date, time of day, compass direction, and location on an aerial photo is useful to minimize the variation from differing field conditions.

Graphical and quantitative analyses require consistent, replicated measurements to reduce variation from field measurement error. Practitioners should write detailed narratives that establish sampling location, frequency, technique, and equipment to be used. They should try to avoid field variables that require subjective judgment. For example, habitat classifications that require considerable professional judgment but which lack consistent, repeatable field protocols should be avoided. In general, the more complex a field variable, the greater the chance for error.

It is advisable to establish a training program before the field season to train

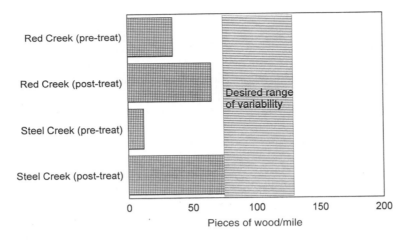

FIGURE 8.1.—An example of a graphical data display comparing frequency of large woody debris before and after restoration to a hypothesized desired range of variability.

field personnel in data collection protocols and techniques. Only when it is clear that monitoring personnel are collecting measurements consistently should they be sent into the field. Quality control checks should be made throughout the field season.

Analyze Data and Report Results

Several useful ways exist to display and analyze monitoring data. Most importantly, a variety of very powerful tools are available if the sample design considerations described earlier have been followed carefully. For some variables, a qualitative approach may be more appropriate, given the complexity or great variation associated with them. For example, it is difficult to accurately quantify bank erosion, but a photograph provides vivid and unambiguous documentation.

Comparative analyses can be simple graphical displays of how data in control and treatment areas differ in the same year and over time. These simple comparisons are often the most visually powerful evidence that change has occurred. It may also be appropriate to compare data from these controls and treatments to objectives from policy documentation. Photographs, bar graphs, and line graphs are particularly powerful ways to show analytical results to nontechnical people and decision makers (Figure 8.1). Graphical analyses should be the first step in any statistical report.

Statistical interpretation may provide further insight that may not be readily apparent from graphical analyses alone. If the study is designed using the principles outlined in Table 8.1, then other statistical options will be available for the analysis. Statisticians should be consulted often during the study design, pilot study, and analysis phases of the project to keep on track. Few statisticians are willing or able to help analyze data from a poorly designed study (Green 1979).

Restoration often fails because data are not analyzed and reported. Monitoring reports can build support for restoration by demonstrating positive environmental change. The results may lead to greater support for restoration and restoration monitoring from supervisory personnel and the public. It is as important to report restoration failure as it is to report success. Determine whether failure was caused by poor design, poor implementation, catastrophic conditions, or poor planning. Practitioners should share information with other restoration professionals, document what worked and what did not, and adapt restoration goals and objectives as new information becomes available.

Technical note on statistics.—Parametric statistics are used to measure data against a set of "assumptions" regarding their use. These assumptions have to do with the distribution of the data, whether the error variation is homogeneous, normally distributed, and independent of the mean. Random or stratified-random samples are a condition for using these analyses as well as nonparametric analyses described below. If data do not meet the assumptions outlined above, then it may be appropriate to statistically transform the data before performing further analyses. Statisticians should be consulted to help with the appropriate transformations of data. Parametric tests that can be used for monitoring data include *t*-tests, analyses of variance, and means tests. The questions asked during the study design phase will guide which of these tests are appropriate.

Nonparametric statistics are used when the assumptions described above are violated for some reason. Random or stratified random samples are still a condition for the use of these tests. Nonparametric tests are not a panacea for poor study design; they are to be used only when the assumptions for parametric tests are violated and cannot be corrected by transformation.

Adapt Goals and Objectives to New Information

Adaptive management is the process whereby management is initiated, evaluated, and refined (Holling 1978; Walters 1986). It differs from traditional management by recognizing and preparing for the uncertainty that underlies resource management decisions. Adaptive management is typically incremental in that it uses information from monitoring to continually evaluate and modify management practices. It promotes long-term objectives for ecosystem management and recognizes that people's ability to predict success is limited by knowledge of the system. Adaptive management uses information gained from past management experience to evaluate both success and failure and to explore new management options.

Monitoring provides the information needed to evaluate management. Monitoring may suggest new approaches or goals for watershed restoration and management. An adaptive management strategy for the restoration practitioner may be to experiment with different types of restoration throughout a watershed while continually monitoring the performance of the measures. By learning from both successes and failures, the restoration practitioner can see which techniques may be most useful and gain insight into which practices best promote recovery. It may be advantageous to conduct several smaller experimental restorations to gauge their effectiveness and system response as a prelude to large-scale projects.

CONCLUSIONS

Watershed restoration monitoring is an important component of aquatic ecosystem management. Resource stewards cannot determine the success or failure of watershed restoration without well-designed and properly implemented monitoring. The call for more efficient government and wise use of public funds will place restoration under increased scrutiny from the public and legislative bodies. The accountability that comes from proficient monitoring will be essential to continued restoration funding.

The challenge to restoration practitioners is to move beyond project implementation into careful analyses and reporting of restoration results. The lack of published restoration monitoring results indicates either that restoration generally is not followed by careful analysis or that restoration practitioners have not widely shared their analyses and findings, or both. Any of these cases is unacceptable. To move forward with well-conceived and well-supported watershed restoration, we must look backward to lessons of the past.

APPENDIX: MONITORING EXAMPLES

Example 1: Introducing large woody debris to improve habitat variety, restore deep pool habitat, and increase juvenile steelhead (rainbow trout) numbers.

Goal: Improve juvenile steelhead habitat to restore runs of summer steelhead.

Key limiting factor: Complex pool habitats created by large woody debris.

Objectives: (1) Increase structurally complex rearing habitat for juvenile steelhead as measured for deep pools and woody debris frequency in the current administrative policy. (2) Increase the numbers of juvenile steelhead to meet downstream migrant numbers defined as optimal in state management plan.

Question to be addressed: Was summer rearing habitat for juvenile steelhead restored?

Implementation monitoring: Number of deep, complex pools.

Measurements.—Numbers of pools per mile; frequency of woody debris after restoration.

Effectiveness monitoring: Was the restoration effective in creating quality summer rearing habitat that produces more juvenile steelhead?

Measurements.—Residual pool depth; cover index; numbers of juvenile steelhead.

Study design:

Control section(s).—No treatment upstream or downstream of treated section(s); no influence from treatment. Sampling unit is two riffle pool sequences, randomly selected in a low-gradient stream segment. Separated from treatments by long distance (300–1,000 yards).

Treatments.—Introduce large woody debris into a randomly selected section, which is two riffle pool sequences in a low-gradient stream segment.

Sampling.—Annual field monitoring of residual pool depth, percentage of complexity, and juvenile steelhead numbers in both the controls and treatments until recovery is established. Each treatment site should have a matched control and replicates of both controls and treatments. Field measurements should be quantifiable and repeatable. (Remember, structural enhancements may move or deteriorate over time.) Monitoring during periods of low flow during summer will allow for the detection of changes in the working efficiency of the structure and suggest when maintenance or further improvement is needed.

Analyses:

Graphical.—Graphical comparison of residual pool depths, complexity values, and juvenile steelhead in the control and treatment sections. Graphical comparison of treatment values and policy standards (if quantitative). Multiyear comparison graphs showing the recovery trajectory of effectiveness-monitoring variables.

Statistical.—Tests to determine whether data meet the assumptions for parametric statistical methods. Transform data to meet assumptions where possible. If transformations do not work, determine appropriateness of nonparametric methods. Parametric methods that may be appropriate include *t*-tests (two-sample paired sites), analysis of variance, and means tests. Consult statisticians to determine appropriate statistical methods.

Example 2: Watershed restoration in a small watershed that provides spawning and rearing habitat for an important anadromous fish stream.

Goal: Restore spawning and rearing habitat for summer steelhead in the subwatershed. Specific restoration goals include reducing fine sediment from roads, increasing complex pool habitat with woody debris, and providing for long-term woody debris inputs from riparian stands that consist currently of mixed hardwoods and conifers of subcommercial size.

Key limiting factors: complex pool habitats created by large woody debris; spawning substrates having minimal fine sediments.

Objectives: (1) Increase structurally complex rearing habitat for juvenile steelhead as measured for deep pools and woody debris frequency in policy standards. (2) Decrease the percentage of fines in spawning gravel to less than 10% during spawning and incubation. (3) Increase the numbers of juvenile steelhead to meet downstream migrant numbers defined as optimal in the state steelhead management plan.

Question to be addressed: Was rearing and spawning habitat for summer steelhead restored in the subwatershed?

Implementation monitoring: Wood additions; miles of roads closed; riparian stand conifer stocking.

Effectiveness monitoring: Was the restoration effective at improving rearing and spawning conditions for summer steelhead?

Measurements.—Numbers of complex pools per mile (residual pool depth, complexity rating); percentage of fines in spawning gravel; emergence survival; numbers of juvenile summer steelhead; long-term woody debris inputs.

Study design (three parts):

1. Wood Introduction

   Control sections. No treatment upstream or downstream of treated sections; no influence from treatment. Sampling unit is two riffle pool sequences, randomly selected in a low-gradient stream segment. Separated from treatment by long distance (300–1,000 yards).

Treatments. Introduction of large woody debris into site in low-gradient stream segment (measurement area is two riffle pool sequences). Treatment may include different debris inputs to vary the complexity.

Sampling. Annual field monitoring of residual pool depth, percentage of complexity, and juvenile steelhead numbers in both the controls and treatments until recovery is established. For each treatment site there should be a matched control and replicates of both controls and treatments. Field measurements should be quantifiable and repeatable. (Remember, structural enhancements may move or deteriorate over time.) Annual monitoring during periods of low flow during summer will allow detection of changes in the working efficiency of the structure and suggest when maintenance or further improvement is needed. Monitoring frequency may be decreased once structures are stabilized.

2. Road Obliteration

Assumptions. Road obliteration typically involves obliterating the existing road grade, removing culverts, regrading to native surface, and replanting. The assumption is that primary sediment delivery occurs via small side drainages that were crossed by the former road and that these drainages delivered the majority of sediment to the spawning tributary.

Controls. Controls exist above the road obliteration and are above the side drainages that delivered sediment from the road. Controls should be placed in spawning riffles under geomorphic conditions in the upstream areas that are similar to those in sites below treatment areas.

Treatments. Treatments are sites below confirmed sources of sediment that were linked to the road. Sites can be placed below side drainages in spawning riffles throughout the treatment area or below the last side drainage that delivered sediment to the spawning stream. To separate treatment effects, sites below the treatment area should be placed in spawning areas that do not have confounding effects from other management sources. It is important to have replicate controls and treatments for each site.

Sampling. Establish sampling protocols to measure fine sediment in spawning riffles that will allow detection of change during the critical spawning-incubation period. Protocols should be measurable and repeatable. Uncalibrated visual estimates of fine sediment are normally unacceptable for detecting change. Preferred are methods that provide repeatable quantitative estimates of sediment size distribution (such as freeze-core sampling or McNeil samplers).

To understand sample variance at the site, a sample protocol needs to be established that takes multiple samples within the riffle in random locations. A small pilot study should help determine the variation between samples. A power analysis will help determine how many samples to take in the riffle to reasonably estimate the particle-size distribution (Green 1979; MacDonald et al. 1991). A lone sample at each control and treatment normally is insufficient to detect a change between control and treatment.

A second part of this sampling may require emergence traps (traps that capture young fish as they emerge from the gravel) to understand embryo survival in the changing gravel conditions. Typically, these are placed behind known nesting locations to estimate the emergent fry that are surviving from each nest. Similar sample-size guidelines to those described for sampling fine sediments may be used when estimating emergence survival.

The temporal component of this monitoring will vary depending on the size of the treatment and climatic conditions. Sediment-rich drainages that have high sediment loads may take decades to respond to treatment. In these cases, it is important to understand the trajectory of recovery and to note whether fine sediment in spawning gravel is decreasing as a result of treatment. Having control sites is particularly important in this case to compare recovery rates between treated and untreated areas. Changes between treatment and controls may be detectable before long-term, large changes become evident in treated sites alone. Monitoring annually for the first few years may be necessary to understand their rate of recovery. Then monitoring frequency can be decreased to every three years or longer, depending on the rate of change.

3. Riparian Silvicultural Restoration

Assumptions. Silvicultural restoration of riparian forest stands refers to planting trees along streambanks to enhance woody debris production, and it may take decades or centuries to fully realize the benefits. In this example, the measure of effectiveness is the amount of wood produced by the stand that will be available to the stream and riparian zone as coarse woody debris.

Controls. Control areas are larger-scale sections (small watersheds and larger) that have not been treated with tree plantings. These may include old-growth forest or other regenerated stands that have not been subject to timber harvest.

Treatments. Treatments are riparian stands subject to planting.

Sampling. Sampling can be a combination of remote sensing (aerial photography) and field methods. The ultimate measure of effectiveness is the contribution of woody debris to the stream channel. However, since time frames will be long, short-term surrogates for woody debris input may be necessary. These could include stand stocking rates (how many trees), canopy closure (how much shade), and stand growth rates. Aerial photography and videography could be used to establish stand densities and canopy closure over intervals of 10–30 years to document stand conditions in the treatment areas. Field measurements of stand stocking rates and canopy closure can be used to verify and supplement remote sensing methods.

Over the long term, woody debris in streams and along their banks can be resampled using basinwide protocols from inventory methods. If full-basin surveys are used, statistical comparisons are unnecessary because the full population (N) is being sampled. Consequently, the total difference between control and treatment areas will be known. If control and treatment sites are subsampled, then it will be necessary to randomize the sample sites and follow sample design recommendations from Table 8.1. Sampling frequency will be relatively infrequent, given the rate of temporal change in vegetative composition (10–30 years).

Analyses:

Graphical.—Graphical comparisons of variables in control and treatment sections. Graphical comparisons of treatment values and policy standards (if quantitative). Multiyear comparison graphs showing the recovery trajectory of effectiveness-monitoring variables.

PART 2

Building Partnerships

    Most watersheds contain land owned or managed by a broad array of private and public entities. It is not unusual to have more than 15 or 20 state and federal agencies involved in various aspects of a watershed's management. The number of private landowners and interests typically is even more diverse. Further, special interest groups may not own land in a watershed, but nonetheless have valid interests in management of its air, water, fish, wildlife, or other natural resources. Bringing these diverse interests together to define and implement a common vision for watershed restoration can be difficult. The three chapters in this section illustrate the partnership process needed to build an effective coalition.

    In Chapter 9, "Changing Roles and Responsibilities for Federal Land Management Agencies," the present and former chief of the U.S. Forest Service join one of this book's editors to detail the varied roles that public lands can serve in watershed restoration. In addition to the lands themselves, public agencies often are able to bring forth people with expertise in hydrology, geomorphology (landforms), community ecology, and other areas of scarce scientific skill.

    Chapter 10, "Building Public and Private Partnerships," continues with the theme of building partnerships from the perspective of a national foundation. The chapter describes how partnerships can successfully compete for outside sources of funding.

    In Chapter 11, "Achieving Private-Sector Involvement and Its Implications for Resource Professionals," the perspective is that of a state fish and wildlife agency biologist. The author describes the efforts of the Missouri Department of Conservation, a leader in developing successful partnerships with the private sector.

    Fortunately, many natural resource managers and landowners share a surprising number of basic watershed-scale interests. Most people recognize the value of clean water, productive soils, healthy forests, and abundant fish and wildlife populations. This common foundation can be expanded through additional ecological understanding, field trips, and open meetings to achieve an even broader appreciation of healthy watersheds.

Perhaps just as important is understanding the relationship between human activities and the loss of attributes that are provided by healthy watersheds, such as clean water. Seldom, if ever, do the actions on any one parcel of land not inflict some reaction on other lands in the watershed. As John Muir pointed out more than a century ago, every time we attempt to pick something out of the landscape, we find it connected to everything else.

The complex ownership and myriad public and private interests within watersheds dictate that no single party or interest can by itself facilitate restoration at the watershed scale. All of us, or at least a good portion, must work together productively to build the relationships necessary to achieve healthy watersheds.

Every generation receives a natural and cultural trust from its ancestors and holds it in trust for its descendants.
—Edith Brown-Weiss, 1990 (in NRC 1992a)

CHAPTER 9

CHANGING ROLES AND RESPONSIBILITIES FOR FEDERAL LAND MANAGEMENT AGENCIES

Michael P. Dombeck, Jack Ward Thomas, and Christopher A. Wood

Approximately 30% of the land in the United States is collectively owned by its citizens. Four federal agencies are responsible for managing more than 90% of these lands—627 million acres. These agencies are the U.S. Bureau of Land Management (USBLM), U.S. Forest Service (USFS), U.S. National Park Service, and U.S. Fish and Wildlife Service. Together, the USBLM and USFS administer 461 million acres. Nearly all of the land managed by USBLM and much of the land managed by the USFS is in the western United States, a result of east-to-west settlement patterns and statehood negotiations.

The economic importance of these federally administered lands is significant. Each year, tens of millions of Americans use public lands for recreation—hunting, fishing, hiking, backpacking, camping, and cycling. Expenditures related to hunting, fishing, and wildlife recreation on USFS lands alone total US$12 billion annually (Fedler 1996). Extractive industries such as mining, ranching, timber production, and oil and gas development provide employment on these lands, supporting the economies of rural communities.

Public lands contain some of the most remarkable cultural, archaeological, and historic sites in the world. They often provide the best—and sometimes last—habitats for rare plant and animal species. For example, Nehlsen et al. (1991) identified 214 stocks (local populations) of Pacific salmon as being at some risk of extinction. The USFS manages land containing habitat for 134 of these stocks, and the USBLM manages land containing habitat for 109 stocks (Williams and Williams 1997). In the eastern United States, lands managed by the USFS provide habitat for 17 threatened and endangered fish species and 41 threatened and endangered aquatic invertebrates.

Multiple-use federal lands managed by the USBLM and the USFS must play a critical role in the restoration of biological integrity and watershed health (J. W. Thomas and H. Salwasser, presented at the annual meeting of the Society for Conservation Biology, 1988). Nearly all USBLM-managed lands are in the western United States where the USBLM and the USFS administer large blocks of land, often contiguous. The USFS manages much of the forested lands in the headwaters, whereas USBLM manages significant acres of land downstream.

Herein, we trace the nation's historic use and policy toward publicly owned natural resources and discuss how multiple-use agencies such as the USFS and USBLM can most effectively support and promote watershed-scale restorations.

HISTORICAL CONTEXT

The United States has passed through four distinct eras of public land and resource management. Each era contained its own legislation, policies, and environmental ethic. The country is now entering a fifth phase, a natural progression from and outgrowth of the previous four.

Briefly overviewing these eras, they are (1) settlement and development of the original public domain (1789–1834); (2) the nation's first venture into public land stewardship with the resource conservation ethic, beginning during Theodore Roosevelt's presidency in the early 1900s; (3) World War II and the ensuing national growth (1941–1962); (4) the era of environmental laws (1962–1990); and (5) the present new era of watershed restoration and collaborative stewardship, signaling emergence of the Aldo Leopold land ethic.

Era 1: Settlement and Development

The first stage was characterized by an all-out effort to settle and "reclaim" the nearly 1.8 billion acres of original public domain. Laws such as the Land Ordinance of 1785 and the Northwest Ordinance of 1787 directed exploration, survey, and disposition of the public domain. To accommodate the needs and demands of the growing nation, the U.S. Treasury "disposed of" immense tracts of land to settlers, corporations, and states.

Between 1789 and 1834, Congress practiced its public land responsibilities with great zeal, passing approximately 375 laws that adjusted the size of public land lots for sale and their payment rates and schedules (Limerick 1987). In 1812, the General Land Office was formed to process land patents and to expedite settlement of the western United States. By 1840, lands from the Atlantic to the

Pacific coasts were settled, culminating with the passage of the Homestead Act of 1862. To applicants over 21 years of age, it granted 160 acres of land, contingent upon 5 years of consecutive residence and land cultivation.

Congress gave large tracts of land to the western states to support schools and school construction, wagon roads, canals, and other public services and facilities. Of the 1.8 billion acres of public domain, about two-thirds was acquired by private citizens, corporations, or states. More than 91 million acres also were granted to railroad companies by Congress. These cessions, sales, and land disposals contribute to the scattered and checkerboard ownership pattern of much of the public domain today.

As the nation expanded westward, Congress passed laws designed to encourage settlement and development of the public domain's vast timber, mineral, and forage supplies. For example, livestock grazing on the public domain was free and unrestricted until passage of the Taylor Grazing Act in 1934. The General Mining Law of 1872 and the Timber and Stone Law of 1878 encouraged extensive development of mineral and timber resources.

Settlers continued to move westward and use the land as they wished, generally with little knowledge of or concern for the health of natural resources. Entire forests were commercially harvested with no effort to reforest cutover lands. Forests also were cleared by settlers for fuel, railroad ties, and agriculture. Rivers and streams were dredged for gold and other precious metals. The waters of many western rivers were diverted from their beds and put to "beneficial use" (irrigation) on the land, leaving previously perennial streams dry during critical times of the year. Dams and other impoundments were built for various purposes. These diversions dramatically altered riverine function, provided significant barriers to fish passage, and often changed natural river dynamics.

Barbed wire was strung across public domain and livestock trespass was common. By the 1870s, federal rangelands were greatly overstocked. In 1887 a severe winter, coupled with malnutrition, killed millions of stressed livestock, bankrupting many cattle companies that were involved in land-damaging and speculative grazing practices (Limerick 1987). Vicious grazing wars among cattlemen and sheep herders broke out over scarce water supplies. As the turn of the century neared, the western public rangelands were broadly degraded.

Era 2: The Resource Conservation Ethic

The presidency of Theodore Roosevelt from 1901 to 1909 signaled the beginning of the second phase in the country's public land policy and the nation's first period of public land stewardship. Four trends were emerging at this time: growing western communities increasingly demanded clean water supplies; public distrust was growing for many of the large timber companies; sustainable timber supplies were needed to replace cutover forests of the Great Lakes region; and the nation generally needed timber products. In response, Roosevelt expanded the forest reserve system initiated by President Benjamin Harrison in 1891 and created the Forest Service (Wilkinson 1992).

In the same breath, Congress both limited and legitimized the creation of federally managed forest reserves through the Organic Act of 1897. The Act stated that

> No national forest shall be established, except to improve and protect the forest within the boundaries, or for the purposes of securing favorable conditions of water flows, and to furnish a continuous supply of timber for the use and necessities of the citizens of the United States.

President Roosevelt and the first chief of the U.S. Forest Service, Gifford Pinchot, used the Organic Act to dramatically expand federal forest reserves. Other public lands were withdrawn from settlement and established as national parks, wildlife refuges, and military bases.

Early in the twentieth century, Congress enacted a number of laws to expand federal control over the use of public lands and resources. Through the Mineral Leasing Act of 1920, Congress directed the executive branch to administer leasing, exploration, and production of minerals resources such as coal, oil, gas, and sodium. In response to the poor condition of the public rangelands and the extended drought that prompted the Dust Bowl during the 1930s, Congress passed the Taylor Grazing Act in 1934. The Act established the Grazing Service, which later merged with the General Land Office to become the U.S. Bureau of Land Management to administer public rangelands.

These laws promoted sustainable use of natural resources, reflecting Gifford Pinchot's resource conservation ethic: national forests should be managed to serve the "greatest good, for the greatest number, for the longest time." Pinchot imbued the U.S. Forest Service with his belief that

> there are just two things on this material earth—people and natural resources . . . Natural resources must be developed and preserved for the benefit of the many and not merely the few. (Pinchot 1947)

Pinchot also brought from Europe the concept of professional forestry in an effort to bring order out of the chaos commonly found on forested lands. During the 1930s, the Civilian Conservation Corps helped to restore degraded public lands, expand recreational opportunities, and construct much of the infrastructure still present on public lands.

Era 3: The Post-War Years

World War II and the ensuing national growth (1941–1962) introduced the third phase of public land management. With the need to provide materials for war, followed by the desire of returning military personnel to have homes and raise families, great energy was focused on production of minerals, livestock, and most importantly, timber. Timber production on national forest land expanded, as did the road network needed to move the trees to the mills. Silvicultural practices (tree management) favored large harvest blocks (clear-cuts). Noncommodity amenities and values such as wilderness, aesthetics, and recreation were perceived as little more than constraints on the production of timber, forage, and minerals.

Era 4: Environmental Protection

The fourth phase of public land management was popularized by Rachel Carson's publication of *Silent Spring* in 1962. This book came to symbolize the country's growing concern over environmental protection. A host of important environmental laws were passed during this era. The 1970 National Environmental Policy Act required public disclosure and citizen involvement in federal land management actions. The U.S. Environmental Protection Agency was formed by President Richard Nixon in 1975. The Clean Water Act was passed in 1972; the Safe Drinking Water Act followed in 1974; and the Clean Air Act was amended in 1977 (originally passed in 1955).

Multiple-use management of federal lands was codified for the USFS and USBLM, respectively, by the Multiple Use Sustained Yield Act of 1960 and the Federal Land Policy Management Act of 1976. The latter recognized the value of public lands to the American people and declared that they would remain in public ownership.

The National Environmental Policy Act, Federal Land Policy Management Act, Endangered Species Act, Wilderness Act, Wild and Scenic Rivers Act, Clean Water Act, and other environmental legislation were clear expressions of the resolve of Congress and the public to protect noncommodity resources such as wildlife and fisheries, wilderness, clean water, recreation, and aesthetic values. Over the past 25 years, these agencies have focused on balancing federal laws that protect or preserve specific environmental benefits or values against laws and congressional directives that emphasize commodity development and production.

No longer can federal agencies manage public lands exclusively for livestock forage consumption, timber, and minerals production. Abiding by Pinchot's resource conservation ethic while implementing environmental laws and regulations has proven a formidable task, and an often controversial one, for multiple-use land management agencies. Thus, we are entering a new era.

Era 5: Watershed Restoration and Collaborative Stewardship: Reemergence of the Land Ethic

It is increasingly clear that federal agencies cannot protect individual resources, be they endangered species or timber and forage, without managing them in the context of larger ecosystems. In the words of Aldo Leopold (1947)

> The practice of conservation must spring from a conviction of what is ethically and aesthetically right, as well as what is economically expedient. A thing is right only when it tends to preserve the integrity, stability and beauty of the community, and the community includes the soil, waters, fauna, and flora, as well as people.

Leopold recognized that humans shape, and in turn are shaped by, the land and its resources. Humans are "members of a biotic team . . . plain members and citizens of one humming biotic community" (Leopold 1949). Experience has proven Leopold correct in that multiple-use agencies, such as the USFS and USBLM, cannot meet the needs of people if they do not first secure the health, diversity, and productivity of the land. As Callicott (1991) noted:

> Human beings are not specifically created and uniquely valuable demigods any more than nature itself is a vast emporium of goods and services. We are, rather, very much a part of nature.

As watershed health declines, lands become less productive. They become less resilient to natural events such as floods, drought, and fire. They become prone to invasions of exotic species (Williams et al. 1997a, this volume). For example, about 5% of all USBLM-managed lands have "serious" weed problems; noxious weeds are rapidly infesting additional acreage, at the incredible rate of 2,300 acres per day (USBLM 1996). Spotted knapweed was first discovered in Montana in 1920; by 1996 it had infested more than four million acres in the state.

Exotic plant species often out-compete native flora and can make the land less productive and more susceptible to devastating fires and soil loss. Countless similar invasions of exotic species in aquatic ecosystems occur annually, although they are hidden from our view and therefore less obvious (Courtenay and Moyle 1992). As a result of these invasions, human uses of degraded watersheds become progressively restricted and management options more limited.

Healthy watersheds provide habitat complexity and diversity, which helps to maintain species diversity (Angermeier 1997, this volume). The relationship among healthy watersheds, biological diversity, and productivity may not always be apparent but can be of immense importance. Tilman and Downing (1994) measured drought resistance of grasslands containing different levels of plant species richness. Those grasslands having the highest levels of plant diversity were found to be more productive during droughts than the less-diverse plots.

The lessons are deceptively simple. Higher diversity results in greater stability, resiliency, and higher productivity. In other words, biological diversity among communities of species can beget stability and productivity, which in turn provides society with potentially higher levels of goods and services from the land.

The growing recognition that sustainable management of natural resources depends on maintaining and restoring ecological processes within and among natural systems is characteristic of the fifth phase of public land management. Ecosystem management is the basis for contemporary USFS and USBLM programs, including watershed restoration (Overbay 1992; USBLM 1994a). Emerging strategies for effective ecosystem management include integration of state and federal technical skills and involvement of communities to define and implement common approaches to management of watersheds.

The history of public land management, whether aimed at settlement of the frontier and commodity development or correcting past resource degradation, was generally spurred by federal legislation. Today, however, local communities and field offices of federal agencies should not look to a future "Watershed Restoration Act" to guide their efforts. No such remedies are likely to restore America's watersheds. In an era of fiscal constraint and given the time frame and geographic scope necessary for restoration (discussed by Ziemer 1997, this volume), legislation alone will not restore the nation's watersheds.

The USFS and USBLM's multiple-use mandates require them to manage public lands in a "harmonious and coordinated" manner "without permanent impairment of the productivity of the land and the quality of the environment." Long-term solutions to watershed degradation, however, cannot overlook the relationship between the health of federal lands and the condition of adjoining state and private lands. A truly comprehensive restoration strategy requires the active participation of all who use, value, and influence how watersheds work and function.

COLLABORATIVE STEWARDSHIP

The USBLM has found the use of balanced and diverse citizen Resource Advisory Councils to be an effective way of involving more people in management decisions. These councils provide a framework within which all who use and value public lands can work together to ensure the land's health and productivity, particularly where publicly owned lands are abundant.

The U.S. Bureau of Land Management created 24 of these councils, each with 10 to 15 members, to represent the interests of three broad constituencies:

1. holders of federal grazing permits, representatives of energy and mining development, right-of-way interests, the timber industry, off-road vehicle users, and developed recreation;
2. representatives of environmental and resource conservation organizations, archeological and historic interests, and wild horse and burro groups;
3. representatives of state and local government, Native American tribes, academicians involved in natural sciences, and the public at large.

Other noteworthy features of Resource Advisory Councils include: (1) all members must be residents of the state or states in which the council has jurisdiction; (2) council members are selected on the basis of their education, training, experience with the issues, knowledge of the geographic area of the council, and their willingness to work together; and (3) at least one member of each council must be an elected official.

As the federal government reduces spending and cuts costs, federal land management agencies such as the USBLM and USFS must harness the many skills possessed by all of these disparate public land interests. The western United States has the nation's fastest-growing population, with the largest proportion living in urban areas. Although Resource Advisory Councils were created by the USBLM, in some states and areas they also provide resource advice to the USFS and the states.

TABLE 9.1.—Examples of resources that the U.S. Forest Service and the U.S. Bureau of Land Management can provide to benefit watershed restoration.

Resource	Example
Technical expertise	Hydrology, plant ecology, soil science, geomorphology (landforms)
Analysis procedure	Cumulative-effect analysis, watershed analysis, monitoring protocols
Collaborative stewardship opportunities	Resource Advisory Councils; public involvement under National Environmental Policy Act
Reference and benchmark habitats	Key watersheds, areas of critical environmental concern, research natural areas, wilderness areas
Resource information	Monitoring data, habitat inventories, riparian condition information, historical data
Experience with broad-scale restoration	FEMAT (Forest Ecosystem Management Analysis Team) Aquatic Conservation Strategy; Interior Columbia River Basin Ecosystem Management Project
Staff and logistical support	Meeting places, clerical staff, facilitators, etc.

As the country continues to change and grow, more demands are placed on public lands. Collaborative stewardship—whether from the watershed associations and coalitions described in this book's case studies or from Resource Advisory Councils—will help to focus on those things that draw us together as a nation of communities.

Restoration and all of the natural and societal benefits that accrue from healthy watersheds must be community-based and community-driven. Because public lands are collectively owned by all Americans, and because of the temporal and geographic scales involved in restoration, federal agencies can and should take a leadership role in coordinating and guiding restoration. Federal land management agencies can support watershed restoration by providing technical expertise, data and analysis procedures, reference habitat information, and other relevant experience (Table 9.1).

Partnerships among state and federal land and resource management agencies, user groups, environmental coalitions, and communities are essential to restoration. Such partnerships are the only way that progress can be made in areas outside the West where multiple landowners and agencies manage intermingled lands and resources.

INFORMATION FRAMEWORK AND KEY WATERSHEDS

Watershed restoration should not proceed without a comprehensive understanding of watershed structure and function. Field offices of the natural resource management agencies are clearinghouses of information and technical expertise in scarce skills such as hydrology, geomorphology, aquatic ecology, range science, and forestry. Through watershed analysis (Ziemer 1997), these technical experts can help watershed organizations and coalitions gain a better understanding of the natural events and human activities that affect and shape watershed health.

Presently, federal agencies are developing watershed analysis procedures for public lands (Regional Interagency Executive Committee 1995a). Specifically, these procedures are for use on public lands west of the Cascade Mountains, within the range of the northern spotted owl, and for other public lands in Idaho, Oregon, California, and Washington that provide habitat for anadromous salmonids (salmon, trout, and chars that migrate to spawn in freshwater, but normally live in the saline ocean or estuaries; FEMAT 1993; USFS and USBLM 1995).

These watershed analyses can provide the technical and scientific underpinnings necessary for recovery of listed species or to offset the need to protect species under the Endangered Species Act. Such actions, however, are only part of the solution. If society is to reap the full social, economic, and aesthetic benefits of healthy watersheds, then the states, local communities, and private landowners should be provided with the incentive and impetus to participate in federal efforts. The USBLM and the USFS cannot dictate management of state lands, nor can they coerce private landowners to participate in restoration. Yet, watershed analysis offers federal and state agencies and private landowners unique opportunities to cooperatively identify and correct sources of degradation.

The USFS and the USBLM have identified key watersheds that presently or potentially provide healthy habitat for anadromous salmonids. This system of key watersheds will be managed to maintain and recover habitat for anadromous salmonids that are at risk of extinction. Because of their ecological importance—and ultimately, their social importance—watershed analyses will be conducted on these public watersheds before further resource development is permitted.

Federal lands that contain key watersheds in good condition can serve as "anchors" for recovery of salmon stocks and maintenance of high water quality. Even those key watersheds that have poorer quality but that have high potential for recovery will be the focus of future restoration. These efforts alone cannot restore anadromous fish stocks, but federal lands can be used to buy time to address other issues that affect salmon abundance, such as hatchery production, passage mortality, and overharvest.

Identifying watershed areas that have the best chance for recovery is an essential first step. The key watershed system was developed in response to the decline of anadromous salmonids in the Columbia River. However, federal agencies across the country should take the initiative to work with local user and conservation groups, Indian tribes, state agencies, and communities. Together, they can prioritize watersheds that merit special protection due to their ecological and social importance or that offer a good potential for recovery.

EDUCATION AND COMMUNICATION

Too often, agency processes and operating procedures belie the relatively straightforward mission of natural resource professionals. Consequently, the effectiveness of these professionals is undermined. For example, the intent of the Endangered Species Act has less to do with long and drawn-out negotiations between federal agencies to determine whether federal action will imperil rare species than it does working with states and communities to conserve the nation's rich biological heritage. Similarly, the intent of the National Environmental Policy Act was not to justify timber sales or produce jargon-filled technical tomes so much as to engage in a frank and open public discussion about the benefits and environmental effects of agency actions.

Translating the benefits of healthy watersheds—and the consequences of watershed degradation—in a manner that nonprofessionals can understand is critical. Professional resource managers cannot be truly effective without good communication skills.

Perhaps one of the greatest challenges facing federal land and resource management agencies is the organization of knowledge and insight into forms (media) that can be readily applied and understood by the general public (Thomas 1979). As Cairns notes (1997, this volume), watershed restoration is unlikely to be initiated, and is unlikely to endure once accomplished, if those who influence watersheds do not support both the restoration and its continuing maintenance. All the technical expertise in the world cannot overcome public disinterest—or worse, distrust—in agency actions. Thus, federal land managers should spend more time with local community leaders, user and conservation groups, state officials, and schoolchildren.

It is insufficient to expend time and money fixing the effects of watershed degradation without addressing their root causes. Addressing root causes is the promise of watershed restoration. For example, it is far more productive to team together to restore a watershed than it is to enhance species by introducing exotics or by using gabions (stone-filled wire baskets) or riprap (a collection of large rocks) to artificially support degraded stream habitats (Dombeck and Williams 1995).

If society is to continue enjoyment of productive watersheds, federal and state land management agencies must redefine how they approach restoration. They must rethink how they communicate the many benefits of restoration to taxpayers, and how they educate land users, interest groups, and local communities. An old Kashmiri proverb reads, "we have not inherited the land from our forefathers; we have borrowed it from our children." No less than their future is at stake.

ACKNOWLEDGMENTS

The authors thank James Clayton, John D. McIntyre, and Russ Lafayette for their reviews and assistance with this chapter.

Problems can become opportunities when the right people come together.
—*Robert Redford, 1987*

CHAPTER 10

BUILDING PUBLIC AND PRIVATE PARTNERSHIPS

Whitney Tilt and Cindy A. Williams

The Beaverkill River, and its sister tributary, the Willowemoc, arise in the Catskill region of New York, a 2-hour drive north of New York City. The Catskills have long been a protected park, due in large part to their role as a municipal water source for New York City. The Beaverkill–Willowemoc is a legendary river system considered by trout aficionados as the "cradle of American fly-fishing." For more than 150 years, fly-fishing legends from Theodore Gordon to Lee and Joan Wulff have called the Beaverkill–Willowemoc their home or have made annual pilgrimages to these waters.

Five hundred miles south, along the spine of the Appalachian Mountains in Tennessee, arise waters with names like North Chickamauga Creek and Horselick Creek. Although not hallowed among American trout anglers, these watersheds are treasured by today's conservationists for their aquatic diversity of darters, madtoms, freshwater mussels, and other species.

In 1991 and again in 1993, drought and habitat degradation within the 260-square-mile Beaverkill–Willowemoc watershed led to fish kills, stressed trout populations, and rising concern about the overall health of the rivers (Rafle 1994). To the south, creeks like North Chickamauga and Horselick fared no better, as decades of acid mine drainage threatened their aquatic heritage.

145

Although these northern and southern watersheds are different in many aspects, their emerging problems have shared a common treatment, and this treatment has not fit the historical pattern of environmental protection. Rather than a top-down, command-and-control response funded by state and federal governments, activities have stayed largely within sight of the watershed. On the Beaverkill–Willowemoc, local activists and Trout Unlimited stepped forward to work cooperatively with state and local governments, highway departments, local zoning boards, and most importantly with the local landowners, shop owners, and citizens (Conyngham and McGurrin 1997, this volume).

For the Horselick and North Chickamauga creeks and other southern Appalachian watersheds, a partnership called the Southern Rivers Council was formed to address immediate conservation needs and to provide local officials and citizens with the information and tools necessary to protect, restore, and manage the watersheds. The partnership is funded with assistance from the U.S. Forest Service, National Fish and Wildlife Foundation, and a variety of other federal, state, and local organizations. The Council's initiative, called Restore Our Southern Rivers, is in initial implementation. Although it is too soon to evaluate the overall benefit to the aquatic resources of these rivers, the initiative remains a good example of conservation partnership.

FROM LARGESSE TO DOWNSIZING

Several chroniclers of the history of the U.S. conservation–environmental movement have described it as occurring in distinct eras or waves (Fox 1981; Wilkinson 1992; Shabecoff 1993). The first wave began at the close of the 1800s as the era of land and wildlife conservation and preservation, characterized by the thoughts and actions of Theodore Roosevelt and Gifford Pinchot. Beginning in the 1960s, Rachel Carson's *Silent Spring* inaugurated the second wave, commencing some 25 years of what arguably could be called the "environmental movement." This second wave spawned an impressive mass of policy and legislation, accompanied by constituent advocacy, congressional lobbying, and frequent litigation. The third wave arose in the mid-1980s during the Reagan Administration, accompanied by such buzz words as "market-based incentives," "regulatory flexibility," and "win–win" (Dowie 1995). Although conservation partnerships are as old as the conservation–environmental movement itself, partnerships are clearly one of the third wave's buzzwords.

The first three waves have been effective in achieving broad changes in public policy, behavior, and attitude. However, their achievements ultimately will fall short of long-term conservation goals unless the efforts are sustained, because there are untold Horselick Creeks in need of immediate and continuing attention across the country. Unfortunately, after years of increasing budgets during the 1970s and 1980s, federal and state agencies in the 1990s are increasingly hard-pressed to fund existing staff, conservation lands, and equipment, let alone new initiatives.

For the foreseeable future, the federal government will be forced to reconcile its budget. Although budgets for natural resource management are a relatively insignificant portion of the federal budget—less than 1%—they are likely to bear a disproportionate burden in paying for a balanced budget. In fiscal year (FY) 1996, for example, the federal budget paid out less than US$.01 for natural resource budgets, while spending $.48 on Social Security, Medicare, and other benefits to individuals, $.16 on defense, $.15 to states and localities, and $.16 on interest payments for the national debt (Office of the President 1995).

Entitlement programs like Social Security and medical payments to individuals collectively compose the largest portion of the federal budget. So, it is not surprising that they also have the largest number of proponents and protectors. These numbers just cited should be viewed as an opinion poll indicating the relative value that society places on natural resource management. They also provide a poignant reminder of where conservation interests sit in the real world—not surprising, considering the size of the watershed restoration constituency relative to the constituency for competing interests (Tilt 1993).

So how will watershed and "ecosystem" interests obtain the necessary resources for river systems and creeks across the land? Three components are essential for effective conservation—education, investment, and partnerships.

EDUCATE AND INVEST

Survey after survey demonstrates that people who are informed and educated about the environment provide greater support for programs and actions that improve environmental quality (Gigliotti 1992; Hausbeck et al. 1992). Yet such support is difficult to tap, because the nation's environmental IQ still appears too low to achieve an environmentally responsible citizenry (Orr 1992). All too often, young and old alike lack the basic factual knowledge required to make environmentally sound decisions at home or work. Environmental literacy is not evidenced by parroting the refrains of *recycle*, *acid rain*, or *global warming*, but by demonstrating an understanding of how the environment functions, how humans fit into the environment, and how their actions affect it.

Effective and proven programs to increase the environmental knowledge of children and adults already exist throughout the country. However, a lack of coordination, sustained investment, and commitment has prevented these programs from achieving their full potential (Tilt 1996). Education is widely recognized as the vital first step that leads to conservation action. Yet, conservation interests that pay lip service to education have repeatedly failed to direct sufficient energy and funding to education. To use a watershed analogy, conservation education has been like a seasonal water course—small, vulnerable, and too often ignored.

Generous government spending is in apparent eclipse, and Congress is preoccupied with balancing the budget. In this difficult fiscal environment, conservationists have come to realize painfully that the federal government lacks the money to buy all the remaining lands necessary to conserve our fish and wildlife.

The Land and Water Conservation Fund (LWCF) is the nation's primary funding source for public land acquisition. It is funded largely by offshore oil-leasing receipts. A look at recent expenditure trends under LWCF is illustrative.

- In 1978, $805 million was appropriated to the U.S. Bureau of Land Management, U.S. Forest Service, U.S. Fish and Wildlife Service, and to the states for land acquisition (U.S. National Park Service, unpublished data).
- For the 5-year period FY1989 to 1993, an average of $276 million per year was provided from LWCF (U.S. National Park Service, unpublished data).
- In FY1996, although the LWCF balance began the year at $10.3 billion, only $138 million—a scant 1.3%—was appropriated (Office of the President 1995). The reason is that LWCF funds are being commandeered for deficit reduction.

Future appropriations remain uncertain as budget concerns lead Congress to divert LWCF receipts to reduce the deficit.

Yet, even with full expenditure of LWCF and other dedicated accounts (such as the Federal Duck Stamp, used to fund acquisitions for the National Wildlife Refuge System), federal funds are insufficient for the conservation task. For example, an estimated 75% of the nation's wetlands are in private ownership. The federal government cannot be expected to buy them all and turn them into national wildlife refuges. Nor, as we are increasingly coming to realize, is such federal largesse a prescription for success. We have learned that lasting conservation achievement is like Tip O'Neil's politics: it must be local and have a strong foundation of community support. The challenge facing resource managers is to ensure that less government spending does not translate into less conservation. What is needed is a movement away from total dependence on federal funds to fix conservation problems and a movement toward shared responsibility and investment—hence the need for *public–private partnerships*.

Identifying funding sources and determining how to glean the money to support a restoration project can be difficult, especially for professionals trained in natural resource management rather than in fund-raising or partnership-building. Although numerous private organizations fund environmental projects, willing donors can be hard to find and worthy projects often exceed available funds. Increasingly, agencies and conservation organizations are recognizing the economic and political benefits of partnership agreements, as are states and local governments. *Partners become shareholders in restoration success* and share the economic cost of doing business by contributing money or in-kind contributions. The result is increased understanding and greater accomplishment than if the project were completed by a single party. Partners also learn to appreciate each other's perspectives, concerns, and limitations, and each party gains ownership in the project while becoming better neighbors.

The National Fish and Wildlife Foundation, a private, nonprofit conservation organization established by Congress in 1984, is proof that shared investment pays dividends. The Foundation works to conserve fish and wildlife resources by providing federal challenge grants, which in turn must be matched by nonfederal funds. The operating premise is simple: the federal government gets leverage for

TABLE 10.1.—National Fish and Wildlife Foundation (NFWF) projects and funding history, 1986 to 1996. Funds are provided by a variety of federal agencies in the U.S. Departments of Agriculture, Interior, Defense, State, and Commerce. Matching funds consist of federal contributions from the National Fish and Wildlife Foundation and nonfederal challenge funds from more than 600 partners.

Year	Number of projects	NFWF federal matching funds ($)	Challenge funds ($)
1986	15	97,164	1,556,859
1987	20	187,160	799,673
1988	57	2,846,734	5,678,138
1989	130	5,142,147	19,242,433
1990	73	2,160,510	6,667,084
1991	143	6,186,834	17,034,768
1992	177	5,748,496	18,486,381
1993	209	5,357,984	17,567,689
1994	298	9,027,380	34,069,240
1995	319	9,912,603	47,283,087
1996	343	13,953,660	45,007,199
Total	1,784	60,620,672	213,392,551

its increasingly tight dollar, grantees get much-needed seed money, and the project gets done faster and more economically.

In awarding challenge grants, the Foundation employs the buy-in strategy. If a project is viable, then other parties should be willing to invest in it, especially local partners who will help the project endure. The Foundation and its conservation partners have consistently proven this axiom. For the period 1986 to 1996, 1,784 grants totaling $60 million in federal matching funds have been awarded to over 600 organizations. For each federal dollar committed, two additional dollars have been raised by the Foundation and its conservation partners. The total exceeds $213 million to conservation practitioners at local, state, and regional levels (Table 10.1). When we consider contributed services and other funds that are attracted to projects as a direct or indirect result of Foundation challenge grants, the return for federal dollars exceeds $400 million.

PARTNERSHIP BASICS

A constant theme in the history of the National Fish and Wildlife Foundation has been partnerships. The Foundation has spent a decade experimenting with them. In part, this experience is reflected in two documents: "Partnerships: Innovative Strategies for Wildlife Conservation" (Trauger et al. 1995) and *Conservation Partnerships: A Field Guide to Public–Private Partnering for Natural Resource Conservation* (Management Institute for Environment and Business 1993).

Broadly defined, a partnership is a collection of entities (often individuals, not necessarily institutions) where each brings to the table some enlightened self-interest mixed with a combination of *time, talent*, and *treasury*. Partners need not agree on politics or the weather, but they are bound by a common interest in the project at hand (Trauger et al. 1995). As for time, talent, and treasury, time should never be underestimated as a valuable asset, for a partnership is hard work and time-consuming. Partnerships are low-technology by nature and there are few, if any, shortcuts. Talent can range from watershed restoration expertise to accounting acumen. Treasury can come from a partner's own pocket or reside in a partner's ability to ask others.

Partnerships must be homegrown. Too often, conservation efforts are developed and implemented by outside parties, without the benefit of local input. Even today, conservation projects are often conducted by well-meaning government agencies and conservation interests without the involvement and ownership of local stakeholders.

There is no single recipe for a successful partnership. Like a favorite recipe, partnerships built with the same ingredients do not necessarily produce the same outcome each time. However, partnerships do share some constants. The following discussion of these constants is adapted from Management Institute for Environment and Business (1993), Tilt (1996), and Trauger et al. (1995).

Warning Label

Like many buzzwords in conservation—from ecosystem management to sustainable development—it is important to first read the warning labels. Partnerships are a tool, and like all tools there are times when they are effective and times when other tools may be more productive. Partnerships are like a living organism—they need constant nourishment and hard work to grow. Egos and turf must be checked at the door. For government agencies, participants need to determine early on that a partnership is the proper path to take or whether a more formal arrangement, such as a contractual agreement, is appropriate.

Worthy Project

A good partnership is founded on a solid conservation need. Some projects lend themselves more readily to partnerships than others. For example, habitat acquisition and restoration are often more attractive to a prospective donor than a research project to collect data for 10 years.

Equity and Participation

The fastest way for a partnership to falter is for the sponsor to treat funding partners differently, based on the perceived value of their contribution, or to withhold project information from them. Before launching a project, the sponsor must know the interest of each partner and why the partner is involved. Partnerships depend on mutual respect and an evenly distributed workload. All partners must be willing to listen openly to ideas advanced by other partners.

Embrace Nontraditional Partners

Given the need for new sources of funding and broader constituencies, conservationists need to embrace new, nontraditional players. The resources that these nontraditional partners can contribute to watershed restoration are enormous. As conservationists, we spend too much time speaking to each other and not enough time reaching out to forge alliances with new partners. A great source of new partners is the rank and file of those perceived to be against natural resource conservation. Once you are successful in converting a former opponent, you now have a powerful ally and persuasive force for converting others to your cause.

Leverage

Leverage is one of the most appealing aspects of a partnership. Examine its appeal from the funder's viewpoint. Funders, be they Congress or local family trusts, are continually bombarded by requests. Each applicant claims that their project is vital and dependent on the funder's attention. Multiply this claim severalfold and even the most conscientious funder grows numb. Alternatively, partnerships provide a perfect platform for joint funding and cost-sharing. Such projects come with an endorsement that other interests view the project to be of sufficient value that they are also willing to invest. Cost-sharing tests the hypothesis that if a project is viable, more than a single donor should be willing to fund it. And cost-sharing has one other important element: constituency. The building of funding partnerships is akin to building constituencies.

Flexibility

Partnerships must be flexible. If the participants bind a partnership too tightly with regulations, it will fail as one or more partners get fed up with red tape and delays. If an agency finds itself overly constrained by regulations, a different relationship may be needed (e.g., contract or grant). For state and federal agencies, the ability to be flexible is one of their greatest challenges.

The paradox facing government agencies is that they are being requested by governors, department secretaries, and the President to be creative and to increase efficiency and effectiveness through partnerships—but meanwhile, rules and policies that govern the conduct of government personnel and financial transactions remain rigid, restrictive, and seemingly unassailable. Clearly, the rules that govern government have not kept pace with this call to partner. A look at some existing legislation, policy, and fiscal procedures is illustrative.

Legislation

Many state and federal agencies lack the direct authority to enter into cooperative arrangements with nongovernmental entities. Others find they can enter into cooperative funding arrangements only on their own lands or within their legislative boundaries. This can be a major impediment for watersheds that encompass a checkerboard of federal, state, and private lands. Such restrictions may prevent partnering outright, or they may allow cooperative programs to be pursued only within an agency's respective arena. As such, legislative restraints may be a major obstacle for watershed restoration projects.

Federal legislation can inadvertently stifle productive partnerships. For example, the Federal Advisory Committee Act was designed to foster openness in government actions. But it can impede the ability of federal agencies to use nonfederal expertise in a timely manner. Two examples where appellants used the Act to thwart conservation measures are attempts by the U.S. Fish and Wildlife Service to develop a status report on the Alabama sturgeon, and the U.S. National Park Service's attempts to remove mountain goats from Olympic National Park. The Federal Advisory Committee Act has been noted by at least one court to be "an uncomfortably broad statute . . . that would, if literally applied, stifle virtually all non-public consultative communication between policy making federal officials and a group of two or more other people, any one of whom is not in government service" (Lein 1994).

Policy

Numerous agreements exist between federal agencies that address how one agency will work with another. It seems baffling that agencies within the federal government require a written document to allow them to "partner" with a sister agency. However, such agreements are necessary because each agency has different mandates and procedural guidelines specific to its mission. Not surprisingly, there are occasions when federal policy endorses a program or strategy that outlines specific activities which run contrary to other agency mandates.

For example, Executive Order 12962 of 7 June 1995 endorsed a "Recreational Fisheries Stewardship Initiative." Unfortunately, the initiative implied the need to favor recreational fishing opportunities over the Endangered Species Act, with the end result of further muddying agency directives. Such a policy also sustained the perceived conflict between recreational fisheries and the conservation of native fishes under the Endangered Species Act. Obviously, achieving balance will be difficult in some cases where endangered species are in direct conflict with sport fish species, many of which are the basis for million-dollar industries.

Fund-Raising

Federal employees and most state employees are officially prohibited from fund-raising in general, and specifically where such efforts will augment their budgets. The U.S. Fish and Wildlife Service and other federal land management agencies have no general authority to "supplement" their appropriations from Congress by actively soliciting gifts of money and materials. On the other hand, these same agencies have a growing number of initiatives that encourage federal agencies to enter into cooperative partnerships with states and the private sector. These partnerships range from "Bring Back the Natives" (native fish restoration) to "Partners in Flight" (neotropical migratory bird conservation).

Such efforts often include cost-sharing and other activities that could be defined as "supplementing" agency budgets. The potential conflict is clear, and the proper avenue for federal employees engaged in such partnerships is poorly defined. A clear policy for such activities should be articulated by the secretaries of the U.S. Department of the Interior, U.S. Department of Agriculture, U.S. Department of Commerce, and U.S. Department of Defense.

Fiscal Controls

Anyone who has managed a federally funded project knows first-hand the procedures and regulations that accompany it. Although many of the controls and provisions can be reduced to Accounting 101, still others threaten to overwhelm partners in a sea of red tape and confusing regulations. When confronted with this maze of financial requirements, ranging from the Davis Bacon Act (establishing wage rate for federally funded construction projects) to Office of Management and Budget circulars, partners who are not accustomed to such procedures may wish they had chosen another profession. On the state level, agencies are typically under a completely different set of restrictions that they must impose in turn.

If cooperating partners are not guided skillfully through this regulatory maze, the partnership's ability to succeed is questionable. For the fisheries biologist, range conservationist, and geographic information system mapper to succeed, there often must be a contracting officer and lawyer willing to help them make it happen.

Case Studies of Public–Private Partnerships

Many of the factors just described found fertile soil in the southern Appalachian Mountains, where restoration of neglected and degraded streams awaited strong community partnerships of the Southern Rivers Council.

Southern Rivers Council.—In the southeastern United States, nine federal and numerous state and local agencies share management of aquatic ecosystems. These aquatic ecosystems encompass the richest and most diverse aquatic faunas and habitats in North America, possessing 490 of the 790 U.S. freshwater fish species. Of these, 21% are extinct or imperiled (Warren and Burr 1994).

Although the Southwest United States contains the greatest number of North American fish species that are listed as endangered or threatened (Williams et al. 1989), the greatest freshwater diversity exists in the Southeast. For example, of the 297 mussel species occurring in North America, 248 of them (84%) occur in the Southeast (Williams et al. 1993). The region is also rich in other aquatic faunas, including aquatic insects, crayfishes, and crustaceans.

Over time, it became apparent that a cooperative partnership should be formed, dedicated to the repair and wise use of aquatic ecosystems in this region. With guidance from the National Fish and Wildlife Foundation, the U.S. Forest Service and other agencies determined that they should develop a mechanism to fund watershed restoration projects that would enhance, repair, and restore selected aquatic ecosystems.

The first meeting of what is now the Southern Rivers Council included representatives from the U.S. Forest Service; U.S. Army Corps of Engineers; Tennessee Valley Authority; U.S. Fish and Wildlife Service; U.S. Geological Survey, Biological Resources Division; Office of Surface Mining; U.S. National Park Service; U.S. Environmental Protection Agency; and the Tennessee Aquarium. Participants were scientists and agency personnel who had some administrative authority and backgrounds in fisheries, aquatic ecology, botany, rangeland conservation, hydrology, and soils. The mission of the Southern Rivers Council emerged: to provide an opportunity and a method for success for on-the-ground restoration projects, beginning with open communication (Pringle et al. 1993; Tangley 1994).

There is no formal membership in the Southern Rivers Council; it simply comprises people who have a commitment to aquatic ecosystems. Anyone who shares the Council's commitment to restoration of aquatic ecosystems is invited to participate. The Southern Rivers Council continues to grow through new contacts, and in February 1995 it began to include representatives from different conservation organizations and universities. State and local government agencies have increasingly joined in the Council's discussion of projects.

Early in its development, the Southern Rivers Council designated a committee to review project proposals prior to their being submitted to the National Fish and Wildlife Foundation for funding consideration. The review committee is made up of one resource professional from each participating federal agency. Between July and November 1994, the Council secured $125,000 from the Foundation for seven projects designed to control the introduction of sediments into river systems and to stabilize riparian areas by using vegetation that, in turn, would lead to improved water quality and physical habitat for aquatic-dependent organisms.

The Southern Rivers Council has much to show for its efforts so far.

- The Council's largest watershed project (in stream miles recovered and dollars spent) has been the remediation of acid mine drainage and restoration in the North Chickamauga Creek (Tennessee) watershed.
- Significant improvement of the Abrams Creek watershed of the Great Smoky Mountains National Park has reduced sediment input and restored habitat for many unique and threatened or endangered fish species. In fact, two extirpated fish species have been successfully reintroduced into the Abrams Creek watershed.
- Sediment runoff caused by flooding and illegal off-highway vehicle roads has been reduced in the Nantahala National Forest (North Carolina) and Daniel Boone National Forest (Kentucky). These projects have protected habitat for several listed and rare species.
- Habitat has been improved for recreational fisheries in the Hiwassee River (North Carolina) and Horselick Creek (Kentucky) drainages (Bowling et al. 1997, this volume).

The Council seeks to involve all parties who are interested in restoring southern aquatic systems and watersheds to participate and submit project proposals. What began as a core of 30 individuals has grown to a coalition of more than 100. As a result of the Foundation's challenge funding and other cost-sharing programs, the Council is soliciting increased participation by private landowners, conservation groups, and state and local governments. In turn, this buy-in strategy has proven critical to building coalitions and expanding the

Council's restoration effort. The Council's continued growth and success depends on maintaining its core membership while expanding to include active participation by additional state and other partners.

North Chickamauga Creek Restoration.—The North Chickamauga Creek watershed is just north of Chattanooga, Tennessee. As in many southern Appalachian locations, coal was once mined on Walden Ridge South, which includes the headwaters of North Chickamauga Creek. Congress passed the Surface Mining Control and Reclamation Act of 1977 to address the heritage of mine-related problems throughout much of the Appalachians from Pennsylvania south to Georgia. These abandoned mine lands continue to be the cause of significant erosion and water quality problems (Udall 1966; Office of Surface Mining 1996). The water quality of North Chickamauga Creek has been severely affected by acid mine drainage, and to a lesser degree by erosion, municipal sewage, and raw sewage.

The Office of Surface Mining identified the North Chickamauga Creek watershed as a potential national model to demonstrate the Appalachian Clean Streams Initiative in 1995. This designation was due in large part to a partnership that combined activism from the Friends of the North Chickamauga Creek Greenway, expertise offered by the Tennessee Valley Authority, and funding connections of the Southern Rivers Council.

Because of acidic water conditions, the North Chickamauga Creek no longer supports a warmwater fishery or provides adequate habitat for many aquatic species that are native to the ecosystem. The goal for the project is to restore the natural pH of the streamwater to within normal limits in the upper 18 stream miles so that the native aquatic communities can become reestablished.

The North Chickamauga Creek project has been funded in two phases: (1) locating mine discharge points and measuring pH values, and (2) continuing reclamation and installing passive water restoration systems. Here is a look at the activities in each phase.

In the first phase, mine discharge sites were identified and prioritized by pH value, discharge velocity, and overall threat to the watershed from the discharge. Anoxic (absence of oxygen) limestone drains and wetlands were designed and constructed to buffer the acidic water that seeps from some of the mines. The drains were constructed by placing a plastic liner in a trench, filling it with coarse limestone rock, topping it with another layer of plastic, and covering with soil. Acidic water was then diverted through the drain in the absence of oxygen. As the now-neutralized water leaves the drain and is exposed to oxygen, the iron particles settle out of suspension and cover the bottom substrate.

Scientists at the Tennessee Valley Authority are experimenting with other methods to treat oxygenated acid mine drainage in the North Chickamauga Creek watershed (Brodie et al. 1993). One system with promise is the successive alkaline-producing system (Kepler and McCleary 1994).

The second phase of this project involves continued land reclamation and installation of passive water restoration systems. Constructed wetlands and alkaline-producing systems were placed on the high-priority sites. The Friends of the North Chickamauga Creek Greenway have raised public awareness of the watershed's condition and are demonstrating the benefits of environmental education that promotes stream restoration. For example, one local high school has begun monitoring water quality below some of the constructed wetlands as their school project. These data can be used to monitor the success or failure of the mitigation efforts and to teach students the importance of maintaining a healthy environment.

In support of this project, the National Fish and Wildlife Foundation provided a $70,000 challenge grant which was matched by $100,000 from the Tennessee Aquarium, the McClellan Foundation, Bowaters Corporation, and the Tennessee Department of Conservation and Environment. The federal partner for this project was the Office of Surface Mining.

Two additional projects have been proposed to monitor the effects of the completed reclamation. The success or failure of the anoxic limestone drains, successive alkaline-producing systems, and constructed wetlands placed in the North Chickamauga Creek watershed will determine if additional mitigation efforts will be funded. Once success has been determined, the technology and partnership structures developed on North Chickamauga Creek can be used in other watersheds affected by acid mine drainage.

Horselick Creek Restoration.—The Horselick Creek watershed is an area of unique biological diversity located primarily on the Daniel Boone National Forest, in western Kentucky. This 40,000-acre watershed is relatively remote and not extensively developed at present. National Forest lands comprise approximately 15,000 acres, The Nature Conservancy owns 2,000 acres, and the remainder of the watershed is in private ownership.

Horselick Creek has many unique features, including tremendous biodiversity of both aquatic and terrestrial species, threatened and endangered species, unique vegetation, caves and other karst features, and a rich troglodytic fauna (cave-dwelling or underground-dwelling) that includes several endemic invertebrates. The most significant populations of two federally listed mussels occur in Horselick Creek—the little-wing pearlymussel and Cumberland bean pearlymussel (D. Biggins, U.S. Fish and Wildlife Service, personal communication).

Impacts to the ecosystem are primarily related to sediment loading from roads, off-highway vehicle trails, livestock, and agricultural practices. Raw sewage from residences in the area is also a problem, as well as some road crossings which prevent fish migration (e.g., culverts, submerged concrete slabs). The goals of the Horselick Creek restoration are (1) to restore water quality; (2) to reduce and eliminate sediment input by closing off-highway vehicle trails and some county roads where possible; (3) to stabilize and vegetate streambanks; and (4) to remove unnatural barriers to fish migration.

The Foundation provided $10,000 in federal challenge funds to be matched by The Nature Conservancy. The U.S. Forest Service and the U.S. Environmental Protection Agency are the federal partners for this project. Efforts are underway to broaden the partnership through inclusion of the Kentucky State Nature

Preserves Commission, Kentucky Department of Fish and Wildlife Resources, Kentucky Department of Transportation, and local landowners.

Publicity for the project has increased local landowner participation, cooperation, and ownership in the restoration efforts. Private landowners can see physical improvements that reduce erosion and increase their property values, creating a strong incentive to assist in recovery of the watershed. One partner demonstrates the need to reach out to nontraditional partners. The local Toyota manufacturing plant graciously has provided 3 years of water-quality laboratory analysis, including periodic testing for insecticide and herbicide residues.

CONCLUSION

It remains a great irony that U.S. residents are more attuned to the decline of distant tropical rain forests than they are to the loss of natural resources in our own backyards. Rare fish, endangered freshwater mollusks, and endemic aquatic insects are poorly known and little appreciated, even when they occur in nearby streams (Williams and Neves 1992). Yet every day we are reminded that each species possesses a unique biochemical makeup and potential contribution to society. For example, some fish eggs and mussels use an underwater glue; certain darters emit a chemical that protects their eggs from fungus; and paddlefish and mussels apparently have natural defense mechanisms that are resistant to cancer and virus (TVA 1995). We have barely plumbed the societal riches contained within this backdoor biodiversity. As natural resource managers, we must work toward a better understanding and increased awareness of our own region's unique habitats and their tremendous biodiversity.

As noted earlier in this chapter, with improved understanding and education comes concern over the degradation of our watersheds and the knowledge necessary for their restoration. Concerned citizens are beginning to form partnerships around the country to improve water quality, restore fisheries, and repair environmental damage from past land use practices. Once such coalitions are formed, increasing capabilities through investment and new partners can help to offset dwindling agency funding for natural resource protection and restoration.

Ultimately, an improved understanding by the public of the value of healthy watersheds also will lend political muscle to support agency spending at levels greater than the paultry 1%, which our natural resources now receive. This renewed commitment, when added to the power of partnerships, will greatly enhance the capabilities and effectiveness of watershed restoration—not only for the Beaverkill–Willowemocs, Horselicks, and Chickamaugas, but throughout the country.

*The significant problems we face cannot be solved at the same
level of thinking we were at when we created them.*
—Albert Einstein, in Cairns 1993

CHAPTER 11

ACHIEVING PRIVATE-SECTOR INVOLVEMENT AND ITS IMPLICATIONS FOR RESOURCE PROFESSIONALS

William M. Turner

Private-sector involvement is critical to broad-scale watershed protection and restoration. Of the watershed area in the United States, 71% is in nonfederal ownership. In many states, the percentage is much higher; 27 states have more than 95% of their watershed area in nonfederal ownership (Table 11.1). In those states that do have a high percentage of government-administered lands, private interest groups representing industry, recreation, and the environment exercise great influence in water and land use decisions.

Past and present actions of both government and private-sector decision makers have resulted in degraded watersheds and associated aquatic ecosystems in the United States. Successful watershed and riverine management will not occur until those who make decisions about water and land use for both privately owned and publicly-owned lands change their perspectives and actions. The opening epigram from Albert Einstein says it best, and to progress beyond our historic level of thinking, we must understand the past attitudes and actions that have created the problems and clearly focus on the changes needed for success.

TABLE 11.1.—Percentage of U.S. watershed acreages that are federally and nonfederally owned. (Data from USGSA 1990.)

State	Federally owned (%)	Nonfederally owned (%)
Northeast		
Connecticut	0.2	99.8
Maine	0.8	99.2
Massachusetts	1.3	98.7
New Hampshire	12.7	87.3
New Jersey	2.4	97.6
New York	0.7	99.3
Pennsylvania	2.1	97.9
Rhode Island	0.3	99.7
Vermont	6.0	94.0
Midwest		
Illinois	2.7	97.3
Indiana	1.7	98.3
Iowa	0.9	99.1
Kansas	0.8	99.2
Michigan	12.6	87.4
Minnesota	10.5	89.5
Missouri	4.7	95.3
Nebraska	1.4	98.6
North Dakota	4.2	95.8
Ohio	1.3	98.7
South Dakota	5.7	94.3
Wisconsin	10.1	89.9
South		
Alabama	3.3	96.7
Arkansas	8.2	91.8
Delaware	2.2	97.8
District of Columbia	26.3	73.7
Florida	9.0	91.0
Georgia	4.0	96.0
Kentucky	4.2	95.8
Louisiana	2.6	97.4
Maryland	2.8	97.2
Mississippi	4.3	95.7
North Carolina	6.3	93.7
Oklahoma	1.6	98.4
South Carolina	4.7	95.3
Tennessee	3.8	96.2
Texas	1.3	98.7
Virginia	6.0	94.0
West Virginia	6.7	93.3
West		
Alaska	67.9	32.1
Arizona	47.1	52.9
California	44.4	55.6
Colorado	36.2	63.8
Hawaii	15.5	84.5
Idaho	61.7	38.3
Montana	28.0	72.0
Nevada	82.7	17.3
New Mexico	33.1	66.9
Oregon	52.4	47.6
Utah	63.8	36.2
Washington	29.0	71.0
Wyoming	48.8	51.2
Total U.S.	28.6	71.4

The book title *Entering the Watershed* (Doppelt et al. 1993) aptly describes the current level to which aquatic ecosystem management has progressed. Fisheries management has evolved from stocking and harvest limits to water quality improvement, and now to watershed management. Traditional instream habitat enhancement and protection have led us to realize that riverine fisheries exist within the context of watersheds and thus are at the receiving end of these complex hydrologic systems. The new watershed perspective mandates that all elements of the system must be considered. Reaching this point culminates decades of land and water conservation, and it leads us into a holistic view of riverine ecosystem management (Leopold 1994; Willard and Kosmond 1995; Williams 1995).

An understanding of fluvial processes and a commitment to the restoration and protection of watersheds are paramount to the restoration and protection of riverine ecosystems (Karr et al. 1983; Heede and Rinne 1990; Schlosser 1991; Bickford and Tisa 1992). Many past failures in river and stream management were actually failures in understanding the physical and biological dynamics of these systems (Karr et al. 1983; Williams 1995).

This lack of understanding has allowed the narrowly focused "reach approach," or site-specific strategy, to proliferate and to direct efforts whether they were intended to restore, protect, or modify. By perceiving a reach (segment) of river as a unit unto itself, the reach approach inappropriately attempts to isolate that reach from the remainder of the system. The problems or opportunities identified within the reach are considered to occur independent of upstream or downstream influences. Conversely, practices within the subject reach are considered to have minimal impact on upstream and downstream reaches.

The reach approach exhibits both a lack of understanding and a disregard for fluvial geomorphic processes (processes by which streams shape the land). This has resulted in decades of channel alteration, landscape modification, and installation of ill-advised instream habitat improvement structures (Brookes 1988; Frissell and Nawa 1992; Rosgen 1994). These efforts have often created many more problems than they have solved (Karr and Schlosser 1978). In contrast, restoration and protection on the scale of an entire watershed is a rational strategy that considers cumulative impact, facilitates interdisciplinary and interagency coordination, and recognizes the vital role of the private sector (Willard and Kosmond 1995).

The question that looms before us is, "How?" In the face of heightened concern about private property rights, an unprecedented move to downsize federal natural resource agencies, and an antiregulatory climate within a sector of the national political structure, how do we solve such a monumental problem? *The answer lies within a partnership of private interests and public agencies.* The private sector is essential to achieving proper watershed management because members have a strong and vested interest. Motivated private individuals and organizations have the incentive to expend the effort to organize, gain political adoption of, and implement innovative management plans (Kusler 1995b). Only with a knowledgeable and empowered private sector, equipped with technical support from public agencies, will watershed restoration and protection be accomplished on a broad scale and far into the future.

The intent of this chapter is to outline ways to achieve constructive involvement by the private sector in watershed restoration and protection. The information presented is derived from other authors and 10 years of experience instituting and developing a statewide stream management program in Missouri (Wehnes 1992).

Discussion of the private sector can be facilitated by focusing on three groups: (1) the general public, (2) conservation advocacy groups, and (3) landowners. The majority of landowners upon which this chapter will concentrate are commodity producers. These include agricultural producers, miners, loggers, commercial fishers, and others who derive a living by extracting commodities from the watershed. Commodity producers are emphasized because of their history of affecting watersheds. Thomas (1985) reported that, in 6 of 10 U.S. Environmental Protection Agency regions, nonpoint sources were the principle remaining cause of water quality problems, and agriculture was the most pervasive nonpoint source in every region. The magnitude of landowner impact shows the important need for their involvement and the need to understand their role and contribution.

Direct decisions regarding water and land use are made by landowners. Indirect influence comes from the general public and conservation advocacy groups, who influence public opinion, legislation, and regulatory decisions. All three of these groups will continue to influence decision making, whether or not natural resource professionals interact with them. However, without resource professional interaction, decisions will be focused on social and economic priorities, and will not be adequately weighted with sound scientific information. Effective communication by knowledgeable resource professionals will positively affect decision making. In fact, the potential is great for resource professionals to be catalysts for enlightened strategies that will replace the status quo and result in healthier aquatic ecosystems.

FUNDAMENTALS OF THE WATERSHED STRATEGY

Effective protection and restoration are rooted in innovative projects and programs that are tailored to the hydrologic, social, and biological characteristics of the watershed. Projects and programs must be compatible with the existing decision making processes for water and land use within a watershed. These processes are often heavily influenced by multidisciplinary viewpoints, multiple political boundaries, tradition, and economic considerations. Because of the complexity of riverine and watershed management, guidance is needed to initiate and develop a successful program. Several published articles have included this type of guiding information: Mahood (1985), Heede and Rinne (1990), Wehnes (1992), Wise (1993), Ticknor (1994), Willard and Kosmond (1995), and Kusler (1995a, 1995b).

This information, plus the author's personal experience, has been used to compile 10 fundamental principles that need to be addressed to coordinate an effective watershed-based program.

1. Authority must be vested in local entities, with full representation of affected community members.
2. The political will to pursue riverine and watershed protection and restoration must be present, or it must be developed early in the project.
3. Many educational needs exist, and it is vital that they be identified, prioritized, and addressed.
4. Clear, well-developed goals should be established and a single authority (i.e., local committee, watershed board, agency) should lead the planning and management effort.
5. A watershed analysis should be conducted using the best available data.
6. Key stakeholders must be understood and their economic and social concerns addressed.
7. Programs and projects must have a strong scientific base that includes adequate trained staff.
8. Clear and frequent communication is needed among resource professionals, project stakeholders, and the general public.
9. Watershed projects should be user-friendly.
10. Project monitoring and evaluation should be ongoing and adjustments made as needed.

All 10 of these principles should be considered essential, and others should be identified case-by-case. There is no significance to the order of the following principles; the order should be determined for each situation.

Principle 1. Authority Must Be Vested in Local Entities

The watershed strategy should be based on the premise that one key to success is to involve local people in setting policy and solving problems. "Local people solving local problems" is a common adage that describes this characteristic. Wehnes (1992), Kusler (1995a, 1995b), and Ticknor (1994) all stressed the need for local authority and full representation from the community. Achieving this level of local involvement will require a fundamental change in the approach used by most agencies.

Key to any program is the role of the administering organization, whether it is government or private. The administering organization generally acts as the ultimate authority and dictates the decision-making process. Based on the hierarchy of authority, programs are commonly categorized as either "top-down" or "bottom-up." For example, most programs based in the federal government are in the "top-down" category, because decisions are made at the higher levels of government. Although some authority to make decisions is delegated to state and local governments, the federal agency retains control. The more authority that is exercised from the "top," the less likely it is to be received favorably at the local level. This "top-down" strategy is suitable for programs with nationwide applications or resources of national interest, but it is poorly suited to addressing complex problems and opportunities at the local level.

In contrast, a "bottom-up" strategy focuses the decision making at the local level. This approach of "local people solving local problems" has been advocated by both government agencies and commodity producers. Speaking for the U.S. Environmental Protection Agency, Thomas (1985) stated that it is only at the state and local level that enough flexibility exists to make the site-specific and source-specific decisions required for success. On behalf of the livestock industry, Coy (1985) made two points in support of locally based programs. First, improved communication is needed among government, farmers, scientists, and the general public. Second, cooperative solutions should be emphasized over government dictum. Within the "bottom-up" strategy planning is conducted at the local level. This gives local people ownership in the plan, because it is a compilation of their ideas that fit their needs.

Upper levels of government do play important support and regulatory roles for a "bottom-up" program. They do so by providing technical assistance, cost-sharing programs, grants, and enforcement of state and federal laws (such as the Clean Water Act, the Endangered Species Act, and state environmental laws). The need for complementary federal programs that enable this approach should not be underestimated. The combination of such federal, state, and local efforts is the key to protecting the nations' river systems (Doppelt et al. 1993).

Events in the St. Louis River watershed in northern Minnesota illustrate citizen acceptance of local authority and rejection of a "top-down" strategy. The St. Louis River Board was formed in 1991 because of concern about the future of rivers in this watershed. The public, local government officials, and Fond du Lac tribal officials are represented on the board. The board was established specifically to formulate a comprehensive management plan for the wise use and environmental protection of the St. Louis River watershed (Hambrock and Murto 1993). The plan includes criteria for (1) protection of critical biological, historical, and archaeological resources, (2) recreational use of the river and adjacent lands, and (3) strong cooperative planning and management practices.

Ironically, 10 years earlier, the Minnesota Department of Natural Resources proposed a similar plan for a subbasin of the St. Louis River watershed. Although public input was sought, the Department's plan failed because it was perceived as state government interference in local affairs. In addition to the St. Louis River initiative, Minnesota has several examples of grassroots planning initiatives that now provide good resource protection (Pauley 1993).

The Watershed Committee of the Ozarks is another excellent example of a successful "bottom-up" strategy to addressing watershed needs. The committee is a nonprofit organization created by the citizens of Springfield, Missouri, to protect the drinking water supply of the Springfield–Greene County community. The committee includes citizens who represent the city, county, water utility, and the general public.

The Watershed Committee of the Ozarks reviews zoning laws, initiated development of a geographic information system for the area, and is the innovator of projects relating to nonpoint pollution, stormwater, and spring-monitoring projects. The committee has addressed educational needs by developing a water resource library, developing a water-testing program in schools, hiring a communication specialist, and producing multimedia presentations. These are all accomplished with funding from local, state, and federal governments (Bullard 1994). Local authority, strong community representation, and the discriminate use of higher-level government expertise and funding are the important attributes that make this watershed program effective.

In contrast, the Locust Creek Riparian Corridor Management Plan, proposed for north-central Missouri, reminds us that all stakeholder groups need to perceive a need for a project. This was a joint federal, state, and county project funded by Section 319 of the Clean Water Act. The project was designed by government planners and technicians, with much input from a small group of landowners. It centered on cost-sharing for streambank stabilization and purchase of riparian corridor easements.

Although landowners were surveyed to determine their concerns and level of interest, they did not act as decision makers. These landowners had a poor understanding of fluvial systems and a low regard for the ability of healthy riparian areas to protect channel stability, floodplain agricultural fields, and water quality. As a result, they did not comprehend a need to protect riparian areas and had not committed themselves to the project. Although improvements could have been made in the project design process, negative landowner attitudes probably would have prevailed, had the project proceeded, which it did not (M. D. Lobb, Missouri Department of Conservation memorandum, 15 June, 1995). Situations such as this exemplify the importance of evaluating participant attitudes and desires at an early stage so the project's direction and level of effort can be adjusted.

Principle 2. Political Will Must be Developed

In many watershed issues, numerous individuals, groups, and government entities must commit their support before significant changes can occur. Without such political support, adversarial views will dominate decision making, and the initiative will either be terminated or significantly altered to render it politically acceptable. In either case, the resource will not receive protection or restoration benefits.

Sampson (1991) noted that solutions to environmental problems often represent as much political pain as the problem itself. In these cases, the politician has little incentive to support the solution, so no action is taken. The political path of least resistance is followed. This scenario supports the concept that citizens lead and politicians follow. Missouri's stream program views this concept in a positive manner by acknowledging that good law is crystallized public opinion. The logical extension of this thinking is that support for protecting and restoring the riverine resource must develop from the grass roots (Wehnes 1992). The presence of strong grassroots advocacy provides incentive for politicians and government agencies to play a positive role.

Natural resource professionals experienced in coordinating politically charged issues know that citizen advocacy is essential to achieving resource protection. The need for citizen advocates is accentuated because most resource professionals are employed by government agencies. As agency employees, their freedom to act fully as advocates for the resource is often impaired by politics and agency policy. This results in citizen mistrust and apprehension toward working with agency staff.

On the other hand, private citizens possess the freedom, and its associated powers, to advocate on behalf of environmental concerns. Citizens must be given a legitimate opportunity to become active participants in issues that often are intimidating. Citizens can be intimidated because they may not be confident in their understanding of the subject, or because they may be inexperienced with the political process. A strategy is needed to alleviate this intimidation and to empower citizens to assume the lead in protecting natural resources. It has been said that knowledge fosters understanding, understanding fosters concern, and concern fosters action. These concepts outline a basic strategy for involving citizens.

Through effective public awareness and by providing opportunities that facilitate citizen involvement, knowledge and understanding can be promoted. If knowledge and understanding are attained, concern and meaningful actions will follow. *How to Save a River* by Bolling (1994) is an excellent book that details the steps vital to organizing a river protection initiative. One chapter of the book is devoted to increasing awareness and involving people.

Fostering knowledge and understanding within the private sector is a formidable and critical task. It requires organization and ample staff capability. One effective method is a canvassing outreach program. This method has been very successful for the Citizens Campaign for the Environment in New York and Connecticut. Canvass outreaching uses volunteers to make one-on-one contact to deliver both verbal and written information. This program informs citizens of watershed issues and guides them on how to effectively communicate with policy makers.

In Suffolk County, New York, outreach canvassing made the public aware of the long-term environmental and economic benefits of comprehensive watershed protection. This led to approval of several legislative initiatives. Outreach canvassing had similar success in the rural area near Albany, New York, and the watershed of Long Island Sound Estuary. This method targets citizens of a watershed in a personal way with information about that watershed. It is a sound marketing strategy.

Missouri's program also has identified objectives for increasing public awareness and citizen involvement. Citizen involvement has been accomplished with the state's Stream Team program. Stream Team is a formal joint venture of the Conservation Federation of Missouri, the Missouri Department of Conservation, and the Missouri Department of Natural Resources. Goals of the program are to educate, foster stewardship, and develop advocates.

Membership in Stream Team is open to all citizens, and in eight years it has grown to 940 teams, representing about 40,000 people. Growth has occurred without promotional campaigns, which is testimony to the latent environmental consciousness present in most of the country. Stream Team is a version of an adopt-a-stream program that aspires to move members beyond traditional cleanup projects and into resource advocacy. Members are asked to ''adopt'' a stream that interests them and are provided a Stream Team inventory booklet that guides them through a qualitative assessment as they canoe or walk their adopted stream.

This program has resulted in many benefits to the riverine resource: greenways have been established, harmful projects have been stopped, restoration projects have been conducted, and political activism has been exercised on key issues. A major accomplishment has been the formation of the Volunteer Water Quality Monitoring Program that trains Stream Team members to collect water quality data for the Department of Natural Resources. Six hundred individuals have been trained and are collecting data on 1,900 miles of stream (J. Bachant, Missouri Department of Conservation, personal communication).

This case history illustrates the potential advocacy strength that lies within the general public. It may be latent, but it is viable and represents opportunities for resource professionals to motivate citizens and to develop the political will that is needed to create positive change in watershed management.

Principle 3. Educational Needs Must be Identified and Addressed

Education is an essential component of effective riverine or watershed management. A basic understanding of the scientific processes affecting these ecosystems is essential to a good understanding of a watershed. Common areas of scientific information required are hydrology, fluvial processes, and biota (Williams et al. 1997a, this volume). However, just as important for an accurate perspective is economic, social, and political information (Preister and Kent 1997, this volume).

In many cases, disseminating such information raises the awareness necessary to develop needed political support. For water and land users, this is often the first information that permits understanding of the impact of their actions on the resource. This basic understanding raises the environmental consciousness of these groups. It helps them to understand *best management practice*, which is the most effective and practical method of preventing or reducing pollution generated by a specific source. Early education can establish the rationale for a best management practice, and it can provide an opportunity for local people to evaluate proposed best management practices and to develop their own ideas. Often, it is the responsibility of participating resource professionals to assess the educational needs of an issue and to coordinate appropriate training. Meeting these educational needs in a timely manner will facilitate the entire process.

A case in point is Missouri's attempt to provide better resource protection against instream gravel mining. The ad hoc group formed to develop new regulations included representatives from regulatory agencies, regulated agencies, the mining industry, private conservation groups, and the Missouri Department of Conservation. Coordination activities soon revealed the need to raise the awareness of gravel mining impact, fluvial processes, and regulatory options.

Groups targeted for awareness efforts were agency representatives and their administrators, industry representatives, and private conservation group members. Each was provided similar information, but the content and presentation method were tailored to each group. This garnered administrative support and reduced contention by neutralizing many issues that were based on misconceptions and false information. Clearing the politically charged atmosphere of misconceptions is a positive step toward reaching a solution.

The physical and biological aspects of fluvial systems are challenging subjects to teach. One reason is that very few people, including resource professionals and engineers, have been exposed to these subjects. The absence of formal instruction leaves a void that is typically filled with misconceptions and false information. Teaching must dispel these erroneous, preconceived ideas and communicate correct information.

A second reason that the aspects of fluvial systems are hard to teach is that many hydraulic, hydrologic, and fluvial geomorphic principles are complex and difficult to grasp. For example, many people do not understand that a channelization project may increase upstream channel instability by increasing stream energy. The ability to understand such concepts is important to gaining public support necessary to restore watersheds.

Innovative methods of communicating complex information need to be used. The use of functioning models that simulate fluvial processes are invaluable teaching tools, because they hold students' attention and transfer the information better. Guy and Denson-Guy (1995) presented the four sensory modalities used in the learning process: visual, auditory, tactile (touch), and kinesthetic (movement). Research has shown that each individual uses all four but is best suited to use one modality. An advantage of functioning models is that they can be used in a format that presents the information in all four modalities, which should improve the educational experience for all. Such models increase the instructor's ability to communicate. Examples of good working models are the model stream (Gough et al. 1990), the Streamlab hydraulic demonstration flume (Newbury 1994), and the sand tank groundwater model (Mechenich 1990).

The model stream (Figure 11.1) is a rectangular water-tight box set at a desired slope and filled with a specific mix of granulated plastic. A stream channel is formed in the plastic substrate, and water is circulated to the head of the model stream. Basic principles of fluvial channels can be demonstrated. This model has proven to be effective with people from many disciplines, age groups, and educational backgrounds. It has been an invaluable instructional tool in critical issues such as channelization, instream gravel mining, and riparian corridor protection.

The Streamlab hydraulic demonstration flume (Figure 11.2) is a miniature flume whose image is projected onto a standard screen with an overhead

FIGURE 11.1.—Model stream used to demonstrate the basic principles of stream channel formation and change.

projector. It is a good instructional tool for demonstrating the varied flow patterns associated with different structures in rivers and streams.

The sand tank groundwater model (Figure 11.3) is designed to teach groundwater characteristics, movement, and potential for contamination. A well-developed functioning model will have an instruction manual and, preferably, an

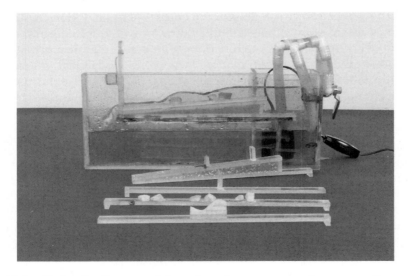

FIGURE 11.2.—Hydraulic flume model used to demonstrate various streamflow patterns associated with in-channel structures.

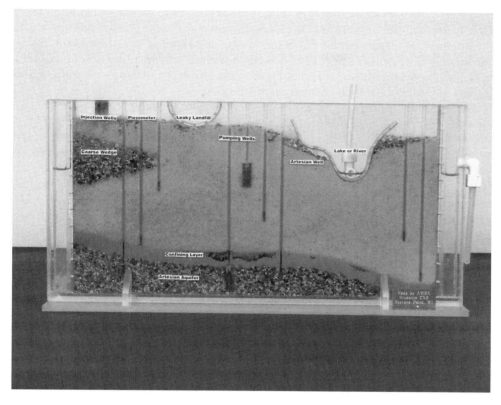

FIGURE 11.3.—Sand tank groundwater model used to demonstrate changes in groundwater aquifers and potential contamination by chemical pollutants.

accompanying videotape to ensure that new concepts are communicated correctly.

Providing technical assistance in development and demonstration of best management practices is another important educational responsibility for resource professionals. This role must be recognized and accepted, because the impetus and expertise often are lacking in the private sector and some agencies to lead in providing technical assistance. Commodity producers often view the development and adoption of best management practices as a staffing and economic cost that will not provide timely economic return, if any. Skepticism over the severity of riverine and watershed problems, and any subsequent need for best management practices, also fosters apathy toward their development. However, if change from the status quo is advocated, then land and water users must have alternative practices that will ensure their economic stability and enhance the resource.

Ideally, best management practices should be developed very early in the riverine or watershed restoration and protection process. Although resource professionals must take the lead, commodity producers and other individuals expected to use these new practices must be closely involved with their development. Ownership in the development of new ideas will promote acceptance by their peers. Working with landowners, on their land, is an important facet of this approach. A well-conceived technical assistance strategy will address pertinent social and economic issues as part of the development and adoption process for best management practices.

The U.S. Natural Resource Conservation Service, formerly the Soil Conservation Service, has exercised this focus on obtaining local input and meeting local needs for many years (Helms 1993). Through this approach, the U.S. Natural Resource Conservation Service has had a strong influence on agricultural practices. The use of terraces, contour farming, and minimum tillage are all significant conservation practices that have been adopted by commodity producers and have increased long-term productivity and profitability. Chaney et al. (1990) described case histories of best management practices for riparian areas, demonstrated on lands in Oregon, Utah, Texas, South Dakota, Nevada, Idaho, and Arizona. These successes are similar, in that solutions were developed locally and proved to be advantageous for both stream health and for livestock producers.

Meeting educational needs does not ensure the adoption of watershed protection and restoration practices. However, ignoring these needs will assure the failure of such efforts.

Principle 4. Clear Goals Must be Developed

Well-developed goals should determine the spatial scale and approximate time line of the project. The scale could be a large watershed, or it could be approached subbasin by subbasin (Ziemer 1997, this volume). Working on a large scale may offer greater efficiency, but it also may preclude the flexibility required to develop site-specific and source-specific solutions. The time line must be realistic so that all participants have reasonable expectations for the time required to achieve expected results.

Setting goals also helps to define real problems, rather than just identifying symptoms that too often are the focus of habitat-improvement projects. A case in point is streambank instability. An eroding bank is often the focus of concern, but it is generally a symptom of a larger problem, such as vertical instability, alteration of the watershed's hydrology, or physical removal of vegetation from the banks. A properly developed goal would address a problem like eroding banks within the broader context of their cause.

This does not imply that a symptom should never be addressed until the cause is identified and corrected. Sometimes the cause cannot be pinpointed, and the symptom (the eroding bank in this example) must be addressed. In other cases, the cause may be apparent, but incurable. A common example is hydrologic change and subsequent channel instability resulting from urbanization. In such cases, identifying the cause is still important because it provides information needed to properly design channel-stabilizing projects and to guide further urban development planning.

Agency resource professionals have important roles in goal development. They can supply technical information, organizational logistics, and support services in the early stages of a watershed project. This allows them to make important contributions while maintaining the position of an objective participant. (This is not meant to imply that the resource professional should attempt to be neutral on significant issues.) Once goals are established, they will guide the entire effort, so it is essential that the goals be developed with sound scientific input of resource professionals. Angermeier (1997, this volume) provides a conceptual framework that should be understood prior to establishing restoration goals for specific watersheds.

A single authority should take the lead in a watershed restoration or protection project. Leadership is essential to retaining organizational structure, continuity, and direction. The lead organization must have both the credibility and the resources to conduct the required tasks. A private organization should take this role, or a leadership group can be formed with representatives of the participating groups. A key characteristic of the lead authority is that it is committed to the full-watershed strategy of land and water management.

Principle 5. A Watershed Analysis Should be Conducted

Watershed analysis provides the essential information needed to write project goals. This basic analysis guides the goal-setting within the watershed context and prevents it from becoming too narrowly focused. Although the level of detail can vary greatly, some common types of information include geology, soil types and erosion hazards, channel development, hydrology, water quality, pollution sources, unique habitats, and biological community information. High-quality maps always are useful and often are critical.

Watershed analysis also provides continuity to information flow as the project progresses and as new participants join. Project documents provide an organized approach to problem identification and act as an educational tool for all interested parties. Ziemer (1997) summarizes the watershed analysis process, its intent, and critical components.

Principle 6. Concerns of Key Stakeholders Must be Understood

Agencies must clearly communicate their intent and have a good understanding of the concerns, perspectives, and capabilities of the stakeholders. Attaining this understanding is often difficult, because skepticism may exist between landowners and agency representatives. Landowners fear the loss of private property rights, and agency staff question the landowners' level of environmental concern and commitment. Such mutual distrust inhibits good communication. However, developing an understanding between these groups provides more accurate perspectives and establishes the credibility needed to produce positive results.

Conducting surveys is a good way for agency personnel to gain a better understanding of landowners. Surveys can be expensive and time-consuming, but they offer valuable insight on landowner perceptions of problems and potential solutions (Wehnes 1992). Surveys also provide basic information on gender, income, location, landowner status, and income commodity.

In central Iowa's Bear Creek watershed, Colletti et al. (1994) used a survey to assess perceptions of water quality, to identify uses of Bear Creek, to determine perceived sources of pollution, to identify conservation practices in use, and to determine the willingness of the watershed's residents to pay for water quality improvement. Survey results were intended to help define best management practices that would be economically feasible, environmentally sound, politically expedient, and socially acceptable. Responses from farmers, absentee landowners, and nonfarmer groups were compared. A key factor in this survey was the use of farm leaders from within the watershed to review and refine the survey. Such a thorough, objective survey provides valuable insight into the attitudes of all watershed residents.

However, surveys are an impersonal contact, requesting substantial information from a cross section of watershed residents. Questions must be carefully worded to reduce biased answers and to reasonably avoid offending those surveyed. Nonrespondents to the survey should not be ignored, for they too are providing another sort of information input. Professional pollsters are recommended for survey development and data analysis.

Understanding the attitudes of watershed stakeholders is important. But it also is important to understand how they make decisions to change land and water use practices. This information can expedite development and adoption of best management practices. Rogers (1962) and Preister and Kent (1997) explain how innovative ideas are accepted into local communities. The diffusion of new ideas depends upon each individual's willingness to try new ideas. Categorizing community members according to their willingness to accept new ideas is called the diffusion model (Rogers 1962). The model consists of five categories: innovators, early adopters, early majority, late majority, and laggards. Although each category is specifically described, there is an assumed continuum among the five categories.

Innovators are critical to the diffusion process because they bring new ideas into the community. Key attributes of the innovator are the ability to apply complex technologies and to cope with a high degree of uncertainty. They also must have the financial resources needed to institute new ideas, and to absorb the loss if they do not work.

Early adopters are more closely identified with the majority, but they are open to new ideas. This position in the social system makes them influential in the widespread adoption of new ideas. Their primary role is to decrease the uncertainty about new ideas and to diffuse information within the community. Early adopters are sought by resource managers to speed the diffusion process.

Early majority group members follow with deliberate willingness, but seldom lead. Members of this group adopt new ideas just before the average member of society does, and they are an important link in the adoption network.

Late majority members are skeptics. They can be persuaded of the need for change, but peer pressure is needed to make them actually change. This group does not facilitate the adoption process.

Laggards are those who base their decisions on tradition, so they are generally the last to adopt new ideas.

An understanding of the inherent differences in how people adopt new ideas can be used to evaluate individuals for different roles in restoration. For example, those who are laggards or late majority members would not be good candidates for testing new best management practices.

Although useful, the diffusion model must be supplemented with knowledge and trust gained from frequent personal contact to fully understand those who will adopt innovations.

Hooks et al. (1983) pointed out that the diffusion model does not adequately consider economic constraints on decision making. The diffusion model implies that access to information is the major factor affecting adoption. It assumes that innovators and early adopters who are given a new idea will convey it to the rest of the community. However, farmers who lack adequate financial and land resources cannot adopt some ideas, regardless of their desire to do so. In addition, commodity producers are motivated to reduce cost and increase income. Thus, best management practices that are contrary to this primary objective will not be considered favorably. Misjudging the importance of economic factors is a common mistake of resource professionals, so economic constraints must be thoroughly evaluated and addressed.

Economic and social concerns of stakeholders, regardless of their background, must be acknowledged in an objective manner, accounted for whenever possible, and continually evaluated for probable effect on project success. Resource professionals must use frequent personal contacts to build trust within the community, to understand citizen needs, and to gain widespread adoption of new ideas.

Principle 7. Programs and Projects Must Have a Strong Scientific Foundation

Establishing a technical base should precede all other activities of a riverine or watershed restoration project. The fact that good decision making is dependent upon good information dictates the need for a strong scientific base upon which to formulate watershed solutions. The best available information is needed regarding the extent of a problem, its cause, and its solution. Proceeding into a project without credible information jeopardizes the project's success because it causes confusion. This leads to participant insecurity and project failure. Credibility loss from scientific incompetence also makes subsequent projects more difficult to promote. A strong scientific base improves agency credibility, allows better understanding of problems, and sets the stage for making good decisions.

An acceptable level of scientific competency should be possessed by all staff members to ensure correct and consistent communication. An efficient method of providing a strong scientific base is through the use of interdisciplinary teams.

Principle 8. Clear and Frequent Communication

Many watershed stakeholders are unfamiliar with watershed issues and concepts. These must be introduced and reinforced regularly. Until familiarity is achieved, public support will be lacking. To achieve effective communication, information must travel in both directions, so resource professionals always should seek input from other participants.

Principle 9. Watershed Projects Should be User-Friendly

Realistically, the logistics of many watershed projects become the responsibility of a government agency and the resource professionals it employs. Tailoring the project to its users will reduce confusion and frustration and produce a positive attitude toward the project. Communicating clear project objectives, providing a streamlined process for permits, and reducing duplication among agencies also will ease frustration. To improve implementation, logistics should be coordinated with representative users.

Principle 10. Project Monitoring and Evaluation Should be Ongoing

Riverine and watershed projects are large and complex, making it difficult to stay focused and on schedule. However, flexibility is also important. The project must retain its focus on goals while being flexible with strategy. Continual monitoring and evaluation are needed to handle unexpected occurrences and to ensure project effectiveness (Kershner 1997, this volume). The evaluation process is facilitated by fully employing the other nine coordination principles because the project will be well-designed and frequent communication will have revealed areas of concern. The time line established during goal development should be adjusted as needed.

These 10 resource management coordination principles facilitate good planning based on sound scientific information, all of which is well-communicated throughout the project. Such projects are tailored to the needs of the watershed and offer greater potential than many traditional regulatory-based programs dictated by government agencies. However, the watershed strategy presents unique challenges that require resource professionals to evaluate their perspectives and priorities.

IMPLICATIONS TO RESOURCE PROFESSIONALS

Resource professionals are critical in watershed restoration and protection, but many of their roles are nontraditional. For example, fisheries biologists traditionally have focused on fish population dynamics, exploitation, and closely related management. Expanding those functions to encompass a watershed perspective greatly increases the complexity and alters the role of the position. Although the watershed strategy rationale is widely accepted, not all professionals willingly accept the changes required. Individuals having the aptitude and inherent attraction to riverine and watershed work should be recruited for these positions.

Attempting to mold other aquatic professionals into these positions is a disservice to them and the resource.

The nontraditional roles induced by the watershed strategy often require expertise in hydrology, geomorphology, economics, water and land use disciplines, and social ecology. Meeting these requirements holds significant implications for resource professionals. The most fundamental challenge they face is the technical preparation to fulfill these functions. Willard and Kosmond (1995) point out that agencies often lack trained professionals who combine an interdisciplinary background with an interest in ecological restoration. Continuing education courses can provide the training, but resource professionals must promote the need for such training.

Beyond being technically prepared, resource professionals need to communicate their knowledge effectively. Significant challenges of public relations and communication are faced in presenting scientific information on watersheds to an audience as diverse—and often polarized—as watershed decision makers. Achieving effective communication requires resource professionals to seriously consider the steps needed to succeed. One option is for lead agencies to hire public relations consultants having rural and urban social science skills. They can help guide the planning process.

Carter (1992) clearly outlined the rationale for fisheries professionals to apply when evaluating communication techniques used in the past to promote watershed aquatic protection and restoration agendas. This experience is based on the 1985 Bay Restoration and Protection Plan and associated programs directed at the Chesapeake Bay and its watershed. Aquatic ecosystem protection cannot be achieved through plans developed solely by fisheries managers and select users. The lack of personal interaction with others in the private sector, such as elected officials and planning–zoning staff in the watershed ensures inadequate representation for the aquatic resource.

When those outside the natural resource disciplines fail to represent the aquatic resource adequately, one should not assume a disregard for the resource. It simply reaffirms the old marketing axiom that personal contact is the most effective form of communication—and resource professionals have the responsibility for making those contacts. Several planning agencies interviewed in the Chesapeake Bay program provide insights for fisheries biologists. These planners explained that they did not have the expertise to factor in fisheries considerations and would welcome input from fisheries biologists, if it could be made specific to a locality, if it were presented in a timely and unbiased manner, and if information were of quality sufficient for decision making.

In addition to increased personal interaction, effective communication is facilitated if resource professionals understand the common factors that make it difficult for aquatic and riparian ecosystem protection to be a convincing argument. Effective ways of dealing with these factors must be developed, because they will arise regardless of the specific issue at hand, whether it is land use in the Chesapeake Bay watershed, instream gravel mining in Arkansas, or riparian corridor restoration in Illinois. The following eight factors were identified by Carter (1992) and Missouri's experience in a statewide stream management program. The argument for ecosystem protection is difficult because

1. individual actions cause cumulative impact, but proving a demonstrable connection between the two is often impossible;
2. the basic concepts of fluvial processes are foreign to most people and difficult for many to grasp;
3. recommendations must often be based on the literature or theory, rather than data collected from the locality of concern;
4. effects on the resource usually cannot be quantified (e.g., in many cases the specific proportion of the biotic community that will be affected by an individual action or cumulative impact cannot be cited);
5. the results of recommended changes in the upper watershed will have to work through a complicated set of ecosystem linkages to affect the aquatic ecosystem;
6. those with a poor understanding of watershed dynamics have unrealistic expectations about the rate at which the resource will recover due to restoration activities;
7. long-term, complex ecological benefits are difficult to make convincing when compared to short-term economic gains; and
8. most often, resource professionals must prove that an action significantly affects the resource, rather than the user having to prove a minimal impact upon the resource.

Developing strategies to overcome these factors is just one challenge facing resource professionals as they assume the role of coordinators and technical advisors in watershed management. Traditional roles must be altered and steps should be taken to prepare professionals for these changes. Professionals in academic, management, research, and administrative positions must adjust their perspectives and priorities.

The implications of the watershed strategy are nowhere more acute than for resource professionals who have agency management responsibility. Strong support from line managers and administrators is needed to carry out successful riverine and watershed restoration and protection. Committing the required trained personnel and entering new areas of coordination and cooperation mandate strong leadership and a commitment to a long-term vision of the watershed strategy's potential.

ACKNOWLEDGMENTS

Gratitude is extended to the Missouri Department of Conservation for supporting this project and to Del Lobb, Richard Wehnes, Jack Williams, Sam Kirby, Brian Todd, John Fantz, David Nolte, Stephen Moran, and Kathie Kinkead-Flores for their valuable contributions.

PART 3

Key Practices

The following six chapters describe innovative procedures that have been successfully used to restore rivers and their watersheds across the country: (1) using the hydropower relicensing process to restore river flows, (2) involving students and school systems in restoration programs, (3) encouraging cities and local communities to see fisheries problems in the context of watersheds, (4) integrating private landowner and federal restoration strategies, (5) building diverse coalitions to address watershed problems, and (6) assembling multidisciplinary scientific teams to restore water quality. These practices were selected because of their broad application to watersheds across the country.

Federal licenses for more than 650 dams will expire during the next 15 years. This will provide unprecedented opportunities to restore more-natural flow regimes in rivers across the country. Chapter 12 follows the Federal Energy Regulatory Commission relicensing process for eight dams on the Deerfield River in New England and demonstrates how this process is being used to reestablish at least partial ecosystem structure and function, fisheries, and recreational opportunities on river systems undergoing the relicensing process.

A lack of public education is often cited as a fundamental cause of watershed degradation. Chapter 13 describes a program in central Oregon that integrates students of all ages into community cleanups, restoration efforts, and monitoring activities. Understanding that watershed scale deterioration can cause problems such as increased flood damage and loss of salmon runs is a critical lesson for members of all communities. Chapter 14 describes the evolution of the thought process of people living in northern California's Mattole River drainage and how they have come to "think like a watershed" to address the many resource problems of their watershed.

Many restoration efforts are stymied by the sheer number and diversity of people and their differing perceptions of what issues are important in their watersheds. How are compatible restoration strategies developed for both private and public lands? What are the common perspectives, if they exist, among farmers, anglers, and public agencies? Often, progress is made in one segment of the community only to see progress stalled in another. Chapters 15 and 16 address these concerns. Chapter 15 describes how restoration efforts on federal lands are being joined with compatible efforts on private lands. Many of the same problems exist, and many of the same ecological principles apply regardless of who owns or manages the land. Chapter 16 describes how environmental advocates, anglers, farmers, and ranchers are all working cooperatively to solve common concerns on the Henrys Fork in Idaho. The need to develop such collaborative partnerships is common throughout watersheds across the country.

Integrating people with diverse expertise, such as hydrology, water quality, engineering, fisheries, and education, the Tennessee Valley Authority is assembling River Action Teams as part of its Clean Water Initiative to monitor and improve water quality in each of the 12 major subbasins of the Tennessee River system. Chapter 17 reports on the model River Action Team for the Hiwassee River and its efforts to gather data, build coalitions, and restore water quality in that watershed.

Now, thanks to a little noticed 1986 provision of federal law requiring that environmental concerns be addressed before privately owned hydroelectric dams can be relicensed, the dams' chokehold on the life of the Deerfield is being released and natural flows and rhythms are being restored. With some 800 dams nationwide up for relicensing in the foreseeable future, what is going on here is being held up as a model for the nation.
—W. K. Stevens, 1995

CHAPTER 12

USING HYDROELECTRIC RELICENSING IN WATERSHED RESTORATION: DEERFIELD RIVER WATERSHED OF VERMONT AND MASSACHUSETTS

Kenneth D. Kimball

Hydropower dams can adversely affect rivers in many ways, including fragmenting free-flowing freshwater ecosystems, impeding the flow of nutrients and sediments, blocking fish migration, destroying streamside habitat, slowing and overheating the river, and compromising flows downstream. Reservoir construction can also catalyze extensive shoreline development within the watershed.

At the turn of this century, electricity became an important source of power, and hydropower became a primary source of electricity. In the northeastern United States, existing hydromechanical-power dam sites were altered and many new dam sites were created to harness the region's water for hydroelectric power. Because of the uneven rate of runoff during the course of the year, large headwater storage reservoirs were also built to capture major runoff events and to provide more consistent year-round flows.

In the early 1900s, there was little public concern for the loss of riverine resources to the proliferation of hydropower dams because many of the rivers were little more than well-landscaped sewers carrying burdens of industrial and domestic sewage. The lower Deerfield River in Massachusetts was no exception: village sewage, a paper mill, and later the country's first commercial nuclear power plant all contributed to degraded water quality and altered water temperature. Though the watershed had many earlier dams, by 1911 a well-engineered and coordinated set of dams that could capitalize on the extensive hydraulic head in the watershed for hydroelectric generation was being constructed without extensive opposition.

In the 1920 Federal Power Act the U.S. Congress recognized rivers as publicly owned resources; private companies could build hydroelectric dams, but they could not buy a river. Finite 30- to 50-year licenses became the compromise, and the public was provided opportunities to review the ownership and terms and conditions of these private leases periodically (Echeverria et al. 1989). On many rivers in the United States, hydroelectric power dams are not owned by the federal government but are under the jurisdiction of the Federal Energy Regulatory Commission (FERC). When a project license expires, the owner must apply for a new license. Until 1993, relicensing was a relatively infrequent procedure that received little public attention. Due to how the federal government originally licensed many of the projects, 160 licenses affecting 237 dams on 105 rivers expired in 1993, or 10% of all FERC-licensed dams nationwide. These 1993 relicensings represented the beginning of an unprecedented wave, with 650 more dam licenses nationwide to expire in the next 15 years. On the Deerfield River one license expired in 1993 that covered 8 of the 10 dams on this river.

Public interest in the relicensing of hydropower operations has increased dramatically since many of these projects were first licensed. The passage of the Clean Water Act in 1972 resulted in dramatic improvements in water quality nationwide and gave a new generation, not accustomed to dirty water, the opportunity to see the inherent value of rivers. In 1986 Congress passed Section 4(e) of the Electric Consumer Power Act. For all new or renewal licenses, FERC's new mandate became "in addition to power and development purposes for which licenses are issued, [FERC] shall give equal consideration to the purposes of enhancement of fish and wildlife (including related spawning grounds and habitat), the protection of recreational opportunities, and the preservation of other environmental quality." Having first contested its responsibility even to conduct environmental impact analyses during the licensing or relicensing process as required under the National Environmental Policy Act of 1969, FERC finally acquiesced under considerable pressure from interventions filed and made requirements outlined in the Act a standard part of the relicensing process for projects with a 1993 expiration date.

A series of fortuitous events increased public interest and participation on the Deerfield River at this time. The paper mill, built in 1888 on the river at Monroe Bridge, Massachusetts, closed in 1984. The Rowe Yankee Atomic Nuclear Power

Plant, which was constructed in 1958 and had increased river water temperatures by up to 15°F, was permanently shut down in 1992. Treatment facilities for domestic sewage had come on-line starting in the 1970s. The first catch–release trout fishery in Massachusetts was established below the Fife Brook Dam in the 1970s, and another section just below the first was created in the 1980s. Paralleling this improvement in water quality and coldwater fish habitat was a scheduled 10-week maintenance in 1987 on dam number 5 in Monroe Bridge, Massachusetts. This maintenance resulted in weeks of raging white water in a section of river that had been a dry channel for 40 years. White-water boaters flocked to the rewatered riverbed to enjoy Class IV white-water runs that were unparalleled in southern New England. In short, the evolution of a dedicated constituency followed the improvement in water quality; hydropower operations became the recognized limiting factor in watershed restoration for the Deerfield River. What laid ahead was the challenge to bring together all the constituencies, with their tangled mix of expectations and conflicting demands on a finite resource.

The process leading up to the multiparty settlement agreement for the Deerfield River actually started in 1987, when the relicensing process started and various stakeholders along the lower reaches of the Deerfield River in Massachusetts realized that the process would provide an opportunity to change the terms of the license to enhance and protect the river. This realization catalyzed the formation of the Deerfield River Watershed Association. Evolving from this movement was the Franklin County Planning Department's lead in the production of a comprehensive river plan for the Massachusetts segment of the river (Rubinstein 1990). Later, a comprehensive plan was developed for the upper reaches of the river in Vermont (VDEC 1992).

In 1990, the Appalachian Mountain Club recognized the opportunity for watershed-level restoration in the FERC relicensing process, an opportunity noted as well by other environmental organizations and natural resource agencies. The Appalachian Mountain Club formed a New England hydroelectric relicensing team with the Conservation Law Foundation, bringing together the technical skills of the former's research department with the legal strength of the latter organization. In 1992, a national Hydropower Reform Coalition formed to restore river ecosystems through the relicensing process and to reform the way FERC licenses all hydropower dams. Members of the coalition steering committee include American Rivers, American Whitewater Affiliation, Appalachian Mountain Club, Conservation Law Foundation, Natural Heritage Institute, New England FLOW (FLOW), New York Rivers United, Michigan Hydro Relicensing Coalition, River Alliance of Wisconsin, Sierra Club Legal Defense Fund, and Trout Unlimited. Where appropriate the Hydropower Reform Coalition seeks in each relicensing

1. improved stream flows and restoration of flows to bypass reaches that have been dewatered;
2. fish passage facilities, where necessary;
3. consistency of projects with water quality requirements;

4. protection of riparian (streambank) and watershed lands;
5. free public access to rivers;
6. river restoration and conservation mitigation funds;
7. sufficient funding for long-term dam maintenance or retirement; and
8. riverwide planning and cumulative analysis.

THE WATERSHED

The Deerfield River flows out of the Green Mountains in southern Vermont, passes through the Berkshire Mountains in neighboring Massachusetts, and then drains into the Connecticut River (Figure 12.1). The watershed drainage encompasses 665 square miles, and the river is roughly 75 miles in length from its headwaters to its confluence with the Connecticut River. The upper reaches in Vermont and northern Massachusetts are primarily forested; in the lower portion of the basin much of the land along the main stem is unforested and agricultural land. The river basin is primarily rural, with little commercial or industrial development except at the river's confluence with the Connecticut River in Greenfield, Massachusetts. Between the first impounded river waters at river mile 66 to the Connecticut River confluence, the river drops from 2,134 to 70 feet mean sea level. The steep topographic relief along with the predominance of erosion-resistant metamorphic rock and generally shallow soils has had a major influence on the river. The result is high runoff during storms; this runoff, when stored and combined with the considerable elevation drop, made for ideal hydroelectric power production. The Deerfield's steep gradient and narrow valleys also meant that relatively small and economical dams could be built.

The 1870 census reported 110 water-powered mills and mill sites on the Deerfield and its tributaries. The first hydroelectric power plant was built in 1897, and in 1910 New England Power Company (NEP) was formed to acquire water rights on the river and to construct new dams. The Deerfield River Project was a model of twentieth-century engineering efficiency, effectively harnessing and impacting the entire basin from the headwaters to its mouth. Under FERC project number 2323, first licensed in 1962, are 8 dams with 15 generating units that have a total capacity of 85 megawatts (MW) and produce about 290,000 megawatt-hours annually (Figure 12.2). This licensed project shares the river with two other hydroelectric licenses: a 600-MW-pumped-storage project combined with a main-stem dam built in the 1970s by NEP and a 3.6-MW project operated by Western Massachusetts Electric Company. In addition to the 25 miles of impounded river reach, the projects have created almost 12 miles of dewatered riverbed through diversions and peak electric power production has affected another 33 miles of river with daily fluctuating flows. In short, the cumulative effect of these projects was not considered during the original licensing of these projects, but under more recent environmental legislation the cumulative effect could not be ignored during relicensing.

In the 66 miles of river below Somerset Reservoir (Figure 12.1), the river provides habitat opportunities for economically important game fish—native and stocked brook trout and nonnative species, such as brown and rainbow trout.

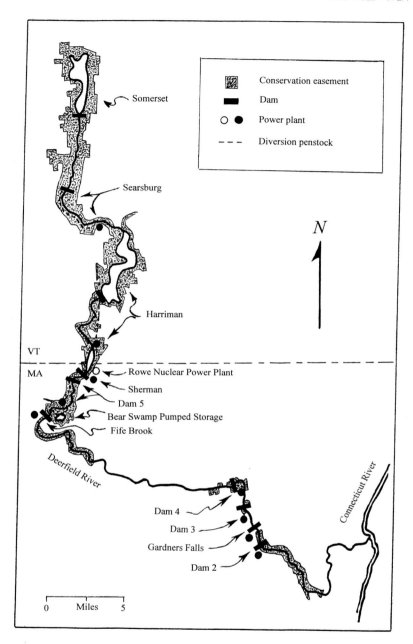

FIGURE 12.1.—Location of New England Power's (NEP) Deerfield River Project developments, Somerset, Searsburg, Harriman, Sherman, and dam numbers 5, 4, 3, and 2 (FERC
project number 2323) and the Bear Swamp Pumped Storage Project and Fife Brook Dam
(FERC 2669), and Western Massachusetts Electric Company's Gardners Falls Project
(FERC 2334). Also shown are projects' lands to be included in the conservation easements
within the Deerfield River watershed. (Modified from NEP maps.)

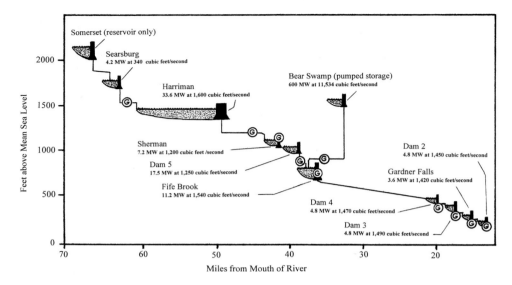

FIGURE 12.2.—Schematic of NEP's hydroelectric projects and reservoirs on the Deerfield River. The location of the generators are identified by "G"; flow is given in cubic feet per second. Lengthy diversions from the natural riverbed for hydroelectric generation occur with the Searsburg, Harriman, and dams 5, 4, and 3. (Figure courtesy of NEP.)

The lower reaches also support smallmouth bass. Federal and state agencies are actively managing the lower Deerfield River basin to restore anadromous Atlantic salmon as part of the Atlantic salmon recovery plan for the Connecticut River basin. The Atlantic salmon disappeared from the Connecticut and Deerfield watersheds in the eighteenth and nineteenth centuries when those watersheds were dammed. The historical record of the size of the Atlantic salmon runs up the Deerfield River is incomplete, but it is known that they traveled up the Deerfield, spawning in the main stem and in some tributaries (Rubinstein 1990).

The federally listed endangered shortnose sturgeon also periodically inhabits the Deerfield's lower reaches. Based on tagging studies it is thought that the shortnose sturgeon probably uses the lower Deerfield River for resting or refuge from high flows in the Connecticut River during April and May as it travels up the Connecticut River to spawn (FERC 1996).

All the reservoirs, though artificial water bodies, now provide a mix of cold- and warmwater fisheries. The largest reservoirs, Harriman and Somerset, are 2,039 and 1,623 acres respectively. Harriman Reservoir has rainbow smelt, brown trout, landlocked Atlantic salmon, and smallmouth bass; Somerset is managed as a stocked brook trout fishery. Comparative studies between the highly regulated Deerfield River, which has variable daily flows, and unregulated rivers showed a reduction in fish community complexity and fewer young fish and more older fish as a result of daily fluctuations (Bain et al. 1988).

The lands immediately adjacent to the river and reservoirs in Vermont are primarily owned by NEP or are part of the Green Mountain National Forest. A large skiing consortium continues to express interest in NEP's lands and water rights to Somerset Reservoir. The lower river basin in Massachusetts is a mix of private ownership, state forest lands, and extensive NEP holdings. Approximately 12 million people live within 100 miles of the river. Recreational use on the Deerfield by the late 1980s was estimated to exceed 500,000 user-days annually at NEP facilities (Kapala 1995), and usage continues to increase.

CONFLICTING GOALS

At the request of resource agencies and environmental organizations, NEP conducted extensive environmental and operational studies on its projects, generating over 30 volumes of information during the preparation and following submission of its relicense application (NEP 1991). New England Power recognized that a renewal of its license on original license conditions would not be acceptable, therefore its relicensing application contained significant proposals for improved flows and recreation and wildlife enhancements. Yet following NEP's relicensing application, 15 separate interventions were filed by state and federal agencies, interest groups, and municipalities. In the eyes of the environmental community, the FERC licensing process contained regulatory responsibilities to insure applicants enhance watersheds relative to the impacts of their operations. But differences between the two states' agendas, between single-interest groups such as the white-water boating community and anglers, and between biologists concerned about the effect of flow and reservoir levels on state-listed species of plants and loon nesting and NEP's interest in the quantity of and daily timing of power production, made achieving every interest groups' expectations problematic. At times the boating and fishing community saw one another as enemies regarding the timing and amounts of water to be released. Each state's river basin plan for the Deerfield provided considerable guidance but also conflicted internally and with the other state's plan. The Federal Energy Regulatory Commission had a track record of being reticent to carry out its own environmental regulatory obligations, therefore indirectly encouraging applicants to take a minimalist approach on environmental mitigation. Conditions were ripe for a risky but classic divide and conquer strategy of the interveners by the applicant in a long, drawn-out litigative process with FERC trying to settle the issues. But NEP's commitment to assume an environmental leadership role, with settlement preferred (Kapala 1995), provided the right environment for successful negotiations.

THE SETTLEMENT PROCESS

Following FERC regulations for relicensing, NEP initiated its application process in 1987 for a 1991 filing. Groups quickly came forward, and early on FLOW, a white-water boating organization of individuals and organizations that included the Appalachian Mountain Club, began a series of negotiations with NEP focused solely on white-water boating issues. Three years and 30 meetings later, NEP and FLOW agreed on a plan, contingent on settlement of issues with other parties. The Appalachian Mountain Club, with a number of parties including FLOW, filed a collaborative intervention in 1991 that covered concerns about water quality, fisheries, white-water boating, enhancement funds, recreation, and watershed land protection. In 1992 and 1993 NEP worked on completing fisheries and other studies ordered by FERC based on requests made by resource agencies and environmental organizations. Building on FLOW's earlier negotiations, the Appalachian Mountain Club met with NEP in late 1992 to see if it was interested in a broader settlement based on all the issues filed in the various interventions by all parties. The answer was affirmative. The Appalachian Mountain Club then contacted the various interest groups and state and federal agencies, set up a settlement process, and took on the coordinating role for the interveners.

In the initial stages there was considerable mistrust—among some of the interveners, some of the interest groups, and NEP. The role government agencies could legally play in a settlement was questioned, particularly in situations in which a degree of confidentiality was needed to encourage innovative problem solving in what was legally a public process. At the same time uncertainty existed within NEP, which was in the middle of a company reorganization plan. However, a relicensing team with the essential direct communications to top corporate management surfaced.

Early negotiation ground rules were developed. To keep the group size manageable, each party selected one representative who could negotiate for his or her group and make the time commitment necessary, although other staff as needed were included. Issues were to be categorized and dealt with individually, but final resolution of them could not occur until the specifics for all issues were known. All issues were on the table but easier issues were dealt with first. The environmental interest groups had legal guidance from the Conservation Law Foundation and NEP from its lawyer; both constructively focused on solutions, not legal roadblocks. Meetings were held frequently to keep the process moving, and groups were allowed to caucus as needed. A watershed mitigation and enhancement approach rather than a state-by-state or special-interest group philosophy was strongly encouraged to allow for innovative and cost-effective solutions that would best benefit society. Maintaining this approach was not easy, with single-interest groups and the states being pressured by their constituencies or legal boundaries. New England Power showed considerable leadership by allowing its Bear Swamp–Fife Brook Project, which is located midriver, to be considered for modification even though this project is not due for relicensing until next century. Modification of this project would mean balancing hydroelectric, fishery, and white-water boating flow needs on a 17-mile reach of the river.

The parties agreed to and tacitly held to a policy of no contact with the press until a settlement was reached. The focus was to be on the facts, not rhetoric. Critical to these negotiations was NEP's willingness to go beyond the studies in its initial application to FERC and use its hydrologic computer model to run some 40 different scenarios to quantify and contrast hydrologic, fishery habitat, and economic impacts. Understanding that any settlement proposal had to be factually sound when presented to FERC, which itself has regulatory boundaries on how license conditions are developed, was also an impetus. Disagreement on methods for determining fishery flows surfaced, and in the end a hybrid of methods including the U.S. Fish and Wildlife Service New England aquatic base flow, instream flow incremental methodology, and professional judgment based on field observations of different flows were employed in the different river reaches. Massachusetts and Vermont state natural resource agencies, because of pending water quality certifications for these projects, attended many of the negotiating sessions but reserved their authority and determinations until their 401 water quality certification processes (as mandated by the Clean Water Act) were completed. Their attendance helped align the final offer of settlement with water quality standards in both states. Following 25 negotiating sessions from the first formal session in early 1993, an offer of settlement was signed and formally submitted to FERC in October 1994. Signatories included Appalachian Mountain Club, American Rivers, American Whitewater Association, Conservation Law Foundation, Deerfield River Compact, Deerfield River Watershed Association, New England FLOW, Trout Unlimited, U.S. Environmental Protection Agency, U.S. National Park Service, U.S. Fish and Wildlife Service, NEP, and from Massachusetts the Division of Fisheries and Wildlife, Department of Environmental Protection, and Department of Environmental Management. The Vermont Agency of Natural Resources attended and supported the settlement process but felt it legally could not sign.

WATERSHED REHABILITATION ACHIEVED IN THE SETTLEMENT

By building on the issues that could be settled, a gradual recognition of what could be achieved evolved into a truly comprehensive watershed-level offer of settlement. New England Power Company's project has a present net value of US$242 million over the requested 40-year license term. The offer of settlement has an estimated value of $27–30 million. The components of the settlement, which reflect both the issues and their resolution by the signatories (NEP 1994), are outlined below.

Flows and Reservoir Level Management

Federal and state agencies and anglers looked to the Deerfield River with high expectations of management goals that would enhance this river as one of the region's premier trout fisheries and provide an opportunity to extend Atlantic salmon restoration. Fisheries rehabilitation was the most costly and complex issue, considering NEP's previous investment in maximizing the river for hydroelectric generation through the use of long diversion penstocks and canals that left over 12 miles of dewatered riverbed. Balancing was also needed between reservoir management objectives and white-water boating interests. Within the Searsburg and Harriman bypass reaches in Vermont, the absence of flows for much of the year had resulted in the colonization of the riverbed by a threatened species, the tubercled orchid. New England Power conducted flow studies to outline the hard choices the Vermont agencies had to make between protecting the plant and rehabilitating the fisheries in these river reaches by means of restored flows. A decision was made to transplant the orchid and restore the fisheries. The settlement offer increases flows in all river reaches and restricts reservoir drawdown; an estimated value of almost $21 million over 40 years is lost from energy sales due to the minimum-flow releases and modifications of schedules at reservoirs (Figure 12.3).

Minimum flows below all dams must be maintained according to standards set in the settlement and range from 12 to 200 cubic feet per second. In the Somerset (6 miles), Searsburg (3 miles), Harriman (4.4 miles), and dam 4 (2 miles) reaches the minimum flows will vary seasonally to consider spawning and incubation needs. At the dam 5 (2.6 miles) and dam 3 (0.4 miles) river reaches, flows will be the lesser of a fixed rate or the inflow. Below Fife Brook (17.4 miles) and dam 2 (9 miles), a year-round flow is guaranteed. The settlement offer flow regime returns water to 12 miles of river that currently receive little or no water at times due to diversions, and it enhances flows in another 33 miles of river over current operating conditions (Figure 12.4). It is estimated that adult trout habitat will be increased from 369,000 to about 5,489,000 square feet throughout the river basin (NEP 1994). Provisions to allow for flexibility during extreme natural precipitation, flood control requirements, ice conditions, equipment failure, and electrical emergencies were adopted. Also, flows in the Somerset reach may be modified in low-flow years to protect Somerset Reservoir resources.

The Somerset Reservoir will maintain a stable level to facilitate loon nesting from ice-out until 31 July. The Harriman Reservoir will be managed to support spawning rainbow smelt and smallmouth bass, with reservoir elevations being stable or rising from ice-out until 15 June. From 16 June through 15 July drawdowns are limited to a foot per day.

Fish Passage

Plans were developed for downstream passage facilities for Atlantic salmon at dams 2, 3, and 4 and will be implemented—as possibly modified by the U.S. Fish and Wildlife Service and the Massachusetts Division of Fish and Wildlife—within two construction seasons of license issuance. A fish stocking program is already in operation. Upstream fish passage will be provided at dam 2 for returning adult Atlantic salmon (Figure 12.5). Following a radio-tagging monitoring program and once specified numbers of Atlantic salmon are found returning to the Deerfield for two consecutive years, NEP will install a permanent upstream trap facility within two construction seasons in accordance with plans approved by the Connecticut River Atlantic Salmon Commission. The estimated cost is $3.2 million for fish passage.

Recreation

Recreational white-water releases are limited to two river reaches. At the dam 5 reach there will be 26 weekend or holiday releases and 6 Friday releases into the bypass for 4 to 5 hours. In the Fife Brook reach there will be 50 weekend and

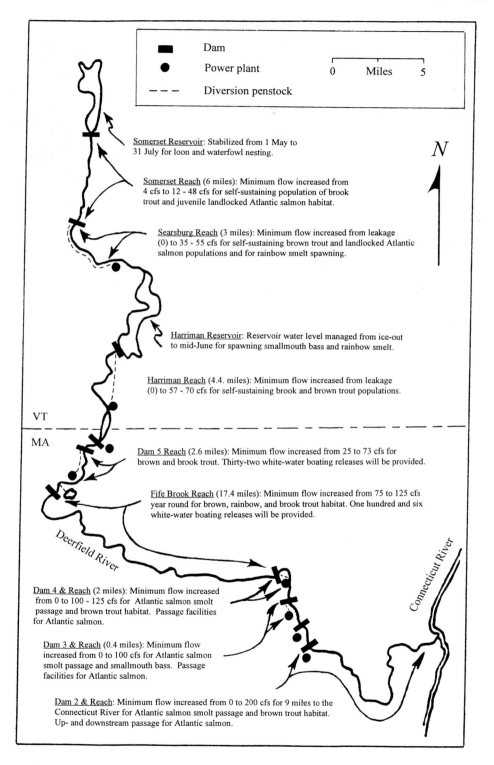

FIGURE 12.3.—Management objectives and enhancement flows (given as cfs) covered by the settlement offer.

FIGURE 12.4.—The Searsburg penstock (foreground) dewaters the natural riverbed (background) for over 3 miles. The settlement agreement proposes to increase the current "dam leakage flow" in the natural riverbed to 35 cfs or inflow during the nonspawning period and 55 cfs or inflow during the spawning season to protect trout, salmon, and aquatic biota habitat.

56 weekday releases, for 3 hours each. A contingency plan exists in the event that low-flow conditions restrict these releases according to schedule. The white-water flows have an estimated cost to the projects of $1.9 million in lost power generation.

New England Power also agreed to provide public access to the water and undeveloped project lands at no cost. In addition they agreed to install, operate, and maintain facilities designated in a basinwide recreation plan. The plan calls for the upgrading of selected picnic areas, boat launches, white-water put-ins, take-outs, and portages, and hiking and ski trails, along with the construction of five camping sites, two picnic areas, two boat launches, and two white-water take-outs, two portage trails, and five hiking trails. The estimated cost of the recreation plan is $1.3 million.

Wildlife Management

Approximately $200,000 of capital cost will be dedicated to waterfowl nesting structures, wetlands management, and other enhancement. Existing beaver habitat in the vicinity of the Somerset Reservoir will be preserved, and nesting structures for wood ducks, osprey, and common loons will be built.

FIGURE 12.5.—Dam 2, the first dam on the Deerfield, currently has no up- or downstream fish passage for the ongoing anadromous Atlantic salmon restoration effort. The settlement agreement provides for downstream passage facilities within 2 years of license issuance, and upstream fish passage will be implemented via a phased approach determined by the number of Atlantic salmon returning to the Deerfield River.

Project Lands

Hydroelectric dam license applicants have the responsibility to provide shoreline buffers on their projects (Code of Federal Regulations, Title 18, Section 2.7), an issue FERC frequently overlooks. Within the Deerfield watershed NEP owned over 18,000 acres, including all the shoreline on the Somerset Reservoir, the majority of it around the Harriman Reservoir, and considerable shoreline along riverine reaches and the smaller downstream reservoirs (Figure 12.6). Protecting watershed and riparian lands has a great effect on the fate of rivers themselves, particularly because hydroelectric operations can place added stresses on riverine ecosystems. The settlement offer grants term easements for the length of the license to qualified government and nongovernment organizations to provide for the continued preservation in a natural state of 18,355 acres of riparian and watershed lands owned by NEP. The intent of the conservation easements is to protect water quality, aesthetics, and the watershed forest. The holders of the conservation easements will be selected by NEP, Appalachian Mountain Club, and Conservation Law Foundation through unanimous decision. New England Power will reimburse each easement holder's reasonable costs for monitoring and enforcing the terms of the conservation easement and give the holder an option to purchase, at fair market value, easement lands that are not required for electrical generation and transmission purposes.

Included for protection are 15,736 acres of project lands in Vermont, which abut the Green Mountain National Forest, and 941 acres of project lands in Massachusetts. Additionally 1,056 acres of lands around the Bear Swamp Pumped Storage Project and an additional 622 acres of nonproject riparian lands are included. New England Power Company will continue to harvest timber from these lands providing (1) riparian lands are protected, (2) visually important vistas are protected, (3) logging is limited on fragile or highly erodible soils, (4) clear-cutting is limited to 20-acre blocks that meet temporal and spatial restrictions and do not have soils rated severe for erosion, and (5) wildlife management considerations will be included in all stand management prescriptions.

Enhancement Fund

There exists a need to provide for currently unidentified and unforeseen programs to meet future needs in the watershed related to project operations over the term of the license. New England Power agreed to establish the Deerfield River Basin Environmental Enhancement Trust Fund in the amount of $100,000 to finance watershed conservation and the planning, maintenance, and monitoring of low-impact recreational and educational projects and facilities. The fund will not be used to carry out the various obligations set forth in other provisions of the settlement offer. It will be administered by a three-member committee—a representative of NEP, a designee of the Secretary of the Vermont Agency of Natural Resources, and a designee of the Secretary of the Massachusetts Executive Office of Environmental Affairs—that shall determine the investment strategy for the fund and the appropriate distribution of available funds each year through unanimous vote.

FIGURE 12.6.—The 2,039-acre Harriman Reservoir shoreline is mostly owned by New England Power and primarily undeveloped. The settlement agreement provides for a "no development" conservation easement on these waterfront lands to protect water quality and aesthetics, and requires the reservoir level to be stable or rising from ice-out until 15 June to protect rainbow smelt and smallmouth bass spawning and early life stages.

Decommissioning and License Term

New England Power acknowledges its responsibility to plan for and collect funds in anticipation of the proper future management of the Deerfield River Project upon retirement from power production. Within 5 years of license approval, NEP will complete a study in consultation with the parties of the settlement and FERC to identify and estimate costs of various options for the retirement of the dams, from continued operation to decommissioning, and the most appropriate likely option will be submitted to FERC for approval.

New England Power will file with FERC an annual certification of financial capability demonstrating that NEP has a tangible net worth at least three times the estimated cost of the project retirement plan. If unable to provide this financial capability certification, NEP will create a segregated trust fund or purchase, post a bond, or provide other means approved by FERC to ensure that the full amount of funds collected will be available at the expiration of the license.

Because of the basinwide level of resource improvement and associated costs, the parties to the settlement offer agreed to support a new 40-year license rather than the standard 30-year renewal.

Adaptive Management

Adaptive management decision-making trees and strategies were incorporated into the Deerfield settlement offer. Riparian and watershed lands to be managed and protected under the conservation easements will be enforced and monitored by a third-party easement holder; NEP will reimburse the third-party easement holder for related costs. Boating interests recognize that drought, electrical emergencies, and other events may prevent strict adherence to the annual white-water release schedule. A river information telephone line will provide information on release schedule changes to the public, and canceled releases will be made up over the next 2 years boating schedule.

For a 36-month period after issuance of the license, NEP will provide flow release data to determine if the new flow regime at dam 2 results in no more than 2 flow transitions per day and not more than 10 per week. The settlement also contained a provision that this dam's operation could be altered if it fails to meet this target. Construction of the upstream fish passage for Atlantic salmon at dam 2 is contingent on the results of a radio-tagging study and return to the Deerfield River of a defined number of fish; the U.S. Fish and Wildlife Service retains authority to prescribe upstream fish passage.

Still unclear is how well the fishery will respond to the enhanced flow regime. Qualitatively, the estimated gains for target fish species are probably reasonably accurate. But the instream flow incremental methodology used to determine flow needs is based on theoretical habitat and more suited for negotiating purposes than for quantitatively monitoring postproject increases in actual fish.

A need exists to improve the instream flow incremental methodology so that it can be used more effectively in determining how effective prescribed flows actually improve fisheries. Environmental impact statement baseline studies are rarely designed for quantitative follow-up research on the efficacy of prescribed management techniques (Naiman et al. 1995). Relative to flow measurements and fishery improvement, the Deerfield case study is no exception, but the enhancement fund concept offers some opportunity for qualitative postlicense follow-up comparisons. Regulatory agencies and FERC need to develop better policy guidelines on the design of preproject fishery studies so that meaningful postlicense studies can be conducted to quantify improvements in the fishery.

CRITIQUE

The settlement offer was submitted to FERC with 401 state water quality certification revisions to supersede parts of NEP's original license application. At the time of this publication the offer was approved in the final environmental impact statement (FERC 1996); new license conditions are expected to be issued soon. The settlement purports to resolve all issues among the signing parties regarding fisheries, fish passage, wildlife, water quality, project lands management and control, recreation, and aesthetics associated with the Deerfield River projects operated by NEP. The Vermont Natural Resources Council elected to challenge the revised settlement offer through an appeal of the state of Vermont 401 water quality certification. New England Power estimated that the additional flows and reservoir management requested by this organization would raise the total cost of energy lost from $20.1 million to over $48 million while achieving an incremental habitat gain for adult trout of approximately 106,027 square feet over the settlement offer. In March 1997, NEP and the state of Vermont signed an agreement whereby NEP agreed to put all of its Vermont project-related lands in a permanent rather than a term conservation easement and gave the state a 6-month option to buy some of the land associated with the Somerset Reservoir and the reservoir itself. In return, Vermont Natural Resources Council dropped its appeal.

The settlement process did not guarantee or achieve resolution to the satisfaction of all parties, but it provided a common voice for the majority of interests, giving both the applicant and participating parties a higher degree of certainty about the terms of the final license to be issued by FERC. As important, it allowed the participating parties to shuffle priorities and balance solutions at the watershed level to maximize the best good, rather than forcing FERC to pass judgment between two warring parties.

Balancing between conflicting interests in watershed restoration also forces parties to determine whether the objective is watershed restoration or rehabilitation. In the Deerfield watershed, restoring the riverine ecosystem's function and structure to its original state could not be accomplished because of the irreversible changes caused by human impacts. It would also have meant no practical settlement process with NEP because of the economic impact. Furthermore, much of the region now favors the continued existence of the major reservoirs, albeit without the major drawdowns. The presence of nonnative fishes, including the brown and rainbow trout, smallmouth bass, and landlocked Atlantic salmon, in the reservoirs is now an economical asset to the region.

Functional ecological rehabilitation—creating a system similar but not equivalent to the natural one—is a realistic alternative in the Deerfield River to accommodate the presence of humankind. This course was followed. Functional rehabilitation can be achieved by returning sufficient flows to currently dewatered bypass reaches. Restoring natural ecosystem structure, that is, the original aquatic species composition in the river, is more problematic. Records of what the river's biota was like before the human-induced degradation occurred are not well documented, with the possible exception of fishes. Species may be extinct. Existing fishery management objectives favor species that are not all native to the river or region, some of which compete with the native fish fauna. Conditions in the settlement offer, however, are designed to rehabilitate some elements of the Deerfield's original biota, for example, anadromous Atlantic salmon and native, self-reproducing brook trout.

The Deerfield relicensing highlights the convergence of a number of factors, including the impending expiration of a hydropower license, a utility company committed to assuming an environmental leadership role, the encouragement of FERC to reach a negotiated settlement, a relatively united coalition of environmental and recreational organizations, and local governments that were willing to participate in the process. This convergence required a low-profile settlement process and no press contacts until completion, a willingness to overcome the tendency for parties to become polarized or be dogmatic, the development of creditable scientific information to sort out options and their implications, and a willingness to share the credit for success. Utility companies are also somewhat unique in that they may pass the costs on to consumers in a way that firms in many other industries cannot; with the ongoing deregulation of the electric industry, however, this difference may be less of a factor in the future.

PROSPECTS FOR THE FUTURE

The prospects of the major elements of the settlement offer becoming either terms in NEP's next license or sidebar agreements with the signing parties are excellent. In August 1996, FERC issued the final environmental impact statement (FERC 1996). The FERC staff recommendations in the final environmental impact statement parallel the settlement offer terms, with several modifications in flow and reservoir-level management to meet additional conditions set in the Vermont 401 water quality certification, which was issued after the signing of the settlement offer. One major exception was that FERC did not adopt the $100,000 Deerfield enhancement trust fund. Though commending the objectives of this fund, FERC cited it impractical to oversee NEP's participation in a fund to carry out future projects and services that may not be within FERC's jurisdiction. The existence of the fund is not jeopardized, however, because FERC reversed itself and incorporated the fund in the final license issued.

On 14 June 1996, FERC also opened for license amendment NEP's Bear Swamp Project (FERC project number 2669) to incorporate the settlement offer terms for enhanced flows in the Fife Brook section and to include 1,257 acres of land in a conservation easement. The final license conditions were being issued when this manuscript went to press.

Although historically reticent toward settlement agreements, FERC, in part out of fear for losing its jurisdictional turf, has shifted dramatically during this decade and now openly encourages such settlements. The Federal Energy Regulatory Commission is offering more guidance in how to craft settlement terms and conditions so that they meet FERC's regulatory needs when issuing new licenses. Settlement negotiations and even collaborative relicensings in New England and across the United States are becoming more common between environmental interest groups, natural resource agencies, and applicants in post-1993 hydro-electric relicensings (Bowman 1996). Examples are ongoing cases in the Androscoggin, Kennebec, and Connecticut river basins in New England with Central Maine Power Company, Union Water Power Company, International Paper Company, and NEP. New England Power has also received numerous calls from other hydroelectric license applicants regarding its perspective on the settlement process and benefits (C. Kapala, NEP, personal communication). The Deerfield settlement process is becoming a nationally recognized example of how watershed rehabilitation can occur amicably, while recognizing the role and value of hydroelectric power. It has received considerable national press, including feature articles in the *New York Times* (28 November 1995) and the *Wall Street Journal* (20 May 1996).

ACKNOWLEDGMENTS

Essential funding for the Appalachian Mountain Club to participate in the FERC hydroelectric relicensing process on the Deerfield River was provided by the Pew Charitable Trust, Jessie B. Cox Charitable Trust, Town Creek Foundation, a Switzer grant through the New Hampshire Charitable Foundation, Gorczyca Charitable Remainder Trust, the Rockefeller Philanthropic Collaborative, an anonymous foundation, and Appalachian Mountain Club's supporting members. The settlement process described resulted from the hard work and dedication of the involved staff of the U.S. Fish and Wildlife Service, U.S. Environmental Protection Agency, and U.S. National Park Service, the Massachusetts Departments of Environmental Protection and Environmental Management, the Massachusetts Division of Fisheries and Wildlife, American Rivers, American Whitewater Association, Conservation Law Foundation, Deerfield River Compact, Deerfield River Watershed Association, New England FLOW, Trout Unlimited, and New England Power.

*There is an irreducible body of knowledge
that all students should know, including
how the earth works as a physical system,
basic knowledge of ecology and
thermodynamics, the vital signs of the
earth, the essentials of human ecology, the
natural history of their own region, and the
kinds of knowledge that will enable them to restore natural
systems and build ecologically resilient communities and economies.*
—David W. Orr, 1994

CHAPTER 13

INVOLVING LOCAL SCHOOL SYSTEMS IN WATERSHED RESTORATION: CROOKED RIVER OF OREGON

David A. Nolte

Five years ago two visionary teachers in the Crook County School District, located in Prineville, Oregon, recognized that education could play an important role in the recovery of the Crooked River basin. Both teachers had grown up in the watershed. Through their own life experiences, they had seen a decline in resource values that were important to them—water quality and quantity and recreational opportunities, particularly trout angling. They recognized that "to have stewardship, we must experience it" (T. Huntley, Crook County High School, personal communication).

As these two teachers shared their educational concepts with local, state, and federal staff, private landowners, and businesses, it was apparent that there was no coordinated effort to provide Crook County students and adults with hands-on, real-world learning experiences in the natural resources. Their challenge, then, was to create a framework in which people from the watershed could come together for a restoration project that also provided meaningful

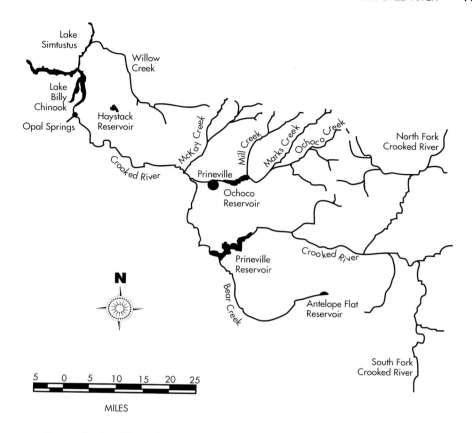

FIGURE 13.1.—Map of the Crooked River watershed in central Oregon.

educational opportunities. This framework would avoid a disjointed watershed recovery process that lacked a long-term foundation in the local community.

THE WATERSHED

The Crooked River watershed (Figure 13.1) is composed of more than 4,300 square miles and is an important tributary to the larger Deschutes River drainage system. The mean annual discharge of the Crooked River into the Deschutes River (at Lake Billy Chinook Reservoir) is approximately 1,131,000 acre-feet. Located partly in the Ochoco Mountains, the watershed is characterized by a semiarid climate with annual precipitation ranging from 10 to 31 inches. Precipitation in the summer may be short-duration, but very active, thunderstorm cells that result in strong hydrologic events.

The drainage includes three dams, Opal Springs, Ochoco Dam (Ochoco

Reservoir), and Bowman Dam (Prineville Reservoir), as well as numerous private impoundments on tributaries. All dams are barriers to fish migration. There are also more than 700 unscreened irrigation diversions currently in the basin (A. Stuart, Oregon Department of Fish and Wildlife [ODFW], personal communication).

The headwaters area of the Crooked River contains bedrock consisting of primarily tuff and lava flows, which results in fairly rugged topography characterized by steep canyons. Much of the soil is very vulnerable to erosion due to the high clay content, steep slopes, and general lack of vegetative cover (USBOR 1992). The Crooked River watershed exhibits the characteristics of a watershed that is not safely capturing, storing, and slowly releasing water. Reservoirs exhibit a high rate of sedimentation. Short-duration thunderstorms result in strong hydrological events, often delivering flows of up to 15,000 cubic feet per second into Prineville Reservoir (Stuart, personal communication).

Historically, the Crooked River supported runs of spring chinook salmon and summer steelhead (Nehlsen 1995). Chinook salmon, first noted by Peter Skene Ogden in 1826 (Ogden 1950), were present in the Crooked River basin until the mid-1960s. Oregon State Game Commission (OSGC 1951) reported that anadromous fishes migrated throughout the entire Crooked River during late winter and spring, when flows were high. Summer steelhead were present until the early 1960s, when upstream and downstream passage were eliminated by the Pelton and Round Butte Dam complex on the Deschutes River. The U.S. Bureau of Commercial Fisheries (draft letter to area engineer, U.S. Bureau of Reclamation, 1966) estimated that the Crooked River basin could produce over 3,000 steelhead and 2,000 spring chinook salmon if water quality and quantity were available. However, accurate estimates of historic population levels are lacking.

Today the principal game species in the basin is the Columbia River redband trout. Other native species present include mountain whitefish, chiselmouth, torrent sculpin, mottled sculpin, and speckled dace. There are other game and nongame introduced species in the basin.

The redband trout found in the Crooked River is part of the Columbia River basin group, which ranges from northern California to eastern Oregon and Washington (Behnke 1992). Locally, Columbia River redband trout in the Crooked River adapt to highly variable stream conditions, and recent genetic studies have suggested that the Columbia River redband trout in Crooked River have diverged from other inland rainbow trout groups (Currens 1994). Inland rainbow trout survive under harsh conditions, such as high water temperatures, lower dissolved oxygen levels, high turbidity, and large fluctuations in streamflow (Behnke 1992). These fish adapt to changes in their habitat both by colonizing new areas of stream reaches previously inaccessible due to drought and by increasing in size and numbers during improved water years (S. Theisfeld, ODFW, personal communication).

The current status of Columbia River redband trout in the Crooked River basin is depressed. This is due, in part, to habitat degradation by the effects of past and present water and land management practices, including dams, livestock grazing, road construction, timber harvest, water withdrawals for irrigation, unscreened diversions, and occasional fish kills by hazardous materials (ODFW 1994).

The majority of the Crooked River watershed lies in Crook County, population 14,111. Land in the basin is split about evenly between private and public ownership. Most of the lower-elevation areas are privately owned, whereas most of the upper elevations and forested areas are publicly owned. About 73% of the land use in the Crooked River watershed is devoted to range, 21% to forest, 4% to irrigated agriculture, and 2% to urban and other uses. The economy of Crook County is based on farming, ranging, manufacturing, milling lumber, and providing outdoor recreation opportunities (USBOR 1992).

There is one state park at Prineville Reservoir and Crook County operates one park at Ochoco Reservoir. In 1988 portions of the North Fork Crooked River and Crooked River main stem were designated as a National Wild and Scenic River; the South Fork Crooked River has been designated for further study as a Wild and Scenic River and is presently within a wilderness study area.

ORIGINS OF THE CROOKED RIVER ECOSYSTEM EDUCATION COUNCIL

In 1988, the Oregon Rivers Council (now Pacific Rivers Council) organized a watershed workshop for residents of the Crooked River area. The workshop provided a forum for a large variety of stakeholders to come together to discuss aspects of watershed health and potential approaches to watershed recovery.

In 1990, two teachers at Crook County High School envisioned a watershed level program based on the Crooked River basin to bring natural resource education to students. By taking students into the watershed for hands-on learning experiences, the students' appreciation of the watershed—and their commitment to protect it—would grow. "What we're trying to do is to allow students to discover the problems we have in the watershed and help them identify and implement solutions. These students are the future," relates Tim Huntley, who, along with Chuck Holliday, originated the concept. Together they envisioned a coordinated program in which experts from the community would provide technical and financial help on specific projects, materials and lectures for the classroom, and fieldwork opportunities for students.

The two teachers began to recruit support for the concept and to involve partners. They first approached the Ochoco Chapter of Trout Unlimited. "We asked the Ochoco Chapter for support mainly because they were active in watershed restoration projects in the basin and were members of the community," Holliday explains. "They had volunteers involved in local restoration projects and through their national office had opportunities for funding support." An added benefit to this initial partnership was the good working relationships the local chapter had established with two major federal land management agencies in the area, the U.S. Forest Service's Ochoco National Forest and the Prineville District, U.S. Bureau of Land Management.

The next step was to recruit and involve these two agencies in the program.

Fisheries biologists from both agencies were approached, and they enthusiastically supported the program. Finally, an Ochoco Chapter member of Trout Unlimited was recruited to serve as fundraiser for the effort. This person, in turn, also enlisted the support of the ODFW. This core partnership provided the platform to launch the Crooked River Ecosystem Education Council (CREEC).

There were several circumstances that facilitated the launch of CREEC. The state of Oregon was undergoing a massive and fundamental change in its education program. Recent legislation known as the Oregon 21st Century Education Act strongly encourages private and public partnerships, business participation, and real-world, hands-on learning in public education. The Crook County School District was in the initial stages of changing the way it educates students, which opened doors for CREEC's education concept to establish itself because it offered both financial and technical support for fostering school–business partnerships that met the intent of the legislation.

Some unique funding opportunities also appeared. In January 1993, CREEC submitted a proposal to Oregon's Governor's Watershed Enhancement Board. This first CREEC proposal was rejected because the Board believed the project lacked sufficient community support. Responding to this criticism, CREEC rewrote the proposal to refine concepts and recruit additional community partners, including Oregon State University Extension Service, Oregon Water Resources, Crook County Soil and Water Conservation District, and several private landowners.

The Crooked River Ecosystem Education Council also requested support from the then-new Bring Back the Natives program, which could provide challenge match funds from the National Fish and Wildlife Foundation by forming a partnership with the U.S. Bureau of Land Management, U.S. Forest Service, and Trout Unlimited. Bring Back the Natives program administrators approved the proposal and have provided funding based upon yearly proposals since 1993: key sustained support over time. The ability to match nonfederal dollars and contributed goods and services greatly facilitated the growth of the CREEC program in the school system. The Bring Back the Natives grant also stimulated CREEC's opportunities with other grants and foundations.

In addition to the Bring Back the Natives funds, additional grants have been received from Trout Unlimited, Phillips Petroleum Corporation, Governor's Watershed Enhancement Board, Wildlife Forever, Coors Pure Water 2000, Thor-lo Corporation (a manufacturer), and the Oregon Department of Fish and Wildlife Restoration and Enhancement Program. Local funding has come from the Crook County School District and everything from bake sales by schoolchildren to donations of cash, professional services, and materials by local businesses and private landowners.

It was obvious to CREEC that an educational partnership needs the full support of the participating school district, teachers, and staff, so the first attempts at marketing the CREEC concept began with the Crook County School District, school board, and teachers. In order to proceed, a common vision and mission statement for the education program was developed by CREEC in early 1992:

[To] provide watershed-based educational curriculum for the Crook County School District that covers classroom lessons, field studies, and work experience environments for all students as well as yearly, renewable watershed rehabilitation projects designed to enhance the Crooked River basin and increase public awareness of the watershed's importance. (Unpublished minutes from January 1992 CREEC monthly meeting.)

Partners of CREEC also made several critical decisions:

1. the program would focus primarily on the school district and secondarily on the community;
2. the program would last a minimum of 10 years;
3. the program would actively and continually seek to recruit additional partners both inside and outside the community; and
4. by brokering partnerships to provide technical and funding support, CREEC would subsidize the watershed level environmental education.

All of the founding partners recognized the need for organizational structure, which was important to support project management, fund-raising, and communication to the community. Several meetings were devoted to creating a council designed to accommodate the exchange of technical and educational information, support project management, and provide a general mechanism for implementation of program activities.

The Crooked River Ecosystem Education Council is organized informally into technical and education committees. Partners recruited into the program may be represented on either one or both committees. At monthly meetings, decisions are made regarding project activities, expenditures, training, outreach, or other program-related activities. Each committee provides necessary technical support, sources of educational materials, and information, such as a grant lead, agency opportunity, or business contact. Monthly minutes are sent to each CREEC partner and each committee member along with an agenda for the next meeting. The school district business manager volunteers time to manage the accounting information and funds for the program.

Outside organizations, agencies, businesses, and citizens are invited to participate in specific activities. For example, the City of Prineville provided personnel to discuss the new water treatment facilities and golf course with Crook County Middle School students during the National Geographic Water Awareness Week. There have been guest lectures by the Oregon State Police on natural resource ethics and a presentation by Ochoco National Forest personnel on geographic information systems. These types of programs for the school district and community provide members of CREEC with information about the latest natural resource management concepts and tools. Experts in forestry, private land issues, rangeland management, and volunteer restoration programs from other parts of the state are recruited to serve as guest lecturers. To date, CREEC partners include the U.S. Bureau of Land Management, U.S. Forest Service, ODFW, Oregon Water Resources, Oregon State Parks, Madras and Culver school districts,

Telephone Pioneers of America, Trout Unlimited, Oregon State University Extension Service, Crook and Jefferson Counties Soil and Water Conservation Districts, Ochoco Irrigation District, and the High Desert Learning Center.

ACTIVITIES OF THE CROOKED RIVER ECOSYSTEM EDUCATION COUNCIL

A principal objective of CREEC is to bring professionals from agencies, businesses, and the local community both into the classroom and out in the field to expose students, teachers, and parents to hands-on, real-world learning situations. For example, in 1993 a Crooked River watershed tour was sponsored for 31 teachers and was held at the Williams Prairie project site. This watershed tour provided guest lectures at the site by the U.S. Forest Service, U.S. Bureau of Land Management, ODFW, Oregon State University, and private landowners. Teaching the teachers about watershed management concepts and the sustainable use of natural resources—both consumptive and nonconsumptive— through interaction with agency experts and the private sector helps leverage partners' contributions and provides partners with a trained labor pool to assist in restoration projects.

Watershed restoration projects in the basin provide a dual opportunity for both education and watershed recovery. Project locations may be at remote reaches of the watershed or in town. For example, a readily accessible urbanized portion of Ochoco Creek, flowing through the City of Prineville, is used as an instream classroom for all students. The Crooked River Ecosystem Education Council has embraced the philosophy that it is important to involve students, teachers, and the community in as wide a range of natural resource management activities as possible. Some are designed to serve as yearly activities, so that teachers and staff can institutionalize the activity, whereas others are one-time events. All grade levels and skills, including special education, home study, and at-risk students, participate in events. This way, teachers have options and a variety of restoration and education opportunities that permit them to fit the CREEC program into their busy schedule.

Restoration of Williams Prairie

This is a restoration activity in partnership with Ochoco National Forest, Trout Unlimited, and ODFW. Objectives focus on restoration of a wet meadow complex that had been acquired by the U.S. Forest Service and is located near the headwaters of the North Fork Crooked River. This meadow area had been altered by livestock overgrazing and other land use practices contributed to severe undercutting by the spring-fed river (Figure 13.2). Restoration of the meadow complex and the elimination of the undercutting will result in improved water quality and quantity and better habitat for Columbia River redband trout. Project work began in July 1993 with Crook County High School students assisting in baseline field surveys that would help locate appropriate sites for restoration activities. Initial restorative efforts focused on a short-term fix to slow down or eliminate the in-channel undercutting through a series of small loose-rock check

FIGURE 13.2.—Student survey teams working at Williams Prairie, Ochoco National Forest, Oregon.

dams and riparian (streambank) fencing. Students and other volunteers assisted in riparian plantings at the site. Cattle are grazed now for only a short duration.

Ochoco Creek Habitat Improvement

Crook County middle and high school students are working annually on restoration of Ochoco Creek where it flows through the city limits of Prineville. Students have been responsible for designing a landscape plan to stabilize eroding and compacted banks, cleaning trash from the creek, planting trees and other riparian vegetation, installing barrier-free fishing and access opportunities, and creating interpretative signs (Figure 13.3). So far, a general landscaping plan has been created by the Crook County High School agriscience students. Several riparian planting efforts along school property were completed by Crook County Middle School students. Yearly creek and community cleanups also have been completed by the middle school students. Crook County High School vocational art students have assisted in construction of a large greenhouse used to propagate native plants, one fishing pier and viewing platform that has handicap access, a stream gauge located on high school grounds, and interpretive signs.

Downstream at Lake Billy Chinook, Culver High School students worked with Cove Palisades State Park personnel and Trout Unlimited to complete a three-panel interpretive display on fish species. The student team adopted a design and fabricated the kiosk for the project.

Restoration of Dixie Meadows

The Crooked River Ecosystem Education Council is supporting a private landowner in the restoration of Dixie Meadows and Lawson Creek, a tributary of

FIGURE 13.3.—In a CREEC-supported project, Culver High School students built an interpretive kiosk of fish species of Lake Billy Chinook Reservoir.

Ochoco Creek. The landowner is interested in eliminating early erosion problems on the meadow, improving riparian vegetation, and developing off-site water sources for livestock. Students are assisting in site mapping by means of field sketches and global positioning satellite technology, photopoint documentation (a photographic record of the project over time at established locations) and riparian plantings. The landowners also supported a CREEC student intern in a summer project.

Ochoco Creek Instream Classroom

Where it flows through the city of Prineville, Ochoco Creek serves as a convenient living laboratory for students at Crook County middle and high schools. Students are responsible for developing management plans, implementing restoration projects, and monitoring and communicating results. Although these are small-scale activities when compared with other restoration projects, the site provides most students with their first introduction to watershed management principles and practices. It also provides students with the fundamentals of project planning, logistics, and publicity. Using CREEC curriculum guidelines and support equipment, teachers can schedule nearby field days at the creek while being supported by local agency professionals. For example, a teacher at the middle school contacted a CREEC partner to request assistance for a discussion about water and watersheds. The partner contacted an archaeologist from the U.S. Bureau of Land Management, a fishery biologist and hydrologist from the U.S. Forest Service, a range conservationist from Oregon State University Extension Service, a private landowner, a representative from the local irrigation company, and a member of Trout Unlimited. It is important to note that each agency provided staff time for preparation as well as presentation. This requires strong commitment and support from agency managers.

Ochoco Creek Stream Gauge

The Governor's Watershed Enhancement Board and the Bring Back the Natives program provided funding to install a stream gauge, which was fabricated and operated by Crook County High School students in cooperation with the Oregon Water Resources Department. State agency personnel serve as mentors for the students, who download data, analyze them, and report results to the Water Resources Department. Two elementary schools, Paulina, located in an outlying region of the watershed, and Ochoco, located in Prineville, were recruited to operate two small weather stations provided by this grant. With this activity, students learn about water management, which is vital to farming and ranching businesses in the basin. Elementary grade level students are introduced to the concepts of weather and the water cycle, which sets the stage for further watershed educational opportunities as they progress through grade levels.

Oregon Department of Fish and Wildlife Restoration and Enhancement Program

The ODFW has been funding restoration and enhancement opportunities through an agency grant program. The Crooked River Ecosystem Education Council has joined the ODFW Restoration and Enhancement Program, Ochoco National Forest, and private landowners to apply for funding to support restoration projects in the Crooked River basin. The Bring Back the Natives challenge grant process is being used to match these contributions to generate additional funds for the project. This funding provides an excellent incentive for private landowners and other stakeholders in the watershed to become involved directly in the restoration process. Current restoration and enhancement projects include (1) fencing Williams Prairie, which helps maintain a grazing allotment on the prairie while improving riparian conditions; (2) providing an ODFW employee to assist in habitat restoration projects for private and public stakeholders and in educational undertakings for CREEC (Crooked River Partnership); (3) providing fencing and fish diversion screens for private landowners, which supports riparian fencing projects and includes fish screens for unscreened diversions; and (4) involving Crook County High School students in construction of recreational facilities at Antelope Flat Reservoir, including a new dock and trail system.

Student Intern Program

In 1994, CREEC launched a pilot program with the Ochoco National Forest and ODFW to hire two high school students to work on a summer research project. The interns were selected through a formal job application and interview process. They completed 2 months of field work under direct supervision of a

FIGURE 13.4.—Crook County High School agriscience classes are assisting the U.S. Forest Service and U.S. Bureau of Land Management riparian restoration program by propagating native black cottonwood cuttings in the high school greenhouse and by planting cuttings on selected streams in the watershed. Construction of the high school greenhouse was an early project of CREEC.

professional biologist and spent 1 month writing a technical report and preparing oral presentations to the sponsoring partners and the community. In 1995, the intern program expanded to three full-time and three part-time students supported by additional partners from Jefferson County Soil and Water Conservation District, Dixie Meadows Corporation, and Oregon Water Resources Department. The program has become an opportunity for the students to gain valuable work experience, learn from professionals, and improve their communication skills. Future intern opportunities include support for a new black cottonwood genetic seedbank program in which native cottonwoods will be collected, typed, cloned, and propagated for future riparian restoration (Figure 13.4).

Watershed Education Curriculum Development

The Oregon 21st Century Education Act provides for Certificates of Advanced Mastery in at least six broad occupational categories, including natural resource systems. The Crook County School District staff, with guidance from CREEC education and technical committees, is developing curriculum guidelines for watershed education. This curriculum is being developed by adapting a wide range of off-the-shelf educational materials from state and federal agencies and private foundations, such as Project Wet, Stream Scene, and Project Aquatic Wild, with other sources of technical and educational materials. Teacher teams were created and funded by CREEC to assist in this curriculum adaptation, more importantly to stage teacher workshops in which the materials were presented directly to the staff (teachers recruiting teachers), and to assist in its implementation.

Fish Fest

This is the largest single event now staged in the school district. The Fish Fest is an annual event sponsored by the Ochoco National Forest in partnership with CREEC. The Fish Fest is an important symbol of the effectiveness of the partnership because it involves local community, state and federal agencies, private citizens, businesses, and outside organizations in the event. In 1995, 1,000 kindergarten to third-grade students and 100 adults attended the event. The Fish Fest includes stations at which students learn about riparian zones, geology, aquatic insects, fish life cycles, and biodiversity (the variety of life and its processes) through presentations, art, interpretative dance, games, and other fun activities (Figures 13.5, 13.6). Crook County High School students are recruited as assistants. In April 1996, a new Fish Fest was "spawned" for Culver School District, which is located near Lake Billy Chinook Reservoir. Over 300 students, in kindergarten through sixth grade, participated, and stations were run primarily by 50 Culver High School students.

ADAPTIVE MANAGEMENT

The diverse partnership and commitment of CREEC has brought together a cooperative effort to develop natural resource awareness and involvement. Both restoration and education efforts are guided by the CREEC technical and education committees. These committees take advantage of partner technical knowledge, experience, existing management plans, and resources in determining which types of activities are appropriate for CREEC.

By meeting monthly and ensuring good communication of partners through monthly minutes, CREEC provides neutral ground, a place for partners to share their knowledge and experience for the benefit of the students and the resource. Decisions made regarding CREEC support of partnership restoration activities, for example, are determined after discussion and vote. Project monitoring ensures completion and provides the necessary feedback to improve upon that type of activity in the future.

Feedback on projects is provided by CREEC members to evaluate activities and make modifications where necessary during the life of the project. Through its technical committee, CREEC has available the technical expertise of biologists, hydrologists, range conservationists, foresters, and other specialists. Restoration projects are monitored by respective agencies by means of redd (nest) counts; riparian, stream, and plant surveys; photopoints; and water quality monitoring. This monitoring provides CREEC with a feedback mechanism that is science based and a means to adapt the knowledge to the task almost immediately. For example, the first year of the pilot student-intern program, adjustments were made in field duties, technical writing requirements, and work schedules based on observations by partners who supervised the interns. These adjustments were later discussed at subsequent CREEC monthly meetings by the education and

FIGURE 13.5.—Brent Ralston, U.S. Bureau of Land Management fisheries biologist, involving kindergartners in discussion with "Tammy and Tommy Trout" alongside Ochoco Creek during the annual Fish Fest.

technical committees, processes and criteria were refined, and new procedures were in place for the next summer intern season.

The local community also provides us with feedback to adjust CREEC tasks. This feedback comes both directly, through parents, and indirectly, for example, as letters to the editor in the local newspaper. A check on CREEC also occurs every time we submit requests for funding through the grant-writing process. This process helps to validate our mission, goals, and objectives.

Educational activities are monitored primarily through CREEC reports. Task reports summarize the accomplishments of a particular activity and aid in adjusting activities to respond to teacher and agency needs and in tracking volunteer involvement. Additional feedback is received from teaching staff and

FIGURE 13.6.—Frances Fish entertains students at the annual Fish Fest.

partners through mail surveys and personal contact during annual workshops and meetings.

In the 4 years that CREEC has been in existence, there have been many tangible results. The instream habitat and bank stabilization project on Williams Prairie has continued in over 20 miles of Crooked River tributaries. Management practices for grazing have been adjusted. Smaller-scale projects involving private landowners are increasing. Most of these include riparian plantings, fencing, and fish screen installation at irrigation diversions and will treat over 15 miles of tributaries. Landowners have been very receptive to involving students in restoration and recovery activities.

Over 2,000 students annually are involved with one or more CREEC activities. The Crooked River Ecosystem Education Council has been an effective tool in breaking down the barriers between schools. In many CREEC activities, high schools work with middle schools and middle schools work with elementary schools. At the state education level CREEC is well recognized as an excellent model for the natural resource career path of the Oregon 21st Century Education Act.

The community response to CREEC has been generally positive. The Fish Fest is an excellent example. More than 100 parents annually assist in this activity, and local businesses donate time, money, and materials.

CRITIQUE

The Crooked River Ecosystem Education Council is entering its fifth year in support of watershed level environmental education and restoration. Over the years it has become clear to CREEC that successful partnerships involve the several attributes.

1. Trust—under the common goals and mission statement, CREEC must foster an atmosphere of trust between partners and the community. Trust is built upon success of the program and commitment. The program began by tackling projects that were very small scale but had a high chance of success. Word-of-mouth success stories spread through teaching staff and school board and outward into the community. Trust is built through constant communication, the opportunity for involvement in any activity by partners and the community, and meeting obligations made in the program.

2. Commitment—partners must demonstrate their commitment to the restoration process and the community and be ready to support that commitment with labor, materials, funding, and technical knowledge. Commitment is the key ingredient necessary to sustain a long-term program. Restoring watersheds and enhancing public perception about the watershed are often slow and long processes. Partners are bringing technical support, funding, and labor to these efforts yearly. Commitment is symbolized by the large amount of volunteer time CREEC partners put into the program—more than 2,500 hours annually—and by financial support from CREEC partners.

3. Perseverance—having a good idea is only as good as the desire and efforts to keep the idea alive in spite of setbacks or problems.

4. Flexibility—all members try to maintain program flexibility to accommodate opportunities and changes. Successful partnerships are adaptable and take advantage of any window of opportunity that may arise. For example, one of the very first funding opportunities came from the Crook County Middle School, which had received information on a grant opportunity from Phillips Petroleum Company, "Phillips Environmental Partnership." On very short notice, CREEC successfully responded to this opportunity. Matched funds from Phillips Petroleum and Bring Back the Natives brought middle school staff into our program.

5. Communication—communication fosters trust and commitment. Frequent communication keeps the community and interested parties informed about CREEC's mission and accomplishments and helps garner additional partners, support, and funding. Monthly meeting minutes with detailed action items are provided to all partners. Press releases are prepared for local newspapers, radio, and television and encourage media coverage of field activities. The Ochoco National Forest has supplied information about CREEC to U.S. Forest Service Region 6, which has highlighted CREEC efforts in their annual publications and reports. Oregon Department of Fish and Wildlife publishes *Oregon Wildlife*, which featured CREEC landowner involvement. Several CREEC partners have written articles for environmental education magazines such as *Clearing* magazine for the Pacific Northwest. Presentations are provided to the Chamber of Commerce, service clubs, and other local organizations.

6. Initiative—in 1995, CREEC partners and students have volunteered well over

2,500 hours for CREEC activities. Tasks are often broken into smaller segments, and partners are encouraged to recruit new participants to assist. An atmosphere of success does tend to spark initiative in others.

7. Leveraging resources—CREEC strives to leverage all resources—people, materials, and funds—to better support the program. Because every cash dollar is difficult to raise, it should be leveraged with another grant, labor, or technical support. For example, a Trout Unlimited Embrace-A-Stream grant may be used as a match to the Bring Back the Natives grant. The same rationale applies to labor. A commitment of funding, time, and technical support from the U.S. Bureau of Land Management may be used to recruit other partners into a particular project.

Strengths

The program has five primary strong points. First, the principal strength of CREEC lies in the diversity of partners and stakeholders involved in education and restoration. A second strength lies in the cultural exchange process that occurs when diverse groups with differing opinions attempt to work together to solve a common problem.

A third strength lies in partnerships. Partners recognize the value of the program and realize the hard work involved in keeping it going. They are committed to the effort and provide technical and financial support.

A fourth strength of the CREEC program is clearly the sustainable funding that the Bring Back the Natives program offers. The restoration and education program generates more than US$50,000–60,000 in cash and contributed services annually to generate $15,000–20,000 of National Fish and Wildlife Foundation funds. This sum does not include the valuable contribution of labor by volunteers.

A fifth strength is the amount of communication and internal marketing between partners and the community. Often, state and federal public affairs people have provided important support in marketing the CREEC program and communicating to the public. In order to sustain program efforts, new partners must continually be recruited: people move, agencies change priorities, and needs vary.

Weaknesses

The Crooked River Ecosystem Education Council is not a true watershed council. It has no legal authority nor is it completely connected to all stakeholders and issues. The often more-difficult socioeconomic issues within the watershed can seldom be addressed. After 4 years, the level of private landowner involvement lags behind expectations. When CREEC meets, almost every idea is "rubberstamped" or approved. Perhaps CREEC needs to focus its efforts more clearly and discuss matters longer. The Council must do a better job of understanding the needs, goals, issues, and perspectives of all partners and find a framework in which to incorporate these perspectives into the program. Public land management agencies do not always accommodate change very quickly.

Their organizational makeup may reflect a high degree of inflexibility, especially in accommodating new roles such as ecosystem management or public partnerships, and this inflexibility can be very frustrating to outside partners.

Finally, most restoration efforts are very small projects—almost "Band-Aid" in nature. The Crooked River Ecosystem Education Council has not taken on larger restoration activities. Larger sources of funding are hard to come by, and expanded levels of activity have been beyond its scope. And, as with any group, while CREEC members share the common goal of watershed restoration and education, how to get there is often debatable. Each partner may have a different need, perception, or goal regarding the process.

PROSPECTS FOR THE FUTURE

As the CREEC program enters its fifth year, several obstacles loom. Due to congressional budget cuts, federal land management agencies are undergoing personnel and funding reductions for key program areas. Likewise, the National Fish and Wildlife Foundation is experiencing the same reductions in funding. Therefore, the ability to sustain community partnerships in watershed restoration may be critically impaired. Watershed restoration, in reality, is a continual and dynamic process. The land degradation and previous management processes that have occurred in some cases over 150 years cannot be simply reversed by a short-term project; 50 or 100 years of restoration effort may be necessary.

One of the goals of CREEC is to expand undertakings with existing partners. For example, CREEC is planning to involve new disciplines within the U.S. Forest Service. In 1996, CREEC started working with the Ochoco National Forest and U.S. Bureau of Land Management to support a genetic seedbank operation for native black cottonwood. This activity will expose students to plant genetics and large-scale nursery operations. Another planned project is to use the technical expertise of Ochoco National Forest silviculturists and fire ecology staff by bringing students to a prescribed fire area located in the Mill Creek Wilderness. The Crooked River Ecosystem Education Council is also looking outside the "pure" natural resource management agencies by directing partnership efforts at local and regional businesses, including local television studios and software development and engineering firms.

If CREEC is to sustain itself, it must continue to add new partners in the restoration process, which will bring in new levels of experience, technical knowledge, contacts, sources of materials, and funding opportunities. Concepts like biodiversity, sustainability (use of natural resources at a rate that resources can be sustained through time without decline), and ecosystem management (integrating science and social values to achieve broadscale ecosystem integrity) will always be a fundamental part of CREEC. The group may want to explore further how to include socioeconomic issues in the process.

As levels of funding and support change, CREEC will be affected. But the fact remains that the work of CREEC must continue: providing students in the Crooked River basin opportunities to work with natural resource professionals, affect change in the watershed through better land management practices and restoration activities, and increase their understandings of ecosystems.

ACKNOWLEDGMENTS

The Crooked River Ecosystem Education Council would like to acknowledge the technical and financial support of the following organizations and programs: the National Fish and Wildlife Foundation, U.S. Bureau of Land Management, U.S. Forest Service, Oregon Department of Fish and Wildlife, Oregon Water Resources Department, Trout Unlimited, Oregon's Governor's Watershed Enhancement Board, Coors Pure Water 2000 Program, Les Schwab, Crook County School District, the community of Prineville, Phillips Environmental Partnership, and Thor-Lo Corporation. The author would like to thank Dean Grover, Amy Stuart, Steve Theisfeld, Dave Heller, Jack Williams, and Dave Young for their support and reviews of this manuscript.

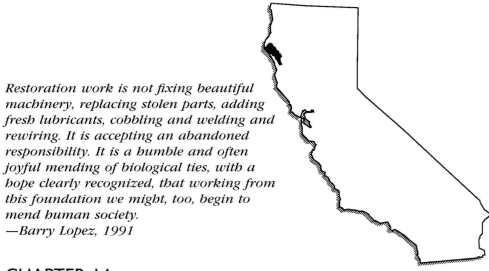

*Restoration work is not fixing beautiful
machinery, replacing stolen parts, adding
fresh lubricants, cobbling and welding and
rewiring. It is accepting an abandoned
responsibility. It is a humble and often
joyful mending of biological ties, with a
hope clearly recognized, that working from
this foundation we might, too, begin to
mend human society.*
—Barry Lopez, 1991

CHAPTER 14

THINKING LIKE A WATERSHED: MATTOLE RIVER OF CALIFORNIA

Seth Zuckerman

In clumps of two or three scattered up and down the banks, teenagers and a few grown-ups are hard at work next to Mattole Canyon Creek this February morning. Armed with digging bars, sledgehammers, and burlap sacks full of willow cuttings, they are replanting the riparian (streambank) forest that once thrived on this tributary of the Mattole River in northwestern California. Students and their teachers will plant more than 8,000 willow cuttings in the week that their high school will devote to this project in lieu of class time. The following spring, these cuttings will sprout and trim the barren gravel banks in green; by the time the younger students have graduated from high school, the willows will provide cover for juvenile salmon and steelhead in the creek and hold the banks in place against high winter flows.

The work these teenagers and their teachers are doing is just one facet of a larger attempt to restore the habitat on this creek, which is part of the larger effort to restore the Mattole watershed in one of the first citizen-initiated watershed restoration projects in North America. Inhabitants of the Mattole basin have been working to improve the ecological health of streams, fisheries, forests, and soils in the watershed for more than 15 years.

Initially inspired by the drastic decline in fall chinook and coho salmon runs, restoration efforts expanded as participants deepened their understanding of underlying ecological and social processes. At first, restoration activists were "thinking like a salmon," and focused on spawning habitat for adult salmonids. Our perspective grew to include aquatic habitat in general, and we were "thinking like a stream." Soon we realized that the entire basin influences instream conditions, so we began "thinking like a watershed." Most recently, we have begun "thinking like a watershed community" by broadening our work to address ongoing land use practices, hoping to prevent further deterioration. This work relies on residents' concerns, observations, and hard work, leavened with technical expertise from nearby research institutions, government agencies, and private consultants.

Although our projects have not always had their intended effect, it has become apparent to us that natural processes of recovery are making themselves felt on the landscape and that we can work in small but significant ways to augment them. Our ongoing enterprise of restoration has made a difference in the way we, as well as some of our neighbors, interact with the landscape and in the ways we conceive of our role here. This change in attitude, from extractor or bystander to helpful participant, is one of the most significant accomplishments of restoration work. By changing land-use practices, this shift in perspective may bring about the recovery of much larger areas than we could work on directly and may prevent damage to other areas that remain biotically intact.

GEOGRAPHIC SETTING AND HISTORICAL CONTEXT

The Mattole River drains a 304-square-mile watershed in the Coast Ranges of northwestern California. The 64-mile-long river empties into the Pacific Ocean near the village of Petrolia, approximately 35 miles south of Eureka (Figure 14.1). River flows are fed by annual precipitation of 60 to 200 inches per year, almost all of which falls between October and May. Unmoderated by a seasonal snowpack, river discharge just upstream of Petrolia (U.S. Geological Survey station number 11469000) fluctuates between an annual low of 20 to 40 cubic feet per second (cfs) and a median high during the 45-year record of 40,000 cfs. The terrain in the watershed is generally steep and is underlain by highly fractured and sheared sedimentary rock, further destabilized by the high rates of tectonic activity in the area (Dengler et al. 1992; California Department of Water Resources [CDWR], Division of Resources Development, unpublished report, 1973). As a result, the terrain is prone to landslides, mass movement, and other forms of erosion. The watershed's natural vegetation, for the most part, is composed of various successional stages of the Douglas fir-hardwood forest (Sawyer et al. 1988). Areas near the headwaters include a substantial component of coast redwood, whereas some ridgetops and hillsides belong to the fescue and oatgrass coastal prairie (Heady et al. 1988).

Human inhabitation began with the Mattole and Sinkyone peoples, whose sustenance depended in large part on the river's salmon runs. With the advent of Euro-American settlement in the 1850s, many native residents died of introduced

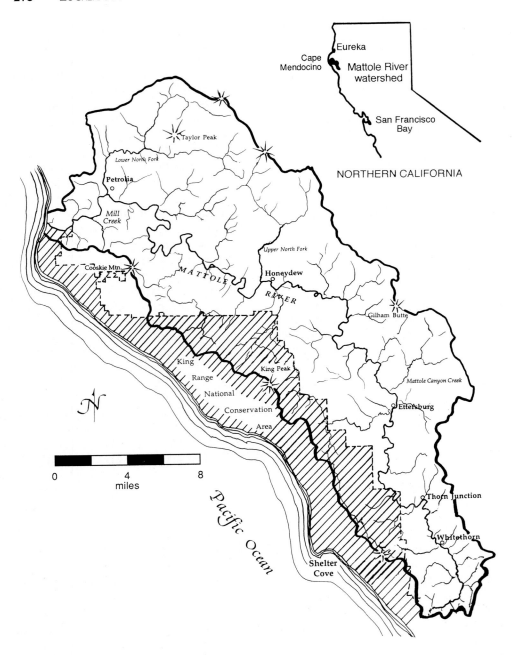

FIGURE 14.1.—Map of the Mattole River watershed in northwestern California.

diseases or were killed, and most of the survivors were displaced to distant reservations. The Euro-American settlers made their living by ranching and other forms of agriculture and continued to harvest a significant sustained yield of salmon. Sporadic resource extraction booms drew on small deposits of petroleum and on peeling the bark of tanoak trees for use in leather tanning.

After World War II, several factors combined to allow large-scale logging of the coniferous forests in the basin, which then covered about three-quarters of the watershed's area. The postwar construction boom raised the value of Douglas fir timber, making logging more attractive. Economic incentives encouraged landowners to cut their trees because they would otherwise be taxed on the value of their standing timber. Finally, the development of crawler tractors made possible road construction in formerly inaccessible areas. Between 1947 and 1988, more than 90% of the old-growth coniferous forest in the watershed was logged, and an extensive network of roads was constructed (MRC 1988). During the logging boom, the two largest floods in recent memory, in 1955 and 1964, brought about major changes in the watershed's stream morphologies. These streams already were destabilized by careless roadbuilding and such forest practices as the skidding of logs down stream channels (CDWR, Division of Resources Development, unpublished report). Three years after the 1964 flood, suspended sediment yields in the Mattole River averaged a staggering 16,370 tons per square mile of watershed area (Kennedy and Malcolm 1977), nearly one-quarter of an inch per year averaged across the watershed's area. A decade later, Griggs and Hein (1980) estimated erosion rates in the Mattole at five-hundredths of an inch per year, still two orders of magnitude greater than the typical rate of soil formation (Wahrhaftig and Curry 1967). Beginning around 1970, some large landowners subdivided their holdings into smaller, 40- to 200-acre homestead parcels, which led to a further increase in the density of the road network. All of these developments overtaxed the streams' ability to transport sediment through waterways, leaving sediment to clog streambeds that formerly provided high-quality spawning habitat.

The effects on salmon runs were severe. Based on surveys by the U.S. Fish and Wildlife Service in 1959 and the California Department of Fish and Game (CDFG) in 1964, the Mattole Salmon Group (MSG 1995) estimated historic spawning runs in the Mattole at 10,000 chinook salmon and 4,000 coho salmon. By the early 1980s, the estimated chinook salmon escapement (number of fish reaching spawning area) had fallen to 3,000, and it continued to drop to a low of 100 in 1990–1991 (Peterson 1996) under the influence of degraded spawning and rearing habitat as well as oceanic factors such as fishing pressures and availability of prey. Nehlsen et al. (1991) categorized the fall chinook salmon in the Mattole as at "high risk of extinction." Coho salmon escapements have fallen similarly and were classified as also facing a high risk of extinction (P. Higgins, S. Dobush, and D. Fuller, Humboldt Chapter, American Fisheries Society, unpublished data). Since then, populations appear to be recovering slowly (Figure 14.2).

In addition to chinook and coho salmon, at least 14 other species of fish have been observed in the lower Mattole since systematic surveys began in 1984

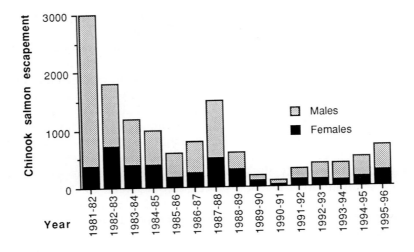

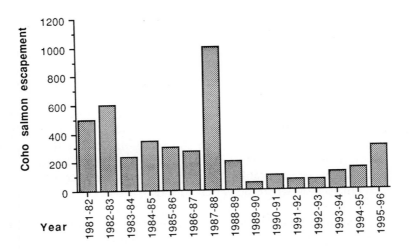

FIGURE 14.2.—Chinook and coho salmon escapements show modest recovery from historic lows. Escapement estimates are based on surveys by the Mattole Salmon Group that were compiled and analyzed by the group's fisheries biologist, Gary Peterson. These data are imprecise, relative estimates and as such are most useful for indicating changes or trends, they point to a decline in salmon runs until 1990 and a gradual increase thereafter (Peterson 1996).

(Table 14.1). Two populations of fish, green sturgeon and spring chinook salmon, known from historical records to have inhabited the Mattole, have not been encountered in the last two decades and are believed to have been extirpated (made locally extinct). A remnant run of summer steelhead was documented in 1996 by the sighting of a dozen of these fish holed up in cold pools (Simpson et al. 1996). Winter steelhead is the healthiest anadromous fish

TABLE 14.1.—Fishes observed in the lower Mattole River, California (derived from Busby et al. 1988; MRC 1995; and MSG 1995). Habitat given as freshwater (FW), marine (M), or freshwater and marine (FW-M). Anadromous species are marked with an asterisk.

Species	Habitat
Chinook salmon	FW-M*
Coho salmon	FW-M*
Steelhead	FW-M*
Pacific lamprey	FW-M*
Green sturgeon[a]	FW-M*
Coastrange sculpin	FW
Prickly sculpin	FW
Western brook lamprey	FW
Green sunfish	FW
Threespine stickleback	FW-M
Surf smelt	M
Redtail surfperch	M
Shiner perch	M
Walleye surfperch	M
Pacific staghorn sculpin	M
Speckled sanddab	M
Starry flounder	M

[a]Extirpated from basin.

stock in the Mattole, with escapements in 1994 and 1995 that may even have exceeded 10,000 fish (MSG 1995).

EVOLUTION OF THE RESTORATION EFFORT

The decline of the Mattole salmon fishery during the 1970s inspired initial restoration efforts in the watershed. Until then, many residents obtained significant quantities of food by fishing for salmon in the river, often with spearpoles, which were sanctioned at that time by local custom although prohibited by state law. The drop in numbers of returning spawners was widely noticed by local citizens. Government agencies did not assign a high priority to restoration of the Mattole because of the river's small size compared with the neighboring Eel and Klamath systems, its distance from urban areas, and its sparse human population, approximately seven persons per square mile. Citizens decided to take matters into their own hands and formed the Mattole Watershed Salmon Support Group, now renamed the Mattole Salmon Group (Salmon Group), to work on behalf of the Mattole's fall chinook salmon and coho salmon. In these early efforts, citizens put themselves in the salmon's scales and tried to ascertain what the fish needed: in effect, trying to "think like a salmon."

Salmon Group Launches Hatchbox Program

The main problem that seemed to face adult spawners was a lack of clean spawning gravels. The Salmon Group's first response, beginning in 1980, was to provide substitute spawning habitat. Being careful to preserve the genetic integrity of the Mattole chinook salmon run, which had never been contaminated with fish imported from other watersheds, the Salmon Group trapped returning spawners, held them until ripe, took eggs and milt from the fish, and incubated the fertilized eggs in small-scale streamside hatchboxes. This was the first grassroots group to be entrusted with such work by the CDFG.

The hatchbox program increased the egg-to-fry survival rate to better than 80%, from less than 15% in the degraded river (House 1990). Simultaneously, the Salmon Group and another community group, Coastal Headwaters Association, trained dozens of residents to survey and monitor streams for spawners, carcasses, and redds (nests). The resident surveyors' observations reinforced their realization that the days of fishing for chinook and coho salmon in the river were over for the time being, but other rewards compensated for that change. "It was incredibly exciting to go from hunter to helper," says Salmon Group board member Michael Evenson, recalling those early days. "None of us were prepared for the glee that came from making a wild animal sacred so you don't take it anymore."

Salmon have been trapped every year since 1980, with volunteers providing about 40% of the labor. Wages and supplies have been paid for with funding from the CDFG, foundations, and private donors. The program has been funded for six seasons by the Salmon Stamp Fund, a self-imposed tax on commercial salmon fishers that is allocated by a committee of trollers and administered by the CDFG (MSG 1995). Chinook salmon have been trapped and spawned every year but two since 1980; coho salmon have been trapped and spawned during approximately half the years the program has been in operation.

Restoration Workers Initiate Instream Projects

From the outset it was clear to the Salmon Group that the native stock hatchbox program could be only a stop-gap measure. No one wanted to midwife every fish born in the Mattole River. In the long term, the salmon needed the good habitat they had enjoyed in the past. Restoration workers began trying to "think like a stream" and undertook projects to make streams more hospitable to fish. For example, with funding from the CDFG, the Salmon Group installed scour structures to create deeper, better-covered pools in which juvenile fish could rear. The Coastal Headwaters Association placed riprap by hand along a stretch of the headwaters where the river was cutting into the banks and sending silt crumbling into some of the best spawning gravels in the river. Other projects sought to stabilize active landslides that were contributing sediment to watercourses or to anchor woody debris to the banks to add complexity to the habitat. Throughout the process, the Salmon Group worked in collaboration with state agencies such as the CDFG, California Conservation Corps, and State Coastal Conservancy, benefitting from their financial support, technical assistance, and contributions of labor.

Groups Broaden Focus to the Entire Watershed

Even as we began to work in the streams, it became clear that the picture would need to be bigger yet. To understand why the streams were in such bad shape, we had only to look up the hill to the slopes that were bleeding soil into the river. The entire watershed mattered, from ridgeline to ridgeline. In the mid-1980s, the Mattole Restoration Council (MRC) was founded to link the various restoration efforts already underway and to coordinate those efforts in the context of the whole basin and the processes that connect the basin's different parts. It was time to "think like a watershed." The MRC includes such varied groups as the Mattole Valley Community Center, a tree-planting cooperative, the Salmon Group, watershed associations organized around tributaries of the Mattole, two land trusts, and a community high school. In addition, more than 100 landowners and residents belong to the MRC as individual members.

Upslope erosion control.—One of the MRC's first ventures, launched in 1987 with funding from the CDFG, was a study of upslope sources of erosion throughout the watershed. Project leaders mapped the watershed into tributary drainages and, for each subbasin, compiled landownership data from the county assessor's office. About 85% of the Mattole watershed is privately owned, so access for studies and projects depends largely on the goodwill and interest of the landowners. By the same token, private landowners' land use practices have the greatest influence on the well-being of the watershed. Project staff recruited residents of each subbasin to map and categorize the erosional features and to help determine which sites were most important to address. Scientists from Redwood National Park, where restoration work had been underway since the park acquired 36,000 acres of cut-over land in 1978, came on weekends to train these citizen surveyors in aerial-photograph interpretation and mapping techniques. The result was a 47-page catalog of erosion sources and prescriptions entitled *Elements of Recovery* (MRC 1989). Advertised to every landowner and resident in the watershed, it sold 3,000 copies and spread the concept among those living in the watershed that problems upslope percolate downhill to influence the streams and the fishes that live in them. Many of the prescriptions identified in *Elements of Recovery* have since been implemented.

Of all the erosional features mapped in *Elements of Recovery*, more than three-quarters were related to roads. The study helped bring home to people of the Mattole watershed that roads change drainage patterns in ways that can seriously destabilize hillslopes and vastly increase sedimentation. Sometimes roads just need to be better maintained; in other cases, the best solution is to remove them. In the King Range National Conservation Area, the MRC closed and rehabilitated three and one-half miles of seldom-used road above one of the most intact tributaries of the Mattole for the U.S. Bureau of Land Management (USBLM). The half-million-dollar federally funded project relies on techniques and expertise developed at Redwood National Park, in which the same size equipment is used to obliterate the road as was used to build it (Zuckerman 1990). Bulldozers and excavators are used to pull up previously laid fill and place it against cut banks or somewhere it will stay put; stream crossings are dug out to their original contours; and as much topsoil as possible is retrieved and spread on the re-created surface. The result is a stabler landscape with much less potential to contribute sediment to the Mattole.

Old growth mapping and preservation.—At about the same time as the erosion study was conducted, the MRC produced a watershed map comparing the extent of old-growth coniferous forests in 1947 and 1988 (MRC 1988; Figure 14.3). Compiled by local residents relying on black-and-white aerial photographs and county and state records, the map showed that 91% of the softwood forest had been harvested since the local Douglas fir logging boom began after World War II. In light of how much had been cut, the MRC called for a moratorium on logging in the ancient forests "until systems of harvest are devised that maintain the structure and characteristics of old growth and all old-growth-dependent species" (MRC 1988). Since then, the MRC has commented on numerous timber harvest plans in the watershed, attempting to reduce the impact of logging. Two of MRC's member groups, Sanctuary Forest and the Mill Creek Watershed Conservancy, are land trusts that have protected significant portions of the remaining old growth identified on the 1988 map by arranging the purchase of groves from willing sellers. Funds have come primarily from state bond issues (Sanctuary Forest) and the USBLM land exchange program (Mill Creek).

Estuary study.—The MRC members have continued to study the watershed and refine the base of knowledge from which we act. In the late 1980s, research conducted for the USBLM by Humboldt State University students and faculty showed that juvenile chinook salmon suffer high rates of mortality in the lagoon that forms at the mouth of the river for a 2- to 4-month low-water period in the summer and early fall (Figure 14.4; Busby et al. 1988). In other river systems, chinook salmon that oversummer in estuarine environments before swimming out to the ocean constitute a disproportionately large share of returning spawners (Reimers 1971). For some reason, perhaps the lack of cool, deep water, riparian canopy, and woody debris, young Mattole chinook salmon were unlikely to survive the summer in the lower river. This lack of good habitat in the lower river was seen as a potential bottleneck to the recovery of the chinook salmon population. A 5-year study commenced, funded by the State Coastal Conservancy, that encompassed the biology, geomorphology, and hydrology of the lower river and extended to consider the effects of upslope conditions on the estuary–lagoon system and its surroundings. Local residents gathered and analyzed the data with the help of technical advisors recruited from natural resource agencies, Humboldt State University, and private consulting firms in the region. Graduate students and USBLM staff continued to monitor juvenile salmon populations in the lower river in collaboration with the Salmon Group.

Our most significant realization was that change is part of the natural order for the lower river, including channel migration, scour, and deposition. Project staff analyzed aerial photographs taken every few years between 1942 and 1992. Findings showed swimming holes now exist where orchards formerly were, and riparian forests stand in the former river channel. This view of the river shifting across the floodplain (Figure 14.5) reminded us of how the river might affect

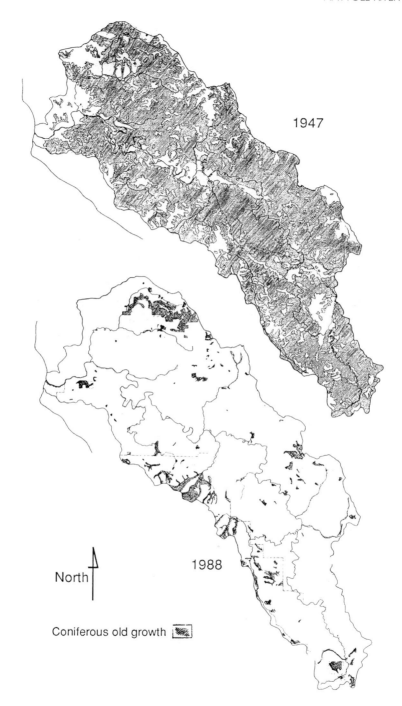

1947

North

1988

Coniferous old growth

FIGURE 14.3.—Nearly all of the watershed's old-growth forests were logged in 40 years. By 1988, the extent of old-growth coniferous forest diminished to just 9% of what it had been in 1947; more has been logged since 1988. Unlogged stands of fewer than 20 acres are not mapped. Because almost all logging was done with tractors, this map implies a tremendous increase in road density in the four decades. (Data from MRC 1988.)

FIGURE 14.4.—The Mattole River estuary as seen in August 1989. Historically, this estuary provided critical rearing habitat for juvenile chinook salmon during the summer. Losses of deep pool areas, shade, and large woody debris have contributed to elevated water temperature and die-offs of juvenile chinook salmon during the past decade. (Photograph by Jack Williams.)

future restoration projects. "The river giveth and the river taketh away," wrote Jan Morrison, then program director of MRC (MRC 1995).

The other landmark realization was that restoration work must be seen proceeding on different time scales simultaneously. Most of the root causes of the watershed's degradation lay far from the estuary. The causes included the entire landscape of cutover forest, active and abandoned logging roads, poorly maintained residential roads, and creeks suffering excessive impacts from livestock (Figures 14.6–14.8). Those causes needed to be addressed for the watershed to heal. At the same time, if we ignored the symptoms, we might lose key elements of biodiversity (the variety of ecosystems and species and the genetic variability within them) while we waited for the healing to work its way downhill—as sources of sedimentation were sutured and stream channels revegetated. Accordingly, we saw value in continuing efforts such as the hatchbox program (now in its 17th year) and initiating rescue rearing, in which the Salmon Group proposes to trap part of the migrating juvenile chinook salmon run and rear the fingerlings over the summer to make up for the lack of good estuarine habitat.

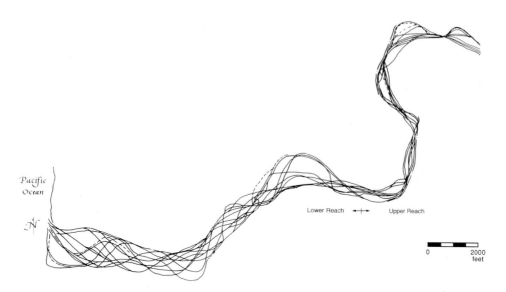

FIGURE 14.5.—Illustration of how the low-flow channel of the lower Mattole River has shifted position between 1942 and 1992. Each line denotes the channel position in 1 year during that period, based on a series of aerial photographs taken every 3 to 9 years. Tightly bunched lines denote little variation during the period of record. Dashed lines represent overflow channels. Note the wider range of variation in the lower reach compared with the upper reach, which is more strongly influenced by bedrock controls.

Residents See Themselves as Members of the Watershed

The erosion survey, the old-growth map, and the estuary study helped in what Jan Morrison called "claiming the watershed," that is, identifying the watershed as a unit of analysis and concern and assuming responsibility for its health. This was a powerful concept because until then no one had collected data on a watershed level or considered the system as a whole. Even the map of old-growth forests, although not strictly aimed at restoration, was significant because it helped more people see themselves as residents of a watershed, not just a state, county, or town.

From the beginning, we knew that the health of the watershed landscape would ultimately depend on the terrain of human consciousness. Most land management activities in the Mattole are unregulated; regulations that do exist, regarding timber harvest for instance, are often spottily enforced. That situation is unlikely to change in the era of underfunded government programs. The fate of the watershed depends on landowners and residents recognizing the value of good stewardship and practicing it on their own lands.

Private landowners control 85% of the land base, with most of the remainder managed for the public by the USBLM (Figure 14.9). Only one-third of the land in the watershed is owned by people who live on it. About one-quarter is ranched (some under absentee ownership), and about one-sixth is owned by four timber

FIGURE 14.6.—King Range Road removal in the upper Mattole River watershed. One of the primary causes of stream sedimentation is erosion from road crossings. Many culverts become plugged by debris during storms, which causes culverts to be blown out and road fill material to be washed into creeks. Photograph shows road being obliterated by the USBLM during August 1995. (Photograph by Thomas Dunklin.)

FIGURE 14.7.—Fill material being removed to restore natural stream channel. (Photograph by Thomas Dunklin of same site in Figure 14.6.)

FIGURE 14.8.—Restored stream channel after removal of road, fill material, and culvert. Riparian vegetation will be reestablished naturally. (Photograph of same sites shown in Figures 14.6 and 14.7, but taken during February 1996 by Thomas Dunklin.)

companies. Roughly one-quarter is owned by people who live elsewhere and use their Mattole holdings as a second home, private wildland, or future retirement home. Almost all of the public land is part of the King Range National Conservation Area, and much of it is under consideration for wilderness status. This diversity of landowners means that we must communicate the need for stewardship and restoration to a wide variety of audiences.

Some of our attempts have been more successful than others. We have come to enjoy widespread support from small landholders, absentee landowners, and a few members of the ranching and logging communities. We have a cooperative relationship with the USBLM. However, during the aftermath of antilogging protests in the region in 1990, known as Redwood Summer, some opposition to restoration arose because restoration seemed to threaten the livelihood of some in the watershed and to conflict with private property rights.

Alliances and comedy.—Two developments helped heal these conflicts. One was the formation in 1991 of the Mattole Watershed Alliance (Alliance), a group that is more broadly representative of the basin's residents and landowners than is the MRC. The Alliance took positions by consensus only and scored early successes by convincing the California Fish and Game Commission to restrict sportfishing in the Mattole to protect the diminished stocks of chinook and coho salmon. Even apart from the policy successes, the Alliance served as a forum for

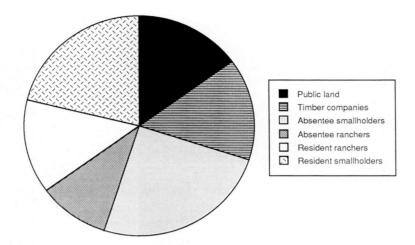

Figure 14.9.—Landownership in the Mattole River watershed. Some 85% of the land in the Mattole watershed is privately owned (all but the black wedge at the top of the pie).

people of diverse perspectives to discuss resource issues and better understand each other's views.

The other development was the production of the musical comedy *Queen Salmon*, which chronicles the response of a community much like the Mattole to rapacious logging, alongside a fish-eye view of the situation as seen by a school of migrating spawners. The writers and many of the cast were long-time participants in the restoration movement in the Mattole and saw the musical as yet another expression of their desire to improve the inhabitants' relationship to the land. The production, which toured three times in 4 years through the Pacific Northwest, also allowed people to laugh at themselves and the conflicts in which they were embroiled. One timber industry employee grinned at intermission, "I think I see myself up there," in the persona of the forester who is in love with a state fisheries biologist but can't seem to break out of his industrial approach to the forest.

Forestry practices.—In recent years, the restoration groups' long involvement in the health of the watershed has gained them standing to influence the way forestry is practiced in the Mattole. One lever came in the form of an opinion from the CDFG that logging in the Mattole watershed should take place only under conditions of "zero net discharge" of sediment (B. E. Curtis, CDFG, in a 1990 memorandum to W. Imboden, California Department of Forestry and Fire Protection). Because all logging produces sediment, this directive means logging operations must include measures aimed at reducing sediment production below current rates to compensate for the erosion timber harvest will cause. Several landowners, including ranchers and timber companies, have enlisted the MRC to design and carry out mitigations for their timber harvests. These mitigations have taken the form of improvements in road design, riparian revegetation, and monitoring of forest and stream conditions before and after timber harvest. Although the logging plans may still not be exactly what the MRC itself would have designed, it has found that working on the plans provides an opportunity to build relationships with land managers and to gather data on ecological conditions. The MRC also has found these collaborations to be an invaluable opportunity to learn from long-time landowners and develop a deeper under-standing of the judgments, perspectives, and pressures affecting their actions.

SIGNIFICANT ASPECTS OF THE MATTOLE RESTORATION EFFORT

Local Citizens Initiate and Direct Work

Watershed restoration as practiced in the Mattole is an expression of the caring and concern that we residents feel for our home place, taking cues from what we see around us. "The voice of the land had been unheard for a hundred years," says the Salmon Group's Michael Evenson. "Some of the people who were drawn here are hearing that voice." For many people, listening has turned into a long-term commitment. "A lot of people have the sense that they're in it for the long haul," says Rex Rathbun, a 77-year-old retired contractor who heads the Mill Creek Watershed Conservancy.

From the beginning, the locals who have directed restoration efforts have sought the knowledge they needed as their work led them to new questions and disciplines. This process has been very satisfying to participants and has helped sustain their efforts over the last decade and a half. "I have studied hydrology and geology willy nilly," says David Simpson, cofounder of the Salmon Group. "There's a shared excitement in learning something and applying it right next door."

That excitement helps provide good data and follow-up, too. It motivates citizen restoration workers who live near the restoration sites to check up on the success of the projects, to spot the need for maintenance before a project unravels completely, and to stay alert to relevant conditions in the field, such as the presence or absence of particular species or changes in water turbidity. "We're out there observing the natural processes; that's what we're going to emulate," says Maureen Roche, a former intensive care nurse who now works with the Salmon Group, logging hundreds of hours a year snorkeling in the river and its tributaries. It is much easier for a resident to find the time for an afternoon hike up a neighboring creek than for an overtaxed agency scientist to schedule a day-long trip out from the county seat.

In nearly two decades of restoration work here, a long series of college students and recent alumni with particular technical expertise have provided assistance. Some came to do research for term papers or masters' theses; others came to try out freshly acquired skills and knowledge in the service of locally directed restoration work. These students have stayed anywhere from 1 month to 17 years and include the Salmon Group's fisheries biologist and a geologist who has worked with the MRC.

Mattole Restoration Council Monitors and Evaluates Projects

In 1991, recognizing that watershed restoration is an evolving discipline in which much of its work is experimental, the MRC adopted a policy that all of its projects would include a plan for monitoring and evaluation. Sometimes this is done through direct measurement, such as tree survival rates for reforestation projects, some of which are sponsored by publishers and catalog-marketing companies wishing to compensate for the trees that are cut down to make the paper they use. Pool depths are measured at instream structures installed to scour the streambed. In other cases, such as channel modification, projects are monitored with photograph and video documentation.

Evaluation is arranged by project coordinators and is used in the design of the MRC's future projects. For example, survival rates in the MRC's Douglas fir reforestation projects have typically ranged from 70 to 80%. When seedlings planted on a harsh, south-facing aspect showed a significantly lower survival rate one year, the reforestation coordinator decided to install shade devices on new trees planted on that site the following year in the hopes of achieving better results.

Most restoration work in the Mattole has addressed very localized problems, such as a particular slide, unstable bank, or a logged hillside where trees never regenerated. Our efforts proceed while land management for the necessary livelihood of the people who live here continues. As a result, there is no control group (a no-intervention treatment for comparison) and we are left to speculate about what would have happened in the absence of restoration efforts versus what actually did occur. Researchers project that it will take 20 to 200 years for badly degraded river systems to recover (Gregory and Ashkenas 1990; Frissell et al. 1993). We aren't expecting any quick fixes. In a much smaller watershed north of here, it took one-quarter of a century of a hatchbox program before the coho salmon population and its habitat were healthy enough to make a go of it on their own (Hight 1996).

We continue to evaluate and learn from the results of our past actions. The MRC revisited projects carried out before 1991 in the lower river and evaluated them as part of the estuary study (MRC 1995). Weather patterns cooperated to make those evaluations as authentic as possible. Just as we were preparing our study for publication in January 1995, the Mattole experienced its highest flows in 21 years, which tested instream structures and vegetation planted next to the channel. The lessons of that flood event have been incorporated into our thoughts about future restoration work. For example, we must plan for lateral channel migration in designing revegetation projects, and a few large structures are more likely to survive high flows than are a passel of smaller ones.

The project on Mattole Canyon Creek described at the beginning of this chapter was also rearranged by the high waters of January 1995. Three miles of stream channel had been engineered into a calculated meander sequence according to a method developed by Rosgen (1994). Revetment structures of logs and boulders were built to guide the water into a channel of calculated sinuosity (curvature) and width. The areas around those structures were planted with native willows to provide riparian cover and anchor the banks. Originally implemented in the summer of 1993, the project needed minor repairs after the moderate flows of the 1993–1994 rainy season, and some structures were strengthened. The January and March 1995 floods, however, deposited up to 5 feet of new gravel and sediments around some of the structures and buried some of the planted willows. The creek jumped its designed channel in its lower reaches, carving a new course across the floodplain (Stemler 1996).

Clearly, this was a situation that needed to be managed adaptively. Project staff from the MRC and scientists from the CDFG, the funding agency, collaborated on an evaluation of the project. The outcome was a strategy that relied on trench planting of larger willow cuttings 8 to 12 feet long. The fate of this project has awakened in the MRC a renewed debate about the value of instream modification when the headwaters are still unstable, as they were on this creek. That debate was informed by a tour of the project given to the board and others active in the MRC by the project's coordinator—just the sort of ongoing education that keeps us on the learning curve, engaged in studying the processes that take place around us.

WHAT THE FUTURE HOLDS

Restoration has taken root in the culture of the Mattole watershed in the last 17 years. Schoolchildren have incubated native salmon eggs in their classrooms for over 8 years and have helped release fingerlings into the river so many times, jokes MRC cofounder Freeman House, "that it has become just another boring thing adults do." The native stock hatchbox program envisioned in 1980 as a 4- to 8-year stopgap measure will likely be retired in 10 years, according to Gary Peterson, the Salmon Group's fisheries biologist. The drop in salmon runs seems to have bottomed out, at least for the time being. With luck, and if external conditions such as ocean fishing and weather patterns cooperate, the hatchbox program will not be needed after that. People call the MRC for advice when storms take away their riverbanks or their roads threaten to fail. We continue doing as best we know how, listening to the watershed around us for our cues, striving to improve as our base of knowledge and experience grows. "After a decade and a half," says Freeman House, "we have only two things to point to: one, a possible improvement in salmon numbers, and two, cultural transformation. Without cultural transformation, those improved fish populations won't last." Effecting the cultural transformation is the biggest task before us—but also the most rewarding.

ACKNOWLEDGMENTS

The author gratefully acknowledges the help of participants in the Mattole projects who were interviewed for this chapter, including Michael Evenson,

Freeman House, Jan Morrison, Rex and Ruth Rathbun, Maureen Roche, and David Simpson. Mattole Salmon Group fisheries biologist Gary Peterson provided invaluable technical assistance. Useful critiques of the draft came from MRC program director Mickey Dulas, geologist Thomas Dunklin, Freeman House, Gary Peterson, and Maureen Roche.

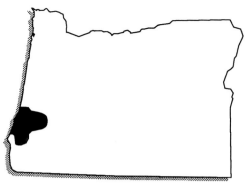

The down turn was cumulative and long term, and the up turn will not happen overnight. It has been long-term historic changes that has led to the decline in salmonids, and we all need to recognize that we are part of the problem.
—Paul A. Heikkila, Coquille River Project, 1995

CHAPTER 15

INTEGRATING PUBLIC AND PRIVATE RESTORATION STRATEGIES: COQUILLE RIVER OF OREGON

William F. Hudson and Paul A. Heikkila

The Coquille River restoration project was initiated to improve habitat for anadromous salmonids and water quality for residents of the Coquille River watershed of southwestern Oregon. The Coquille River basin, like other Pacific Northwest basins, has experienced declines in salmonid habitat and water quality due to land use activities that started in the late 1800s. Just as the declines occurred over a long period, restoration will require a long-term commitment to improve land management practices and upland, riparian (streambank), and aquatic habitats (Lichatowich 1989; Bilby and Ward 1991). Because land ownership patterns and social and political factors in the watershed are diverse and complex, restoration requires cooperative involvement and coordination among the various interests in the watershed and necessitates an ecosystem approach (i.e., one that looks at all components of the whole basin regardless of political or ownership boundaries).

Restoration efforts on federal lands administered by the U.S. Forest Service and U.S. Bureau of Land Management (USBLM) are guided and supported by the Northwest Forest Plan (FEMAT 1993), which includes a strong restoration component aimed at repairing riparian function (e.g., providing shade to the stream channel, stabilizing streambanks, and mitigating erosion from upland areas) and reducing sediments from forest road systems. The Bring Back the Natives program is another partner to the Coquille River restoration project. Bring Back the Natives is a challenge grant program sponsored by the National Fish and Wildlife Foundation, the USBLM, the U.S. Forest Service, and Trout Unlimited that seeks to restore native aquatic species and watershed health through partnerships formed among federal and state agencies, industry groups, conservationists, and local residents. Through such programs, linkages are formed between public and private land restoration efforts. Recognizing the need for partnerships among private, state, and federal land managers and funding limitations, those involved in Bring Back the Natives developed a network of people capable of addressing all aspects of watershed restoration—from the educational and social aspects to those of technical design and implementation of on-the-ground restoration.

COQUILLE RIVER WATERSHED

The Coquille River is the largest coastal river that has its origin within the Coast Range in Oregon. It is bordered by the Coos and Umpqua river basins to the north and the Rogue River basin to the south (Figures 15.1, 15.2). The Coquille River drains approximately 1,059 square miles (677,760 acres) and has four major tributaries: the north, east, middle, and south forks (Table 15.1). The lower river is tidally influenced from the mouth at the town of Bandon to the town of Myrtle Point, 40 river miles upstream.

The Coquille River basin is located between two large areas of consolidated federal ownership, the Siskiyou National Forest to the south and the Siuslaw National Forest to the north. Federal agencies manage approximately 30% of the watershed. Lands administered by the USBLM are primarily interspersed with private lands, creating a checkerboard ownership pattern across the landscape. Even with its fragmented landscape pattern, the Coquille River basin can provide crucial bridges for floral and faunal dispersal between adjacent and less fragmented ecosystems and landscapes.

Within the Pacific Border province, vegetative communities of the Coast Range are generally distinct from those of the Klamath Mountains, although overlap occurs. Western hemlock is the dominant forest species throughout the Coast Range and is most common in soils of Tyee sandstone parent rock (Hemstrom and Logan 1984). It is restricted, except for isolated pockets, to the windward side of the coast crest and north of the Rogue River in the Klamath Mountains. Port Orford cedar is the dominant forest species in the northern Klamath Mountains and extends north into the Coast Range inland of Coos Bay (Atzet and Wheeler 1984).

Historical Perspective

Human activity in southwest Oregon dates back thousands of years. Native Americans hunted waterfowl, deer, elk, and other animals, harvested salmon, and

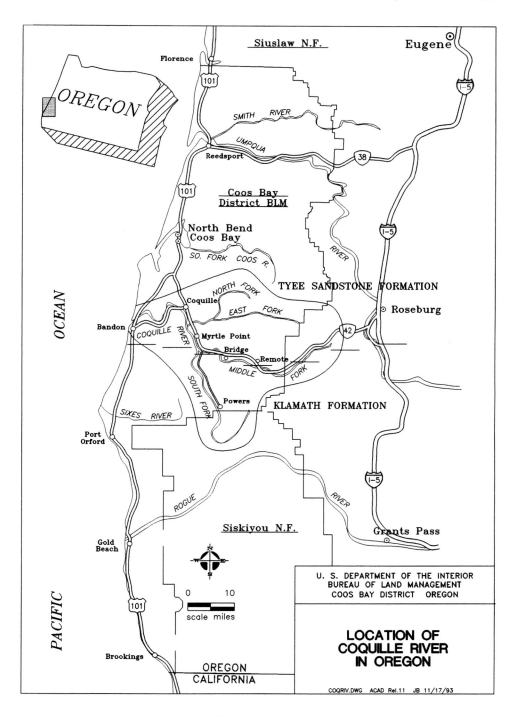

Figure 15.1.—Location and general features of the Coquille River watershed in southwestern Oregon.

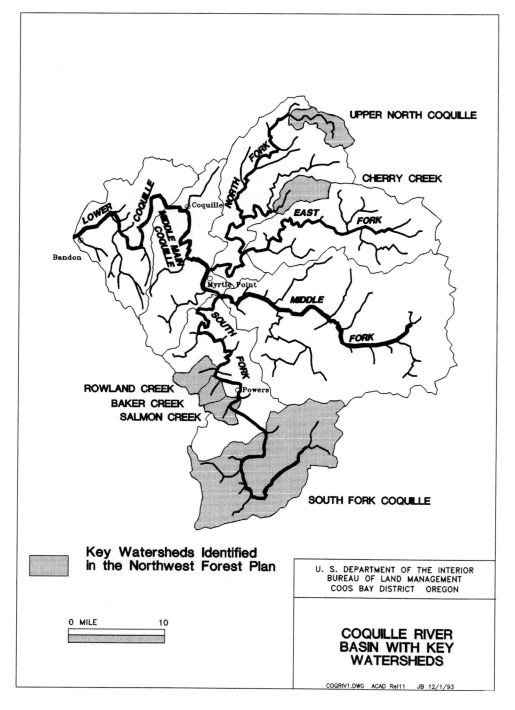

FIGURE 15.2.—The Coquille River and major tributaries. Key watersheds were identified as part of the Northwest Forest Plan (FEMAT 1993).

TABLE 15.1.—Basin area and land ownership (given in approximate stream miles[a] and percentage of ownership) for the Coquille River watershed (USBLM 1994b).

Basin	Acres	Ownership in miles (%)		
		U.S. Bureau of Land Management	U.S. Forest Service	Other
Lower river and mainstem	138,240	0	0	153 (100)
North Fork	98,560	97 (37)	0	166 (63)
East Fork	86,400	118 (51)	0	113 (49)
Middle Fork	197,760	97 (22)	0	344 (78)
South Fork	156,800	7 (5)	72 (48)	74 (48)
Total	677,760	319 miles	72 miles	850 miles

[a]Given for third-order streams and greater; first- and second-order streams do not contain significant populations of fish. Furthermore, the database is inadequate to determine total length of first- and second-order streams.

gathered plants throughout the Coquille River basin. Subsequent arrival and settlement by Europeans caused drastic changes to the landscape in a very short period.

The Coquille River estuary was an ideal place for early Euro-Americans to settle. Although treacherous, the bar at the mouth of the river was navigable. The estuary provided access to 40 miles of navigable river and adjacent flat bottomland. Surveys conducted in the mid to late 1800s described extensive marshes and wetlands that were later diked, drained, and converted to agricultural lands (Benner 1989). Historically, such bottomlands provided critical rearing habitats for juvenile salmonids.

As settlements grew and expanded upriver, timber was harvested from floodplains and later from hillslopes. The vastness and value of the timber resource attracted sawmills to provide wood for shipbuilding and exportation. Rivers and streams were the major source of transportation for moving huge conifer logs to mills and shipping facilities along the estuary. Splash dams were built within every major tributary by 1900 and were operated for several decades to flush rafts of logs downriver (USEPA 1991). This practice eliminated many miles of productive spawning and rearing habitat for salmon, trout, and other fishes as the log drives ripped out riparian vegetation and scoured stream bottoms.

The Coquille River has a long history of gill netting and commercial fishing, which supplied salmon for several local canneries and for export. This industry contributed to the depletion of most salmon stocks in the basin and in the region. According to U.S. Army Corps of Engineers records, exports of salmon from the Coquille River between 1882 and 1918 averaged roughly 38,000 fish per year (USEPA 1991). The historic productivity of the Coquille River can be appreciated from these records, which represent exports alone and not total annual catch or run size.

Status of Fish Species and Population Trends

The Coquille River presently supports populations of coho salmon, spring and fall chinook salmon, winter steelhead and resident (nonanadromous) rainbow trout, resident and sea-run (anadromous) cutthroat trout, and numerous non-salmonid fish species. Table 15.2 identifies 57 species within the basin (ODFW 1992). This list includes chum salmon, which were historically abundant but now are exceedingly rare. Nehlsen et al. (1991) reported stocks of spring chinook salmon, coho salmon, and sea-run cutthroat trout in the Coquille River to be at high or moderate risk of extinction as a result of habitat degradation, overharvest, and other factors.

Coho salmon in California, Oregon, and Washington (including Coquille River stocks) were petitioned in 1993 for listing under the U.S. Endangered Species Act. Spawning data show a clear decline in spawning escapement (number of fish reaching spawning area) of coho salmon over the past 20 years with stable, but reduced, populations during the last 10 years. Recent stable populations may be a result of reduced harvest on stocks (PFMC 1996). Spring chinook salmon in the Coquille River basin are threatened by a combination of high prespawning mortality, caused by high water temperatures during the summer, poaching, and genetic effects of hatchery manipulation of the very small population (Oregon Department of Fish and Wildlife [ODFW], unpublished data, presented at the Southwest Regional Fish Management meeting, 1993). Annual spawning escapement of spring chinook salmon is estimated at approximately 200 adults, most of which return to the South Fork Coquille River. Fall chinook stocks in the Coquille are depressed from historic levels, although recent data show spawning populations have increased and stabilized over the last several decades (ODFW 1995). Severe declines in fall chinook populations probably occurred in the era of splash damming, before spawning data were collected. Winter steelhead in the Coquille River, proposed for federal listing in July 1996, have had runs below their 20-year average 7 out of 10 years from 1981 to 1990 (Nickelson et al. 1992b). Winter steelhead returns in the Coquille basin are composed largely of hatchery stocks, with the exception of mostly wild stock in the Middle Fork Coquille River. Cutthroat trout exhibit great genetic diversity between isolated populations throughout the Coquille River basin. These populations have been adversely affected by hatchery releases and the degradation of complex pool habitat (ODFW, unpublished data). Native populations of Pacific lamprey are classified as vulnerable by the Oregon Department of Fish and Wildlife (ODFW, unpublished data).

Factors causing decline.—Readily erodible, deep, unstable soils in the basin coupled with widespread human disturbances have contributed to degradation of fish habitat. Disturbances have been associated with settlement, agricultural development, and timber harvest.

Road construction in the Coquille basin has greatly increased sediment delivery to streams. Nearly every square mile of the landscape has been affected by road construction, and nearly every tributary stream has been constrained by a road along the narrow valley bottom. The average road density ranges from a high of 4.5 miles per square mile in the main-stem drainage to a low of 3.5 miles per square mile in the south fork (Interrain Pacific 1996). Disturbances that have encroached upon and removed riparian vegetation have resulted in the loss of large coniferous trees from stream channels and consequently the ability of the riparian area to supply much-needed large woody material to the streams (Jones and Grant 1996).

Presently, only a fraction of stream miles in the basin are capable of functioning as spawning or rearing habitat for anadromous fishes. Stream channels in many of the lower valleys are deeply eroded down to the bedrock and disconnected from floodplains as a result of land clearing, channel straightening, splash damming, removal of large wood and trees from channels, and ditching, draining, and diking throughout the watershed.

Streams throughout the basin lack instream structure and channel complexity (diverse habitats). Harvest of large conifer trees from riparian areas, and removal of logs from channels, depleted streams of large woody material that is needed to form complex fish habitat. Historical records on the removal of large wood, known as snags, from the tidal and lower section of the Coquille River indicate the presettlement conditions of the basin. Between 1899 and 1923, more than 8,000 snags were removed from the river to improve navigation (USEPA 1991). The volume of large wood at the mouth of the Coquille River declined by 60% between 1970 and 1985 (Benner 1989). These natural large-wood accumulations once provided abundant complex and deep pool habitat for all aquatic species.

Episodic landslides and debris torrents (rapid movement of a large quantity of wood and sediment down a stream channel) are common in the sedimentary rocks of the Coast Range, whereas large-scale slumps (sudden collapse) and earth flows (downslope movement of soil and weathered rock) are more common in the Klamath Mountains. The Klamath Mountains section has very old sedimentary and volcanic rocks that may be locally metamorphosed. Both ranges weather rapidly and produce fine-grain sediments that erode into and degrade stream channels. Logging and road building accelerate the rate of debris slides (downslope movement) and debris flows (instream movement) well above predisturbance levels (Bisson et al. 1992; Megahan et al. 1992). Such activities have resulted in an increased proportion of scoured headwater stream channels and sediment buildup in lower stream reaches. One of the most common and detrimental effects of increased sedimentation has been the loss of habitat complexity in streams, particularly the loss of deep pools preferred by many species (Bisson and Sedell 1984).

Water Quality

Ten years of fixed-station monitoring along the Coquille River have indicated that dissolved oxygen, temperature, and fecal coliform (colon bacteria that indicate fecal contamination) levels frequently have not complied with standards adopted by the Oregon Department of Environmental Quality. Nonpoint sources of pollution are difficult to assess, but sediment, removal of shade-producing vegetation, fecal coliform bacteria, oil, and toxins all have contributed to deterioration of water quality in the Coquille River. Data collected by agencies, industry, and the Coquille Watershed Association will assist in identifying stream reaches for potential rehabilitation projects.

The most intensively monitored water quality attribute has been temperature.

TABLE 15.2.—Fish species occurring in the Coquille River system, Oregon. Habitat is noted as freshwater (FW), estuarine (E), or freshwater and estuarine (FW–E). Anadromous species are marked with an asterisk. Data are unpublished collection records of U.S. Bureau of Land Management Coos Bay District office and Oregon Department of Fish and Wildlife (ODFW 1992).

Common name	Habitat
Lampreys	
Pacific lamprey	FW–E*
Western brook lamprey	FW
Sturgeons	
Green sturgeon	FW–E*
White sturgeon	FW–E*
Herrings	
American shad	FW–E*
Pacific herring	E
Anchovies	
Northern anchovy	E
Carps and minnows	
Speckled dace	FW
Suckers	
Largescale sucker	FW
Bullhead catfishes	
Brown bullhead	FW
Smelts	
Eulachon	FW–E*
Surf smelt	E
Trouts	
Brook trout	FW
Chinook salmon	FW–E*
Chum salmon	FW–E*
Coho salmon	FW–E*
Cutthroat trout	FW–E*
Rainbow trout	FW
Steelhead	FW
Cods	
Pacific tomcod	E
Livebearers	
Western mosquitofish	FW
Silversides	
Jacksmelt	E
Topsmelt	E
Stickelbacks	
Threespine stickelback	FW–E
Pipefishes	
Bay pipefish	E
Scorpionfishes	
Black rockfish	E
Bocaccio	E
Copper rockfish	E
Quillback rockfish	E
Yellowtail rockfish	E
Greenlings	
Kelp greenling	E
Lingcod	E

TABLE 15.2.—Continued.

Common name	Habitat
Greenlings	
Rock greenling	E
Whitespotted greenling	E
Sculpins	
Brown Irish lord	E
Buffalo sculpin	E
Cabezon	E
Coastrange sculpin	FW
Pacific staghorn sculpin	FW–E
Prickly sculpin	FW–E*
Temperate basses	
Striped bass	FW–E*
Sunfishes	
Black crappie	FW
Bluegill	FW
Largemouth bass	FW
Surfperches	
Pile perch	E
Redtail surfperch	E
Shiner perch	E
Silver surfperch	E
Striped seaperch	E
Walleye surfperch	E
White seaperch	E
Gunnels	
Saddleback gunnel	E
Wolffishes	
Wolf-eel	E
Sand lances	
Pacific sand lance	E
Righteye flounders	
English sole	E
Sand sole	E
Starry flounder	E

In 1996, based upon available temperature data, large portions of the basin were identified as having degraded water quality. Historically, removal of shade-producing riparian vegetation, particularly along the middle and lower reaches of the river, has resulted in chronic high water temperatures during summer months. In 1992, seasonal maximum temperatures in the South Fork Coquille River increased steadily in a downstream direction. Maximum temperatures increased from 69.4°F near the headwaters at Buck Creek to 71.8°F at the National Forest boundary to 81.5°F near the Broadbent Bridge (ODEQ 1992), a distance of approximately 20 miles. Similarly high temperatures have been recorded in the middle fork (77°F) and north fork (72.3°F) (ODEQ 1992).

Because salmonids are stressed by temperatures exceeding 64°F for extended periods, much of the river is not suitable for juvenile fish rearing in the summer. Even in deeper pools, salmonids can be subject to severe thermal stress and high mortality. An improved understanding of the river's temperature regime and plans to reduce summer temperatures are crucial to the recovery of salmonid populations (P. Blake, Oregon Department of Environmental Quality, personal communication).

Social Setting

Private agricultural interests own much of the low-elevation areas within the estuary and along the main stem of the Coquille River. It is in the lower-elevation valley bottoms where the most complex stream and river habitats historically occurred. The higher-gradient streams in upland forested areas are on a mixture of private (timber company) and public (federally administered) lands.

The major sources of employment in the basin remain the timber industry, commercial fishing, and agriculture. Timber and fishing industries have been in decline, resulting in depressed local economies. Tourism revolving around recreational salmon fishing is also reeling from the closure of coho salmon fishing and greatly restricted ocean chinook salmon fishing designed to protect the weaker coho salmon stocks in the ocean, where stocks of both salmon species are mixed.

ISSUES AND GOALS

Partnerships developed through Bring Back the Natives and other watershed health initiatives (discussed below) have identified the seven objectives.

1. Promote watershed planning for coordinated resource management.
2. Seek partnerships with a variety of individuals, organizations, corporations, and agencies interested in restoration of the aquatic environment and salmonid stocks.
3. Obtain funding or in-kind services from partners to conduct restoration. Use challenge grants and services to secure additional funds from the National Fish and Wildlife Foundation and other partners.
4. Collect information for basin assessment and watershed analyses to direct watershed management and restoration effectively and efficiently.
5. Use existing data to restore riparian zones, instream habitat, wetlands, and the health and genetic integrity of fish stocks.
6. Resolve problems related to road and upslope erosion and developments that have prevented or restricted fish passage.
7. Provide education and information on aquatic resource issues.

RESTORATION EFFORTS

A basinwide working group has been formed to improve habitat for salmonids and water quality within the watershed. The partners' participation has been enthusiastic, voluntary, and sincere. The working group meets every other month or more frequently as needed. Current partners include timber companies such as Georgia-Pacific Corporation and Menasha Corporation, Oregon Department of Environmental Quality, ODFW, and ODFW's Salmon Trout Enhancement Program volunteers, Oregon State University Extension Service, Sea Grant, Trout

Unlimited, Thor-lo Corporation, U.S. Forest Service, USBLM, and many others. All partners are members of the larger Coquille Watershed Association (CWA) and many serve on the CWA's technical advisory committee. The CWA has a charter and a watershed action plan and works by group consensus.

Restoration efforts are guided by the partners, Bring Back the Natives funding guidelines, the aquatic conservation strategy described in the Northwest Forest Plan (FEMAT 1993), and the CWA's technical advisory committee. The technical advisory committee includes federal, state, and private participants with expertise in fisheries, hydrology, forestry, water quality, wetlands and riparian habitats, soils, and engineering.

Private lands constitute approximately 70% of the basin, with over 1,000 individual landowners living adjacent to the river and its tributaries. Private lands are the priority restoration areas because these lands exhibit the greatest need for restoration and because private lands often occur in valley bottoms, which historically were some of the most valuable fisheries habitat. In the past, few funds were available for enhancement work on private lands. The group consensus has been to identify opportunities on private lands that (1) build on federal efforts at restoration and (2) improve factors that limit fish production. Projects that meet these criteria are implemented through the CWA or private partners and are submitted to the National Fish and Wildlife Foundation for matching funds. Matching funds provide an important incentive to the private sector to become involved in restoration.

Restoration efforts by the CWA have been undertaken in areas where there is good potential for recovery and which provide important habitat for early life history stages of fishes in the lower reaches of the watershed. The types of projects being implemented are designed to address specific limiting factors for salmonids, improve water quality, advance early ecological stages of the riparian community to a later condition, and restore historic ecological functions. This approach makes fiscal sense and complements federal efforts to restore and protect key watersheds. The CWA also is coordinating with its partners in the development of a whole basin assessment and a restoration plan for the Coquille that builds on the federal watershed analysis effort on public lands.

THE AQUATIC CONSERVATION STRATEGY

On publicly owned lands managed by the U.S. Forest Service and USBLM, the Aquatic Conservation Strategy was developed to guide the agencies in restoring and maintaining the ecological health of watersheds and the aquatic ecosystems contained within them (FEMAT 1993). The strategy has the following four primary components:

1. *riparian reserves*—the lands along streams for which special standards and guidelines direct land use;
2. *key watersheds*—refugia of crucial watersheds for at-risk fish species;
3. *watershed analysis*—a procedure for conducting analysis of the geomorphic and ecological processes in a watershed that provides guidance for land

management planning, restoration, and monitoring (see Ziemer 1997, this volume); and

4. *watershed restoration*—a comprehensive long-term effort to restore aquatic ecosystems and watershed health.

Key watersheds have been designated on federal lands in the upper reaches of the Coquille basin (Figure 15.2). There has been a deliberate, coordinated effort to plan and implement restoration projects with interested private landowners adjacent to and within the key watersheds in order to expand the restoration effort on federal public lands. The Jobs in the Woods program, intended to provide work for former timber industry employees, funds many of the restoration activities on federal lands (Doppelt 1997, this volume).

Watershed Restoration Activities

Inventories.—Data collection and information sharing among partners for the basin assessment and watershed analyses are considered to be essential in developing on-the-ground projects. Habitat inventories have been funded and conducted by private timber companies to determine stream habitat conditions on their lands. Federal agencies have also conducted habitat inventories on public lands. Habitat inventories indicate present baseline aquatic conditions and through analysis can reveal factors limiting fish production in the basin. Through the CWA, a basinwide assessment was conducted on over 250 culverts to determine if fish passage problems existed for adult and juvenile salmonids.

An inventory of the basin's tide gates and channelized streams was conducted in 1996. Fish distribution surveys have found areas that have human-induced passage problems which are preventing or limiting passage of salmonids to existing or historic spawning streams. The CWA is currently assessing these areas to prioritize and determine appropriate corrective actions.

Fish passage.—Culverts that prevent or hinder fish passage have been identified by the CWA inventory and scheduled for replacement or modification to improve migration for both adult and juvenile fish (Figure 15.3). In 1995, initial funding repaired or replaced five high-priority culverts on private lands identified by the technical advisory committee of the CWA. Additional culverts will be modified or replaced as funding permits. On federal lands, seven culverts that were barriers to upstream fish movement were replaced in 1994, which restored fish passage to 9 miles and improved access to another 2.5 miles. Georgia-Pacific Corporation, working with ODFW, constructed a fishway (fish ladder) with a 17-foot rise to a culvert on Baker Creek, a tributary and key watershed on the South Fork Coquille River. The fishway has resulted in the reestablishment of coho salmon and steelhead in over 2 miles of stream. After having been absent from Baker Creek for over 30 years, these species were observed spawning in the first season after construction.

Instream work.—Riparian areas that have been adversely affected by past activities may take 200 years or longer to restore to full ecological function if projects depend on natural regrowth of trees as sources of large woody debris in

FIGURE 15.3.—Typical culvert that has been installed to replace existing culverts that were barriers to fish movement. Note baffles that provide flows conducive to movement of both juvenile and adult fishes. These culverts have been used in the federal Jobs in the Woods program and by the Coquille Watershed Association.

stream channels. Instream habitat modification can provide immediate cover and suitable spawning and rearing habitat while longer-term management of riparian habitat is addressed (House 1996). In the Coquille River basin, boulder weirs, logs, boulders, and rootwad complexes have been used to enhance streams that lack functioning, self-maintaining riparian areas to supply the needed large-wood component to the channel.

Instream projects in the Coquille River basin have included placing large woody material in Giles and Hudson creeks and the construction of boulder weirs on the upper north fork, all on Menasha Corporation lands. Menasha has also donated materials for instream projects on public and other private lands (J. Carr, Menasha Corporation, personal communication). In 1996, boulder and log structures were added to increase complexity at six sites in Hudson Creek on USBLM lands. Also in 1996, treelining was completed on Rowland and Sandy creeks on USBLM lands. Treelining is the practice of pulling three or four mature trees from riparian or upland areas into the stream channel. The trees act as key structures that trap additional large woody debris during periods of high flows. Twenty-one treelining structures were added to a one-mile stretch of Rowland Creek. Such efforts are undertaken only where adequate numbers of trees will remain in riparian areas to provide shade and other ecological functions.

Wetland restoration.—The construction of off-channel ponds and other wetland habitat projects will provide overwintering habitat for salmonids, especially coho salmon. Present and future projects will address tide gate and dike removal to reestablish marsh wetlands that are rich in nutrients and provide feeding areas for salmonids. Many of these areas are currently used as pasture land. Tide gates can be modified to facilitate or improve passage for both adult and juvenile salmonids.

Genetic analysis.—The ODFW, National Marine Fisheries Service, and Oregon State University are examining cutthroat trout populations above secondary migration barriers (a falls occurring upstream of an initial barrier falls on a tributary) and anadromous populations below barriers to determine genetic differences between the populations. The study is also investigating differences among isolated cutthroat trout populations. Interestingly, isolated resident populations of rainbow trout were discovered above some secondary barriers in the Coquille basin. Generally in the Coast Range of western Oregon, these fish are anadromous and therefore are considered to be steelhead.

Livestock management and riparian restoration.—On private lands along the lower Coquille River tidal zone, over 25 miles of agricultural riparian areas have been planted with native conifers, such as western red cedar and spruce, native hardwoods, or a mixture of both, and fenced to keep livestock from eating or trampling the seedlings (Figure 15.4). Fences may be a temporary measure and may not be needed after the trees become well established. On federal lands, 24 acres of riparian areas along Rowland, Baker, and Salmon creeks have been replanted at a density of 436 trees per acre.

Erosion and sediment control.—Eight miles of road on USBLM lands were closed in 1995 to reduce stream sedimentation in the Middle Fork Coquille River drainage. Another 5 miles of road on USBLM lands in the South Fork Coquille River drainage were closed in 1996. More than 130 culverts on federal land were improved in 1995 and 1996 to reduce sedimentation. Also in 1996, a survey of roads and logging landings on private lands was completed to assess conditions and develop plans for improvements needed to avoid failures that result in erosion. Improvement of roads and landings should result in reduced sediment delivery to streams.

The CWA has contracted a hydrologic evaluation of bank erosion in the lower 5 miles of the South Fork Coquille River. Bioengineering techniques have been used in two areas of high erosion on agricultural lands. Bank sloping and plantings were used in both areas to stabilize severe bank erosion (Figures 15.5, 15.6). One of the sites was also treated with Enkamat, an erosion control mesh through which planted species are allowed to grow; rock barbs were employed at the other site to protect the toe of the wash. Early indications reveal bioengineering techniques are working as planned.

Education.—Landowners who participate in restoration activities on their private streams develop personal commitment to and ownership of the projects,

FIGURE 15.4.—Private lands along the East Fork Coquille River that have been fenced and planted with conifers to help restore riparian vegetation. Note Vexar tubing to protect seedlings from deer browsing.

especially, for example, when salmon, trout, and steelhead return to a previously inaccessible stream through a newly constructed culvert or fishway. Personal ownership of restoration projects by private landowners promotes concern for environmental protection of their investment. Concerned citizens and private landowners who take personal interest in improving land management practices can provide the best land stewardship and watershed health.

Early on, the CWA recognized the need to inform the public about restoration activities. Educational activities to date include tours given across the watershed to regional and local public interest groups; development of a slide presentation; distribution of pamphlets describing the effort and opportunities in the Coquille basin; production of a video describing issues and objectives of the restoration effort; and a video showing implementation of restoration projects. The partnership also developed an informative display about the effort, which was seen by over 2,000 people at a recent county fair. Educational efforts have generated support among public interest groups, the local community, and landowners who wish to initiate restoration of their lands.

Monitoring and evaluation.—Monitoring determines if projects were properly implemented and if they are effective, and it provides insight on changes needed for the future. Cooperative restoration of the Coquille watershed has

FIGURE 15.5.—Typical bank erosion that occurs along all forks of the Coquille River where riparian vegetation is lacking to hold soils.

been underway for only a few years. The changes that caused the decline in anadromous fish stocks and water quality occurred over long periods. Progress in improving the health of the ecosystem, reversing the declining fish stocks, and improving water quality will also take a long time.

Monitoring will need to be conducted at appropriate intervals to measure detectable changes. Thus far the effort is too new to determine what efforts need to be modified. The technical advisory committee of the CWA is actively reviewing current literature and research for improved restoration techniques that may be implemented.

Presently, the monitoring effort within the basin is coordinated among private landowners, state agencies, and federal land management agencies. Riparian and stream habitat conditions, water temperature, adult and juvenile fish numbers, and fish species distribution are a few factors and indicators on which data are being collected to establish baseline conditions and trends throughout the basin. Preproject and postproject site evaluations are being conducted as projects are implemented. Implementation monitoring will be completed within the first season of projects to determine whether the project was implemented as designed and if preliminary results indicate progress toward desired future conditions. Monitoring requires a long-term commitment and is essential to success of a project.

FIGURE 15.6.—Same area shown in Figure 15.5 after banks have been modified to reduce slope. Enkamat matting has been applied to provide short-term stabilization following grass seeding and planting of native willow cuttings.

CRITIQUE

Strengths

The level of cooperation among partners has flourished since the inception of the Coquille River basin restoration project. The experience of people from private timber companies, public interest groups, regulatory agencies, state agencies, and federal land management agencies all working together on a project has been gratifying for all involved. Many factors help make the effort work, including developing trust at a personal level, respecting one another's views, working in a neutral environment, staying on the positive side of issues and working toward positive outcomes, having personal commitment, having partners who are experienced and have a high level of professional expertise, and enjoying community support. Coquille Watershed Association meetings are structured with an agenda, a group facilitator, and a chairperson to move the discussions through the business items. Meetings are conducted to make efficient use of the partners' time and are informal so that partners feel comfortable in participating. The facilitator ensures that everyone has an opportunity to contribute.

Weaknesses

This is an active effort that requires consistent funding to achieve watershed restoration across a large basin in which there exists many issues and opportunities for restoration. Without a secure source of funding the effort will fail. Thus far, funding for federal and private lands has been consistent, but the partners worry that funding sources may dry up before significant restoration is achieved. Many private landowners would be unable or unwilling to become involved with restoration work on their property without the incentive of the cost being shared. The commitment required of people who have full-time jobs is extraordinary. Time restraints and work commitments spread partners thin and could easily hinder effectiveness. Partners worry that the public may become frustrated and abandon support or begin to look for quick fixes.

PROSPECTS FOR THE FUTURE

Current inventory and survey efforts in the basin continue to reveal opportunities for restoring ecosystem structure and function. Many surveys have been conducted, and more are planned of streams to determine factors limiting habitat and water quality and to identify artificial barriers for migratory fishes. Surveys are also planned of roads and landings, which often are sources of sedimentation affecting aquatic organisms.

A tremendous amount of needed restoration work has been identified in this 1,059-square-mile basin. Restoration of federal lands in the upper- and mid-elevation reaches continues through the Northwest Forest Plan. The CWA and all its partners provide a companion restoration mechanism on private lands. Both efforts are long term and will continue as more partners become involved and more people learn of the benefits of restoration. The partnerships have been formed. The infrastructure, along with the skilled technical advisory committee to implement the restoration, is in place. All that is needed is continued sources of funding to take the effort forward into the future.

ACKNOWLEDGMENTS

The Coquille River Bring Back the Natives partners acknowledge the following organizations for their technical or financial support (or both) of the project: National Fish and Wildlife Foundation, Coquille Watershed Association, Georgia-Pacific Corporation, Menasha Corporation, Oregon Department of Fish and Wildlife, Oregon Department of Environmental Quality, Oregon's Governor's Watershed Enhancement Board, Thor-lo Corporation, Trout Unlimited, U.S. Natural Resources Conservation Service, Coos County Soil and Water District, U.S. Forest Service, and U.S. Bureau of Land Management. For their support and review of the document the authors thank Steve Wickham and Mike Grifantini, Georgia-Pacific Corporation; Max Yager, U.S. Forest Service; Jim Muck, Oregon Department of Fish and Wildlife; Pam Blake, Oregon Department of Environmental Quality; Bob Kinyon, Coquille Watershed Association; Joe Cone, Oregon State University Sea Grant; Bob House and Ron Wiley, U.S. Bureau of Land Management; and the volume editors. Use of trade names or manufacturer's names in this chapter does not imply endorsement.

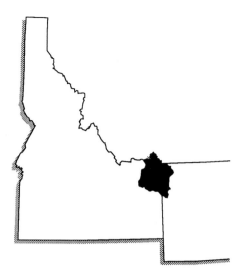

The differences are not so great between environmentalists and farmers. A big wall has been built. First we helped people peek over, and now we're tearing it down.
—Ron Ard, farmer and Henry's Fork Watershed Council member, 1995

CHAPTER 16

BUILDING A COLLABORATIVE PROCESS FOR RESTORATION: HENRYS FORK OF IDAHO AND WYOMING

Robert W. Van Kirk and Carol B. Griffin

Despite its scenic vistas, abundant recreational opportunities, and adjacency to Yellowstone and Grand Teton national parks, the Henrys Fork watershed (Figure 16.1) is best known for rainbow trout fishing and seed potatoes. A billboard on Highway 20 south of the farm town of Ashton, Idaho, welcomes visitors to "the world's largest seed potato producing area." Twenty-five miles to the north, fly-fishing-related businesses, a complete collection of insect-specific fly patterns, and an entire lifestyle have been built around the mystique of fishing for rainbow trout on the flat water of the Henrys Fork.

Rainbow trout and seed potatoes have much in common. Both are nonnatives brought to the Henrys Fork watershed a century ago by Euro-American settlers. The rainbow trout were originally imported for commercial fish farming operations and so, like the potatoes, were a food product. Rainbow trout displaced Yellowstone cutthroat trout, the only trout native to the watershed. The potato tuber provided food for Euro-American people, who displaced the native Shoshone, who, in turn, had eaten the tuber of the native camas plant. Rainbow

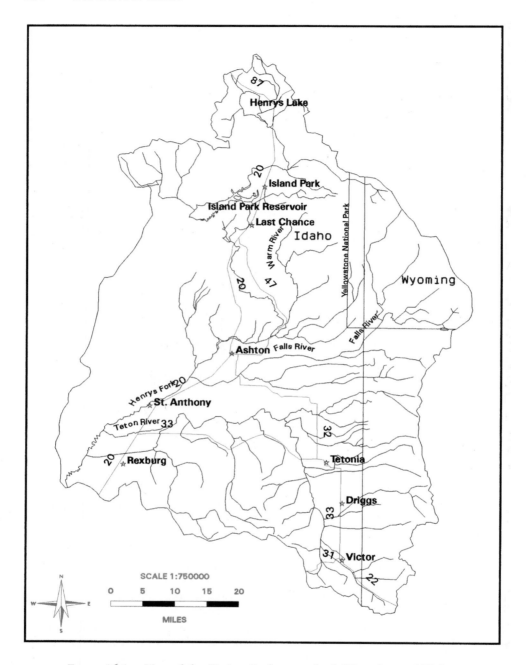

FIGURE 16.1.—Map of the Henrys Fork watershed, Wyoming and Idaho.

trout and seed potatoes both thrive in the watershed because of its cool climate and abundant spring water. The U.S. Bureau of Reclamation built storage reservoirs for farmers; the Idaho Department of Fish and Game (IDFG) stocked rainbow trout for anglers.

Fishing and farming coexisted peacefully in the watershed until the 1970s. About this time, anglers became vocal advocates for wild trout management, riparian (streambank) protection, and consideration of fish in water management decisions. In a scenario played out in many parts of the western United States during the past 25 years, the struggle for control over natural resources in the watershed pitted "rich, out-of-state anglers" against third- and fourth-generation local farmers, loggers, and ranchers. A secondary conflict evolved between wild trout advocates and local anglers, many of whom favored the harvest opportunity afforded by hatchery stocking. This conflict created further animosity between commodity interests and fly anglers, who were portrayed as outsiders attempting to lock up the natural resources of the watershed for their own benefit. Government agencies were caught in the center of this struggle as they tried to accommodate the desires of both commodity and recreational interests at the local level while, higher up, pressure from all sides politicized management decisions.

This conflict escalated throughout the 1980s, eventually peaking in 1992. As a result, the political and social climate prior to 1993 pitted each of the watershed's user groups against each other. In the absence of cooperation and consensus, the health of the watershed's fisheries and lands began to decline. The formation of the Henry's Fork Watershed Council (HFWC) in 1993 created a new cooperative atmosphere for the coordination of watershed management and restoration projects. Fittingly, the HFWC's cofacilitators are the primary advocates for rainbow trout and seed potatoes: the Henry's Fork Foundation and the Fremont-Madison Irrigation District.

In this chapter we describe the development of a multiorganizational framework within which cooperative natural resource management and watershed restoration can occur. Because issues and conflicts in the Henrys Fork watershed are similar to those found in other parts of the country, the evolution of the HFWC and the collaborative environment it has created has application to other organizations seeking to build broad-based partnerships for watershed restoration in the face of government downsizing and conflicts over resource use.

The original spelling of the lake and river name was, appropriately, the possessive "Henry's," as can be seen on early maps (e.g., USBOR 1939). However, due to the recent cartographic trend of dropping the apostrophe in possessive place names, the official modern names are "Henrys Lake" and "Henrys Fork" (e.g., Whitehead 1978), although the Henry's Fork Foundation and the Henry's Fork Watershed Council use the apostrophe. The Henrys Fork is also referred to as the north fork of the Snake River (e.g., IDFG 1958), acknowledging that the river is a major fork that joins the main Snake, or south fork, southwest of Rexburg, Idaho.

FIGURE 16.2.—Falls River, the Teton Range, and grain fields in the lower Henrys Fork watershed. (Photograph by Robert Van Kirk.)

THE HENRYS FORK WATERSHED

Physical Setting and Human Influences Prior to 1860

The confluence of the north and south forks occurs on the Snake River Plain, a 50- to 70-mile-wide crescent of lava covering much of southern Idaho (Hackett and Bonnichsen 1994). The plain was formed by the gradual eastward migration of a hot spot of volcanism that is currently centered under Yellowstone National Park. At the northeastern edge of the plain is the Island Park caldera, formed by the collapse of a large volcano two million years ago (Whitehead 1978). The Henrys Fork originates in and around this caldera, primarily from springs emerging from the contact zone between the rhyolite of the Yellowstone Plateau and the older basalt of the caldera floor (Whitehead 1978). To the west of the Yellowstone Plateau, the watershed boundary is formed by the Centennial Mountains and to the south by the Teton Range.

Prior to settlement by Euro-Americans, the 4,000- to 6,000-foot-elevation zone of the watershed was primarily grasslands and shrub steppes. Dominant species included wheatgrasses, needlegrasses, Idaho fescue, big sagebrush, rabbit-brushes, bitterbrush, and serviceberry (Marston and Anderson 1991). The grasslands and shrub steppes have been largely replaced by irrigated crops of potatoes, barley, wheat, and canola (Figure 16.2).

As precipitation increases with elevation, the sagebrush steppes give way to Rocky Mountain juniper, limber pine, and Douglas fir. Dense stands of lodgepole

FIGURE 16.3.—Henrys Fork of the Snake River immediately below Island Park Dam, with the Centennial Mountains in the background. (Photograph by Robert Van Kirk.)

pine cover the Island Park caldera and Yellowstone Plateau (Figure 16.3), where elevations range from 6,000 to 8,000 feet and annual precipitation ranges between 20 and 60 inches. Riparian vegetation consists of cottonwoods at low elevations and willows at higher elevations (Marston and Anderson 1991).

The species composition and age structure (number of individuals of each age in a population) of a particular patch of vegetation at any point in time is dependent on the history of disturbance, primarily fire, at that site. Prior to Euro-American settlement, intervals between fires ranged from 10 to 25 years on the grasslands and steppes, 20 to 50 years in the transitional forests, 100 years in the lodgepole pine forests, and up to 400 years in the high-elevation forests (Marston and Anderson 1991; USFS 1996a). Although individual patches of forestland in the watershed today may look the same as they did to early settlers, the overall appearance of the landscape is probably much different because fire suppression and timber harvest over the past century have changed the regime of disturbance (USFS 1996a).

Euro-American Settlement and Natural Resource Development:
1860 to 1940

Richard Leigh, the watershed's first Euro-American resident, settled in the Teton Valley in 1860, 12 years before the creation of Yellowstone National Park (Brooks 1986; Green 1990). While Leigh made a living from hunting, trapping, and guiding, another early settler, Gilman Sawtell, was one of the first people to raise domestic cattle in the area and to establish a commercial trout fishery. Sawtell settled on Henrys Lake in 1868, and by 1877 when General Howard passed through the area in pursuit of Chief Joseph and the Nez Perce, Sawtell had established a commercial fishery on the lake (Brooks 1986; Green 1990). Subsequent settlers also discovered the abundant Henrys Lake trout, and each winter between 50,000 and 100,000 pounds were harvested, frozen, and shipped to markets in Butte and Salt Lake City (Arbuckle 1900; Stephens 1907; Brooks 1986; Green 1990). In 1891, a Henrys Lake resident reported that "I can catch from 5 to 100 pounds a day and sometimes a good many more when I find a place that hasn't been fished out" (Henrys Lake Collection, Idaho State Historical Archives, unpublished data).

The fish that supported this commercial fishery were the native Yellowstone cutthroat, or black-spotted, trout (Trotter 1987). Although it is difficult to pinpoint the earliest date of nonnative fish introduction, it is known that Joe Sherwood established a commercial rainbow trout hatchery at Henrys Lake in 1891 and that in 1893 George Rea was operating a hatchery in Shotgun Valley that was supplied with brook and rainbow trout (Brooks 1986; Green 1990). By 1900, there were at least 37 commercial fish operations in the watershed (Arbuckle 1900). The U.S. Commission of Fish and Fisheries reported that only half as many cutthroat trout eggs were taken in annual operations at Henrys Lake in 1900 as had been taken in 1899 because of a "scarcity of fish, 50 tons of trout having been taken from the lake . . . the previous winter" (USCFF 1901).

Although fish and game laws were passed by the territory and later the state of Idaho, they were essentially unenforceable until the IDFG was created by an act of the 1899 state legislature. The IDFG succeeded in stopping the commercial exploitation of Yellowstone cutthroat trout in the watershed by 1910 but at the same time promoted the distribution of nonnative brook trout (Arbuckle 1900; Stephens 1907). State Game Warden W. N. Stephens reported that brook trout "seem to thrive and grow in our mountain streams . . . better than our native fish . . . [The brook trout] is considered the best of all the trout family and its propagation should be encouraged in every way possible" (Stephens 1907). Stephens's own department entered the fish propagation business in 1908, and Stephens stated that "the fish culture work . . . will keep the streams well stocked with the finest species of fish and will insure an opportunity for all who come to catch a mess of trout . . . [T]he replenishment of the streams will not only afford the residents of the State pleasure, but will attract many nonresidents and this will help in the development of Idaho" (Stephens 1909).

Many of those nonresidents were well-known actors, politicians, and authors who fished the waters of the Island Park area as guests of the Harrimans, Trudes, and other landowners (Brooks 1986; Green 1990). Private fishing clubs formed in the upper watershed around the turn of the century as fly anglers from around the country learned of fishing opportunities on Henrys Lake and Henrys Fork.

To keep up with the increasing demand for recreational angling on the Henrys Fork, the IDFG intensified its fisheries management activities in the watershed. In 1920, the headwaters of the Henrys Fork from Big Springs to the Henrys Lake outlet confluence were closed to fishing to protect spawning trout (Thorp 1919), a closure that remains in effect today. A newly acquired and expanded IDFG hatchery at Ashton provided 40,000 brook and 262,000 rainbow trout for planting in the watershed during 1923 and 1924 (Thomas 1925). By 1939, the IDFG was planting nearly 2,000,000 trout, mostly rainbow trout, in the watershed each year (IDFG 1940).

While out-of-state anglers, many from nearby northern Utah, enjoyed fishing for cutthroat trout, then cutthroat–rainbow trout hybrids, and finally rainbow trout in the upper part of the watershed, a local economy centered on farming, ranching, and logging was building up around them. Starting at the mouth of the Henrys Fork in 1879 and spreading up the valley from there, Mormon farmers from Utah settled the lower watershed and began to build irrigation systems as they had in Utah 50 years earlier (Carter 1955; Brooks 1986; Reisner 1993). Construction of a railroad line between 1905 and 1908 from St. Anthony, Idaho, to West Yellowstone, Montana, significantly increased transportation support for both commodity and tourism industries. The new rail line served numerous ranchers who shipped cattle to and from summer pastures in Island Park and logging companies that began cutting timber on the Targhee National Forest when it was created in 1908 (Brooks 1986; Green 1990; USFS 1996a).

As the need for irrigation infrastructure grew, farmers formed canal companies and irrigation districts, which were assisted by federal programs designed to promote the West's agricultural development. The first company to construct a major storage reservoir in the watershed was the North Fork Reservoir Company (D. Rydalch, North Fork Reservoir Company, personal communication). The company completed a dam on Henrys Lake outlet in 1923 that raised the level of Henrys Lake approximately 15 feet and provided 90,000 acre-feet of storage (Fremont-Madison Irrigation District, unpublished system management program, 1992). In what was likely the first action to mitigate the effects of irrigation development in the watershed, the IDFG established a hatchery on Henrys Lake in 1924 to compensate for lost natural reproduction in tributaries flooded by the lake expansion (Thomas 1925).

The Fremont-Madison Irrigation District was formed in 1935 and entered into a contract that year with the U.S. Bureau of Reclamation to construct storage reservoirs in the Henrys Fork watershed. In 1939, the U.S. Bureau of Reclamation completed Grassy Lake Dam on a tributary to Falls River and the much larger Island Park Dam on the Henrys Fork about 25 miles downstream of Henrys Lake (Fremont-Madison Irrigation District, unpublished data). With a capacity of 135,000 acre-feet, Island Park Reservoir inundated approximately 15 miles of the Henrys Fork and over 30 miles of tributaries, including Shotgun Creek, a large spring creek that had supplied water for George Rea's turn-of-the-century fish farm (USBOR 1939). The dam also altered the Henrys Fork flow regime (Figure 16.4) and blocked fish access to Big Springs and other spawning and rearing areas upstream. The only major tributary to the 27-mile stretch of the Henrys Fork between Island Park Dam and 114-feet-high Upper Mesa Falls is the Buffalo River, and fish access to it was blocked in the mid-1930s by a small hydroelectric dam

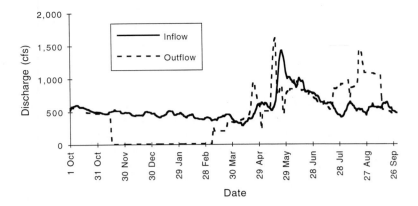

FIGURE 16.4.—Hydrographs showing reconstructed inflow and actual outflow of Island Park Reservoir in cubic feet per second (cfs) for water year 1950. Inflow was reconstructed from reservoir content and discharge data and represents the discharge that would have been present in the river without the dam. Dam management in 1950 is representative of management from 1939 until 1970.

just above its mouth (Rohrer 1983). In contrast to the concern expressed by the IDFG over the impacts of Henrys Lake Dam a decade earlier (Thomas 1925), published IDFG biennial reports did not express any concern over loss of stream habitat or spawning and rearing areas due to construction of Island Park and Buffalo River dams. Ironically, it was the 20 river miles immediately below Island Park Dam that would become some of the most famous rainbow trout fly-fishing water in the world during the 1970s, when popular angling author Ernest Schwiebert (1979) wrote, "The Henrys Fork may be the finest trout stream in the United States."

Natural Resource Management, Economics, and Sociology since World War II

Compilation and analysis of existing data indicate that the excellent rainbow trout fishing Schwiebert wrote about was much better during the 1970s than it had been during the preceding decades due to a favorable combination of water and fisheries management (Van Kirk 1995, 1996). Prior to the early 1970s, Island Park Reservoir was filled by completely shutting the dam gates on 15 November and not opening them again until the reservoir was full, usually in March (Figure 16.4). This management practice left no water in the Henrys Fork for the one-quarter mile between the dam and the confluence of the Buffalo River. Below the Buffalo, only one-third of the river's natural flow was present (U.S. Geological Survey, unpublished data). Low winter discharge at Island Park Dam has been cited as a major factor limiting growth of rooted aquatic plants (Shea 1991; Vinson 1991) and survival of juvenile rainbow trout (Smith and Griffith 1994; Griffith and Smith 1995). In addition, rainbow trout populations were decimated twice prior to 1970 by chemical treatments designed to rid the reservoir and river

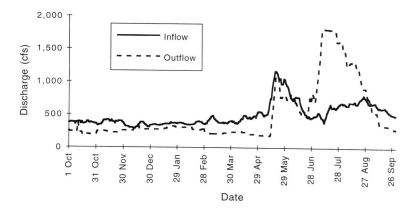

FIGURE 16.5.—Hydrographs showing reconstructed inflow and actual outflow of Island Park Reservoir in cubic feet per second (cfs) for water year 1991. Inflow was reconstructed from reservoir content and discharge data and represents the discharge that would have been present in the river without the dam. Dam management in 1991 is representative of management since 1970.

of nongame fish (Jeppson 1967, 1969). In 1958 and 1966, Island Park Reservoir, its tributaries, and the Henrys Fork from the dam downstream to Ashton were treated with rotenone (a commonly used fish poison) and then restocked with rainbow trout (Rohrer 1983). Rotenone projects in 1979 and 1992 were limited to the reservoir and upstream tributaries.

Beginning in 1970, winter minimum discharges from Island Park Dam were increased (Figure 16.5). Ruth Shea (Idaho State University, unpublished data) theorizes that this change in dam operations created favorable conditions downstream for the development of dense beds of macrophytes, which increased river depth (Vinson 1991; Vinson et al. 1992) and may have contributed to better winter survival of juvenile trout. From the 1920s through 1977, the IDFG stocked catchable-size rainbow trout into Island Park Reservoir, the Buffalo River, and the Henrys Fork below the dam (IDFG 1973; Jeppson 1973; Ball et al. 1982). An emerging theory is that the combination of increased winter flows, increased macrophyte density, cessation of rotenone treatments in the river, and consistent input of hatchery fish to mitigate loss of access to spawning and rearing habitat above Island Park and Buffalo River dams was responsible for developing the excellent fishing of the 1970s (Van Kirk 1995, 1996).

Surveys conducted by IDFG documented increasing angling pressure from out-of-state fly anglers below Island Park Dam as the result of the good fishing, a change in fishing regulations, and publicity provided by articles, such as that by Schwiebert (1979), in national fly-fishing publications. In 1973, 78,241 angler-hours—45% from nonresidents—were recorded on the 10.4-mile section of the Henrys Fork from Osborne Bridge upstream to Island Park Reservoir (Jeppson 1973). Only on the Harriman Ranch (formerly the Railroad Ranch) section, which had fly-fishing-only regulations, did nonresidents dominate the angling effort. In

1978, IDFG implemented special regulations on 16.3 miles of the Henrys Fork from Island Park Dam to Riverside Campground that consisted of a slot limit of three trout under 12 inches and one trout over 20 inches and tackle restrictions of only artificial flies and lures with a single barbless hook (Rohrer 1983). The fly-fishing-only regulation continued on the Harriman section upon gift of the ranch to the state in 1977 (Rohrer 1983; Green 1990). By 1981, total effort on the 10.4-mile section had increased to 86,103 angler-hours, with 77% provided by nonresidents (Rohrer 1983).

Although Jeppson's (1973) assessment of the rainbow trout population below Island Park Dam reflected his knowledge of spawning and rearing habitat limitations and the role of stocked fish in maintaining the sport fishery, it is unlikely that many of the fly anglers who enjoyed the good fishing of the 1970s were aware of habitat or other problems that would eventually lead to a decline in rainbow trout numbers below the dam. In 1981, more than 90% of the anglers fishing the Henrys Fork from Island Park Dam to Osborne Bridge rated the fishing as either excellent or good (Rohrer 1983) and thus had little reason to be concerned with habitat restoration. However, IDFG research on Henrys Lake showed that between 1951 and 1972 the catch rate had declined 50% and the average weight of creeled fish had declined from 2.7 to 1.8 pounds (IDFG 1973). In 1976, only 41% of Henrys Lake anglers rated the fishing as either excellent or good (Coon 1978). Furthermore, diversion and siltation of tributaries had made the fishery more dependent on hatchery stocking (IDFG 1975). Siltation was caused primarily by livestock grazing around Henrys Lake (Rohrer 1982), although salvage harvest of lodgepole pine in the Targhee National Forest had accelerated erosion and degraded streams throughout the upper Henrys Fork watershed (USFS 1996a). By 1980, fly anglers who fished Henrys Lake and other waters above Island Park Dam had begun to realize that the quality of their recreational experience was being jeopardized by management of the area's natural resources primarily for commodity production, setting the stage for the conflict that would occur as the watershed's economy began to shift from agriculture and forestry to one dominated by tourism and retirement and investment income.

This economic shift has been documented in the group of rural Idaho, Montana, and Wyoming counties that surround Yellowstone National Park (Power 1991; Rasker et al. 1992), and it is also occurring, albeit more slowly, in the Henrys Fork watershed, which primarily consists of Fremont and Teton counties in Idaho. In 1980, farming accounted for 25% of the 4,348 jobs in Fremont County, whereas in 1992, farming supplied only 18% of 4,350 jobs. Meanwhile, retail trade and services increased from 25% of Fremont County jobs in 1980 to 30% in 1992 (IDC 1994). The percentage of total retail trade that is tourism related is estimated at 20 and 23%, respectively, for Fremont and Teton counties. Growth in the Fremont and Teton county economies has come primarily from increases in farm revenues and retirement and investment income. The tourism industry in the watershed is just developing, and recreational service needs are generally not being met at present (IWRB 1992).

If the Henrys Fork watershed continues to follow the trend of the greater Yellowstone area as a whole, the actual contribution of agriculture to the local economy will remain fairly constant but will continue to decline in relation to retirement and investment income and tourism and recreation-related services (Power 1991). This economic and social shift will continue to create the potential for conflicts over natural resource management and land use in the watershed.

EVOLUTION OF COOPERATIVE NATURAL RESOURCE MANAGEMENT

Angling Groups, River Conservation, and Political Conflicts: 1981 to 1991

The Henrys Lake Foundation.—The first watershed restoration projects in the Henrys Fork basin were implemented around Henrys Lake by the IDFG and the Henrys Lake Foundation (HLF), which was formed in 1981 by a group of fly anglers concerned about declining numbers and size of fish in the lake (Brooks 1986; Prange 1995). Restoration projects have included construction of riparian fencing along tributaries, improvements to irrigation structures, removal of fish passage barriers, and revegetation of riparian areas. The HLF has been successful in its restoration efforts because it has chosen to work cooperatively with both the IDFG and private ranchers around the lake and because it has a narrowly focused mission: to protect and enhance the trout fishery of Henrys Lake. The HLF's support of creel and gear restrictions has been in conflict with anglers who do not fish with flies and prefer more liberal harvest opportunities. In general, though, the HLF has avoided political conflicts. Restoration projects and improved livestock grazing management have been the major focus of the HLF, in part because the IDFG management direction on Henrys Lake has been to improve reproduction of wild Yellowstone cutthroat trout in the tributaries (Rohrer 1982; Brostrom and Watson 1988). However, the preferred fish of Henrys Lake anglers are nonnative brook trout and first-generation cutthroat-rainbow trout hybrids, which are produced at the Henrys Lake hatchery (Coon 1978).

The Henry's Fork Foundation.—In 1984, a group of fly anglers initially concerned with riparian degradation on the Henrys Fork below Osborne Bridge formed the Henry's Fork Foundation (HFF). The group was also concerned with the general consensus among anglers that rainbow trout numbers in the river below Island Park Dam were declining. Population estimates made by IDFG corroborated this consensus; the rainbow trout population in the Box Canyon section immediately below the dam had declined from almost 19,000 fish in 1978 (Coon 1978) to 15,000 fish in 1982 (Rohrer 1983). The HFF's first project was to construct fencing to protect the streambank from cattle grazing along the mouth of Box Canyon and downstream 12 miles to Pinehaven. Unlike earlier riparian fencing on Henrys Lake tributaries and on the Teton River, the HFF's project did not involve private landowners and was implemented on U.S. Forest Service and Harriman State Park land. Thus the HFF did not have any personal connection with local landowners when it began its 1987 campaign to change fishing regulations below Island Park Dam from the slot limit to catch and release. Despite public opposition to the regulation changes, the HFF was successful in lobbying the Idaho Fish and Game Commission to implement catch-and-release fishing in 1988. The price for this success was increasing distrust and dislike for the HFF among local residents, who viewed the catch-and-release outcome as proof that out-of-state fly anglers, led by the HFF, were determined to lock up the river for their own use.

The HFF increasingly found itself at odds with the majority of local residents as it entered more legislative and political battles. In addition to the catch-and-release campaign, the HFF successfully lobbied to protect stretches of the Henrys Fork and its tributaries from hydroelectric and irrigation developments (Mickelson 1994). In all of its political campaigns, the HFF opposed local agricultural and development interests. A positive outcome of one of these adversarial battles was the formation in 1986 of the Island Park Hydroelectric Project Advisory Committee, of which the HFF is a member. Because of the constructive dialogue that occurred in the committee among Fall River Rural Electric Cooperative, government agencies, and recreational interests, the power plant, which began operation in 1994, was built with aeration and monitoring equipment that has improved water quality below the dam.

Meanwhile, rainbow trout numbers in the Henrys Fork continued to decline. In 1989, the year after catch and release was implemented, the rainbow trout population in Box Canyon was only 5,000 fish (Gamblin et al. 1993a), and by 1991 the population had fallen to 3,000 fish (Gamblin et al. 1993b). Current research suggests that the decline in the Henrys Fork rainbow trout population between 1978 and 1991 was due to the cessation of stocking in 1978, a significant decline in the aquatic plant community throughout the 1980s due to an increase in the number of wintering trumpeter swans (Shea, unpublished data), and degradation of spawning and rearing habitat (Van Kirk 1995, 1996). However, reasons for the rainbow trout population decline were not known at the time by the HFF, which operated under an adversarial model of environmental advocacy. Under this model, interest groups line up on opposite sides of a given issue and lobby government agencies and legislatures, which have the authority to make decisions. General beliefs based loosely on generic science (e.g., cattle degrade streambanks or catch-and-release regulations improve trout populations) provide the basis for lobbying activities. The HFF was not involved in cooperative efforts. Events of 1992 changed this.

Formation of the Henry's Fork Watershed Council:
1992 to 1993

In 1989, Marysville Hydro Partners, Inc., received Federal Energy Regulatory Commission approval to build a small hydroelectric plant on the Falls River and enlarge an existing irrigation canal to carry water to the plant. Construction began the following year amid allegations from local citizens that proper public review procedures were not followed in granting the Federal Energy Regulatory

Commission license and that the environmental assessment contained inaccurate information. In June 1992, a construction accident caused failure of the canal and sent 17,000 tons of sediment into Falls River. Later that summer, as Island Park Reservoir was drained for irrigation, IDFG took advantage of the nearly empty reservoir to carry out a rotenone project. In early September, as the remaining reservoir pool was being drained, the river began to cut down into sediment accumulated on the reservoir bottom. Over a period of 2 weeks, more than 50,000 tons of sediment were transported from Island Park Dam into the Henrys Fork (ERI 1995).

Meanwhile, a citizen committee that included recreationists, hydroelectric developers, county commissioners, and irrigators had just completed a contentious, 3-year debate over recommendations to be incorporated into the Henrys Fork Basin Plan by the Idaho Water Resource Board (IWRB 1992). The plan protected 195 miles of streams in the watershed from further irrigation or hydroelectric development and was adopted unanimously by the state legislature in 1993. After the controversies surrounding the Marysville hydroelectric project and the Henrys Fork Basin Plan, people were growing weary of confrontations over natural resource management issues in the watershed. The 1992 sediment spill angered and frustrated many watershed users, who believed that it could have been avoided by better communication among government agencies. Amid mistrust and skepticism, some proposed that a cooperative, watershed-based approach was needed to develop constructive solutions to natural resource management problems in the Henrys Fork watershed.

Throughout the spring and summer of 1993, citizens, scientists, and government agency personnel met to discuss possible cooperative approaches to watershed management. The general consensus was that a single organization was needed to coordinate management activities among dozens of government agencies and interest groups in the watershed. Agency representatives agreed that such a group might be most effective if it were led by citizens. The two groups most at odds over the basin plan, the Fremont-Madison Irrigation District and the HFF, volunteered to work together and lead the new organization, the Henry's Fork Watershed Council (HFWC).

Coordination of Natural Resource Management

The HFWC was chartered by the 1994 Idaho legislature as a "grassroots, community forum which uses a nonadversarial, consensus based approach to problem solving and conflict resolution among citizens, scientists, and agencies with varied perspectives" (State of Idaho 1994). The HFWC's charter identified four duties: (1) promote cooperation in resource studies and planning that transcends jurisdictional boundaries; (2) review, critique, and prioritize proposed watershed projects; (3) identify and coordinate funding for research, planning, implementation, and long-term monitoring programs; and (4) serve as an educational resource about the Henrys Fork basin (State of Idaho 1994). The HFWC is organized into three component groups: (1) a citizen's group of local community members representing commodity, conservation, and community development interests; (2) an agency roundtable of representatives from federal, state, tribal, and local entities with land and resource management jurisdiction in the watershed; and (3) a technical team of university, agency, and independent scientists from varied disciplines. Council mailings are sent to over 100 people, and attendance at the all-day meetings averages 50. The HFWC has no formal membership or appointed positions; all HFWC meetings are open to any person who wishes to attend. Proposals sponsored by an individual, group, or agency are presented to the entire HFWC. The sponsor answers questions, and then participants separate into component groups to evaluate the proposal using the 10-point watershed integrity review and evaluation (WIRE) criteria (Appendix 16.1).

Two facilitators, one from the HFF and one from the Fremont-Madison Irrigation District, encourage HFWC participants to use the WIRE criteria to evaluate proposals. Detailed suggestions are written down and reported back to the entire HFWC, which then uses a consensus-based model of decision making to determine whether to endorse the proposal, endorse it if the sponsor makes improvements, or suggest that substantial revisions are needed before reevaluation. At the end of the meeting, the HFWC reconvenes to discuss the day.

Projects and programs that meet the WIRE criteria may receive a letter of endorsement and possible funding from the HFWC's watershed fund. The fund was established by the Idaho Division of Environmental Quality using mitigation money collected from Marysville Hydro Partners following the Marysville canal accident. Approximately US$25,000 of the $100,000 given to the HFWC has been allocated. Concern about the rate of spending prompted the HFWC in 1995 to develop new funding guidelines that limit the amount of money a project can receive to $5,000 or 20% of total project costs, whichever is less. These guidelines also stipulate that money will be allocated to approved projects twice a year to facilitate prioritization of projects that have been approved during the previous six months. The HFWC tracks the status of approved projects by requiring sponsors to present a report at the annual state-of-the-watershed conference held in October.

An evaluation of the HFWC's efforts by the Northwest Policy Center (NPC 1995) identified seven accomplishments, many of which have also been achieved by other watershed-based groups. First, and "perhaps its greatest accomplishment," (NPC 1995) the HFWC has enabled previous adversaries to work together on a variety of natural resource issues in the Henrys Fork watershed. The increased communication among government agencies and citizen and business groups has enabled potential sponsors to develop proposals cooperatively and constructively before proposals are brought to the HFWC.

A second HFWC accomplishment is the education of its participants on natural resource management issues. The National Research Council (NRC 1992a) and Doppelt et al. (1993) cite the lack of understanding about stream and river dynamics as a cause of inadequate protection and restoration policies. The HFWC has invited agency personnel to give educational presentations about water rights, irrigation delivery systems, reservoir management, and endangered species habitat management. These presentations and those of the sponsors are critical to informed decision making.

Another HFWC accomplishment is the development of the WIRE process (Appendix 16.1) to focus participants' attention on the specifics of a proposal instead of on blanket endorsement or rejection. Although many participants have a preconceived notion that a given proposal should be endorsed or rejected, facilitators urge them to use the WIRE criteria to evaluate the proposals. This results in specific feedback to the sponsor about aspects of the proposal that could be improved.

Fourth, HFWC meetings are open to anyone who wishes to attend. A consistently diverse group of participants helps the HFWC provide a broader perspective on public and private natural resource problems and potential solutions than any single interest group could. This diversity also helps make people aware of other interests and, hopefully, more sensitive to competing demands regarding natural resource management.

Fifth, HFWC serves as an important venue for local, state, and federal agencies to present proposals and obtain feedback from the public. The HFWC is structured to involve the public in ways that are different from those of traditional public hearings. Rather than a shouting match pitting individuals against agency personnel and each other, the HFWC emphasizes shared learning in a nonconfrontational manner. Agencies have brought several types of activities to the HFWC, including scoping of potentially controversial issues such as the U.S. Bureau of Reclamation's proposal to transfer title of the Island Park Dam to the Fremont-Madison Irrigation District and the U.S. Forest Service's Targhee National Forest Plan revision. The U.S. Forest Service has also directed groups applying for special-use permits to present their proposals to the HFWC as a way of gathering public input (USFS 1996b).

A sixth accomplishment is the HFWC's emphasis on community building and the consensus-based model of decision making as a method of encouraging people to work for better natural resource management. Although discussions may become tense at times, the HFWC's process is based on mutual respect so that participants do not engage in personal attacks. Participants have agreed to disagree but to do so in a nonadversarial manner.

Finally, the HFWC has used monetary incentives to encourage private landowners to improve natural resource management. Private land proposals focus on farming and ranching since these are the dominant land uses in the watershed. The HFWC has approved proposals for improving irrigation delivery, implementing rotational grazing management, constructing fences, and improving irrigation management near Henrys Lake and for preventing shoreline erosion and protecting riparian areas on private rangeland around Henrys Lake.

Most participants would agree the HFWC is fulfilling its legislatively mandated duties, which consist of coordinating projects and funding, reviewing proposals, and educating the state legislature and the public (State of Idaho 1994). Because its mission is to serve as a community forum, the HFWC has a process orientation rather than an ecological vision (i.e., promotes coordination rather than restores habitat). There is no question that the HFWC has succeeded in improving the relationship among conflicting interests, but it is too early to determine if its efforts have made a measurable improvement in watershed health.

CRITIQUE OF THE HENRY'S FORK WATERSHED COUNCIL APPROACH

Challenges Facing the Henry's Fork Watershed Council

There are a number of challenges facing the HFWC. First, the Northwest Policy Center (NPC 1995) evaluation found that many participants do not believe the HFWC has been truly tested. The proposals that the HFWC has evaluated have not provoked deep or widespread disagreement among participants, nor have there been any proposals that cause conflict between the cofacilitators. Participants are concerned that while the HFWC needs to take on more controversial topics, there may be some reluctance to do so because of the fear of breaking down the cooperative relationships that have been created. As the issues increase in contentiousness, which may happen as the HFWC diversifies, there may be increased difficulty in reaching consensus.

Another criticism of the HFWC is that there is no formal process for prioritizing proposals that come before it (NPC 1995). The time allotted to a proposal does not always bear a direct relationship to its significance, its complexity, or the amount of money requested. There is an informal process by which members of the facilitation team set meeting agendas based on participant interest and sponsor desires. The HFWC generally responds to rather than solicits proposals, which makes it difficult to ensure that the HFWC is addressing the most important restoration needs first or getting the most improvement in watershed health for its money.

Two frequent criticisms of the consensus-based approach to decision making are that it can lead to deadlock and can result in a decision aimed at the lowest common denominator. The University of Colorado (1996) suggested that one of the keys to success in maintaining participation in watershed groups is to establish a decision-making structure that will prevent deadlock. One individual who disagrees with the will of the majority, especially new participants to a group, could break down the entire process. Carpenter and Kennedy (1988) have suggested that in such circumstances the entire group can choose to override the objections of one dissenting individual. At its first meeting, the HFWC decided that consensus for its purposes means ''general agreement'' and not unanimity. As the HFWC diversifies, it increases the possibility of encountering fundamentally irreconcilable differences of opinion.

Because the HFWC has no legal authority and all of its participating agencies and entities retain any and all legal authority they possess, there is always the possibility that a sponsor could implement a project that the HFWC did not endorse, but this has yet to happen. Participants are just beginning to appreciate that agencies must manage public resources by means of both science and public desires, and thus it is possible that an agency may choose to implement a project that the HFWC did not endorse. Because agencies have the legal responsibility for managing natural resources in the watershed, losing their participation would decrease the significance of the HFWC's role in coordinating natural resource management. A related problem is that because HFWC review of proposals is voluntary, projects routinely are implemented without an HFWC presentation; so far this has not eroded trust among participants.

Some concerns have been raised about whether all interests participate in the HFWC. Most rank and file Fremont-Madison Irrigation District members do not attend HFWC meetings because they believe that their opinions are represented by the board members who attend as facilitators (S. Steinman, Fremont-Madison Irrigation District member, personal communication). Likewise, HFF members do not attend because they believe the HFF facilitators are representing them or because they live in other parts of the country. However, by definition, facilitators should not represent the views of their constituents, and thus both facilitating organizations should encourage their members to attend HFWC meetings. There is little participation by regional and national environmental groups because these groups have few members in the watershed and because there are no nationally contentious issues (e.g., spotted owls or salvage logging) in the watershed (NPC 1995). Increased participation by regional or national groups of all types would help address the concern that the interests of residents outside the watershed are not being represented because these people cannot attend HFWC meetings. Some agencies may also represent the interests of nonresidents. For example, the U.S. Forest Service has a congressional mandate to manage the national forests for the benefit of all U.S. citizens. Finally, even though the HFWC intends to include all interests by inviting anyone to attend meetings or receive mailings, realities of time and space will always prevent some interested people from attending HFWC meetings.

Financing is another difficulty the HFWC may face in the future. At the current pace of spending, the HFWC will run out of money in 4 years, potentially limiting future interest. Inconsistent with the University of Colorado's (1996) finding that all watershed groups struggle to find adequate funding, the HFWC has not yet faced that problem because initial money came from the Marysville Hydro Partners fine. The HFF has begun seeking new funding sources for the HFWC, but the HFWC as a whole has not tackled the funding issue. It is possible that agencies, legislatures, and private corporations will help fund the HFWC in the future.

Management by Means of Collaboration

The HFWC's approach to collaboration and problem solving is increasingly viewed as an alternative to top-down, overly bureaucratic, natural resource management systems. The River Network (1995), a national nonprofit organization designed to help people organize to protect and restore rivers and watersheds, developed a strategic plan that identifies as a goal having an active citizen watershed council in every one of the country's 2,000 watersheds by 2020. "These efforts have arisen for a variety of reasons, including pressure to reform how public agencies interact with the public; a growing concern over the long term effectiveness of traditional interest group advocacy tactics; and a movement among resource managers toward more complex ecosystem or watershed management approaches" (NPC 1995). In its evaluation, the Northwest Policy Center (NPC 1995) noted that, "The [Henry's Fork Watershed] Council is one of a group of leading experiments in new, more responsive, and potentially more effective ways to manage and maintain healthy ecosystems while integrating the needs and desires of people." John (1994) coined the phrase "civic environmentalism" to describe these local, bottom-up, decentralized approaches to natural resource management. Clearly, there is a movement to abandon traditional command-and-control environmental management in favor of a new approach, but there are at least two aspects of the watershed approach that require careful consideration before it is implemented in a given watershed.

Using a watershed as the unit for natural resource management is arguably more useful than using current political boundaries, but many physical and ecological processes do not respect watershed boundaries. It is critical that watershed-based groups recognize that many natural resource management decisions made inside a given watershed will affect resources and people in other geographic areas. For example, changes in irrigation water delivery systems in the Henrys Fork watershed affect groundwater levels in a different part of the Snake River Plain 200 miles away (Hackett and Bonnichsen 1994). A second example is management of trumpeter swans, which winter on the Henrys Fork in large numbers but which breed throughout the northern Rockies (Shea 1991). A third example, discussed by the HFWC, is the difficulty in modifying management of Island Park Reservoir because, like many reservoirs in the western United States, it is part of a multiwatershed system that supplies irrigation water to a large geographic area. Another issue faced by many groups is interbasin water transfers. The National Research Council (NRC 1992b) suggested that all interbasin transfer decisions should include people in the watershed of origin, suggesting the need for cooperation across watersheds on this and many other issues.

Because issues, stakeholders, and environments differ from watershed to watershed, each individual watershed-based organization must necessarily determine its own goals, processes, and participants. Differences in goals among watershed organizations dictate that their funding, composition, structure, and processes will vary. As outlined in the Swift River Principles (Coyle 1995), it is apparent that there is no one standard approach to creating watershed organizations that is universally applicable.

THE FUTURE OF RESTORATION IN THE HENRYS FORK WATERSHED

Watershed Restoration

In general, state and federal agencies are unlikely to increase their restoration efforts substantially in the Henrys Fork watershed. Agencies such as IDFG have little land management authority and thus have a limited ability to implement on-the-ground restoration projects unless they partner with private landowners or other entities. Where opportunities arise for such partnerships, the HFWC may increase the effectiveness of agencies with limited land management authority. Land management agencies such as the U.S. Forest Service have more opportunity to undertake restoration, but the trend of reducing government spending is likely to decrease funds available for restoration. Where there are restoration opportunities on public lands, the HFWC may again help by facilitating cooperative funding agreements to help willing agencies carry out restoration projects they would not be able to fund themselves. The most likely scenario for increased agency involvement in watershed restoration is listing of a species under the U.S. Endangered Species Act. Particularly if the species requires aquatic or riparian habitat, a listing could require agencies to undertake active watershed restoration.

Given the limited role of agencies in restoration and the coordination role of the HFWC, the burden of designing and implementing large-scale restoration in the Henrys Fork watershed has fallen on nongovernmental organizations, primarily the HLF, HFF, and The Nature Conservancy. Because the HLF is concerned only with Henrys Lake and its tributaries and lacks financial and staff resources, it is not in a position to undertake an inventory of the physical and biological attributes of the entire watershed, which is necessary to prioritize restoration needs and design a watershed-wide restoration program. Likewise, The Nature Conservancy is primarily concerned with restoration in and around its 1,400-acre Flat Ranch on Henrys Lake outlet, purchased in 1994 to provide a model of how cattle and healthy riparian and aquatic habitat can coexist on a working ranch.

This leaves the HFF to serve as the primary force behind large-scale restoration in the watershed. Until 1995, the HFF's mission was to preserve and protect the unique recreational, wildlife, and aesthetic qualities of the Henry's Fork. Recently the HFF developed a new mission statement: "to *understand*, *restore* and protect the unique fishery, wildlife and aesthetic qualities of the Henry's Fork of the Snake River" (emphasis added). This new mission statement reflects the HFF's commitment to understanding watershed processes and conditions as a precursor to restoring and protecting the watershed. Ongoing research includes hydrology, fish population dynamics and habitat requirements, nutrient dynamics and water quality, and structure and function of macrophyte and aquatic invertebrate communities.

However, the main research focus for the next 4 years is an assessment of current habitat conditions and aquatic organism populations throughout the watershed. Results from the assessment will be analyzed along with historical information and data from relatively pristine reference streams in the watershed. Restoration projects will then be prioritized and aimed at aiding long-term recovery of natural processes rather than conducting short-term habitat improvement on a single section of river. Monitoring will determine the success of restoration and improve future restoration efforts. This plan of assessment, analysis, prioritization, implementation, and monitoring constitutes the first attempt at a comprehensive restoration program for the Henrys Fork watershed.

Although the HFWC is not likely to be directly involved in research and analysis, individual restoration projects will be brought by the HFF to the HFWC for endorsement and possible funding. The collaborative atmosphere created by the HFWC will certainly aid the HFF in its research, implementation, and monitoring. Even though the HFWC itself is not the leader of the restoration effort, it makes possible the coordination and cooperation among agencies, landowners, and nongovernmental organizations necessary to carry out large-scale restoration of the Henrys Fork watershed.

Native Trout Restoration

Prior to the 1994 discovery of lake trout in Yellowstone Lake (Kaeding et al. 1996), there was little concern about the viability of the Yellowstone cutthroat trout. In the Henrys Fork watershed, only a few agency biologists, the occasional angler who appreciates fishing for native trout, and those involved with Henrys Lake restoration had given any thought to the status of native Yellowstone cutthroat trout. In the case of Henrys Lake, the attention given to cutthroat trout was aimed at maintaining the cutthroat–rainbow trout hybrid fishing opportunity. The HFF concerned itself exclusively with rainbow trout, even as it began to expand its view beyond the 20 miles of river below Island Park Dam. Both the HLF and HFF were motivated to improve habitat and restore watersheds almost exclusively by the desire to improve the quality of their fishing, which primarily consisted of the opportunity to catch large rainbow trout, cutthroat–rainbow trout hybrid, and brook trout. All these nonnative species have contributed to the decline of Yellowstone cutthroat trout. However, since 1994 there has been a greatly increased awareness among both the fisheries management community and the general fly-fishing public that native fishes of all types are at risk of going extinct (Butler 1996; Gierach 1996).

Yellowstone cutthroat trout are either extinct or at risk in over 85% of the Henrys Fork drainage, exclusive of the Teton River, where healthy populations are estimated to occur in 74% of its watershed (USFS 1996a). However, rainbow trout are present throughout the Teton River drainage and thus pose a constant threat of hybridization, and recent surveys of headwater tributaries in the upper Henrys Fork watershed have discovered only brook trout in streams that a decade ago were believed to contain Yellowstone cutthroat trout (J. Griffith, Idaho State University, unpublished data; D. Delany, U.S. Forest Service, personal communication). The general consensus among biologists familiar with the Henrys Fork watershed is that healthy populations of pure Yellowstone cutthroat trout probably exist outside the Teton River only in a few tributaries originating in and around the remote southwest corner of Yellowstone National Park. Although some Henrys Lake tributaries may contain genetically pure Yellowstone cutthroat trout, the presence of rainbow trout in the lake for the past 100 years and the hatchery hybrid program make it likely that almost all of the Henrys Lake Yellowstone cutthroat trout contain at least some rainbow trout genes.

The cutthroat trout is an IDFG species of concern and a U.S. Forest Service sensitive species. Accordingly, both of these agencies have management policies designed to protect populations of Yellowstone cutthroat trout, and both are engaged in research and inventory efforts to gain more information about its status in the Henrys Fork watershed. A U.S. Endangered Species Act listing of Yellowstone cutthroat trout, which is possible in the next few years, would dramatically change agency actions toward cutthroat trout restoration. However, because nongovernmental organizations are currently the most active proponents of habitat restoration and trout population enhancement, the big question is whether increased attention on native trout will motivate the HLF and HFF to become strong advocates for restoration of native species as part of their watershed restoration efforts, even though recovery of Yellowstone cutthroat trout may necessitate abandoning nonnative sport fish favored by anglers.

CONCLUSION

Whether the HFF decides to restore native Yellowstone cutthroat trout at the expense of rainbow trout, the farmers of the Fremont-Madison Irrigation District are extremely unlikely to give up growing seed potatoes and other irrigated crops. In fact, because the Henrys Fork watershed, for a variety of reasons, has lagged behind the rest of the greater Yellowstone area in being converted into a recreation and tourist area, planning and zoning commissions, economic development agencies, and newly formed land trusts in the watershed are seizing the opportunity to promote agriculture, both to maintain economic stability and to preserve open space. As long as there are advocates for both agriculture and trout, be they rainbows or cutthroats, there will be competing demands for water and other resources in the Henrys Fork watershed. The long-term viability and relevance of the HFWC will improve the chances that watershed restoration and improvements in natural resource management will occur and that anglers and farmers can once again coexist peacefully in the Henrys Fork watershed.

ACKNOWLEDGMENTS

Preparation of this chapter was supported by a grant from the National Fish and Wildlife Foundation that was matched by private and corporate contributions. The authors thank Russ Thurow and Cindy Deacon Williams for helpful suggestions that greatly improved the content and readability of the chapter.

Appendix 16.1: Watershed Integrity Review and Evaluation (WIRE) Criteria

1. Watershed perspective.—Does the project employ or reflect a total watershed perspective? The project demonstrates an understanding of the relationships that exist among

a. physical parameters of watershed (soil formation and other geologic processes);
b. surface and ground water resources (headwaters and lowland resources);
c. biological components (organisms associated with the watershed);
d. ecological communities and processes (forests, meadows, riparian zones, migration corridors, nutrient cycles, and predator–prey relationships);
e. human communities (towns, transportation corridors, historic and archeological sites, economies); and
f. climatic factors (weather patterns and air quality).

2. Credibility.—Is the project based upon credible research or scientific data?

a. The project demonstrates use of scientific principles and procedures (rather than strictly a response to political agendas or impending crises).
b. The project clearly cites references or current research results to support its approach or meets research goals and objectives set by the HFWC.
c. The project has undergone appropriate regulatory processes.
d. The project's goals and approach are clear and understandable by the general public.

3. Problem and solution.—Does the project clearly identify problems and propose workable solutions that consider the relevant resources?

a. Based on scientific evaluation, the project demonstrates that problems exist.
b. The project contributes toward the maintenance, enhancement, or restoration of specified resources to proper functioning condition.
c. Cumulative effects of project strategies have been considered.

4. Water supply.—Does the project demonstrate an understanding of water supply?

a. The project describes the quantity, quality, timing, and source(s) of water involved.
b. The project considers potential effects on water interests within and beyond the Henrys Fork watershed.
c. The project demonstrates an understanding of watershed dynamics and regional water policy.

5. Project management.—Does project management employ accepted or innovative practices, set realistic time frames for project implementation, and employ an effective monitoring plan?

a. The project sets reasonably achievable objectives with measurable results.
b. The timeline for project implementation is clear and has contingency plans.
c. A monitoring plan is in place to evaluate the project effectively.

6. Sustainability.—Does the project emphasize sustainable ecosystems?

a. The project recognizes the natural limits of the resources involved.
b. The project helps to ensure the sustainability of the ecosystem for future generations.
c. The project recognizes the importance of maintaining the basin's biological diversity, preventing the need for species listing under the U.S. Endangered Species Act.

7. Social and cultural perspective.—Does the project sufficiently address the watershed's social and cultural concerns?

a. The project provides or maintains educational, recreational, and cultural opportunities.
b. The project considers community welfare, health, and safety needs and local lifestyles in its design.
c. An understanding of ongoing social change and its costs and benefits to local communities is demonstrated.
d. The project considers the development pressures being experienced by the basin.

8. Economy.—Does the project promote economic diversity within the watershed and help sustain a healthy economic base?

a. The project creatively supports a sustainable basin economy.
b. It is clear who benefits from and who is sharing the costs of the project.
c. The project employs local labor, materials, and expertise where realistic and available.

9. Cooperation and coordination.—Does the project maximize cooperation among all parties and demonstrate sufficient coordination among appropriate groups or agencies?

a. The project uses the expertise and talents of local citizens, agencies, and scientists and outlines how communication among these interests will be maintained.
b. The project transcends political agendas and jurisdictional boundaries.
c. The project maximizes efficiency among agencies and is coordinated with other activities in the watershed or subwatershed.

10. Legality.—Is the project lawful and respectful of agencies' legal responsibilities?

a. The project complies with federal, state, and local laws and regulations, including the National Environmental Policy Act and the U.S. Endangered Species Act.
b. The project respects vested water rights and protects the beneficial consumptive and nonconsumptive uses of water established by law.
c. The project points out any conflicts in legal mandates and suggests any needed changes in laws or regulations.
d. The project recognizes both public and private property rights in its design.

Never doubt that a small group of thoughtful, committed citizens can change the world; indeed, it's the only thing that ever has.
—Margaret Mead

CHAPTER 17

USING RIVER ACTION TEAMS TO RESTORE WATER QUALITY: HIWASSEE RIVER OF NORTH CAROLINA, GEORGIA, AND TENNESSEE

David L. Bowling, Jr., Terry S. Chilcoat, Janice P. Cox, James R. Hagerman, Christopher D. Ungate, and Gary G. Williams

The mission of the Tennessee Valley Authority (TVA), as stated in the TVA Act of 1933, is to provide for the "unified conservation and development of the Tennessee River system." The Tennessee River drains a 41,000-square-mile watershed that covers portions of seven southeastern states. The river has more than 30 major reservoirs operated by the TVA for navigation, flood control, water quality, power production, recreation, and other purposes.

In 1992, in response to numerous requests from recreational users, lakeshore homeowners, and the general public, the TVA Board of Directors adopted the goal of making the Tennessee River the cleanest and most productive commer-

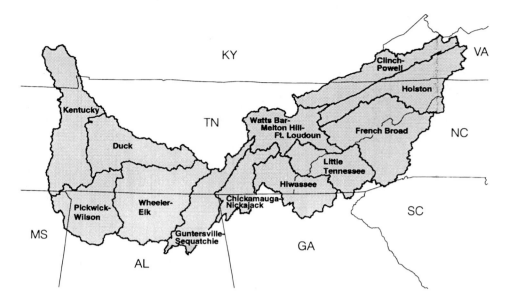

FIGURE 17.1.—The 12 River Action Team subwatersheds of the Tennessee River valley.

cial river system in the United States. The TVA's Clean Water Initiative was created to carry out this mission.

The Clean Water Initiative developed an innovative approach, placing a River Action Team (RAT) in each of the Tennessee River's 12 major subwatersheds (Figure 17.1). These small, self-directed teams of water resource and community specialists gather information, help build coalitions, and coordinate resource protection and recovery efforts. The Hiwassee RAT was the first to take the field and initially served as a model for the RATs that followed.

THE HIWASSEE WATERSHED

The Hiwassee River originates in the southern Appalachian highlands of Georgia and North Carolina. The river basin covers some 2,700 square miles in all or parts of 11 counties in Georgia, North Carolina, and Tennessee. Over 4,000 miles of stream drain the basin. There are eight TVA impoundments in the watershed, and the river flows into a ninth at Chickamauga Reservoir on the Tennessee River itself. About 75% of the Hiwassee River drainage area is forested and about 18% is pasture. Twelve towns and five major industries rely on the Hiwassee for their water.

The basin is largely rural. The population of about 230,500 is projected to grow to approximately 250,000 by the year 2000 (USDOC 1990). There is only one town, Cleveland, Tennessee, with a population above 15,000. Employment is weighted toward the manufacturing and service industries, with a secondary emphasis on transportation, agriculture, and forestry. As of 1993, per capita income hovers around US$15,000, about 70% of the national average, and the

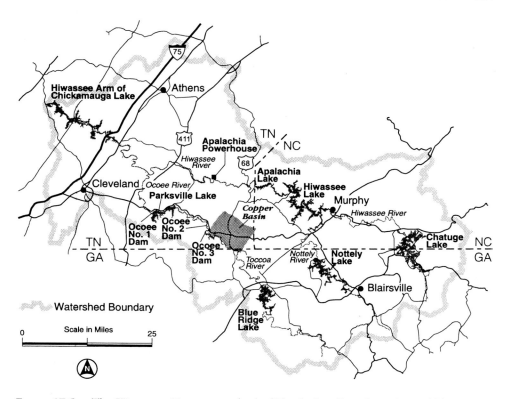

FIGURE 17.2.—The Hiwassee River watershed of North Carolina, Georgia, and Tennessee.

unemployment rate is near 7.5%. An apparently significant factor in the economic and social future of the region is a growing emphasis, particularly in the upper basin, on real estate (second home) development and the recreation and tourism industries.

The Hiwassee basin is divided naturally into three sections: a forested upland, a narrow middle section where the rivers have cut canyons through sandstone ridges, and a region of farmland and industry on the lower reaches of the river (Figure 17.2).

The Upper Valley

The upper, southeastern portion of the Hiwassee basin is a region of woodland, streams, and lakes. Almost 90% of the area is forested, and much of it is U.S. Forest Service land (Chattahoochee National Forest in Georgia and Nantahala National Forest in North Carolina). The topography is mountainous, and the streams drain through narrow, high-gradient valleys.

The most striking hydrologic features in the upper watershed are the five TVA impoundments. Because these reservoirs drain forest underlain by thin soils and insoluble metamorphic rock, their waters are extremely soft and have very low concentrations of plant nutrients, especially nitrogen and phosphorus. These clear, nutrient-poor reservoirs are naturally infertile, but they can respond dramatically to any increase in nutrient input, making them vulnerable to eutrophication as the lands in their watersheds are developed for agriculture or homes. The streams of the upper watershed are primarily high-gradient coldwater habitat. These extremely soft-water streams have base flow pH measurements ranging from 5.8 to 6.8 and are very vulnerable to acidification. Brook trout, the only salmonid native to the southern Appalachians, is of special importance in this portion of the watershed to the Hiwassee RAT and its partners. Prior to 1900, the range of brook trout probably extended down into the foothills to an elevation of about 1,600 feet (Bivens 1985), covering as much as one-third of the total Hiwassee River watershed. After the turn of the century, the introduction of rainbow and brown trout, as well as widespread logging, road building, and other land disturbances, contributed to the extirpation (local extinction) of most of the brook trout populations in the Hiwassee drainage (Bivens 1985). The team recently documented brook trout populations in 30 streams (Williams 1996). The combined drainage area of these 30 streams is only 26 square miles, representing less than 4% of the historical distribution. These streams are restricted to elevations above 2,500 feet, where small first- and second-order streams are susceptible to droughts, flash floods, anchor ice, and other extreme conditions. Twenty-six of the brook trout populations are small and disjunct and are isolated in small headwater streams where there is little possibility of replacement by introduced trout.

The Middle Valley

The middle valley begins at the tailwaters of Apalachia Dam and extends through the ridgelines that form the last outliers of the Blue Ridge physiographic province. The topography is steep and mostly wooded, and nearly half of the land is within the boundaries of the Cherokee National Forest in Tennessee. There are no major population centers in the middle valley, and the region is sparsely settled.

The middle Hiwassee is dominated by coolwater streams that tend to support a higher number of fish species (species richness) than either the coldwater streams of the higher elevations or the warmwater streams of the more-developed lower watershed. Coldwater habitat persists in the main river because of release of waters from the hypolimnion (the deepest, coldest layer) of the Apalachia Reservoir; this release provides one of the premier tailwater trout fisheries in the southeastern United States. The Hiwassee River in the middle valley is also classified as a State Scenic River in Tennessee and was considered for federal wild or scenic status.

This middle section of the watershed has areas of great natural beauty, which attract tourism and outdoor recreation, but it also has significant environmental problems. The Copper Basin is located in the drainage of the Ocoee River, a large tributary of the Hiwassee River. About 50 square miles of the Copper Basin were denuded by copper smelting before the turn of the century. Siltation, acid, and metals from the basin all but destroyed aquatic life in long stretches of the Ocoee, and siltation has reduced storage capacity in three small TVA impoundments. Thanks to a major cooperative effort, beginning in 1984 most of the denuded area left unreclaimed by earlier efforts has been revegetated, but contaminated sediments and groundwater inflows continue to limit aquatic life in the Ocoee. Nevertheless, part of this same stretch is rated as one of the top-10 white-water attractions in the country. In 1994, commercial outfitters operating on the Ocoee took more than 210,000 visitors rafting down the river, and an additional 23,000 descended in their own watercraft. These visitors contributed an estimated $23 million to the local economy. An artificial slalom course recently completed on this stretch of the Ocoee was the site of white-water slalom events during the 1996 Olympic Games.

A unique reach of the Hiwassee in the middle valley also poses an environmental management challenge. During normal reservoir operation for power production, the TVA diverts the Hiwassee River at Apalachia Dam into an artificial tunnel that leads to Apalachia Powerhouse. At the powerhouse, the water is returned to the riverbed after passing through the hydropower generators. The 12.4-mile cut-off reach of the natural channel between Apalachia Dam and Apalachia Powerhouse usually carries only inflows from small tributary streams. The TVA's monitoring of biological indicators shows that the overall ecological health of this section of the Hiwassee River is only fair, despite the fact that most of the tributary streams draining this area are in good ecological health. Several of these streams carry heavy sediment loads. The sediment is deposited in the shallow, braided channel of the cut-off reach, affecting bottom-dwelling organisms and young fish. Nevertheless, the cut-off reach supports a good smallmouth bass fishery and a variety of rare species, including the federally listed endangered Cumberland bean mussel, tan riffleshell mussel, and Ruth's golden aster. Because the flows from Apalachia Powerhouse tend to limit upstream colonization by benthic organisms into the cut-off reach, this area could have unique value as a sanctuary where native mussels can evade the nonnative zebra mussel, which is decimating native mussels in other U.S. river systems.

The Lower Valley

The flatter land of the lower valley is the most densely populated and heavily farmed area in the Hiwassee watershed. There is no national forest land, and a number of industries and several small towns are located along the tributaries. The greater population and more intensive land use make the lower Hiwassee streams vulnerable to siltation from land and streambank erosion and to pollution from urban runoff.

OPERATION OF THE HIWASSEE RIVER ACTION TEAM

The Hiwassee RAT was created in 1992 to perform a specific function: to bring together stakeholders in a cooperative manner to increase the effectiveness of watershed restoration. In its role as a facilitator, the team is able to address needs and issues that are often not effectively addressed by single-focus programs, including

1. holistic resource management, which focuses on resource needs rather than restricting efforts to certain types of pollution problems, technologies, or resources, political or jurisdictional boundaries, or legislative requirements;
2. development and promotion of strategies that balance human use of the resource with resource integrity to achieve economic and resource sustainability;
3. consideration of the needs and expectations of all stakeholders (resource users, landowners, and cooperators) to increase the likelihood that voluntary solutions will be widely implemented and accepted; and
4. proactive protection of high-quality or rare aquatic resources.

The Hiwassee team's initial decisions on where to focus financial and technical resources set the tone for its early efforts and continue to influence resource allocation 3 years later. Initial decisions about how to target resources set priorities based on three criteria:

1. protecting and improving "high-value" (ecological or human use) resources;
2. solving problems that have a significant negative effect on the resource or that constitute a critical threat to the sustainability of the resource; and
3. targeting projects that were likely to be successful as cooperative ventures.

Several projects that clearly met these criteria were initiated during the first year. In some cases, however, the team lacked current information on the status of resources and needed a better understanding of the interests, goals, and limitations of potential cooperators. Consequently, a significant portion of the total team effort during the first 2 years was invested in obtaining baseline information for the basin.

An Information Base to Support Decisions

In order to learn more about the condition of watershed resources, the team conducted stream bioassessments at over 170 sites in the watershed. In the past, the TVA, like many other state and federal agencies, relied on the measurement of physical and chemical parameters to monitor water quality. As a pilot team, the Hiwassee RAT decided to invest a larger portion of its resources in biological monitoring. The anticipated benefits of biological monitoring were fourfold.

1. Although the interpretation of physicochemical monitoring is relatively straightforward, it provides a very incomplete picture of something as complex as ecological health. The team's original intent, therefore, was to use biological monitoring to measure ecosystem response to a wide range of interacting water quality and habitat stressors.
2. Given that the biological community integrates the effects of stressors over a period of time, the team anticipated that it would be more cost-effective to conduct biological assessments (which required 2 to 4 hours of team effort per site and had no associated laboratory costs) than to conduct detailed storm runoff and base flow water quality monitoring.

3. The team expected the baseline bioassessments to provide a benchmark for tracking and measuring the success of watershed restoration activities.
4. Conducting bioassessments would provide a mechanism for the team to build firsthand knowledge about the watershed, which was invaluable in supporting the team's credibility with cooperating agencies and resource users.

The team's bioassessments were based on a modification of the index of biotic integrity (IBI) originally introduced by Karr (1981). The index is based on the assumption that fish community features change in a predictable manner with stream degradation (Leonard and Orth 1986). In some streams (e.g., coldwater streams with naturally low species richness, high-gradient streams with bedrock substrate, and streams slightly enriched relative to the naturally occurring low fertility), Karr's original IBI proved difficult to apply, and metrics had to be replaced, deleted, or added to reflect regional features better. To provide data from another trophic level in addition to the fish community, the team also assessed the benthic macroinvertebrate community. The techniques outlined by Plafkin et al. (1989) were used and results reported as an EPT index (the total number of distinct taxa within the pollution-sensitive Ephemeroptera [mayfly], Plecoptera [stonefly], and Trichoptera [caddisfly] orders of aquatic insects).

At the end of 2 years, the team had modified IBI scores for all major streams in the watershed. Twenty-nine percent of the sites rated poor, 44% rated fair, and 26% were good or excellent. With few exceptions, poor ecological health scores in the Hiwassee watershed appeared to be attributable to nonpoint sources and land use practices rather than point source discharges.

Becoming knowledgeable about the watershed also required that the team begin cultivating relationships with potential cooperators in the public and private sectors. Although adequate planning is necessary for the long-term success of any project, the team found that day-to-day cooperative efforts depend upon developing personal relationships with technical staff in other agencies, community leaders, other stakeholders, and the general public. These relationships open the door to a level of communication, trust, accountability, and stability that cannot be achieved by any level of office preparation.

The Hiwassee RAT continues to spend a significant amount of time gathering information and developing partnerships with stakeholders, resource users (e.g., communities, industries, environmental groups, and recreation groups), and other agencies. The goal is to develop cooperative ventures that pool technical and financial resources to solve water resource problems while meeting the economic, social, regulatory, and environmental needs of the stakeholders.

A Cooperative Effort to Develop a Watershed Strategy

Issues that the team is addressing can be grouped into several categories, each of which calls for different strategies (Table 17.1). Although the team has long-term strategies for dealing with certain problems in identified locations, there is no written charter or overall watershed plan. In some cases, it is clear that future activity will be grounded in organized, chartered groups such as the Hiawassee River Watershed Coalition (discussed below). In other cases, partic-

Table 17.1.—Major issues, goals, and strategies identified by the Hiwassee River Action Team.

Issue	Goals	Strategy
Protect high-quality mountain streams in middle and upper valley from small-scale but widespread development	Ensure development does not overtax watershed Maintain and enhance communities of native species	Identify high-quality and high-risk resources Develop local coalition to champion protection efforts Educate stakeholders on how land-use activities affect the aquatic resource
Improve quality of streams throughout the watershed that have been affected by agricultural activities—especially livestock rearing and destruction of riparian vegetation	Target best management practices to improve stream ecological health to fair or better Protect reservoirs from excessive productivity, hypolimnetic dissolved oxygen depletion, and sediment loads	Target sites with the greatest potential for improvement Locate cost-sharing funding from other sources Implement targeted best management practices Monitor completed projects to verify effectiveness
Improve ecological health of Tennessee Valley Authority (TVA) reservoir tailwaters	Maintain dissolved oxygen minimum of 5 mg/L for cool waters and 6 mg/L for cold waters Sustain flow Address other problems on an as-needed basis	Allow TVA Reservoir Releases Improvements (RRI) program to take the lead Support RRI activities by addressing water quality problems that would diminish the effects of RRI mitigation projects
Improve quality of lower valley streams that have been affected by urban runoff	Improve stream ecological health to fair or better Ensure recreation areas meet bacteriological standards	Develop and nurture local coalitions that will champion improvement projects
Return lands and streams of the Copper Basin to productive use	Revegetate denuded lands Control sedimentation in Ocoee hydropower projects Control nonpoint sources of toxic mine waste runoff	Form partnership with industries, National Resources Conservation Service, and local landowners to complete revegetation Form partnership with industry to identify and address problems of nonpoint source mine waste Verify efficacy of land treatment activities with stream bioassessment
Achieve an equitable balance between conflicting resource uses	Include stakeholders in decisions about conflicting resource uses Base decisions on good data	When appropriate, collect information needed to support good management decisions Provide a forum for stakeholders to provide input to resource management decisions

ularly where the success of restoration activities depends on their acceptance by individual landowners who may be skeptical of government agency involvement in decisions that affect the use of private land, a less-formalized approach appears more likely to be successful. In any case, a comprehensive watershed plan devised by only one or a few agencies is unlikely to be accepted by all the landowners and cooperating agencies who have a stake in the outcome.

In cases in which different interest groups have diverse visions of how the resource should be managed, the strategy of the team has been to ensure that accurate information is available for decision making and that stakeholders have an opportunity to voice their concerns and have a place in the decision-making process.

Two federal agencies, the U.S. Forest Service and the U.S. Natural Resources Conservation Service (USNRCS; formerly the Soil Conservation Service), have been the major cooperators with the TVA on restoration efforts in the Hiwassee watershed. Both are natural partners for restoration projects. The U.S. Forest Service administers a substantial amount of the land in the basin, and the USNRCS has been working with farmers in the area for many years.

Other natural cooperators include conservation organizations, state game and fish management agencies, state water pollution control management agencies, and state transportation departments. Counties, municipalities, educational institutions, industries, special interest groups, volunteers, public service organizations, and members of the general public are also valued partners. Although the team is officially composed of only TVA personnel, partners from these organizations and groups are instrumental in decision making and project development.

An encouraging development in the process of building partnerships has been the organization of the Hiawassee River Watershed Coalition. This citizen-based coalition was formed in 1994 to give local guidance and coordinate water quality activities in the cluster of counties along the North Carolina and Georgia border. Voting members of the steering committee include representatives of the three soil and water conservation districts (quasi-governmental organizations that operate on a county basis) and four county governments in the upper watershed. The U.S. Natural Resources Conservation Service, U.S. Forest Service, Tennessee Valley Authority, Georgia Department of Natural Resources, North Carolina Wildlife Resources Commission, and other agencies have nonvoting representatives and provide technical support. The coalition has been particularly effective in coordinating multistate and multiagency groups in targeted water quality projects. If the coalition continues to be successful, it could serve as the model for similar organizations in other parts of the Tennessee Valley.

WATERSHED RESTORATION EXPERIENCES

Once the need for a restoration or protection project has been demonstrated through monitoring for biological health or for water quality standards designated by the state for specified water uses, most projects go through several developmental phases before a full-fledged restoration project is initiated. The rationale is explained as follows.

1. Not all worthy projects can be attempted at once. The team narrows the initial list of potential projects by looking at resource and staffing constraints, conflicting demands on cooperators, and planning considerations such as economies of scale. This reduces the list of potential projects to a point at which strategic planning for individual projects can be undertaken.
2. In some instances, the cause of the problem is not clear or there is insufficient information to allow accurate targeting of activities. The first stage of the project then becomes gathering strategic information.
3. In cases in which there are questions about the effectiveness of available technologies, the activity may develop into a demonstration effort or be transferred to research and development experts.
4. When there is limited support from cooperators and stakeholders, the team must first build a coalition of supporters who will contribute financial or technical resources to the effort before the project can proceed.

Only when these hurdles have been passed does the team begin large-scale investment of resources. In any given year, the five-person team (two environmental engineers, one environmental scientist, one biologist, and one education specialist) has had 20 to 30 projects of various types underway. The projects discussed below provide an example of the team's approach to each of the six categories of issues (Table 17.1) addressed by the Hiwassee RAT.

Protection of High-Quality Resources: Shuler Creek

Shuler Creek drains approximately 13 square miles of predominantly forested land in the middle valley of the Hiwassee in North Carolina. A gravel road, owned by the North Carolina Department of Transportation but lying within Nantahala National Forest, runs parallel to the creek along much of its length. After damaging floods in the winter of 1993–1994, emergency repairs to the road resulted in about 1 mile of streambank being covered with bare, unconsolidated dirt fill. Because of its gradient, Shuler Creek was effectively transporting the sediment from the eroding streambank into a section of the Hiwassee River that harbors a variety of rare mussel species.

Complications arising from the split ownership of the land and the road made a simple project surprisingly difficult. However, the team eventually brought together a cost-share grant from the National Fish and Wildlife Foundation, in-kind services (heavy equipment and operator) from the North Carolina Department of Transportation, and technical assistance from the U.S. Forest Service and the North Carolina Wildlife Resources Commission. The Hiwassee RAT then contributed its own labor and materials to help build rock-filled timber crib structures to stabilize 1,050 linear feet of critically eroding streambank (Figure 17.3).

Subsequently in 1996, the U.S. Forest Service administered the contract, North Carolina Department of Transportation again contributed heavy equipment, and

FIGURE 17.3.—Construction of streambank stabilization structures on Shuler Creek, North Carolina.

the Hiwassee RAT designed a bioengineering solution (primarily rootwads driven laterally into eroding streambanks) that will help stabilize another 600 feet of streambank. The Hiwassee Chapter of Trout Unlimited has agreed to collect stormwater runoff samples in the creek to measure the effectiveness of the project.

Control of Agricultural Nonpoint Sources: Chatuge Tailwater

The Hiwassee RAT is presently working with a number of partners on projects designed to reduce the effects of nonpoint source pollution on the tailwaters of Chatuge Dam, North Carolina. The Clay County Soil and Water Conservation District, the USNRCS, and individual landowners were instrumental in encouraging landowners to build structures and adopt practices that would help control runoff problems from dairy farming and beef cattle production activities. To date, 12 landowners have agreed to participate, adopting such practices as fencing cattle from watercourses, providing watering lanes (fenced pathways to the stream that limit access to the streambank) and hardened water crossings (paved or otherwise surfaced fords that protect the streambed), preserving riparian vegetation, and demonstrating an artificial wetland to treat waste from a dairy farm. The Hiawassee River Watershed Coalition was a valuable partner in carrying out these projects.

FIGURE 17.4.—Streambank stabilization work on the Nottely River, North Carolina.

Mitigation of Downstream Impacts of Reservoirs: Nottely Tailwater

While the TVA's Reservoir Releases Improvements program addressed the low dissolved oxygen and intermittent flow concerns in the tailwater below Nottely Dam, the Hiwassee RAT was able to concentrate on a severe streambank erosion problem that threatened to eclipse the potential ecological recovery expected from the enhancements of flow and dissolved oxygen. In 1993, the TVA, the Cherokee County Soil Conservation Service, the Southwestern North Carolina Resource Conservation and Development Council, and local landowners in North Carolina met to identify solutions to the water quality problems on the Nottely River. In April 1995, bank stabilization work on the Nottely River began. To date, seven of the worst eroding sites on the river have been stabilized, and adjoining farms have adopted measures to reduce erosion from overland runoff into the river (Figure 17.4). At least five more sites will be completed by 1997.

Control of Urban Runoff: South Mouse Creek

Some of the streams in the lower Hiwassee valley—South Mouse Creek in Bradley County, Tennessee, for example—are negatively affected over such a long distance by a variety of sources that restoration to pristine conditions is not feasible. The Hiwassee RAT is developing a project with local industries to draw attention to the problems and educate landowners on what can be done. Possible solutions include development of a greenway along the urban reaches of the creek.

Reclamation of the Copper Basin: Cooperative Revegetation

More than 50 years of open-air smelting of copper ore, coupled with logging and stump removal for smelter fuel, open-range cattle grazing, and indiscriminate burning of pasture land, resulted in the destruction of vegetation on about 32,000 acres (50 square miles) in the Copper Basin before the turn of the century. The resulting soil erosion, which ran unchecked for over a century, led to the loss of most of the topsoil and much of the subsoil on about 23,000 acres (36 square miles). The eroded soil has essentially filled Ocoee dam 3 reservoir and is building a large delta in the upstream end of Parksville Reservoir (Ocoee dam 1). Furthermore, the Ocoee River downstream of the Copper Basin has elevated concentrations of dissolved zinc and copper. The sediments in Ocoee dam 3 reservoir, immediately downstream of the Copper Basin, are markedly enriched with arsenic, cadmium, copper, lead, and zinc; however, the degree to which sediments are presently serving as either a source or a sink of metals in the water column has not been determined.

The creation of this problem preceded today's level of environmental awareness, regulation, and pollution control technology by nearly a century. Efforts to revegetate the Copper Basin began over 50 years ago, but early work was not very successful due to technological limitations and the harsh conditions in the basin. An intensive cooperative effort was initiated in 1984, and the Hiwassee RAT was able to leverage funds that multiplied the rate of reclamation activities fivefold in 1992, the team's first year of operation.

Improvements in the condition of the land, along with improvements in wastewater treatment, are beginning to be reflected in the condition of the streams of the Copper Basin. Although few fish have moved back into the basin's streams, the number of pollution-sensitive aquatic insects has doubled in some streams since the 1980s. This is an early and encouraging sign of ecological recovery. The Ocoee River itself is also showing signs of recovery. Although the river is still in poor ecological health, the number and diversity of fishes and benthic (bottom dwelling) insects is increasing. Some areas of the river that were completely devoid of aquatic life in the 1980s now support pioneering benthic insect species and several species of fish.

The cooperative reclamation project depends on partnerships among the Tennessee Valley Authority, Boliden Intertrade Ag, the U.S. Natural Resources Conservation Service, Bowater, Inc., and a number of private landowners. Today, only 15% of the original problem area (4,800 acres out of 32,000 acres) remains to be treated. This partnership plans three more years of intensive reclamation efforts followed by 2 years of maintenance work to ensure that the vegetation is self-sustaining.

Equitable Balance of Conflicting Resource Uses:
A State Scenic River

The 23-mile reach of the Hiwassee River downstream of Apalachia Dam was the first State Scenic River designated by Tennessee. This section of river contains the cut-off reach discussed above. In addition, directly downstream of Apalachia Powerhouse, the river supports superior rainbow and brown trout fisheries. The 11-mile reach below Apalachia Powerhouse is very popular with rafters, canoeists, and float anglers, attracting about 43,000 boaters in 1994. This section of the Hiwassee River was recently evaluated for addition to the federal Wild and Scenic River system, but the proposal was not adopted because of lack of public support.

Flows in this stretch of the river depend on hydropower generation, but the peak demand for hydropower does not necessarily coincide with the times when either rafters or fly anglers would like flows scheduled. Other factors also may compound conflicts over release schedules. For instance, the temperature of hydropower releases in late summer is marginal for the trout fishery in the tailwater because the cold water in the reservoir is sometimes depleted by late summer. Commercial interests, for example an entrepreneur who is proposing to raise trout in holding pens in Apalachia Reservoir, would also like to have access to the cold water that is in such short supply during late summer. Managing the resource to balance conflicting uses (e.g., hydropower production versus recreation) is a difficult challenge. In some cases, the Hiwassee RAT has been able to increase user satisfaction simply by working with small user groups to ensure that they understand how resource management for one use affects management for other legitimate uses.

Along with partners from other resource management agencies, the team is also developing an action plan to monitor and protect the unique aquatic resources of the cut-off reach. In the first phase of this activity, the partners have turned to the watershed for answers. A volunteer network to collect sediment samples during storm runoff events has been in place since 1994. Preliminary results from this effort were used to target a more intensive sampling project on one of the major tributaries in this reach, Turtletown Creek. A cooperative project to control erosion in the Turtletown Creek watershed is now being developed.

Monitoring and Adaptive Management

The team's objective is to measure the ultimate success of projects in terms of improvements in ecological health, as measured by indices such as IBI, or suitability for use, as measured by compliance with water quality standards. Therefore, tracking projects requires resource assessments on a regular basis. Modified IBIs or other rapid bioassessments are performed annually on all streams affected by ongoing projects. The bacteriological quality of waters at most developed recreational sites is routinely monitored by other agencies. However, the team often arranges to monitor sites—such as swimming areas that do not have developed recreational facilities—that are not being monitored by others.

The team has not conducted routine chemical water quality monitoring. However, special water quality studies are undertaken when needed to define the nature and extent of a problem better so that appropriate restoration efforts can be planned. For example, water quality sampling is being done in the streams of the Copper Basin to determine whether heavy metal pollution, as well as unfavorable habitat features, are limiting recovery of the aquatic community.

The team is also testing single-stage samplers (a set of suspended sediment samplers that automatically collect samples during the rising stage of a storm) for suspended sediment monitoring. This inexpensive, low-technology method for evaluating sediment loads from storm runoff is proving useful for targeting subwatersheds for more detailed analysis, such as by aerial inventory. The sediment samples can be collected easily by volunteers after the runoff event, providing another opportunity to involve the public in cooperative projects.

Response to New Information

New resource assessment information is reviewed during the development of the team's annual business plan. The team can, and sometimes does, redirect efforts in response to new assessment data, but changes are usually at the level of fine tuning rather than dropping entire projects. More often, major changes in focus result from the opening or closing of a window of opportunity, such as when partners propose a new and unexpected approach or cooperative effort. Team consensus is always sought in cases involving a major redirection of effort.

Since its inception, the Hiwassee RAT has undergone two major evolutions that affect the way it approaches its work. The first change was a move away from the very objective mechanism that was used to define priorities for initial efforts. In the first year's plan, a detailed matrix was developed to evaluate each potential project in terms of (1) value of the resource, (2) significance of the impact to be addressed, and (3) probability of successful resolution through a cooperative venture. In subsequent years, a less-structured, more-intuitive process has proved adequate. Second, the team is now putting much greater emphasis on combining resources with a number of partners rather than assuming the lion's share of the financial burden. Given the funding constraints that government agencies are operating under today, the team believes that efforts to pool resources are crucial to making government more effective in the environmental arena.

CRITIQUE

The TVA's approach to solving water quality problems on a watershed basis through the use of interdisciplinary teams has proven to be successful. According to a report issued by Water Quality 2000 (1994), a coalition of 70 national organizations charged with developing a national water quality agenda, it should be "expanded, promoted, and replicated in other watersheds." The TVA's Clean Water Initiative also won a Hammer Award from Vice President Gore's office for its efforts in reinventing government. The TVA's River Action Teams were recognized specifically for addressing the causes rather than the symptoms of pollution and for their efforts to make government work better and cost less.

Strengths of the River Action Team Approach

The Hiwassee RAT approach to problem solving has proven to have a number of advantages, some of them due to team structure and others to the focus on a narrow geographic area.

1. Operating as a small, self-directed team allows a rapid response to evolving or newly discovered problems and opportunities. The Shuler Creek work discussed above is a good example. The streambank erosion problems were discovered in April, funding sources were pulled together by August, and the first year's construction work was completed by the end of September.
2. The team's decisions about where to focus resources take into account resource-oriented factors (such as the inherent value of the resource, the reversibility of the problem, or the presence of sensitive species) rather than being driven solely by windows of opportunity and political exigencies.
3. Having a self-directed team assigned to a limited geographical area on a long-term basis generates an understanding of—and commitment to—the resource among team members. It also enhances the establishment of very productive one-on-one cooperative relationships with stakeholders and staff from other agencies.
4. The team pursues holistic solutions that address sources rather than just symptoms. For example, the team's approach to controlling sedimentation impacts on threatened and endangered mussels in the Hiwassee cut-off reach involves moving upstream into the watershed to control erosion at the source rather than trying to manipulate the flow regime or channel geomorphology to route sediment through the system more effectively.
5. The team places a high value on projects that prevent problems *before* they occur.
6. Leveraging funds through cooperative interagency ventures allows the team to undertake large-scale projects that it would not be able to undertake on its own.

Weaknesses

The first few years of operation have revealed some weaknesses in the team's initial strategy. Every watershed is different, however, and other RATs have not necessarily experienced the same problems.

1. The team has not yet achieved broad public input into team projects. Workable methods to solicit input in a form that can be used to conduct project planning are still being developed.
2. The team does not always have an effective strategy for increasing the voluntary use of best management practices (structural or managerial land treatment practices devised by the USNRCS to help control the introduction

of pollutants from agricultural practices into the water resources, e.g., rotational grazing and waste lagoons), especially in areas where per capita income is far below the national average and farm income is very low.

3. Ecological health, as measured by a modified IBI, is one of the primary measures that the team had planned to use to track effectiveness of projects. However, this biological assessment tool is still under development and is not well accepted by all cooperators. Further, the team's experience to date with the modified IBI has given some unanticipated results (e.g., lower IBI scores in streams where best management practices have been installed and higher average IBI scores in streams with disrupted riparian vegetation). Although the modified IBI may still be useful for tracking long-term recovery, at the very least another tool will have to be developed for tracking short-term progress. The team's annual performance measures are presently process oriented (e.g., number of best management practices installed and number of critically eroding acres treated), which can make it difficult to terminate ineffective projects or to make midcourse corrections.

PROSPECTS FOR THE FUTURE

As facilitators of cooperative efforts, RATs can concentrate on (1) managing resources holistically, that is, focus on overall resource needs rather than restricting efforts to certain types of pollution, resources, technologies, political or jurisdictional entities, or legislative requirements; (2) promoting strategies that balance human use of the resource with resource integrity to achieve resource sustainability; (3) factoring in the needs of all stakeholders to increase the likelihood that voluntary solutions will be accepted and implemented; and (4) protecting resources before uses are negatively affected or ecological integrity is degraded. Assigning self-directed teams of water resource professionals to specific watersheds has proven to be an efficient, flexible, and effective method of addressing water quality problems in the Tennessee Valley.

The TVA originally planned to establish self-managed teams of five to eight water resource professionals in each of the 12 subwatersheds in the Tennessee Valley. However, full implementation of this plan awaits full funding and greater focus on the water quality problems of the valley. In the interim, the Clean Water Initiative will continue to work toward its long-term goal of improving the beneficial uses of water resources in the valley and transferring the responsibility for sustaining these improvements to the user public by the year 2015.

PART 4

Case Studies

Each watershed has a unique set of restoration needs, human resources, opportunities, constraints, and land ownerships. It is probably also true that no problem is unique. We can all learn from the hard work, successes, and failures of others.

In this section, case studies of watershed-scale restoration efforts are presented from around the country. A variety of restoration projects were chosen. Each is broad in geographic scale, strong on vision, and long on effort. Chapter 18, the case study of work on the Anacostia River, chronicles efforts in one of the most heavily urbanized watersheds in the country, which includes much of Washington, D.C., and adjacent suburban Maryland. Other case studies follow from agricultural lands in Iowa (Chapter 19), forestlands of the Pacific Northwest (Chapter 20), rangelands of the Great Basin (Chapter 21), and large river systems of the Catskill Mountains (Chapter 22), the Rocky Mountains (Chapter 23), and southeast Florida (Chapter 24). These case studies were not chosen because they are all great success stories—although some definitely are—but because they all provide important experiences, methodologies, and lessons that can be helpful for those involved in restoration efforts in almost any region of the country.

Watershed restoration projects would be considerably easier if we had dozens of successful examples to follow. Unfortunately, most restoration efforts are too narrowly focused, too short term, or otherwise suffer from some flaw that reduces their value as a general model for others to follow. Some simply have not been operational long enough to know whether the efforts will be productive in the long term. Finding successful restoration efforts is a challenge; even defining what constitutes success can be difficult. For many watersheds, the process of degradation has proceeded for many decades, if not many generations. It is a complex, long-term process to restore watershed health. Can the coalitions described in these case studies remain as effective and cohesive forces over the span of decades? What new problems will be uncovered tomorrow that we have not a clue exist today?

Perhaps nowhere is the value of a watershed-scale, long-term perspective more clear than in the Fish Creek case study in Oregon. For years, the U.S. Forest Service was engaged in a noble effort to restore instream structure and habitat complexity to Fish Creek on the Mt. Hood National Forest after decades of road building and logging. Hundreds of rootwads, boulders, and logs were placed in the stream to improve fish habitat. Chapter 20 describes what happened to all the years of work when the watershed was hit by a major flood in 1996. Many parts of the landscape unraveled, dislodging instream structures and sending them far downstream.

It is important that we learn from the past. No efforts have been entirely successful. But, if we are to learn from our mistakes and profit from our collective experience, we must chronicle our work, monitor our progress, and report our findings. Increased communication and dialogue are key to benefitting from the experiences of others.

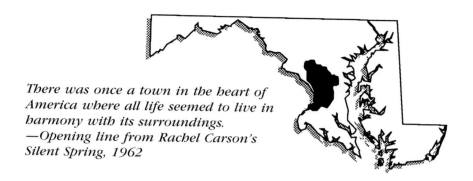

There was once a town in the heart of America where all life seemed to live in harmony with its surroundings.
—*Opening line from Rachel Carson's Silent Spring, 1962*

CHAPTER 18

RESTORATION IN AN URBAN WATERSHED: ANACOSTIA RIVER OF MARYLAND AND THE DISTRICT OF COLUMBIA

David L. Shepp and James D. Cummins

The Anacostia watershed is a largely degraded urban ecosystem located in suburban Maryland and the District of Columbia (Figure 18.1). The Anacostia has often been called "the forgotten river" (Bandler 1988) because, prior to 1987, its decline never received the attention that did its parent in the Washington metropolitan area, the Potomac River. However, a concerted and focused effort to restore the Anacostia watershed began over a decade ago. Since that time, local, state, regional, and federal government agencies, as well as environmental organizations, businesses, and dedicated private citizens have contributed significant resources toward protecting and restoring as much of the watershed ecosystem as possible. Formal cooperation between government agencies came with the 1987 signing of the Anacostia Watershed Restoration Agreement and the formation of the Anacostia Watershed Restoration Committee (AWRC). Members of the AWRC include the District of Columbia, Montgomery and Prince George's counties in Maryland, the state of Maryland, and the U.S. Army Corps of Engineers (USACOE) as the federal representative and liaison. In

Figure 18.1.—The Anacostia River in the District of Columbia.

addition to the AWRC members, the Metropolitan Washington Council of Governments and the Interstate Commission on the Potomac River Basin provide staff support to the AWRC. Also, the Anacostia Watershed Citizens Advisory Committee was recently created by the AWRC to provide a formal avenue for citizen input and information.

The history of ecological degradation in the Anacostia watershed is long and varied. More than two centuries of intense use and alteration have been compounded by an expanding human population and associated changes in land use and land cover. The loss of important forest and wetland habitats, alterations of natural drainage patterns and streamflow, increases in nonpoint source pollution, and discharges of combined (sanitary and storm) sewer overflow and industrial waste all have contributed to the decline of the ecological health of the watershed.

Initial efforts of the AWRC involved the identification and coordinated implementation of restoration projects throughout the watershed. The problems were abundant, and many demanded immediate attention. Stormwater retrofitting (of existing systems and for needed systems) and watershed restoration inventories were developed for each of the three local jurisdictions within the watershed: the District of Columbia and Montgomery and Prince George's counties in Maryland. These inventories identified about 450 projects to correct existing environmental problems and protect and enhance existing environmental resources throughout the watershed. The AWRC members concurrently recognized the need to establish a framework to guide and channel a long-term restoration commitment and effort. The vision was for a comprehensive, ecologically based restoration of the watershed which, by 1991, took shape in the form of a six-point action plan (MWCOG 1991).

Approximately 28% of the 450 identified restoration projects have already been implemented. The successes experienced to date required the identification of problems and associated solutions, coordination of programs, and the mobilization of significant government resources. A key ingredient in the success of the Anacostia program has been partnerships: some 60 government agencies are partners in the restoration effort. In its initial years, the driving force behind the effort was primarily local, state, and regional governments. As the restoration effort has evolved and broadened, active participation by the federal landowners has increased and become a priority objective of the AWRC. Collectively, federal agencies, including the U.S. National Park Service, the U.S. Environmental Protection Agency (USEPA), the U.S. Department of Agriculture (USDA), and the U.S. Department of Defense, own and operate facilities that constitute approximately 15% of the total watershed area. The importance of the restoration effort has been recognized by both the U.S. Congress and the White House (Graham 1994). In 1995, the Anacostia watershed restoration effort was selected as a National Ecosystem Management Model by a White House interagency task force (Interagency Ecosystem Management Task Force 1995).

WATERSHED SETTING

The Anacostia watershed is an ecologically and physically diverse system that extends into two physiographic provinces and three political jurisdictions. The fall line, a geologic transition zone that delineates the steeper-gradient Piedmont province from the lower-gradient Coastal Plain province, roughly mirrors the Montgomery–Prince George's counties boundary. The Piedmont province, with its moderately thin metamorphic soil associations, is characterized by relatively narrow and steeply sloped valleys as compared with the Coastal Plain province, which contains deeper, sedimentary soil complexes and features flatter valleys and broader, meandering streams. The Anacostia watershed comprises three major drainage components: the Northwest Branch, the Northeast Branch, and the tidal drainage. The Northwest and Northeast branches are nontidal stream systems, and their confluence, at the head of tide in the vicinity of Bladensburg, Maryland, forms the tidal Anacostia River (Figure 18.2). The tidally influenced portion of the watershed lies primarily within the District of Columbia and consists of the freshwater tidal river and its floodplain, as well as small Coastal Plain streams that flow directly into the tidal river. The majority of the streams that flow into the Anacostia in the District are partially or completely enclosed in storm sewer systems. The tidal reach of the Anacostia River is 8.4 miles in length, from the confluence of the Northwest and Northeast branches downstream to the Anacostia's mouth at the Potomac River. The Anacostia joins the Potomac approximately 108 miles upstream of Chesapeake Bay.

The 1990 population of the 170-square-mile Anacostia watershed was approx-

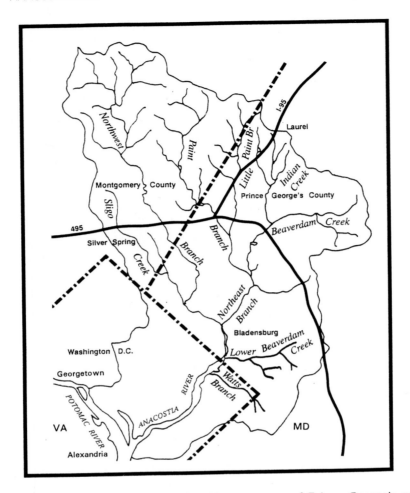

FIGURE 18.2.—The Anacostia watershed in Montgomery and Prince George's counties, Maryland, and the District of Columbia.

imately 804,500 people, yielding an average density of 4,732 people per square mile. Population within the watershed is projected to increase to 838,100 by the year 2010 (P. DesJardin, Metropolitan Washington Council of Governments, unpublished data). The latter figure corresponds, on average, to an additional 200 people per square mile.

The Anacostia has been recognized as one of the nation's prime examples of the problems typically associated with an urbanized watershed. The Anacostia's waterways are significantly affected by conversion of natural drainage networks to hydraulically efficient enclosed and channelized systems; by increased runoff and urban pollutants from the watershed's extensive impervious surfaces; by channel erosion and associated aquatic habitat loss from land-use transition; by sediment-bound toxins from motor vehicles, electrical transformers, and past application of persistent pesticides; by nutrients from excessive and poorly timed lawn fertilizer applications, combined sewer overflow episodes, atmospheric deposition, and pet wastes; and by tons of unsightly floating trash and debris. For three consecutive years, 1993 to 1995, American Rivers, Inc., designated the Anacostia as one of North America's most endangered and threatened rivers (American Rivers 1995). American Rivers reported that the Anacostia River, as it ran within one-and-a-half miles of the U.S. Capitol, "typified the appalling conditions of America's urban waterways."

Historic information on ecological conditions and fisheries of the Anacostia is, unfortunately, limited. Available accounts show that although heavily degraded in modern times the watershed's past reveals a dramatically contrasting picture. The American Indian village of the Nacostines (the root name of the Anacostia) is mentioned as being "thickly settled and the waters there were favorite fishing resorts" (McAtee 1918). Henry Fleet, who visited the area in June 1632, related "The Indians in one night commonly will catch thirty sturgeons in a place where the river is not above twelve fathoms broad. And as for deer, buffaloes, bears, turkeys, the woods do swarm with them, and the soil is exceedingly fertile" (Neill 1871). At one time, commercially valuable species were abundant. In the 1800s, the Anacostia port of Bladensburg boasted a sizable river herring (alewife and blueback herring) curing and packing industry, and fishermen worked the tidal Anacostia for Atlantic sturgeon, American shad, and striped bass near the Washington Naval Yard (Bandler 1993).

The existing degraded condition of the Anacostia watershed reflects more than two centuries of land-use transformation. Many of its current problems were set in motion during the colonial and early American periods when its original forests and meadows were converted to farms, principally for the production of tobacco. This activity, in the absence of today's understanding of sound agricultural practices, led to massive erosion and sedimentation of the river and many of its tributaries. As a case in point, Bladensburg (see Figure 18.2) initially served as the principal ocean-going colonial port for the Washington area *prior* to the development of the port of Alexandria, Virginia. It was only *after* Bladensburg and the tidal river silted in that the ports of Georgetown and Alexandria emerged as major international ports for the region. Also, during the 1800s the new capital city brought growth, and growing pains, to the Anacostia watershed. The capital's first sewer lines were primitive and carried untreated human wastes directly to the Anacostia River. River fringe wetlands were initially used as landfills—for convenience and in attempts to combat real, or perceived, health problems. Later, following the turn of the century, wetlands were used as disposal zones for sediment dredged from the main channel for navigational purposes and filled to create more buildable land in an increasingly crowded metropolitan setting.

A second distinct period of significant ecological impact began in the mid-1900s following World War II, when the human population in the watershed began to grow exponentially, as did the intensity of land development. Farms were converted to residential developments and country roads to commercial

TABLE 18.1.—Estimated total pollutant loads (pounds per year) to the Anacostia River and its tributaries from nonpoint sources (stormwater runoff), combined sewer overflow, and facilities holding national point discharge elimination system (NPDES) permits (data from Warner et al. 1996). Total suspended solids are in tons per year.

Parameter	Nonpoint sources	Combined sewer overflow	NPDES facilities
Nitrogen	809,710	88,860	6,112
Phosphorus	112,580	27,930[a]	1,391
BOD$_5$[b]	2,915,700	977,460	5,390
Arsenic	850		<1
Chromium	1,250		<1
Copper	3,720		1
Lead	740	4,316	2
Zinc	21,070	3,808	3
Hydrocarbons	803,600		1,955[c]
Total suspended solids	45,890	2,330	10

[a]Total inorganic phosphorus.
[b]Five-day biological oxygen demand.
[c]Oil and grease.

corridors as the nation's capital grew and expanded toward the Maryland cities of Baltimore and Annapolis. Flooding, exacerbated by these land-use changes, led to engineered solutions such as stream enclosure and channelization. In the absence of effective stormwater controls, accelerated channel erosion and related habitat degradation occurred. In addition to the physical changes to the watershed associated with this growth, nonpoint source pollution, runoff of potential toxins, discharges of combined sewer overflows, and, to a lesser extent, industrial waste further contributed to the decline in the ecological health of the Anacostia. As such, the Anacostia's problems are not only long standing but complex.

Land use in the Anacostia watershed today follows the general pattern of other metropolitan areas, with the most dense development concentrated near the urban center. The average imperviousness of the entire watershed is 22.5%. Nonpoint sources are the dominant form of pollution in the watershed, typically constituting 75 to 90% of the total pollutant loads to the river and its tributaries (Warner et al. 1996; Table 18.1). Between 80 and 90% of the annual nutrient loads (nitrogen and phosphorus) in the watershed are from nonpoint sources, with combined sewer overflows contributing a majority of the remainder. Eighty-five percent of the lead received by the river each year is believed to result from combined sewer overflow. Conversely, 85% of the annual zinc loadings is attributed to nonpoint sources. Additionally, the Anacostia receives approximately 48,000 tons per year of total suspended solids, which on average equates to 0.44 tons per acre per year. Ninety-five percent of the total suspended solids loads originate from nonpoint sources such as construction activities, solids washed off of paved areas, and instream erosion of receiving channels in the form of bank erosion and substrate scour (Warner et al. 1996).

Residential development is the single largest land use in the Anacostia watershed, constituting more than 43% of the watershed area. The "undeveloped" category, covering 31% of the watershed, is primarily forest and parkland.

The industrial and manufacturing base in the watershed is confined to 4% of the land area. This land use is predominantly light industry and is concentrated primarily in the tidal Anacostia area. Sand and gravel mining activities are generally isolated to a single subwatershed, Indian Creek, found in the upper reaches of the watershed.

Average imperviousness in the individual subwatersheds ranges from 11% in the Beaverdam Creek subwatershed (owned predominantly by the USDA and operated as the Beltsville Agricultural Research Center) up to 48% in the highly developed subwatersheds. A moderate correlation ($r^2 = 0.62$) exists between population density values and imperviousness in the subwatersheds of the Anacostia, with imperviousness typically increasing 4 to 5% for each additional 1,000 people per square mile (Warner et al. 1996).

Twenty-five percent of the Anacostia watershed (approximately 27,700 acres) remains forested. Unfortunately, nearly 50% of the forest tracts in the watershed are smaller than 12 acres and have limited potential for optimal species diversity and wildlife utilization. It should be noted that although their species diversity potential is limited, the actual use of these small tracts is critically important because they typically offer the only available habitat in the majority of the developed areas of the watershed. Most of the subwatersheds in the Anacostia retain between 20 and 30% of their total area in forest, with an adequate riparian cover along 35 to 50% of their stream miles. The Beaverdam Creek subwatershed, mentioned above, contains more than twice this typical percent forested area (71%), and more than 87% of its stream miles retain adequate forest cover. At the other extreme, 10% or less of the Sligo Creek subwatershed and the District of Columbia's Northwest Bank and tidal Anacostia areas are forested, and only 20% of the stream miles of Lower Beaverdam Creek and Northeast Branch retain adequate riparian cover (Warner et al. 1996).

In total, 3,208 acres of wetlands remain in the Anacostia watershed, constituting just under 3% of the total watershed area. It is estimated that more than 4,000 acres of nontidal wetlands have been lost from the Anacostia watershed due to both the suburban development boom of the last five decades and earlier urban development and agricultural activity. Additionally, more than 90% of the nontidal wetland acreage loss has occurred in the Coastal Plain portion of the watershed (Warner et al. 1996). The loss of tidal wetlands in the watershed has been dramatically more extensive than that observed for nontidal wetlands. The USACOE (1994) estimated that 2,600 acres of freshwater tidal wetlands have been destroyed in the Anacostia watershed between Bladensburg and the confluence with the Potomac River. Many of the wetlands were filled by the USACOE during a time when policies favored wetland removal for human health concerns (Dr. Walter Reed reported that Fort Meyer and Washington barracks led the Army in incidences of malaria for 1897 [Gordon 1987]). Even with the recent restoration of Kenilworth Marsh (representing approximately 32 acres), less than 100 acres of freshwater tidal wetlands currently exist. Moreover, the total area of remaining tidal wetlands is approximately 180 acres, constituting a loss well in excess of 90% of the originally occurring tidal wetlands from the watershed.

THE RESTORATION EFFORT

Complementing the development of the Anacostia Watershed Restoration Agreement in 1987, the creation of the AWRC, and the foundation provided by the retrofitting inventories for the District of Columbia and Montgomery and Prince George's counties, are the broad goals for restoring the Anacostia watershed presented in the six-point action plan (MWCOG 1991). The AWRC adopted the framework for an ecologically based restoration of the watershed as found in this plan, which identified six broad watershed restoration goals that address the major problem areas in the watershed. The following section individually identifies those goals and for each summarizes the problems and proposed restoration strategies and highlights examples of progress to date.

Reduce Pollutant Loads

Goal one of the six-point plan is to dramatically reduce pollutant loads delivered to the tidal river to improve water quality conditions measurably by the turn of the century.

Problem.—The tidal Anacostia River has some of the poorest water quality in the Chesapeake Bay system. It receives a substantial annual load of urban pollutants, sediment, and debris, it experiences combined sewer overflow events, and its dissolved oxygen levels frequently violate water quality standards. Its sediments contain toxics (such as PCBs, petroleum hydrocarbons, trace metals, and pesticides) and are enriched with excess nutrients.

Strategy.—Sharply reduce the number of combined sewage overflow events and stormwater pollutant loadings. Effectively control increased stormwater loadings from new development. Remove trash and floatable debris now trapped in the estuary and its tributaries. Prevent future trash and debris deposition. Evaluate and address the problem of toxic sediments in the tidal river.

Progress.—Initiatives to date to reduce pollutant loads and improve water quality conditions are described below.

- The District of Columbia and Montgomery and Prince George's counties have instituted stringent erosion and sediment and stormwater management controls for all new development (several hundred urban best management practices have been implemented since the mid-1980s).
- The District of Columbia, Montgomery and Prince George's counties, the state of Maryland, the USACOE, and the USEPA have undertaken the installation of stormwater retrofits, including both new stormwater controls for previously uncontrolled development and the modification of existing stormwater controls to enhance their pollutant removal performance. Approximately 159 stormwater retrofits have been proposed, and approximately 50 projects are currently in planning, design, or construction.
- The D.C. Department of Public Works and the USEPA installed an innovative swirl concentrator facility to reduce the combined sewer overflow from the

largest combined sewer system in the Anacostia. The D.C. Environmental Regulation Administration and the USEPA, via the Hickey Run Comprehensive Pollution Abatement Program, directed the Metropolitan Washington Council of Governments to develop the first subwatershed action plan for the Anacostia and to develop a prototype storm drain tracing system for petroleum hydrocarbons for Hickey Run, Anacostia's most degraded subwatershed.

- The Washington Suburban Sanitary Commission, a regional water utility, instituted a sanitary sewer rehabilitation program for aging sewer lines in the Anacostia's tributaries.

- The D.C. Environmental Regulation Administration and the USEPA directed the Interstate Commission on the Potomac River Basin to develop a remedial action plan to mitigate the contaminated sediments in the main stem tidal Anacostia.

- The D.C. Department of Public Works, the Prince George's Department of Environmental Resources, the Prince George's Maryland–National Capitol Park and Planning Commission, the USEPA and the Interstate Commission on the Potomac River Basin developed floating-trash management initiatives for the main stem and its larger tributaries. The District of Columbia and Montgomery and Prince George's counties supported citizen's initiatives to include stream cleanups and "Don't Dump" catch basin stenciling, which identifies a catch basin's connection to the Anacostia watershed.

Protect and Restore Ecological Integrity of Streams

The second goal is to protect and restore the ecological integrity of Anacostia streams to enhance aquatic diversity and encourage a quality urban fishery.

Problem.—Dozens of miles of stream habitat have been severely degraded by poorly controlled stormwater runoff and, in some cases, by engineered channel "improvements." Urbanization has profoundly altered the flow, water quality, geometry, and ecology of these streams, many of which possess only a fraction of their original natural biological diversity.

Strategy.—Apply stormwater retrofits to control runoff and hence restore a dynamic equilibrium to the receiving streams; protect remaining supporting habitat; apply stream restoration techniques to improve habitat in the most degraded streams; and effect land-use controls and stringent stormwater and sediment practices at new development sites, prioritizing the most critical subwatersheds.

Progress.—Initiatives to date to protect and restore the ecological integrity of streams in the watershed are described below.

- The USACOE, Montgomery County Department of Environmental Protection, Prince George's County Department of Environmental Resources, the Maryland State Department of the Environment, the Metropolitan Washington Council of Governments, and the Interstate Commission on the Potomac River Basin have initiated or completed eight major urban stream restoration

FIGURE 18.3.—Wheaton Branch, a tributary of the Anacostia's Sligo Creek, before (above, 1989) and after (next page, 1993) restoration. (Photographs by John Galli, Metropolitan Washington Council of Governments.)

projects, improving approximately 15 miles of degraded habitat in Sligo Creek, Northeast Branch, Northwest Branch, and Paint Branch (Figure 18.3).

- In the upper Paint Branch subwatershed, a naturally reproducing brown trout population exists in the Anacostia's highest-quality stream; several efforts have

been undertaken to protect this valuable resource. Trout Unlimited, the Maryland–National Capitol Park and Planning Commission, and the Maryland Department of Natural Resources have worked to expand the pool habitat in a prime spawning tributary. The USDA's Beltsville Agricultural Research Center has initiated stream restoration efforts in the lower portion of Paint Branch. A multiagency task force, including the Montgomery County Department of Environmental Protection, the Maryland–National Capitol Park and Planning Commission, the Maryland State Department of Natural Resources, the Interstate Commission on the Potomac River Basin, and the Metropolitan Washington Council of Governments, developed an upper Paint Branch management plan that was approved by the AWRC. The Montgomery County Council recently voted to adopt the majority of the plan's recommendations and took steps to purchase 248 acres of critical riparian property and identified another 152 acres of critical buffer and recharge areas for future purchase.

- The Maryland–National Capitol Park and Planning Commission supported the reintroduction by the Interstate Commission on the Potomac River Basin and the Metropolitan Washington Council of Governments of 17 species of freshwater fishes into a restored portion of Sligo Creek. All reintroduced species, both pollution tolerant and intolerant, are surviving.
- The U.S. National Park Service, in conjunction with the Interstate Commission on the Potomac River Basin, has initiated efforts to restore streams in Greenbelt National Park that drain into the Northeast Branch.
- The Prince George's County Department of Environmental Resources and the Maryland State Department of the Environment have undertaken initiatives to restore Brier Ditch and other Northeast Branch tributaries.
- The D.C. Environmental Regulation Administration, the USEPA, the USDA, and the Metropolitan Washington Council of Governments have initiated efforts to restore Hickey Run, which flows through the National Arboretum.
- The USACOE is restoring lost pool and riffle habitat within flood control project authorization zones in the Northeast and Northwest branches.

Restore Spawning Range of Anadromous Fishes

The third goal of the six-point action plan is to restore the spawning range of anadromous fishes to historical limits.

Problem.—For centuries, certain fish species (Atlantic menhaden, yellow perch, river herring [blueback herring and alewife], shad [American shad and hickory shad], white perch, and striped bass) have annually migrated from the Atlantic Ocean and Chesapeake Bay up into the freshwater nontidal Anacostia tributaries to spawn. By the 1970s, the historical annual migration of these anadromous fish species had been interrupted by as many as 25 unintentional and artificial fish barriers (primarily along the lower portion of the Anacostia).

Strategy.—Remove key fish barriers to expand the available spawning range for anadromous fishes, and improve the quality of the their spawning habitat. Once habitat is expanded, assist the anadromous fish community to imprint genetically on their newly-opened territory to encourage the return of future generations.

Progress.—Fish imprinting efforts—the manual transportation of fish to upstream habitats in order to imprint unique stream chemistry—were performed in 1991 with the aid of students from local schools. As of fall 1996, six priority fish barriers have been removed or modified. The first of these was done by private industry, three were accomplished through the USACOE's Environmental Restoration Project, and two fish passages were installed through Maryland's Watershed Habitat Enhancement Initiative. This work has had significant results: for the first time in many decades, alewife again migrate up the Northeast Branch to areas beyond the Washington beltway (interstate highway 495, Figure 18.2) and many stream miles beyond prerestoration limits.

Increase Quantity and Quality of Wetlands

Goal four of the plan is to increase the natural filtering capacity of the watershed by sharply increasing the acreage and quality of tidal and nontidal wetlands.

Problem.—Wetlands historically have been an integral part of the self-cleansing system of the Anacostia watershed, as well as key wildlife and waterfowl habitat. When the restoration began in 1987, over 98% of the once-extensive tidal wetlands and nearly 75% of the watershed's nontidal wetlands had been destroyed.

Strategy.—Accept no further net loss of wetlands in the watershed. Restore the ecological function of the existing degraded wetland areas. Create several hundred acres of new tidal and nontidal wetlands.

Progress.—Initiatives to date to increase the acreage and quality of wetlands are described below.

- The D.C. Department of Public Works, the USEPA, the U.S. National Park Service, and the Metropolitan Washington Council of Governments initiated efforts to restore Kenilworth Marsh, a freshwater tidal system in the District of Columbia. Their efforts were successfully merged with a nearby ongoing USACOE dredging project on the main stem Anacostia, resulting in the creation of 32 acres of freshwater marshlands, which represents the largest freshwater tidal marsh restoration project in the nation.
- The Maryland State Department of Natural Resources asserted new authority in 1992 to protect nontidal wetland areas; the department is also evaluating ways to transfer wetland mitigation requirements between sites, so that mitigation requirements can be used to expand watershed-wide restoration efforts.
- The Montgomery County Department of Environmental Protection has retrofit several of its stormwater management ponds with fringe wetland plantings; both emergent and submerged species have been successfully incorporated.
- Through USACOE's Anacostia Feasibility Study, designs have been undertaken to restore Kingman Lake, a system similar to the Kenilworth Marsh and located in the same river reach. Lessons learned from the Kenilworth experience will be transferred to this project (approximately 46 acres of emergent wetland are planned).
- Also identified in USACOE's Anacostia Feasibility Study is the creation of an additional 30 acres of emergent river fringe wetlands. They are planned for the

shoreline of the Anacostia main stem, in the vicinity of Kingman Lake. Though originally part of USACOE's Anacostia Feasibility Study project scope for the District of Columbia, due to funding considerations this element will probably be phased into part of future initiatives.

- The Prince George's County Department of Environmental Resources, in cooperation with the USDA's Beltsville Agricultural Research Center and the Washington Metropolitan Area Transit Authority, constructed 19 acres of nontidal wetlands.
- The U.S. National Park Service and the Interstate Commission on the Potomac River Basin teamed to develop a one-half acre freshwater wetland in Greenbelt National Park.
- Overall, approximately 138 acres of constructed wetlands have either been completed, or are currently under construction within the Anacostia watershed.

Expand Forest Cover

Goal five is to expand the range of forest cover throughout the watershed and create a contiguous corridor of forest along the margins of its streams and river.

Problem.—Nearly 50% of the forest cover in the Anacostia basin has been lost due to agriculture, and later, urbanization. The extensive losses have occurred in the forest cover along the streambanks and riverbanks, where riparian vegetation plays a critical role in maintaining stream temperature and water quality, preventing streambank erosion, and providing aquatic and terrestrial habitat.

Strategy.—Reduce the loss of forest cover associated with new development and other activities by local implementation of the 1991 State Union Forest Conservation Act. Extensively reforest suitable sites throughout the basin. Reforest ten linear riparian miles over the next three years with the ultimate goal of an unbroken forest corridor from the tidal river to the uppermost headwater streams.

Progress.—Initiatives to date to expand forest cover and create a contiguous forest corridor along streams and the river are listed below.

- Local and regional agencies, including the D.C. Forest Council, Montgomery County Department of Environmental Protection, Maryland–National Capitol Park and Planning Commission, Prince George's County Department of Environmental Resources, and the Metropolitan Washington Council of Governments, have completed reforestation projects affecting an estimated 50 acres. Much of the impetus has come from forest mitigation requirements created by county tree ordinances, buffer criteria, and the Chesapeake Bay Critical Area Program.
- The District of Columbia, through its urban forester (hired in 1991), is exploring options with federal landowners to reforest approximately 2.7 miles of riparian zone along the Anacostia main stem.

- The U.S. National Park Service has supported several citizen-based local tree planting efforts within its Anacostia Park system.
- The USDA's Beltsville Agricultural Research Center has initiated efforts to plant approximately 4 acres of riparian habitat within its jurisdiction.
- The USACOE plans to improve riparian habitat along several miles of its Northeast Branch floodway authorization zone.
- The Maryland Department of Natural Resources assigned a forester to the Anacostia watershed in 1993, and the forester has coordinated the planting of more than 2,000 trees. The forester is also active in public outreach activities.
- The Metropolitan Washington Council of Governments, in conjunction with D.C. Cares, has organized eight tree maintenance events in the watershed.
- The USEPA's Chesapeake Bay Program, in coordination with the AWRC, the U.S. Fish and Wildlife Service, and the D.C. Department of Consumer and Regulatory Affairs, is committed to funding another 8 acres of watershed riparian reforestation.
- The Metropolitan Washington Council of Governments has worked with the Earth Conservation Corps to collect native seeds from local trees. The seeds were propagated for use in the watershed through the National Tree Trust program.
- The National Arboretum has modified its mowing policy to promote the regrowth of a natural buffer along a major tributary to Hickey Run.
- The U.S. National Park Service has modified its mowing policy to promote a natural buffer along portions of the tidal river in the District of Columbia.
- Many civic associations and environmental groups have enthusiastically been planting trees throughout the District of Columbia and Montgomery and Prince George's counties.

Increase Public Awareness and Involvement

The sixth goal of the six-part plan is to make the public aware of its key role in the Anacostia cleanup and increase citizen participation in restoration activities.

Problem.—Watershed residents generally are unaware of the stream system where they live. They do not understand their connection to their streams and the associated ecosystems. For the Anacostia, a better and sustained appreciation of the watershed by its roughly 804,500 residents is crucial to the success of long-term restoration and protection efforts.

Strategy.—Raise public awareness about the problems of the Anacostia River and associated, ongoing restoration efforts. Seek active public support and sustained commitment and involvement. Educate the public about the watershed system and the citizen's role in reducing urban pollution. Encourage a grassroots network of citizens to participate in a variety of restoration initiatives.

Progress.—Initiatives to date to increase public awareness and participation are listed below.

- The Interstate Commission on the Potomac River Basin has developed a strong public outreach program. Its public education and participation program has reached more than 60,000 people since its inception in 1988. The program reaches the public through the efforts of five sub-basin coordinators, publications focusing on sub-basin problems, newsletters, and Anacostia information packets produced in partnership with the Chesapeake Bay Trust. Sub-basin coordinators have been instrumental in the formation of several citizens watershed groups and have enlisted other citizens in "stream team" programs run by Montgomery and Prince George's counties. Sub-basin coordinators give presentations, organize cleanups, and work on various restoration projects related to stream stewardship. They have worked closely with the Metropolitan Washington Council of Governments in setting up an Anacostia Watershed Citizens Advisory Council. In addition, many of the more than 60 agencies involved in the cleanup effort also have instituted public outreach programs.
- The Washington Metropolitan Council of Governments developed a Small Habitat Improvement Program designed to enlist volunteers to implement small-scale restoration projects.
- The Maryland–National Capital Park and Planning Commission has committed itself to public education through its nature centers and Anacostia Visitors Center at the Port of Bladensburg.
- The Anacostia Watershed Society, the major citizen-based private nonprofit organization devoted to restoring the Anacostia River, has effectively mobilized many of the local communities. Volunteers have removed hundreds of tons of debris from the river and its tributaries, sending a very positive message of caring to those who simply observe the activity. The society organizes river sojourns, community action days, and a variety of other events that have brought national attention to the plight of the Anacostia River.

FUTURE DIRECTION

Ultimately, restoration of a watershed requires making decisions about implementation while balancing the environmental, technical, financial, and political issues unique to each individual project. Additionally, each individual project should be viewed within the larger context of its subwatershed's needs. In an urban watershed, these decisions become amplified by the greater number of people potentially affected by each project. Although this larger constituency can be beneficial in budgetary arguments in favor of a project, involving necessary stakeholders in the decision-making process may increase the complexity and time required for any project-related decision. Another reality in urban watershed restoration is that the costs of a project's design and construction and, particularly, any required real estate, are typically much greater in urban settings.

As the Anacostia watershed restoration effort concludes its first decade, many new challenges have accompanied its early successes. This section identifies major areas requiring additional effort to ensure the restoration's continued success and provides a brief discussion of ongoing initiatives that have been undertaken by the AWRC.

Citizen involvement.—Effectively integrate citizens into the AWRC process. As mentioned previously, the Anacostia watershed restoration effort is somewhat unusual in that the initial impetus for the effort stemmed directly from local, regional, and state governments, not from the grassroots citizenry as is more typically the norm. As such, citizens were not directly involved with the AWRC from the outset. Recognizing this important missing component of the restoration effort, the AWRC charged the Metropolitan Washington Council of Governments' staff with structuring the Anacostia Watershed Citizens Advisory Committee and providing administrative assistance to the committee, once established. In spring 1996, supported by the AWRC, the Anacostia Watershed Citizens Advisory Committee held its first meeting. The advisory commmittee's purpose is to provide citizens residing within the watershed a direct line of communication with the AWRC regarding restoration issues. Each of the three local governmental entities selected three individuals to serve as committee members for the first term. The chair is handled on a rotating basis and is passed from jurisdiction to jurisdiction.

Private-public partnerships.—Identify and develop private-public partnerships. Over 450 restoration projects have been identified to date for the Anacostia watershed, and approximately 28% have been either completed or are in progress. The remainder will require substantial financial resources to implement. In 1996, the AWRC realized that in order to sustain the restoration effort new private-public partnerships must be pursued and established. The AWRC members signed a memorandum of agreement for the purpose of establishing a blue ribbon panel on resources for continued restoration. The panel's mission is to produce a report that identifies potential public and private partnership opportunities that would help to meet and financially support achievement of the goals of the Anacostia restoration initiative.

Restoration targets and ecological indicators.—Develop specific and quantifiable, ecologically based restoration goals and associated targets with which to measure restoration progress. As previously mentioned, the restoration goals of the six-point action plan represent broad restoration concepts. Now that the restoration is underway, with major financial resources committed to projects, a mechanism for quantitatively assessing progress is needed. In an effort to gauge restoration progress toward those broad goals, a series of measurable ecological indicators and associated restoration targets, specific to each indicator, was needed. To fulfill this requirement, the District of Columbia's Environmental Regulation Administration, the USEPA's Chesapeake Bay Program office, and the AWRC charged the Metropolitan Washington Council of Governments staff with developing a system of ecological indicators and restoration targets for the Anacostia restoration. This effort will be conducted in concert with the local jurisdictions and other agencies involved in the Anacostia restoration. The vehicle for this cooperative effort is the Anacostia Watershed Technical Oversight Subcommittee. This initiative is currently being undertaken by the Metropolitan

Washington Council of Governments as part of a larger ongoing Anacostia Watershed Special Study.

Monitoring network.—Develop and maintain a viable, balanced monitoring network to provide data for the ecological indicators and restoration targets. As mentioned above, ecological indicators and associated restoration targets will be developed for the purpose of quantitatively assessing restoration progress. In order to use this system of indicators and targets, a watershed-wide system of monitoring will be required. As a part of the Anacostia Watershed Special Study, a long-term monitoring program is being developed in conjunction with the member jurisdictions and many other involved agencies via the technical oversight subcommittee. It is anticipated that this monitoring program will consist of a scientifically balanced approach that includes physical, chemical, and biological components.

Remaining restoration gaps.—Close major gaps in the existing scope of the restoration effort. Although the restoration effort currently focuses upon numerous areas of restoration need, two major gaps remain: combined sewer overflow and toxic sediments. Both of these needs represent major efforts, and both are focused upon the tidal river in the District of Columbia. Combined sewers in the Anacostia contribute approximately 6% of the total watershed annual pollutant load (5,500,000 pounds per year of total nitrogen, total phosphorus, lead, zinc, biological oxygen demand, and total suspended solids) from four major source areas. The District of Columbia Water and Sewer Administration is pursuing a comprehensive combined sewer overflow abatement program for all of the combined sewer areas within the District (to include the Anacostia, Potomac, and Rock Creek drainages) to meet the goals of the National Combined Sewer Overflow Policy (Warner et al. 1996). This initiative will require additional characterization monitoring and computer modeling to guide water resource managers toward the optimal approach for solving this major issue. The other major gap in the ongoing restoration effort regards contaminated sediments in the tidal portion of the river. The Anacostia watershed has been designated as a region of concern for toxic contamination by the USEPA's Chesapeake Bay Program. Elevated levels of contaminants including total hydrocarbons, chlordane, DDT and its metabolites, lead, and PCBs, have been consistently observed over time in various monitoring surveys throughout the tidal river in the District of Columbia (Velinski and Cummins 1996). Resultant health advisories (no consumption) are currently in effect within District waters for human consumption of bottom-dwelling fishes, such as catfishes, eel, and common carp. Advisories (less than one-half pound per adult per week) are also in effect for sport fishes, such as largemouth bass and sunfishes (bluegill, longear sunfish, pumpkinseed, and redbreast sunfish). A regional action plan for managing toxins in the sediments has recently been developed by the Interstate Commission on the Potomac River Basin for the District of Columbia Environmental Regulation Administration (Gruessner 1996). The plan represents a first step in managing this problem; it features an overview of the problem, volumetric estimates of contamination, a discussion of potential remediation options, and associated cost estimates. Management efforts are currently hampered by the absence of information regarding the existing sources of contaminants, both within the District of Columbia and upstream from Montgomery and Prince George's counties. Efforts are currently underway to identify the existence of any pertinent data in the upstream jurisdictions. Similar to the previously discussed combined sewer issue, sufficient monitoring to characterize the input of toxicants into the system adequately must first be collected. Then modeling efforts to describe the active fate and transport mechanisms for these compounds must be developed *prior* to undertaking any large-scale comprehensive management initiatives. In the shorter term, small-scale remediation pilot measures, such as capping (the physical isolation of contaminated sediments by means of a medium such as sand), may be undertaken to determine the potential feasibility of physically isolating in-situ contaminants.

Integration of federal resources.—Explore, identify, and create pathways for cost-effective integration of federal programs and initiatives with relevance to the Anacostia watershed restoration. As a part of the long-term funding strategy that is to be developed by Metropolitan Washington Council of Governments staff as a component of the Anacostia Watershed Special Study, various avenues to optimize federal involvement and financial support will be investigated. Metropolitan Washington Council of Governments staff is currently working in partnership with USACOE to identify problems and recommend solutions for federal facilities within the watershed (federal property constitutes approximately 15% of the total watershed area). This effort is a part of a congressionally mandated Anacostia Federal Facilities Impact Assessment project. In addition to working with the individual federal landowners in the watershed, efforts by the AWRC are underway to integrate ongoing programs, currently existing within various natural resource management agencies, to optimize the use of federal personnel and financial resources toward achieving the restoration of the watershed.

Congressional support.—Pursue and maintain a closer working relationship with the local congressional delegation. Discussions are ongoing with USACOE and USEPA staff to identify existing legislative authorization that could be helpful in directing resources to the long-term restoration of the Anacostia watershed. Once a comprehensive review of existing authorities is conducted and compiled, the AWRC will initiate a series of discussions with the local congressional delegation to solicit their ideas, legislative support, and assistance for the numerous remaining restoration initiatives.

A comprehensive restoration plan.—Develop a comprehensive restoration plan representing authorship and input from all stakeholders in the Anacostia watershed restoration. As a final component of the Anacostia Watershed Special Study, the Metropolitan Washington Council of Governments is charged with the development of a comprehensive restoration plan for the Anacostia watershed. Input from the local jurisdictions and various stakeholders will be solicited to obtain widespread authorship and endorsement of the plan. Existing problems and needs will be prioritized to guide implementation strategies. The plan is envisioned to function as a working, living document to guide the focus and priorities of the restoration effort. It is expected that over time and as changing situations in the watershed dictate, the plan will be revisited and updated to reflect the dynamic nature of the restoration effort.

CONCLUSION

As the restoration effort enters its second decade, it has experienced success and associated growing pains. It has outgrown its infancy and currently exists as a maturing effort. In many areas, with the notable exceptions of the combined sewer overflow and toxic sediments problems, the issues are understood and more than 450 individual projects have been identified to restore the watershed. Even within the current major gaps of the restoration effort, the general parameters of the problems have been described.

A huge early revelation in the restoration process has been an appreciation of not only the *scope* of the problems but also the *time* required to bring about the change tantamount to a meaningful restoration of the watershed. Fortunately, time is relatively plentiful; however, the financial resources required to implement projects and to undertake the basic research needed to define unexplored areas and to devise and use analytical and predictive tools to optimize our solutions for those problems, are in increasingly short supply. The major bottleneck in the pace and momentum of the restoration, now in its middle years, is funding. With the contracting financial climate we have experienced throughout the nation in the mid-1990s, implementation has slowed and, consequently, momentum has declined.

The current national commitment to restoring and protecting watersheds and associated ecosystems, as evidenced by numerous federal initiatives (e.g., the Chesapeake Bay Restoration, the Great Lakes Program, and the Everglades Restoration Project), was recently reinforced at the executive level with the formation of a White House Interagency Ecosystem Management Task Force (Task Force). The Task Force was established in 1993 to carry out Vice President Gore's mandate "for the agencies of the federal government to adopt a proactive approach to ensuring a sustainable economy and a sustainable environment through ecosystem management." The Task Force formed a working group, which conducted case studies to learn about ecosystem efforts to date, to identify barriers to implementing the ecosystem approach, and to identify ways the federal government could assist in overcoming those barriers. Seven areas were selected as case studies, including the Anacostia River watershed.

The Task Force's report (Interagency Ecosystem Management Task Force 1995) provided its perspective of the Anacostia Watershed Restoration effort:

Of the seven case studies, the Anacostia River watershed most represents the vision of local and state governments. Federal agencies were perceived as facilitators and implementors of local goals through design and funding of projects and through technical assistance. However, some federal activities in the basin were viewed as not supportive of the goals. Because federal agencies did not participate in vision setting, they may not have modified their priorities in accordance with the vision. Some interviewees said the plan does not provide a comprehensive vision for restoring the watershed, but they agreed that it *does* provide an effective beginning to focus action. Indeed, the key role of any vision may be to provide a general guide for moving diverse entities in a common direction.

The Task Force report makes a powerful argument for focused federal assistance, in the form of substantial human and financial resources, in support of the Anacostia watershed restoration effort. It is clear that federal participation in the restoration effort is central to its continued success; it is also clear that the Anacostia watershed restoration effort is an important national prototype and field laboratory for urbanized watersheds and those undergoing urbanization. Several factors serve to underscore the importance of providing sufficient federal resources to ensure that the restoration effort can move steadily and scientifically forward: (1) the restoration effort is well established and broadly based and has a proven record of getting projects done; (2) the AWRC has a proven record of applied experimentation through the design, implementation, and monitoring of numerous watershed restoration prototypes and procedures, ranging from stormwater management to stream restoration to watershed assessment, and (3) the AWRC also has a proven record of effective evaluation, documentation, and public dissemination of information relative to the range of watershed restoration initiatives being pursued.

The twin pillars critical to ensuring the long-term success of the restoration are human and financial resources. If we can build and maintain a broadly based coalition of citizens, environmental groups, all levels of government, and the private sector, we should be able to translate that energy into congressional support and federal legislative authority. The Anacostia watershed restoration effort has been designated as a National Ecosystem Management Model on the strength of its success to date. It is critical that sufficient federal resources are directed and applied in a planned, scientifically based sequence to sustain the effort and maintain and expand this unique example of urban watershed restoration—not only for the Anacostia, but for other similar urban watersheds throughout the nation.

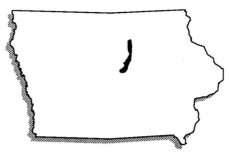

Between the watersheds and at distances of two or three miles from one another were little clear brooks with banks of black sod, their waters flowing on floors of bright colored glacial pebbles; . . . They were beautiful little brooks, so clear, so over-arched with tall grasses and willows, so plaided with the colors of the pebbles in the sun, so dark and mysterious in the shade; with secret pockets under the soddy banks for the shiners, pumpkinseeds, dace, chubs, and other small fish which populated the pure waters. . . . All those beautiful brooks are now forever gone. They were such lovely streams to us children . . . : but they were like delicate flowers, too tender for the touch of humanity.
—Herbert Quick, 1925, recalling the Iowa prairie of the 1860s

CHAPTER 19

WATERSHED RESTORATION AND AGRICULTURAL PRACTICES IN THE MIDWEST: BEAR CREEK OF IOWA

Thomas M. Isenhart, Richard C. Schultz, and Joe P. Colletti

The Western Corn Belt Plains ecoregion, which covers most of Iowa and parts of surrounding states, can be characterized as extensive cropland on level to gently rolling dissected glacial till plains, hilly loess plains, and morainal hills with broad smooth ridgetops (Griffith et al. 1994). This landscape has been largely converted to agricultural uses; more than 80% of most counties of the ecoregion are dedicated to corn, soybeans, and forage for livestock (Burkhart et al. 1994). Nearly two-thirds of the native hardwood forests have been converted to row crop agriculture or pasture. Modification of the local and regional hydrology has been an essential part of this land use conversion. Annual cultivation has reduced soil quality, lowering rates of infiltration and increasing surface runoff. Creation of extensive networks of subsurface tile drains, excavation of surface drainage ditches, and channelization of many perennial streams has facilitated the conversion of nearly all prairie and wetland acreage to agricultural uses and contributed to many off-site, downstream problems of water flow and quality.

This large-scale modification of regional hydrology and terrestrial and aquatic ecosystems has had profound effects on the biological integrity of the surface waters of the region. Menzel (1983) reviewed the natural structure and function of stream ecosystems of the corn belt region with special reference to the impacts of past and present agricultural management practices. He concluded that effects on water quality were not the sole problem, but that aspects of water quantity, habitat structure, and energy transfer were also often profoundly affected by agricultural land use practices. This alteration of the physical, chemical, and biological processes associated with the water resource has dramatically reduced the species composition and diversity of the aquatic ecosystems and the functional organization in the region (Karr 1991).

A challenge for resource managers in such modified landscapes is the development and implementation of restoration-based management approaches that build upon traditional soil and water conservation and pollution control efforts. One promising approach to increasing the effectiveness of efforts to protect soil and water quality while also enhancing the integrity of the terrestrial and aquatic systems is the creation or restoration of landscape buffer zones (NRC 1993b). For example, riparian vegetation, particularly the vegetation bordering smaller streams and tributaries, is now recognized as an important resource that should be protected to serve as a sink for sediments, nutrients, and pesticides, to protect the streambank from erosion, and to reduce excessive runoff into stream channels (NRC 1993b). However, most of the considerable body of evidence confirming the ecological value and effectiveness of riparian zones as sinks for nonpoint source pollution has come from existing vegetated riparian zones (Lowrance et al. 1984, 1985; Peterjohn and Correll 1984; Jacobs and Gilliam 1985; Cooper et al. 1987; Lowrance 1992; Osborne and Kovacic 1993; Castelle et al. 1994; Hill 1996). Little information is available for restored or constructed riparian buffer systems in extensively modified agricultural systems, particularly for buffer systems in the midwestern United States (Osborne and Kovacic 1993).

Superimposed on the physical aspects of landscape restoration efforts are the social, political, and economic questions associated with land ownership patterns in the region. About 92% of the land in Iowa is in private ownership (Iowa Agricultural Statistics Service 1994). Although most landowners seem to agree that landscape buffers would function to improve water quality, most also indicate a need for shared private and public responsibility associated with the voluntary establishment of such buffers (Colletti et al. 1994). Thus, for implementation to occur on a watershed scale, restoration models must remain flexible enough to fit the objectives of private landowners yet be scientifically robust enough to attain governmental and public acceptance.

This chapter describes a watershed management approach for the environmental enhancement of intensively modified agricultural landscapes in the Midwest. An explicit goal of the project described is to develop a riparian management system that has broad-scale applicability to watersheds in the midwestern agroecosystem. This is being accomplished by designing a management system with several components, each of which can be modified to fit local landscape conditions and landowner objectives. Specific objectives of these components are to intercept eroding soil and agricultural chemicals from adjacent crop fields, slow flood waters, stabilize streambanks, and provide wildlife habitat and alternative, marketable products. Additionally, such management systems may improve local aquatic systems by restoring ecological functions associated with the riparian zone. This ecological restoration may be brought about by reducing discharge extremes to modify the flow regime, improving structural habitat, and restoring energy relationships by adding organic matter and reducing temperature and dissolved oxygen extremes.

DESCRIPTION OF THE BEAR CREEK WATERSHED

The Bear Creek watershed in central Iowa is a small (26.8 square miles) drainage basin, typical of the region in terms of landscape and land use. The watershed is located within the Des Moines Lobe subregion of the Western Corn Belt Plains ecoregion, one of the youngest and flattest ecological subregions in Iowa (Griffith et al. 1994). In general, the land is level to gently rolling and has a poorly developed stream network. This region was once part of the vast tallgrass prairie ecosystem, which was interspersed with wet prairie marshes in topographic lows and gallery forests along larger streams and rivers. Soils of the region were primarily developed from glacial till and alluvial, lacustrine, and wind-blown deposits. Present land use in the Bear Creek watershed is typical of the region: over 87% of the land area is devoted to row crop and pasture agriculture.

The presettlement landscape and drainage history of the Bear Creek watershed is being described using original land survey notes (approximately 1847) and accompanying field plat maps (K. Anderson, Iowa State University, unpublished data). Early county atlases, original drainage district maps, and historical accounts by early settlers provide a historical perspective of the dramatic changes in watershed hydrology and the modification of the prairie, wetland, and riparian ecosystems that have occurred since European settlement of the region. The townships through which Bear Creek flows were originally surveyed in 1847. These surveys suggest that prior to settlement, the watershed was "rolling prairie" with "first rate soil" and a substantial portion was "low and marshy." Native forest was limited to the Skunk River corridor into which Bear Creek flowed. The upper portion of the watershed was characterized as low, wet prairie connecting more defined marshes and would likely have contained intermittent or seasonal water flow.

Subsequent changes have dramatically altered watershed hydrology and have resulted in the change of the upper watershed from a low, wet prairie landscape with slowly moving water to one with a well-defined perennial stream and numerous tributaries. Conversion of the land from native vegetation to row crops, extensive subsurface drainage tile installation, and ditch dredging have resulted in substantial stream channel development and incision. Records suggest that artificial drainage of marshes and wet prairies in the upper reaches of the Bear Creek watershed was completed by about 1902, with ditch dredging completed shortly thereafter. While the main stream system appears to have remained about the same since that time, significant channelization continued into the 1970s. Modern stream systems also indicate development of intermittent flow drainages throughout the watershed. Ground surveys show that these are typically grass waterways associated with agricultural row crops.

Such dramatic changes in regional hydrology and vegetation have had substantial impacts on water quality and the biotic integrity of terrestrial and aquatic ecosystems. The alteration of headwater fish communities in the midwestern states after agricultural development has been well documented (Menzel et al. 1984; Karr et al. 1985). Numerous fish species adapted to conditions of clear water, firm substrates, and lush aquatic vegetation have been widely decimated and replaced by ecological generalist species (organisms having broad environmental tolerance ranges and relatively unspecific resource requirements) that are tolerant of degraded habitat conditions, such as turbid, warmer waters, and that have wide functional flexibility (the ability to use a wide range of resources), especially for food and reproductive requirements (Menzel et al. 1984; Liang 1995). Collections of the fish community made in Bear Creek from 1991 to 1993 (Liang 1995) demonstrate this shift to be the case in this watershed. Fish populations are dominated by minnows (Cyprinidae), with 11 species represented. Three sucker species (Catostomidae), three bullhead catfish species (Ictaluridae), and three sunfish species (Centrarchidae) were also collected, though in much fewer numbers than the minnows. The johnny darter was well represented. As a group, the dominant species found in Bear Creek may be characterized as ecological generalists. Many of the dominant species are omnivores that feed on invertebrates, algae, and organic detritus. They also commonly have an extended reproductive period and are able to use various substrates and structures for reproduction (Liang 1995).

Similarly, invertebrate community structure of Iowa headwater streams modified by agriculture demonstrates a relatively low diversity, dominated by functional groups (groups composed of different species that perform a similar ecological function within an ecosystem) adapted to collect a wide range of organic matter or to scrape algae attached to rocks or other substrates (Barnum 1984; Menzel et al. 1984). Invertebrate density and biomass (the summed mass of individuals in a given volume or area) can be relatively high in these streams, particularly within those groups adapted for using the predominant food resource, fine particulate organic matter (Barnum 1984). These communities must also be adapted to the wide swings in temperature and dissolved oxygen characteristic of these stream systems.

Land ownership in the Bear Creek watershed is typical of the region; nearly all of the land is privately owned. Public lands are limited to road rights-of-way and parcels owned within the small community of Roland.

RESTORATION APPROACH

Restoration efforts in the Bear Creek watershed were begun in 1990 by the Agroecology Issue Team of the Leopold Center for Sustainable Agriculture and the Iowa State University Agroforestry Research Team. Other stakeholders in this effort have included the local farmer-owned cooperative, Iowa Department of

Natural Resources, U.S. Department of Agriculture, U.S. Environmental Protection Agency, and Pheasants Forever.

A first step in the project was to conduct a watershed-scale assessment of land use, condition of the riparian zones, and stream and groundwater quality. This information was combined with geographic information system data and computer modeling to develop vulnerability maps for the watershed to identify critical stream reaches where modified management might be expected to reduce the impact of nonpoint source pollution on Bear Creek. A survey of watershed residents and landowners was also conducted at the initiation of the project to determine the level of concern for water quality problems, identify acceptance of a riparian management system, and quantify the value placed on improvement of surface and groundwater quality (Colletti et al. 1994). Survey respondents indicated a concern for nonpoint source pollutants and a desire to improve water quality in Bear Creek. Insights gained from survey results are combined with research relating to the effectiveness of riparian systems to guide watershed-level planning for soil conservation and water quality improvement.

The second stage of the work has been the actual development and establishment of the riparian management system model along stretches of Bear Creek and evaluation of the model system's effectiveness in reducing nonpoint source pollution. The riparian management system consists of three major components (Figure 19.1): (1) a constructed, multispecies riparian buffer strip, (2) soil bioengineering technologies for streambank stabilization, and (3) constructed wetlands to intercept and process nonpoint source pollutants in agricultural drainage-tile water. In addition, rotational grazing systems that limit livestock access to the creek channel are being demonstrated. Restoration efforts to date include collaborative studies and demonstrations in the upper half of the watershed. This work was initiated along a 0.6-mile length of Bear Creek on the Ron and Sandy Risdal farm. Subsequently, a buffer strip system has been planted along an additional 1.8 miles of Bear Creek on three farms upstream from the original site.

Multispecies Riparian Buffer Strip

The general multispecies riparian buffer strip layout consists of three zones (Figures 19.2, 19.3). Starting at the creek or streambank edge, the first zone is a 33-foot-wide strip of four to five rows of trees, the second zone is a 12-foot-wide strip of one to two rows of shrubs, and the third zone is a 21-foot-wide strip of native, warm-season grass. Fast-growing native trees, such as willow, poplar, silver maple, and green ash, are planted nearest the stream to provide a functioning multispecies riparian buffer strip in the shortest possible time. Where site conditions permit, slower-growing species, such as northern red oak, bur oak, swamp white oak, or black walnut can be planted in the outer rows. The trees provide perennial root systems for streambank stabilization and long-term nutrient storage close to the stream.

Shrubs are included in the design because they have permanent roots and because they add habitat diversity. Their multiple stems also function to slow

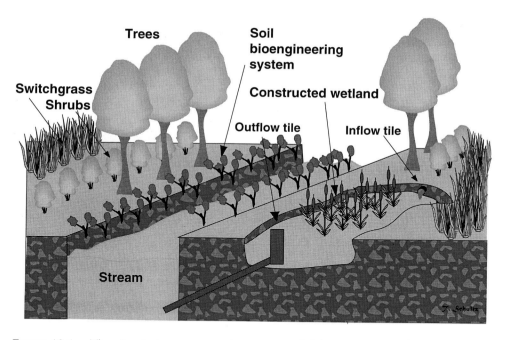

FIGURE 19.1.—The riparian management system model integrates a multispecies buffer strip, streambank stabilization technologies (soil bioengineering system), and constructed wetlands.

flood flows. The mixture of species used includes ninebark, redosier, and gray dogwood, common chokecherry, Nanking cherry, hazel, and nannyberry. Other shrub species can be used, especially if they are native and provide the desired wildlife habitat and aesthetic characteristics.

The grass zone functions to intercept and dissipate the energy of surface runoff, trap sediment and agricultural chemicals in the surface runoff, and provide a source of organic matter for soil microbes that can metabolize nonpoint source pollutants. A minimum width of 20 feet of switchgrass is recommended because it produces a uniform cover and has dense, stiff stems that provide a highly frictional surface to intercept surface runoff and facilitate infiltration. Other warm-season grasses, such as big bluestem and Indiangrass, and native forbs may also be included within the mix. Because of its growth habit, switchgrass should be used where surface runoff is most severe (Dabney et al. 1993).

The multispecies riparian buffer strip model presented here prescribes a zone of trees, shrubs, and prairie grass that, in total, is 66 feet wide. Although these species combinations and width provide a very effective buffer system, they are not the only combinations that can be effective. Site conditions (e.g., soils and slope), desired buffer strip biological and physical function(s), landowner objectives, and cost-share program requirements should be considered in

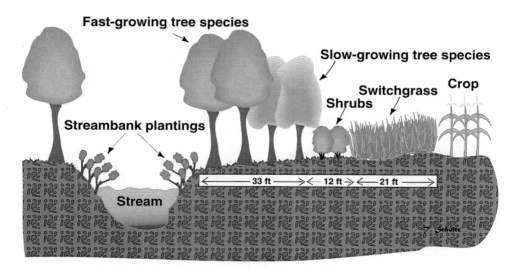

FIGURE 19.2.—The multispecies riparian buffer strip model includes tree rows closest to the stream, next to the trees, shrubs, and then a strip of switchgrass adjacent to the cropland.

specifying species combinations and placement. At a minimum, a native grass community with a width of 20 to 30 feet is needed to meet the objectives of a basic, functioning system.

Monitoring and ongoing research.—One of the best ways to demonstrate the results of a restoration project is to keep a visual record. Figure 19.4 shows two photos taken from the same location at different times. They illustrate the dramatic changes in the structure of the plant community that can take place in a short period. The diverse buffer strip vegetation serves as a physical barrier to both water and wind movement, provides wildlife habitat, and dramatically changes the aesthetic impressions that visitors have of the site.

A major focus of the Bear Creek watershed restoration project is to demonstrate the capability of the buffer strip to reduce nonpoint source pollution impacts on surface and groundwater and to provide wildlife habitat (Schultz et al. 1995). For example, the ability of the buffer strip to retain sediment is being assessed using a combination of passive collectors (Daniels and Gilliam 1996) and runoff simulation studies. These studies demonstrate that the 21-foot-wide switchgrass component of the buffer strip is capable of reducing sediment contained in runoff from nearly 1,000 parts per million to less than 250 parts per million, a 75% reduction in sediment load.

The potential of the buffer strip to reduce chemical loading to Bear Creek is assessed by using piezometers (measures elevation of water table) and soil suction lysimeters (measures percolation and removal of soluble constituents) to monitor nitrate and atrazine (an herbicide) moving within groundwater and vadose zone (zone between the land surface and the water table) water through

FIGURE 19.3.—Cross section of the multispecies riparian buffer strip at the Bear Creek, Iowa, riparian management site. Photograph was taken in July 1994, the vegetation's fifth growing season.

the buffer strip. Results indicate that nitrate concentrations in the vadose zone are much lower across the buffer strip than within the adjacent, cultivated field (Figure 19.5). Whereas concentrations of nitrate-nitrogen within the vadose zone of the cropped field will vary from year to year depending upon crop rotation, average concentrations measured within the vadose zone nearest the stream have never exceeded three parts per million in monitoring conducted between 1994 and 1996. In contrast, concentrations of nitrate-nitrogen measured in the vadose zone within a field cropped to the stream edge showed no reduction nearer the stream. In areas where shallow groundwater contributes significantly to streamflow, this buffering function is important in reducing nitrate loads to the surface water. Atrazine concentrations are similarly reduced across the buffer strip compared with the adjacent, cultivated fields.

Additional research is investigating changes, both in soil quality parameters potentially regulating water movement and in microbial activity that have occurred since the cropped or pastured lands have been planted with buffer strip vegetation. These efforts indicate that the establishment of the perennial vegetation of trees, shrubs, or native grasses has increased soil infiltration capacity, lowered soil bulk density (ratio of the mass of dry soils to the bulk volume of the soil), and improved the quantity and quality of soil organic matter in only 5 years since establishment. Such improvements in soil quality serve as an indicator of the soil's structural and biological integrity, which in turn are related

FIGURE 19.4.—Bear Creek riparian management site. Top photograph shows site in March 1990, prior to buffer strip establishment. The land on the right side of the stream had been in cultivation and the land on the left had been grazed. Bottom photograph shows same site in June 1994. Notice the rapid growth of the riparian vegetation and the dramatic improvement in the condition of the streambanks after only five seasons of riparian management.

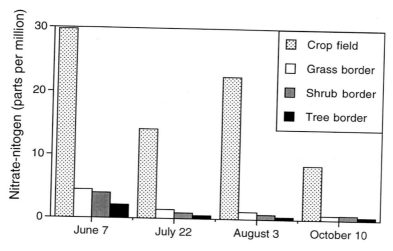

FIGURE 19.5.—Nitrate-nitrogen concentration within the vadose zone of the cropped field (corn) and within the three zones of the multispecies riparian buffer strip on several dates in 1995.

to the status of certain degradative processes and to environmental and biological plant stress (Parr et al. 1992).

The habitat value of the buffer strip was assessed by comparing bird species composition within a section of the buffer strip with a contiguous channelized section of stream that had little riparian cover. Results demonstrate that buffer strip establishment has dramatically increased bird species diversity in the short period since planting. On average, only 4 species per day were observed on the channelized stream section whereas 18 species per day were found within the 4-year-old buffer strip. A two-sample comparison indicated that this difference in the number of species was significant ($P < 0.05$). The total number of bird species observed throughout the study period was also much greater within the buffer strip. Within the buffered stream reach, 30 species were observed, compared with only 8 species along the channelized stream section. Although community indices are not widely accepted as the sole indicators of habitat complexity, these observations strongly suggest that establishment of the multispecies buffer strip positively influences bird species diversity.

Streambank Stabilization

Several authors have estimated that greater than 50% of the stream sediment load in small watersheds in the Midwest is the result of channel erosion (Roseboom and White 1990). This soil usually consists of small silt and clay particles that are ultimately deposited in rivers, lakes, or backwater areas, choking them with sediment and diminishing their value as habitat for fishes and aquatic macroinvertebrates (Frazee and Roseboom 1993). This problem has been worsened by the increased erosive power of streams that have been channelized and have lost riparian vegetation. The typical solution is to buttress blocks of concrete, rock, wood, or steel along the stretch of the bank that is eroding (Frazee and Roseboom 1993). Such solutions are costly to build and maintain and provide little aquatic habitat.

An alternative streambank stabilization technique is the use of locally available natural materials, such as willow or other live plant material, often in combination with revetments of rock or woody material, such as eastern redcedar. These techniques are often referred to as soft engineering or soil bioengineering (Coppin and Richards 1990). Vegetative means of stabilization protect streambanks in several ways (Klingeman and Bradley 1976). First, the root system helps hold the soil together and increases the overall bank stability by its binding network structure. Second, the exposed vegetation can increase the roughness resistance to surface flow and reduce the local flow velocities, causing the flow to dissipate energy against the deforming plant and away from the soil. Third, the vegetation acts as a buffer against the abrasive effect of transported materials. Fourth, close-growing vegetation can induce sediment deposition, which reduces stream sediment load and reestablishes the streambank.

Several different soil bioengineering techniques have been employed in the Bear Creek watershed. These include the use of willow harvested as posts or stakes in late winter while still dormant and driven into the bank, bundles of live willows (fascines) partially buried along the slope, and biodegradable erosion control fabric anchored with willow stakes or redosier dogwood on bare slopes. Alternatives used to stabilize the base of the streambank include rock and anchored dead plant material, such as eastern redcedar or bundled silver maple. Figure 19.6 illustrates the dramatic improvement in streambank stability that can be achieved over just several months. These bioengineering solutions are very effective and less expensive than traditional streambank stabilization techniques.

Constructed Wetlands

A characteristic of many parts of the upper Midwest is the presence of an extensive network of subsurface tile drainage that has facilitated the conversion of many wetland acres to agricultural uses. Such tile drains often are the primary cause of increases in stream discharge and provide a direct path to surface water for nitrate or other agricultural chemicals that move with the shallow groundwater. In such instances, constructed wetlands integrated into new or existing drainage systems may have considerable potential to remove nitrate from shallow subsurface drainage (Crumpton et al. 1993).

To demonstrate this technology, a small (2,900 square feet) wetland was constructed on the original project site (Risdal farm) to process drainage-tile water from a 12-acre cropped field. The size and shape of the wetland were designed to fit into the 66-foot-wide buffer strip. The wetland was constructed by excavating a low area near the creek and constructing a small berm. The subsurface drainage tile was rerouted to enter the wetland at a point that maximizes residence time of drainage-tile water within the wetland (see Figure 19.1). A simple, gated structure at the wetland outlet provides control of the water level maintained within the wetland. Cattail rhizomes collected from a

FIGURE 19.6.—A streambank stabilization structure on Bear Creek. Top photograph taken shortly after completion in May 1995. Prior to stabilization, the streambank was vertical with exposed soil. Techniques employed were rocks for toe stabilization, erosion control fabric to protect the exposed slope, and dormant willow posts. Bottom photograph shows same site in July 1995. Notice the rapid establishment of grasses and the growth of the willows.

FIGURE 19.7.—Wetland constructed on the Risdal farm on Bear Creek. Photograph taken in August 1994, a few months after construction. Notice the aquatic vegetation has spread rapidly within the wetland basin. The wetland berm was planted with native prairie vegetation. Water control structure can be seen in the foreground. Buffer strip vegetation can be seen in the background.

local marsh and road ditch were planted within the wetland. Growth during the initial season was dramatic. The cattails spread rapidly throughout the wetland and many achieved heights in excess of 6 feet. Big bluestem, Indiangrass, gray-headed coneflower, and black-eyed Susan were planted on the constructed berm to provide vegetation diversity. Establishment of these native grasses and forbs was very successful; all species flowered by the second year. Figure 19.7 shows the rapid establishment of vegetation within the first months after construction.

Monitoring and ongoing research.—Water samples at the inlet and outlet of the wetland are collected using automated water samplers. In 1995, inflow concentrations of nitrate-nitrogen in the drainage-tile water fluctuated between 7 and 11 parts per million (Figure 19.8). In contrast, outflow concentrations were lower during most times. Notable exceptions were during May and early July when large precipitation events increased drainage-tile discharge substantially and reduced the residence time of nitrate-laden waters within the wetland. Precipitation during the months of July and August 1995 was substantially below normal. As a result, outflow from the wetland ceased by 1 September, and the wetland was dry shortly thereafter.

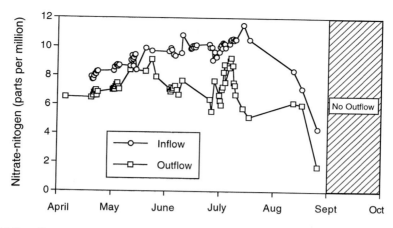

FIGURE 19.8.—Concentration of nitrate-nitrogen in wetland inflow and outflow during 1995. The field being drained was in soybeans.

Experimental studies have demonstrated the considerable capacity of fresh-water wetlands to remove nitrate (Crumpton et al. 1993) and have confirmed that denitrification is the dominant removal process for externally loaded nitrate (Isenhart 1992). These studies also have demonstrated that wetlands containing a large amount of standing vegetation and a large buildup of decaying plant litter will be more efficient in nitrate removal, given sufficient contact time between nitrate-laden water and the substrate. This indicates that the Risdal wetland will likely become more efficient at removing nitrate after several growing seasons and may take several years to reach a steady state with respect to its nitrate removal capacity.

The ratio of the wetland surface area to the area of crop field drained for the Risdal wetland is 1:180. Previous studies suggest that a mature, 1-acre wetland could remove significant amounts of nitrate from waters draining approximately 100 acres of corn to which moderately high nitrogen levels have been applied (Isenhart 1992; Crumpton et al. 1993). The Risdal wetland is, therefore, below this benchmark ratio, and it remains to be seen what the maximum retention capacity for nitrate will be once the wetland reaches maturity.

Rotational Grazing Systems

The gently undulating, highly fertile soils of the Midwest allow development of large crop fields. Because of the value of this land for crop production, livestock grazing is often conducted in a narrow belt within the riparian zone, where cultivation with large equipment may be difficult. As a result, livestock have a dramatic impact on streambanks, riparian zone vegetation, and surface water quality. The use of rotational grazing, with controlled access to the stream, can reduce these livestock impacts. We have just begun to study this last component of the riparian management system model. Comparisons of streambank stability are being made among paddocks (enclosed pasture areas) that totally restrict livestock access to the channel, paddocks that provide very limited direct access to the channel, and paddocks where complete access is allowed during the short periods of grazing. Restricted cattle access should improve water quality and provide better instream habitat.

ADAPTIVE MANAGEMENT

Development of the riparian management system within the Bear Creek watershed is an ongoing process incorporating several actions. The initial concept and design was developed by an interdisciplinary team of scientists and was based on the emerging recognition that rivers and their floodplains are so intimately linked that they should be understood, managed, and restored as integral parts of a single ecosystem (NRC 1992a). Recognizing this, the interdisciplinary team decided to initiate a watershed management project on a real-world agricultural basin and to manage the landscape by using native plant communities to create or restore riparian buffer zones that would complement field-scale practices to reduce nonpoint source pollution. An idealized model that was hypothesized to accomplish the functions of natural riparian zones of the region was developed. Buffer system functions are assessed through experimental work with intensive process monitoring. As more of the process mechanisms and rates are identified, modifications to the model are made and incorporated into subsequent buffer zone establishment.

The social acceptance of the riparian management model is assessed through the use of surveys, focus groups, and one-on-one information exchange. A better understanding of landowner objectives and economic considerations has resulted in numerous variations of the model system. What initially began as just the buffer strip component of the system now includes the three other components: streambank stabilization, constructed wetlands, and rotational grazing. This flexibility is designed to encourage adoption of the management practices by satisfying landowner goals and concerns as well as fitting specific biogeophysical conditions of the site. For example, the buffer strip component of the model can be modified by using different species combinations and by varying the width of each zone. Although such variation in design may not be optimal for water quality or wildlife benefits, the flexibility is important if it means that a landowner is accepting the concept. After the landowner has had experience with a smaller system, he or she may be willing to increase the size and effectiveness of the buffer or add additional system components.

Technology transfer efforts are geared toward quickly getting the results and information into the hands of landowners and natural resource professionals. This is accomplished through on-site tours, field days, self-guided walking tours, videos, and extension bulletins. Other methods of information dissemination include presentations at meetings of natural resource professionals, conservation groups, and local civic organizations, articles in local newspapers and trade publications, and publications in refereed journals. Local ownership of the restoration effort is encouraged through the development of voluntary citizen action teams that assist in buffer strip establishment, water quality monitoring, and construction of wildlife nesting boxes. Finally, training workshops are being organized for agricultural and natural resource professionals to help disseminate the information and validate results.

A challenge that remains is the uncertainty about the level and extent of restoration needed to effect change at the watershed level. With limited resources and a mosaic of privately owned land, restoration efforts have to be targeted at willing landowners within critical areas of the Bear Creek watershed. On-going modeling efforts will assist in making predictions about the extent of restoration required to improve the biotic integrity of the aquatic resources.

To date, the Bear Creek project lacks an accurate accounting of all benefits associated with the riparian zone management efforts (e.g., aesthetics, wildlife enhancement, and filtering of nutrients and sediments). Farmers and citizens in the watershed clearly prefer a shared responsibility for implementing riparian best management practices. With additional details on system function and valuation of extra market benefits, such as nonpoint source pollutant reduction and wildlife habitat and aesthetic enhancement, landowners will be more willing to install riparian best management practices, especially if cost share and technical assistance are available. Our research and demonstration program stresses voluntary adoption versus regulatory approaches of buffer strip installation. Regulation usually sets rigid parameters that do not apply well to the wide range of conditions encountered.

PROSPECTS FOR THE FUTURE

Prospects for the future can be viewed from a watershed restoration perspective as well as a research and model development perspective. The former is specific to the restoration of the Bear Creek watershed. The research team is in the process of identifying the critical reaches of Bear Creek at which to target restoration, and work will continue to identify willing landowners and sufficient cost-share opportunities. Most landowners do not want to install the systems themselves but are willing to rely on consultants. Thus there is a need to identify restoration consultants well skilled in the design, installation, and maintenance of these systems. Hands-on training workshops are being organized with the sponsorship of nongovernment organizations such as Trees Forever and government organizations such as the Natural Resources Conservation Service and the Iowa Department of Natural Resources.

From the research and model development perspective, several pieces of information must be developed or refined. Especially critical is the establishment of rates for critical biological and physical processes of the system, for example, the relative importance of plant uptake and denitrification as the major mechanisms of nitrate uptake (Hill 1996). It will be important to generalize information gained within specific riparian zones to the watershed level. Also needed is a detailed assessment of the ability of the system to improve the biotic integrity of the aquatic resources. To date, established riparian management systems in the Midwest are not extensive enough or old enough to assess changes in factors regulating the biotic integrity of the aquatic ecosystems. These factors include the ability of the management systems to regulate instream temperature and dissolved oxygen concentrations through shading, to provide energy resources through annual inputs of organic matter, and to provide instream habitat diversity. One of the major research goals for the Bear Creek project is to restore enough miles of stream length so that these assessments can be made.

Riparian management models will need to be integrated with farm planning models to coordinate conservation measures required to reduce nonpoint source pollution. Also, further definition of the human constraints on management system adoption will allow for the targeting of education efforts or for the need to change the model to address socioeconomic and political realities.

Ultimately, it is intended that ongoing work within the Bear Creek watershed will contribute to a comprehensive agricultural watershed management strategy for the midwestern corn belt. To accomplish this, the project must act as a demonstration site for landowners, a demonstration and training site for natural resource managers, and a research site for scientists developing and testing a riparian management system model under real-world conditions.

ACKNOWLEDGMENTS

Support for this work is from the Leopold Center for Sustainable Agriculture and from the Iowa Department of Natural Resources through grants from the U.S. Environmental Protection Agency under the Federal Nonpoint Source Management Program (Section 319 of the Clean Water Act); the U.S. National Research Initiative Competitive Grants Program, U.S. Department of Agriculture Award 95-37102-2213; the Agriculture in Concert with the Environment program, jointly funded by the U.S. Department of Agriculture Cooperative State Research Education and Extension Service and the U.S. Environmental Protection Agency under Cooperative Agreement 91-COOP-1-6592; and Pheasants Forever. This is a contribution of the Iowa Agriculture and Home Economics Experiment Station, Ames, Project 3209 and supported by Hatch Act and state of Iowa funds.

We must ... not divorce the stream from its valley.
—*H. B. N. Hynes, 1975*

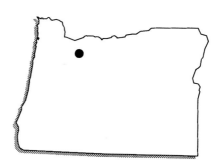

CHAPTER 20

FISH HABITAT RESTORATION IN THE PACIFIC NORTHWEST: FISH CREEK OF OREGON

Gordon H. Reeves, David B. Hohler, Bruce E. Hansen,
Fred H. Everest, James R. Sedell, Tracii L. Hickman,
and Daniel Shively

The decline of anadromous salmonids in the Pacific Northwest of the United States is attributable to a suite of factors that includes overexploitation in sport and commercial fisheries, habitat alteration, migration barriers, variable ocean conditions, and influence of hatchery practices (Nehlsen et al. 1991). A combination of these factors is generally associated with the depressed status of almost every population. The factor most associated with the decline of individual populations is habitat alteration, which includes a decline in the quantity and quality of freshwater habitat (Nehlsen et al. 1991).

Habitat in streams used by anadromous salmonids in the Pacific Northwest has been simplified as a consequence of many human activities (Hicks et al. 1991b; Bisson et al. 1992). Simplification includes the loss of habitat, quality, diversity, and complexity. Such changes have occurred as a result of past as well as more recent activities and are common throughout much of the range of anadromous salmonids in western North America (Sedell and Luchessa 1982; Hicks et al.

1991b). Lawson (1993) argued that the arrest and reversal of the continuing decline of the quantity and quality of freshwater habitat are imperative if populations of anadromous salmonids are to be protected from further decline or local extinction.

Responsible population and habitat management agencies have implemented a multitude of efforts to increase natural populations of these fishes. Habitat restoration programs are integral parts of these recovery programs. Much of the habitat restoration effort has been directed at restoring physical elements, such as wood and boulders, that form and create in-channel (see Reeves et al. 1991a for examples) and off-channel habitats (Peterson 1982).

Results from such efforts have been mixed (Reeves et al. 1991a). Frissell and Nawa (1992) reported most structures placed in streams in southwest Oregon had high failure rates. Fontaine (1988) found no response of juvenile steelhead to the placement of boulders and large wood in an Oregon stream. In contrast, Brock (1986) and Moreau (1984) observed increases in local populations of juvenile steelhead following placement of similar structures in northern California. Numbers of juvenile (Peterson 1982; House and Boehne 1985) and smolt (House 1996) coho salmon have increased following habitat restoration efforts.

Reeves et al. (1991a), in an extensive review of fish habitat restoration efforts for anadromous salmonids in the Pacific Northwest, argued that proper monitoring and evaluations should be an integral component of restoration efforts. Although examples of proper monitoring and evaluation are cited above, they are exceptions rather than the rule. Even rarer are long-term monitoring programs of restoration efforts. In this paper, we (1) describe a long-term program to monitor the response of juvenile anadromous salmonids to habitat restoration efforts; (2) quantify physical and biological responses to habitat restoration in Fish Creek, Oregon; and (3) place fish habitat restoration efforts in perspective to watershed-scale flood events that occurred in November 1995 and February 1996. Preliminary results have been reported in a series of earlier publications (Everest and Sedell 1984; Everest et al. 1986b, 1987, 1988; Reeves et al. 1990).

THE FISH CREEK WATERSHED

The Fish Creek watershed lies in north-central Oregon on the west slope of the Cascade Mountain Range and drains into the upper Clackamas River (Figure 20.1). The watershed is 13 miles long, approximately 6 miles wide, and covers 66 square miles. Main-stem Fish Creek is a fifth-order stream (Strahler 1957). Elevation varies from 4,600 feet in the headwaters to 820 feet at the mouth. Gradient generally exceeds 5%, but averages 2% in the lower 4 miles. The terrain is steep and mountainous; bluffs in the lower canyons are typical of the Columbia River basalt formation. Valley bottoms are typically narrow with incised channels and narrow floodplains. The steep gradient and volcanic geology result in the stream having a predominately riffle environment with boulder substrate. Annual flow varies from about 18 cubic feet per second (cfs) in late summer to more than 3,500 cfs in winter. The hydrograph is dominated by rain and snowmelt. Water

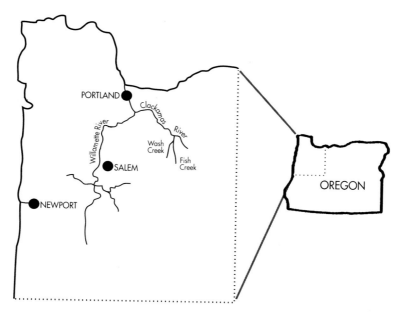

FIGURE 20.1.—Location of Fish Creek, Oregon.

temperatures are within the range favorable to anadromous salmonids, ranging from near 32°F in winter to about 68°F during most summers.

One major tributary, Wash Creek, is a fourth-order system (Strahler 1957) that originates in the southwest portion of the Fish Creek basin and enters Fish Creek at mile seven (Figure 20.1). The Wash Creek basin covers 14 square miles and has a main-stem length of 5 miles. Habitat in Wash Creek is steep boulder riffles in a narrow, incised channel. Average summer flow is 11 cfs.

The Fish Creek watershed has been subjected to intensive land management activities. Over 41% of the basin has been subjected to timber harvest and associated activities, which began in 1944 and peaked in the 1980s (USFS 1994). Much of the riparian area was subjected to salvage logging prior to the mid-1980s, when a new forest plan that restricted such activities was developed. There are currently 140 miles of roads (2.1 miles per square mile) in the watershed. The system has experienced several debris flows (movements of materials in intermittent or small, nonfish-bearing perennial upslope channels) and landslides as a result of these factors. Additionally, the channel was subjected to an intensive debris-clearing program following a 100-year flood in December 1964.

The Fish Creek watershed supports populations of summer and winter steelhead in its lower 10 miles, which includes the lower 3 miles of Wash Creek (Figure 20.2). Coho salmon distribution is limited to the lower 3 miles of Fish Creek (Figure 20.2). Resident (nonanadromous) rainbow and cutthroat trout are found in upper portions of the basin, primarily above barriers to anadromous

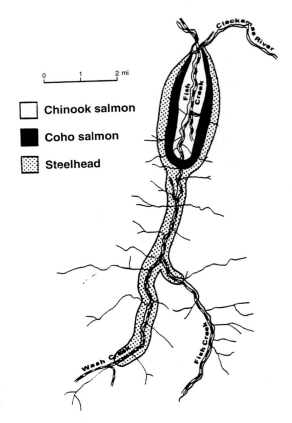

Chinook salmon

Coho salmon

Steelhead

FIGURE 20.2.—Distribution of anadromous salmonids in Fish and
Wash creeks, Oregon.

fishes. Adult and juvenile chinook salmon were found in the lower portion of
Fish Creek in early years of the study (Everest et al. 1985); however, since 1986
we have not observed any. The reason for the decline is not known at this time.

HISTORY OF HABITAT RESTORATION WORK

In 1959, it was estimated that pools formed 45% of the habitat in the range of
anadromous fishes in Fish Creek (USFS 1994). Large wood was abundant and
contributed to the formation of many pools. A subsequent survey of the system
following the December 1964 flood found that pool habitat was reduced to 27%.
This decrease was attributed to large wood redistribution within and movement
out of Fish Creek. Additional wood was removed from 1965 until the early 1980s
through salvage logging (USFS 1994). By 1982, pools constituted 11% of the
habitat. Loss of wood also contributed to the incision of the channel to bedrock
and loss of side- and off-channel areas.

Fish Creek was also affected by other land management activities. Removal of
streamside vegetation during timber harvest along tributaries increased water
temperatures (USFS 1994). This resulted in an increase in water temperatures in
main-stem Fish Creek. Monitoring at 12 sites throughout the upper Clackamas

River basin from 1993 to 1995 found that Fish Creek had the highest summer low-flow temperatures (D. Shively, U.S. Forest Service, unpublished data). An extensive road network was constructed throughout the basin in conjunction with timber harvest. Roads located in the floodplain limited the stream's ability to meander across the valley bottom. Tributary roads exacerbated the natural instability and increased the number of debris torrents originating in steep headwater channels (USFS 1994). Poorly designed stream crossings created migration barriers at three tributaries, eliminating 2 to 3 miles of habitat for anadromous salmonids.

The focus of the habitat restoration program in Fish Creek was to increase the amount and complexity of pool habitat for summer and winter rearing and the amount of spawning habitat for all anadromous salmonids and to rehabilitate riparian vegetation to increase shading and decrease water temperatures. Design, execution, and evaluation of the restoration effort began in 1981 as a cooperative agreement between the Mt. Hood National Forest, Estacada Ranger District, and the Pacific Northwest Research Station, Corvallis, Oregon, both of the U.S. Forest Service. The project was initially conceived as a 5-year effort and was financed with National Forest System funds. The restoration program and companion evaluation effort were expanded in 1983 when the Bonneville Power Administration entered into agreement with the Mt. Hood National Forest to cooperatively fund work in Fish Creek.

During the first 3 years of the program, techniques used were considered prototypes. The goal was to determine which techniques were most effective in achieving the desired conditions. Treatment and techniques evaluated included (1) boulder berms, built to collect spawning gravels and provide rearing areas; (2) development of off-channel and side-channel areas, designed to provide additional rearing habitat; and (3) introduction of large trees into the channel by means of explosives (Everest and Sedell 1984; Everest et al. 1985).

None of these techniques, with the exception of development of one limited off-channel area, were considered promising enough for broad application in Fish Creek (Everest et al. 1987). Evaluation of boulder berms, most of them constructed with on-site material, found that they successfully collected gravels and created pools. However, there was a loss of winter habitat as a result of the removal of larger material from stream margins. Berms built with angular quarried rock showed greater durability and reduced on-site disturbance, but they were fairly expensive to build because of the high cost of hauling material. The rocky substrate of the margins along Fish Creek made it difficult to place explosive charges under the rootwads of trees and avert damage to the rootwads. Consequently, rootwads left on blasted trees did not provide an anchor to keep trees in the stream channel.

Drawing on experience gained in the prototype phase, a different approach to habitat restoration was adopted in 1986. The approach was to treat areas in the lower and middle portions of Fish Creek intensively, particularly along stream margins (Figure 20.3). Structures built between 1986 and 1988 were combinations of logs and boulders anchored together with cable and epoxy and located

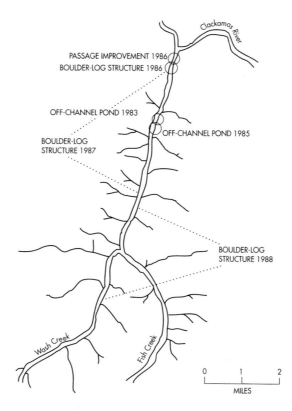

FIGURE 20.3.—Location and type of restoration efforts in Fish Creek, Oregon.

along the streambank (Figure 20.4). More than 500 structures were created during this period at a cost of US$240,000. Large trees used in these structures were obtained from adjacent riparian areas. Trees selected were generally located more than 115 feet from the channel in order not to diminish significantly the potential future recruitment of large woody debris and current shade conditions. Boulders were brought to the sites from outside Fish Creek. The primary purpose of these structures was to increase the frequency and complexity of pools, thus benefiting all species and age-classes of anadromous salmonids throughout the year. Active construction of structures ended in 1988.

Off-channel ponds were created on ancient floodplain terraces on both sides of the lower end of Fish Creek (Figure 20.3). The pond on the east side of the stream was constructed in 1983 by building a 10-inch-diameter gravity-feed pipeline from Fish Creek, which delivered about 9 gallons per second of water (Everest et al. 1985). A small stream near the northeast corner of the pond was diverted into the pond by building a wall with sandbags. The pond, an area formerly dry in summer, was approximately 1 acre, and its depth varied from 8 to 50 inches. A two-way trap was located at the outlet to monitor adult movement in and smolt movement out of the pond. In 1985, 13 cubic yards of gravel were placed in the inlet channel just above the pond. It was hoped that

FIGURE 20.4.—Photograph of large wood and boulder structure that was placed in Fish Creek, Oregon.

adult coho salmon would spawn in this area and provide recruitment for the pond. In 1985 a similar pond was constructed on the west side of Fish Creek, just above the east-side pond (Figure 20.3; Everest et al. 1986b).

From 1985 to 1988 coho salmon fry were placed in the ponds in the spring because of the low numbers of adult coho salmon returning to Fish Creek. The fry were the offspring of wild coho salmon from the upper Clackamas River that had been captured at North Fork Dam and spawned at the Eagle Creek Fish Hatchery, located on the lower Clackamas River. It was believed that these fish would accelerate the increase in coho salmon numbers in Fish Creek. After stocking the 1988 brood was found to be infected with bacterial kidney disease *Renibacterium salmoninarum* (Reeves et al. 1990). The ponds were dried in 1989 and 1990 to minimize establishment of the disease in Fish Creek. The ponds were restocked in April 1991; however, survival was poor and no additional supplementations of the ponds were made.

MEASUREMENT OF HABITAT CONDITIONS AND FISH POPULATIONS

Evaluation of habitat restoration efforts in Fish Creek have focused on (1) quantifying the amount of habitat available to anadromous salmonids in late summer; (2) estimating the number of juvenile anadromous salmonids in late summer; and (3) estimating the number of smolts leaving in the spring. Two approaches have been used to estimate the amount of late-summer habitat. From 1982 to 1984, five 0.3-mile sections of stream, one on Wash Creek, one on upper Fish Creek above its confluence with Wash Creek, and three on main-stem Fish Creek between its confluence with Wash Creek and its mouth, were sampled each year. These reaches were believed to be representative of habitat conditions in Fish Creek. Surface area and volume of habitat units in each reach were measured with a tape to estimate available habitat for the reach. To estimate the total available habitat, reach estimates were summed and the results extrapolated to the rest of the basin accessible to anadromous salmonids. Refer to Everest and Sedell (1984) and Everest et al. (1985) for more detail.

Beginning in 1985, total habitat available to anadromous salmonids was estimated using the procedure of Hankin and Reeves (1988). This technique generated a more accurate estimate of total available habitat and was considered more statistically valid. For these surveys, five habitat types—pools, riffles, glides (slow, shallow runs), side channels, and beaver ponds—were identified (Bisson et al. 1982). Detailed descriptions of yearly protocols are contained in Everest et al. (1986b, 1987, 1988) and Reeves et al. (1990).

Beginning in 1982, estimates of the total number of juvenile anadromous salmonids residing in Fish Creek in late summer were made annually. From 1982 to 1984, estimates were made by sampling individual habitat types at eight different locations in areas accessible to anadromous fishes. Population estimates were made in 36 habitat units (one in each pool, riffle, side channel, and alcove or beaver pond at each of the eight sites). In smaller habitats, estimates were made using backpack electroshockers and the multiple-pass population estimator (Zippen 1958). In larger areas, counts were made by divers with masks and snorkels. The total number of fish counted was considered the population estimate; no attempt was made to account for possible diver bias. These estimates were then extrapolated to the basin based on previous estimates of available habitat.

Beginning in 1985, the method of Hankin and Reeves (1988) was employed to estimate juvenile anadromous salmonid numbers. This technique improved the statistical validity of population estimates and allowed for the construction of confidence intervals around the estimates. More detailed descriptions of the methodology are contained in Everest et al. (1986b, 1987, 1988) and Reeves et al. (1990).

Fish captured by electroshocking for the population estimates were anesthetized in tricaine methanesulfonate and weighed and measured. We compared the annual mean (average) weight of each species and age-class for the period before versus after habitat restoration to determine possible responses in growth. Data from 1982 and 1983 were excluded from the analysis because of questions about their accuracy.

Numbers of anadromous salmonid smolts migrating from Fish Creek were estimated beginning in 1985. From 1985 to 1988 smolts were captured in a modified Humphrey trap (see Everest et al. 1988 for description); beginning in

1989, a revolving helix-screw trap was employed (see Reeves et al. 1990 for description). The trap was positioned 0.2 mile upstream from the mouth of Fish Creek by means of cables attached to trees on the bank. A rock berm constructed immediately upstream of the trap opening directed flow and fish into the trap. The trap was operated for the duration of the smolt migration, which was generally from mid-March through mid-June.

Smolts leaving the off-channel ponds at mile 2 and mile 2.2 (Figure 20.3) were monitored from 1985 to 1988. A rotating-drum screen at the outlet of the pond diverted migrants into a trap box adjacent to an adult access ladder. Smolts leaving Fish Creek and the off-channel ponds were weighed and measured before being released into Fish Creek. See Reeves et al. (1990) for more details.

We compared the estimated number and mean annual lengths of smolts leaving Fish Creek for the periods before and after restoration (but not including the floods in November 1995 and February 1996) to determine possible responses. Only data on the amount of habitat and estimated smolt numbers leaving Fish Creek are available for 1996. These 1996 data were considered by themselves in order to assess the initial effects of the floods. We excluded coho salmon smolts originating from the off-channel ponds in our analysis because they could have biased the effects of the primary in-channel work. As stated previously, the production of these ponds was reduced because of the introduction of diseased fish. We believe that following cessation of annual stocking the production from these ponds was minimal.

We did not directly count the number of adults that returned to Fish Creek to spawn; to do so would have required a tremendous amount of money and time. Instead, we used counts of adults passing North Fork Dam on the upper Clackamas River as an index of adult returns.

RESPONSES TO RESTORATION EFFORTS PRIOR TO FLOODS

Changes in Stream Habitat

The total amount of habitat available to anadromous salmonids in Fish Creek varied from 164,220 square yards to 294,470 square yards during the study (Table 20.1). However, the relative composition of fish habitat in Fish Creek changed between 1985 and 1995. The amount of pool habitat increased steadily from 1988, the second year of the intensive restoration effort (Table 20.1). The continued increase in pool habitat following structure placement is most likely attributable to the lag in pool formation. All structures did not immediately create pools; rather, scouring of the bed, which results in pool formation, may take more extended periods. By 1995, pools constituted 39% of the total habitat area compared with 11% in 1982 at the start of the restoration program.

Concurrent with the increase in pools has been a decrease in glides (Table 20.1). Glides increased in relative area between 1987 and 1989 but have declined steadily since. Many of the structures were located in glides which, consequently, were lost as pools formed. A possible reason for the initial increase in glides (between 1987 and 1989) was the conversion of riffles, particularly low-gradient riffles, to glides during early phases of pool formation.

TABLE 20.1.—Estimated amount of habitat (given as area in square yards and percent of total area) before (1982–1988) and after completion (1989–1995) of habitat restoration program and after floods (1996) in Fish Creek, Oregon.

	Habitat type								
	Pools		Glides		Riffles		Side channels		Total area
Year	Area	Percent	Area	Percent	Area	Percent	Area	Percent	
1982	22,050	11	a		165,620	86	5,080	3	192,750
1983	24,920	8	a		262,140	89	7,410	3	294,470
1984	22,920	10	a		193,230	87	6,360	3	222,510
1985	31,520	18	25,130	15	112,060	65	3,080	2	171,790
1986	29,250	15	32,720	16	136,710	69	0	0	198,680
1987	34,820	21	29,000	18	100,400	61	0	0	164,220
1988	38,910	20	41,160	21	110,870	57	3,200	2	194,140
1989	43,510	19	50,240	22	130,840	58	610	<1	225,200
1990	48,660	25	27,200	14	114,660	60	910	<1	191,430
1991	58,270	27	28,460	13	124,270	58	1,710	<1	212,710
1992	61,760	35	12,630	7	100,380	57	810	<1	175,580
1993	70,270	34	11,010	5	121,170	59	1,830	<1	204,280
1994	76,210	39	6,660	3	110,040	57	1,170	1	194,080
1995	81,640	39	7,970	4	120,290	57	940	<1	210,840
1996	52,528	33	12,714	8	93,171	58	1,177	1	159,590

[a]Did not attempt to identify this habitat type in this year.

The amount of riffle habitat in Fish Creek has remained relatively constant during this study (Table 20.1). It continues to be the primary habitat in Fish Creek, constituting 57% of the total habitat area in 1995.

Changes in Coho Salmon and Steelhead Populations

Coho salmon—The mean annual number of juvenile coho salmon after restoration but before the floods (i.e., 1989–1995) was 41.8% less compared with the mean for the period before completion (1982–1988; Figure 20.5A). Mean number of coho salmon smolts leaving Fish Creek following restoration (i.e., 1990–1995) was 12.7% more than the mean for the period before completion (i.e., 1985–1989; Figure 20.5B). Neither of these differences in the means were statistically significant (for juveniles, $t = 0.685$, $P = 0.5$; for smolts, $t = 0.318$, $P = 0.76$). Please note that the periods for comparisons vary with life history stage. The effects of restoration on smolts would not be apparent until at least a year following restoration.

The situation was reverse for the mean number of coho salmon smolts leaving the upper Clackamas River system (Figure 20.6A). Numbers of anadromous salmonid smolts leaving and adults entering the upper Clackamas River are monitored at the North Fork Dam located below Fish Creek. The mean annual number of coho salmon smolts leaving the upper Clackamas River in the period following restoration in Fish Creek was 19.8% less than the mean for the period before completion of restoration. Similar to the differences in Fish Creek, these means were not statistically different ($t = 0.792$, $P = 0.44$).

The mean annual return of coho salmon adults to the upper Clackamas River

A. Juvenile Coho Salmon

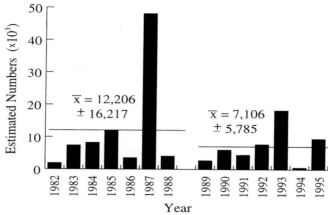

B. Coho Salmon Smolts

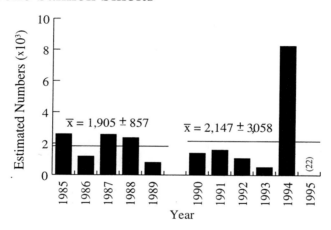

FIGURE 20.5.—Estimated number of coho salmon (**A**) juveniles in and (**B**) smolts leaving Fish Creek, Oregon, before and after completion of habitat restoration but prior to floods of November 1995 and February 1996. Means ($\bar{X}$) and standard errors ($\pm$) are given for the two periods.

also declined in the period following completion of habitat restoration in Fish Creek compared with the period before restoration (Figure 20.6B). Mean annual adult returns to the upper Clackamas River declined 5.1% for the period following restoration efforts in Fish Creek compared with the mean for the period before. The difference in adult returns was not statistically significant (t = 0.137, P = 0.9).

The responses of juvenile and smolt coho salmon in terms of size to habitat restoration efforts differed. On average, juveniles were 14.8% longer for the period following restoration compared with the period before (Figure 20.7A).

A. Coho Salmon Smolts

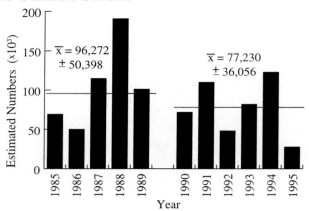

B. Adult Coho Salmon

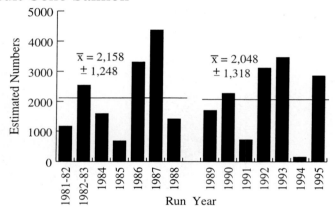

FIGURE 20.6.—Estimated number of coho salmon (**A**) smolts leaving and (**B**) adults returning to the upper Clackamas River, Oregon, before and after completion of habitat restoration in Fish Creek. Means ($\bar{X}$) and standard errors ($\pm$) are given for the two periods.

The annual mean length of juveniles following restoration was significantly greater ($t = -2.48$, $P = 0.03$) than the mean for the period before. Smolts on average were 6.8% longer in the period following restoration compared with the period before (Figure 20.7B). This difference was not statistically significant ($t = -1.266$, $P = 0.236$).

Steelhead—The different age-classes and life history stages of steelhead responded differently to habitat restoration efforts in Fish Creek. Mean estimated annual number of young of the year declined by 53.2% for the period following completion of habitat restoration (Figure 20.8A). The difference of the means before and after restoration was statistically significant ($t = 5.37$, $P = 0.01$).

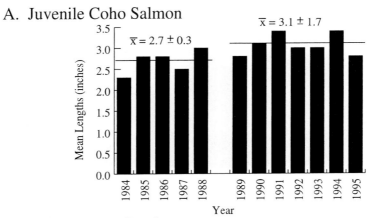

A. Juvenile Coho Salmon

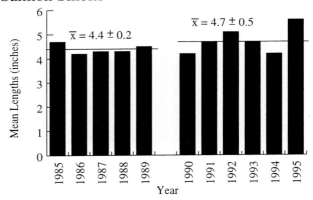

B. Coho Salmon Smolts

FIGURE 20.7.—Mean length of coho salmon (**A**) juveniles captured in and (**B**) smolts leaving Fish Creek, Oregon, before and after completion of habitat restoration but prior to floods of 1995–1996. Means ($\overline{X}$) and standard errors ($\pm$) are given for the two periods.

In contrast, mean annual estimated numbers of age-1 steelhead (Figure 20.8B) and steelhead smolts (Figure 20.8C) exhibited relative increases following completion of habitat restoration but before the floods. The mean for age-1 fish increased 11.7% and that of smolts 27.7% following restoration. The means before and after restoration were not statistically significant for either group, however (for age 1, $t = 0.951$, $P = 0.35$; for smolts, $t = 0.827$, $P = 0.43$).

The mean annual estimated numbers of steelhead smolts leaving the upper Clackamas River (monitored at the North Fork Dam) remained relatively constant during the period before the floods (Figure 20.9A). The mean estimated numbers for the period before versus after completion of habitat restoration in Fish Creek were not statistically different ($t = 0.035$, $P = 0.98$).

The mean annual estimated number of steelhead adults returning to the upper Clackamas River following completion of habitat restoration efforts but prior to

A. Age-0 Steelhead

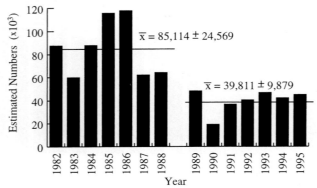

B. Age-1+ Steelhead

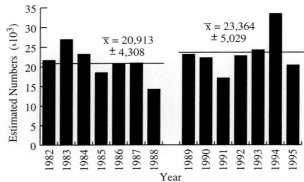

C. Steelhead Smolts

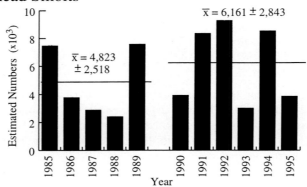

FIGURE 20.8.—Estimated number of steelhead (**A**) age 0 in, (**B**) age 1 in, and (**C**) smolts leaving Fish Creek, Oregon, before and after completion of habitat restoration but prior to floods of 1995–1996. Means ($\overline{X}$) and standard errors ($\pm$) are given for the two periods.

A. Steelhead Smolts

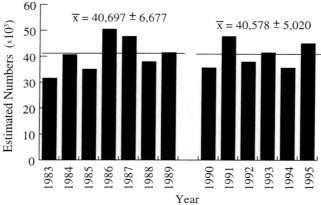

B. Steelhead Adults

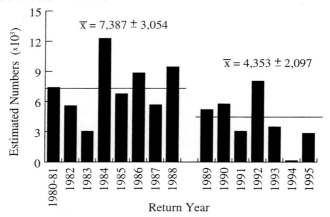

FIGURE 20.9.—Estimated number of steelhead (**A**) smolts leaving and (**B**) adults returning to the upper Clackamas River, Oregon, before and after completion of habitat restoration effort in Fish Creek. Means ($\overline{X}$) and standard errors ($\pm$) are given for the two periods.

floods in Fish Creek was 41.1% less than the mean for the period before completion (Figure 20.9B). The difference between the means was not statistically significant ($t = 1.787$, $P = 0.12$).

The response of the size of steelhead to habitat restoration in Fish Creek varied with age-class. Age-0 and 1 fish were on average 12.5% and 4.1% larger, respectively, in the period following restoration but prior to floods compared with the period before restoration (Figure 20.10A, B). For both age-groups, the difference between the two periods was statistically significant (for age 0, $t = -3.97$, $P = 0.03$; for age 1, $t = -2.20$, $P = 0.05$). Like coho salmon, the mean length of steelhead smolts was relatively larger (3.2%) in the period following restoration compared with the period before (Figure 20.10C). This difference was not statistically significant ($t = -1.430$, $P = 0.186$).

EFFECTS OF THE 100-YEAR FLOOD

Fish Creek experienced major flooding in November 1995 and February 1996. The February 1996 storm was considered to be a 100-year event. Damage to stream channels and adjacent slopes was severe. Many landslides in the Fish Creek watershed delivered large quantities of sediment directly to the stream system (Figure 20.11). The U.S. Forest Service inventoried 236 landslides in the watershed. Many tributaries had major debris torrents (rapid movement of a large quantity of wood and sediment down a stream channel). The inventory attributed 42% of the landslides to timber harvest, 34% to roads, and 24% to nonmanagement-related causes (Shively, unpublished data). More than 15 miles of principle tributaries (second- and third-order streams) were scoured by debris torrents. Fish Creek migrated laterally in several locations throughout its floodplain, washing out many segments of the main river valley road and contributing large quantities of sediment to the system (Figures 20.12, 20.13).

More than 50% of the habitat restoration structures were destroyed (Figure 20.14). Forty-nine percent of the structures present within the Fish Creek watershed in 1995 were exported entirely out of the basin by storm events and associated scours.

There were only moderate changes in the habitat composition of Fish Creek following the floods (Table 20.1). Percent of pools declined to 33%, similar to what was observed in earlier years of the post-restoration period. Concurrently, there was no change in percent of habitat in riffles.

Changes in the estimated number of smolts leaving Fish Creek were larger than habitat changes. In 1996, it was estimated that 106 coho salmon smolts and 1,018 steelhead smolts migrated from Fish Creek. This was the lowest number of steelhead smolts observed during the study and represented an 83.5% decline from the post-restoration period mean (Figure 20.8C). The 1996 estimate for coho salmon smolts was 5% of the post-restoration period mean (Figure 20.5B).

DISCUSSION

The large wood and boulder structures created pools along the length of Fish Creek used by anadromous salmonids and achieved the physical objective of the restoration effort relatively quickly. Studies of habitat restoration efforts in other systems have reported similar responses in physical conditions (House and Boehne 1985; Gowan and Fausch 1996).

Reeves et al. (1991b) argued that changes in smolt numbers are the best indicator of the effect of habitat restoration efforts for anadromous salmonids. Mean numbers of coho salmon and steelhead smolts were relatively larger in the period after restoration (but prior to the floods) compared with before restoration. However, neither of these differences were statistically significant, even though increases were relatively large for steelhead (27.7%).

Lack of statistical significance even in the face of relatively large differences

A. Age-0 Steelhead

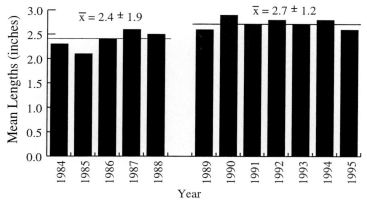

B. Age-1+ Steelhead

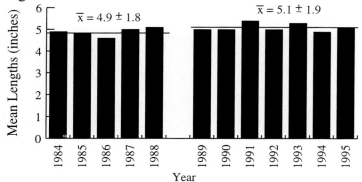

C. Steelhead Smolts

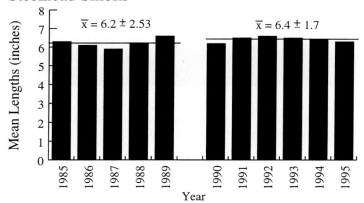

FIGURE 20.10.—Mean length of steelhead (**A**) age 0 and (**B**) age 1 captured in and (**C**) smolts leaving Fish Creek, Oregon, before and after completion of habitat restoration but prior to floods of 1995–1996. Means ($\overline{X}$) and standard errors ($\pm$) are given for the two periods.

FIGURE 20.11.—Landslide along Pick Creek as a result of February 1996 storm event. This landslide and many others delivered large amounts of sediment into Fish Creek. (Photograph taken 16 February 1996 by Dan Shively.)

between means appears to be primarily due to the large variation among annual estimates for both the pre- and post-restoration numbers for most life history stages. The inherent variability in populations of anadromous salmonids requires that evaluations of restoration programs extend for relatively long periods. Reeves et al. (1991a) suggested that evaluations should be conducted over a minimum of two generations of the fish species of interest; for steelhead and

FIGURE 20.12.—High flows from February 1996 storm scoured out roads constructed along Fish Creek, contributing additional sediment and debris to the stream. (Photograph by Dan Shively.)

coho salmon in Fish Creek this would be 6 to 10 years. Post-restoration evaluations in Fish Creek were conducted for 7 years (i.e., 1989–1995) and continue today; even 7 years of monitoring was insufficient to detect statistical differences.

It is difficult to assess the biological response to the restoration effort in Fish Creek prior to the floods. Statistically few measures were significant, although relative differences between means for the pre- and post-restoration means were relatively large, particularly for steelhead. The mean numbers and size of age-1 fish and smolts were larger following restoration than they were before. Because all of these measures tend toward increases and because, in terms of numbers, there was essentially no difference in numbers leaving the upper Clackamas River before and after restoration, we believe that the restoration effort in Fish Creek at least appeared to have been successful for steelhead until the floods.

Conclusions about coho salmon are more equivocal. Mean estimated number of coho salmon smolts was relatively larger following restoration (but before the floods) compared with the period before restoration. It appears that this difference was attributable primarily to the large estimated number observed in 1994 (Figure 20.5B). In 3 of the 5 years in the pre-restoration period, estimated smolt numbers were greater than all years in the post-restoration period, except for 1994. Numbers of juveniles appeared to have declined following restoration. Consequently, at present we cannot conclude with any certainty that the

FIGURE 20.13.—Multiple timber harvest and road-related landslides initiated a 3-mile long debris torrent along Third Creek that washed out the main road paralleling Fish Creek. (Photograph taken at the confluence of Third and Fish creeks by Dave Hohler.)

restoration effort in Fish Creek appeared to have been successful for coho salmon.

It is possible that factors operating at spatial scales larger than the Fish Creek watershed may have confounded the evaluation of the habitat restoration effort on coho salmon. The parallel decline in estimated numbers of juvenile coho salmon in Fish Creek and smolts in the upper Clackamas River suggest that perhaps some factors operating at a basin scale were influencing these populations. Gowan and Fausch (1996) found that trout populations in Colorado varied concurrently over an 8-year period. Platts and Nelson (1988) reported concordance in variation of trout populations in streams in Idaho, Nevada, and Utah. The greatest concordance was with streams that were closest together.

There are some factors specific to Fish Creek that could, at least partially, be responsible for the apparent decline of juvenile coho salmon. Structures in the main stem may have failed to create suitable rearing and spawning conditions for coho salmon. Numbers of juvenile coho salmon were initially relatively low in Fish Creek and declined despite attempts to increase available habitat quality and quantity. The relatively steep nature of Fish Creek makes it marginal habitat for coho salmon (Reeves et al. 1989). Coho salmon numbers increased following habitat restoration efforts in lower-gradient, less-constrained streams in other parts of Oregon (e.g., House and Boehne 1985; Nickelson et al. 1992a) and in Washington (e.g., Peterson 1982). The off-channel ponds, particularly the one on

FIGURE 20.14.—Instream habitat improvement structures suffered a high failure rate in Fish Creek as a result of high flows during November 1995 and February 1996. This log structure was torn apart and deposited into the riparian zone, others were exported downstream and out of the watershed. (Photograph taken April 1996 by Dan Shively.)

the east side of Fish Creek, were the best potential sites for coho salmon production. However, they were only a small portion of the total available habitat in the system, and their productivity was compromised by the introduction of bacterial kidney disease in 1988. Additionally, spawning habitat in the ponds was limited and never used by adult fish to any extent.

Another possible factor influencing the decline in juvenile coho salmon numbers following habitat restoration efforts in Fish Creek was the decline in adult returns. The magnitude of the declines in juvenile coho salmon in Fish Creek was nearly eight times greater than the decline in adult returns to the upper Clackamas River. Consequently, it could be argued that it is unlikely that the decline in juvenile coho salmon in Fish Creek following restoration was attributable to any substantial degree to the decline in adult returns to the upper Clackamas River. This lack of correlation would be the case if the decline in spawners was relatively even throughout the upper Clackamas River. If the decline in adults to Fish Creek, which has marginal habitat for coho salmon, was disproportionately large compared with other streams, then the decline in juvenile numbers could have been at least partially attributable to declining adult returns.

The decline in returns of adult steelhead could have contributed to the decline of age-0 steelhead in Fish Creek following habitat restoration but before the floods. Estimated mean numbers of age-0 steelhead in Fish Creek and adults returning to the upper Clackamas River declined by relatively large amounts, 53.2% and 41.1% respectively, following completion of habitat restoration in Fish Creek. We assumed that adult returns to the upper Clackamas River represented returns to Fish Creek. Numbers of age-0 fish could decline if fewer eggs were deposited in Fish Creek.

Another factor that could explain the decline of age-0 steelhead in Fish Creek for the period following habitat restoration was the change in habitat conditions. Fast-water habitats such as riffles and glides are used by age-0 steelhead (Hartman 1965). These habitats had the highest densities of age-0 steelhead before completion of the restoration effort (Reeves et al. 1990). The amount of fast-water habitat declined and pools increased as a result of restoration. It appears that the capacity for age-0 steelhead declined as a result.

Habitat changes resulting from restoration in Fish Creek appear to have favored older age-classes of steelhead. Hartman (1965) found that age-1 steelhead used pools and generally were bottom oriented. Pools had the highest densities of these fish prior to completion of restoration (Reeves et al. 1990). Numbers and size of age-1 steelhead also could have increased because of the complex nature of the pools. Many of the pools created as a result of restoration had wood and boulders. Pools present before restoration were relatively simple because of the lack of wood. Increased complexity would have increased visual isolation among individuals. Keenleyside and Yamamoto (1962) found that densities of Atlantic salmon could be increased by providing greater visual isolation among individuals. Increased complexity may also have increased food production and availability and reduced energy expenditures for food capture and intraspecific (within a species) interactions. The combination of increased carrying capacity and quantity and quality of pools likely contributed to the increased numbers and size of older steelhead in Fish Creek following completion of the restoration effort. This trend is likely to reverse because of the loss of pool habitat caused by the floods.

The increased complexity of pools may have also contributed to the increased numbers of older age-classes of steelhead by increasing survival. Estimated mean numbers of age-1 and smolt steelhead increased following restoration despite decreasing numbers of age-0 fish in Fish Creek and decreasing numbers of adults returning to the upper Clackamas River. Complex pools could have provided refugia against high flows in winter. Grunbaum (1996) found that age-1 steelhead used pools in winter in streams with flow and water temperature regimes similar to those of Fish Creek in other parts of Oregon. Alternatively, pools could have provided refuge during summer low flows and increased numbers of fish entering the winter period.

The increased size of younger age-classes of coho salmon and steelhead may be attributable to several possible factors. The increased size of age-0 steelhead and juvenile coho salmon following restoration could be a compensatory response to the reduced numbers. Negative correlations between abundance and growth of sockeye salmon (Johnson 1965) and brown trout (LeCren 1965) have been observed. Several examples of similar responses in other fishes are cited by Weatherly and Rogers (1978). Increased amounts of wood may have provided the same benefits as those described above for older age-classes of steelhead.

The habitat restoration effort in Fish Creek and its accompanying long-term evaluation have many important lessons for other habitat restoration programs. First is the importance of evaluations. From the outset, the evaluation was conducted concurrently with the restoration effort. There was almost immediate feedback on prototype structures and projects. We were able to consider a range of possible approaches before settling on final designs. This form of adaptive management undoubtedly contributed to the successes of the restoration program—the change in habitat and the increases in size and number of steelhead smolts and juveniles—and allowed information to be applied to other stream systems in a relatively quick time period.

Advice from persons in a variety of disciplines was continually solicited during the implementation phase of the restoration effort in Fish Creek. Field reviews of the project designs were very productive. These reviews uncovered potential problems and used the experience of restoration experts before plans were finalized. People who had little restoration experience also contributed significantly: they generally provided a test of our conventional wisdom by challenging our assumptions.

Knowledge of previous conditions in Fish Creek allowed establishment of realistic goals and objectives for the restoration program. Unfortunately, an assessment of earlier conditions was not done at the start of the project. Initial assessments, based on professional opinions of various fish biologists, were that spawning gravels were limited and efforts should be directed at increasing available spawning habitat of all species. These efforts were not successful. Final habitat goals were established after we determined conditions in Fish Creek prior to degradation from documents of agencies responsible for resource management in the watershed. Such information is often available for stream systems. It may take time and effort to find and study such information, but the effort is imperative to increasing a program's chances for success. Developing a picture of previous conditions also increased the likelihood that structure designs and materials were appropriate for existing conditions and project objectives.

The criteria for assessing the success of a restoration effort should be established at the start of a project and need not be based solely on statistical differences. It would be difficult to argue that the restoration effort in Fish Creek was successful from a statistical perspective. Only one of the increases in estimated numbers was statistically significant. However, the percent increases in mean estimated numbers were relatively large (i.e., >10%). Unfortunately, we did not establish a priori what expected increase would indicate success. Such criteria could be established by examining results from other studies or conducting an analysis to identify limiting factors such as was done by Reeves et al. (1989). Other criteria, such as changes in growth or survival rates, could also be established (Reeves et al. 1991b), given that the logic and rationale for those criteria are presented. Explicit establishment of criteria and goals can prevent unnecessary debates about the outcomes of the effort.

Different species or age-classes of fish should not be expected to respond in the same manner to changes resulting from habitat restoration (Hartman et al. 1996). In our evaluation, mean annual estimated numbers of age-1 and smolt steelhead increased whereas numbers of coho salmon and age-0 steelhead declined. Habitat changes resulting from restoration efforts may create conditions favored by one species or age-class while, concurrently, habitats favored by other species or age-classes may be lost. Recognition of this possibility is important when establishing objectives and expectations for any restoration program.

The floods in November 1995 and February 1996 had only moderate effects on the relative amounts of habitat types available in Fish Creek. The large number of structures and the large material in them may be partially responsible for this. Structures that remained in the watershed helped trap organic debris and sediment delivered to the channel by landslides and debris torrents. Given the large amount of sediment and wood delivered to the channel by the floods, we expect that habitat composition should return to preflood levels in a short time.

Biological effects of the floods were greater and will likely be expressed for several years. Smolt production in 1996 was well below the average for the period before the flood. We expect this trend to continue until younger age-classes recover. Given that there is potential for habitat to recover relatively quickly, we would expect biological production to recover to levels observed before the floods.

Ultimately, the successful restoration of fish habitat is dependent on the restoration of in-channel and upslope conditions (Hartman et al. 1996). We believe that past land management activities contributed to the large number of upslope failures and ultimately exacerbated the effect of high flows on in-channel conditions in Fish Creek. In hindsight, we realize efforts should have been made to address upslope conditions more aggressively while doing in-channel restoration in Fish Creek.

Failure to consider large-scale watershed conditions may in the long-term override the desired effects of small-scale manipulations of in-channel conditions (Jones et al. 1996). The purpose of upslope efforts should be twofold. First, roads need to be removed or stabilized and road areas revegetated in order to reduce future landslides and debris torrents. Second, upslope and riparian vegetation needs to be re-established, including large trees, so that the proper elements are in place for the next flood or other catastrophic disturbance (Reeves et al. 1995).

ACKNOWLEDGMENTS

Several people have contributed to the evaluation effort in Fish Creek. D. Hellar, U.S. Forest Service, initiated and funded the restoration program and the evaluation effort. G. Haugen, U.S. Forest Service, secured funding from the Bonneville Power Administration for early parts of the study. J. Moreau, U.S. Forest Service, provided financial assistance more recently. R. Diebel, U.S. Forest Service, also provided help and personnel in recent years. Several others from the

Bonneville Power Administration and the U.S. Forest Service Mt. Hood National Forest and Pacific Northwest Research Station have provided assistance with the publication and distribution of reports and with field work and data analysis. Finally, we would like to dedicate this paper to the memory of Carl McLemore. Carl was a fish biologist at the Pacific Northwest Research Station and worked on this project from it inception to 1987. He was responsible for data collection and analysis during the first few years and devoted a tremendous amount of energy to those tasks.

Their flavor was excellent—superior, in fact, to that of any fish I have ever known. They were of extraordinary size—about as large as the Columbia river salmon—generally from two to four feet in length. From the information of Mr. Walker, who passed among some lakes lying more to the eastward, this fish is common to the streams of the inland lakes.
—*John C. Fremont's (1845) description of Lahontan cutthroat trout from Pyramid Lake.*

CHAPTER 21

WATERSHED RESTORATION AND GRAZING PRACTICES IN THE GREAT BASIN: MARYS RIVER OF NEVADA

Laura A. Gutzwiller, Randy M. McNatt, and Roy D. Price

The Lahontan cutthroat trout is the only salmonid native to the Lahontan basin of northern Nevada, northeastern California, and southeastern Oregon. The Lahontan cutthroat trout was federally listed as an endangered species in 1970 under the U.S. Endangered Species Act and was reclassified as a threatened species in 1975 to facilitate management and allow regulated angling. Historically, Lahontan cutthroat trout occurred throughout the Truckee, Walker, Carson, Quinn, and Humboldt river systems, including Lakes Tahoe, Pyramid, and Walker (La Rivers 1962). This native trout may have occupied as much as 2,210 stream miles of the Humboldt River drainage of northern Nevada (Coffin 1983) but currently occurs in only 318 miles (USFWS 1995). Historically, Lahontan cutthroat trout may have occurred in over 200 stream miles of the Marys River subbasin, a 520-square-mile watershed that drains into the Humboldt River (Figure 21.1); however, only 69 miles are currently occupied (USFWS 1995).

The Marys River subbasin is still considered to have some of the best native Lahontan cutthroat trout populations and habitat remaining in the Lahontan

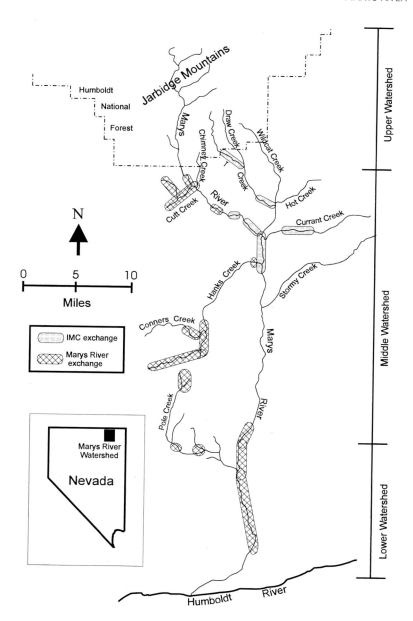

FIGURE 21.1.—Map showing upper, middle, and lower sections of the Marys River watershed in northeastern Nevada and showing stream segments acquired via land exchanges. The Marys River exchange placed 47,000 acres under the U.S. Bureau of Land Management (USBLM). The Independence Mining Company (IMC) exchange placed 4,133 acres under USBLM jurisdiction.

basin. As recently as the 1960s, the Marys River was considered a trophy class fishery (NDCNR 1963). However, due in large part to improper livestock management practices, trout habitat throughout the river system has deteriorated, leaving only a small percentage of the system with suitable habitat. Consequently, only remnant populations of Lahontan cutthroat trout persist.

In 1983, the Nevada Division of Wildlife (NDOW) prepared a fishery management plan for Lahontan cutthroat trout in the Humboldt River basin (Coffin 1983). This plan provided recommendations and guidelines for the protection and enhancement of the species and its habitat. The NDOW, U.S. Bureau of Land Management (USBLM), U.S. Forest Service (USFS), and U.S. Fish and Wildlife Service (USFWS) are all signatories to the plan.

Despite the agreement, management coordination among these federal and state agencies continued to be sporadic. The Marys River Riparian/Aquatic Habitat Management Plan (USBLM 1987a) provided specific guidance for maintaining and improving the quality of riparian and aquatic resources on public lands along the middle reaches of the Marys River and its tributaries. Management plans for grazing allotments (USFS 1989) revised management practices on lands administered by the Humboldt National Forest within the Marys River subbasin to reduce impacts of livestock on riparian and aquatic resources. On private lands, management and its resultant outcome varied. Some landowners were managing their irrigation, haying, and grazing activities to the benefit of riparian and stream resources. The activities of other landowners were detrimental because willows were removed from streambanks, hay was cut up to the water's edge, and livestock grazed season long each year. Fragmented land ownership patterns, as well as local resentment of federal and state governments and their regulations, stymied attempts to manage at a watershed level.

In 1991, the Marys River land exchange between the USBLM (Elko and Las Vegas districts) and Olympic Management, Inc., provided an opportunity for a more comprehensive approach to management of the Marys River subbasin (Figure 21.1). Approximately 47,000 acres of private land in Elko County, primarily within the Marys River watershed, were exchanged for 660 acres of public land within the Las Vegas city limits. As a result of this exchange, the USBLM acquired almost 55 miles of stream in the subbasin, including approximately 6,700 acre-feet of water rights on the Marys River (USBLM 1990a).

Although the environmental assessment for the land exchange mandated the USBLM only to coordinate with its Multiple Use Advisory Committee to develop a management plan for portions of the acquired lands, the opportunity for comprehensive management and the high level of concern about and interest in the area prompted the USBLM to initiate a watershed approach to management of the Marys River subbasin. The Marys River Project provided an opportunity to demonstrate the values and benefits of a healthy, functional watershed and to determine management options to achieve that end. Project planning incorporated the concerns of the USBLM, USFS, NDOW, USFWS, and others interested in enhancing management of the subbasin (Table 21.1).

The Marys River Project concept fit the Bring Back the Natives initiative, a

TABLE 21.1.—Marys River Project participants.

American Fisheries Society, Western Division	Nevada Mining Association
American Forestry Association	Nevada Wildlife Federation
Barrick Goldstrike Mines, Inc.	Newmont Gold Company
Elko County Conservation Association	Northeast Nevada Naturalists
Elko District Multiple Use Advisory Council	Sierra Club
Hawks and Bellgardt Marys River Ranch	Tabor Creek Cattle Company
Independence Mining Company	The Wildlife Society
Marys River Ranch	U.S. Bureau of Land Management
National Fish and Wildlife Foundation	U.S. Fish and Wildlife Service
Nevada Cattlemen's Association	U.S. Forest Service, Humboldt National
Nevada Division of Forestry	Forest
Nevada Division of Wildlife	White Horse Associates

cooperative national effort by the National Fish and Wildlife Foundation, the USBLM, and the USFS to restore the health of aquatic ecosystems and their native species. The Bring Back the Natives approach emphasizes interagency cooperation and revised land management practices to restore native aquatic species and watershed health. A project proposal was submitted in early 1991, and since then Bring Back the Natives has been a valuable tool for increasing funding for restoration.

THE MARYS RIVER WATERSHED

The Marys River originates in the Jarbidge Mountains of northeastern Nevada and drains south to the Humboldt River. The small, high-gradient headwater streams in these rugged mountains flow through aspen or cottonwood forests to the meandering lower reaches of the Marys River. Elevations range from 5,250 feet at the point where the Marys River joins the Humboldt to over 10,500 feet atop Marys River Peak in the Jarbidge Mountains.

The Marys River watershed is one of the more important headwater tributaries of the Humboldt River; its average annual water yield to the Humboldt is 32,000 acre-feet. Snowmelt runoff within the watershed normally begins in April or May and peaks during May in most streams. No storage reservoirs are located on the Marys River or its major tributaries, but several irrigation diversions remove substantial amounts of water when flows are sufficient (USBLM 1990a).

Native vegetation of the area is typical of the Great Basin. In the uplands, mixed stands of low sagebrush and big sagebrush dominate in association with Thurber needlegrass, Indian ricegrass, bottlebrush squirreltail, bluebunch wheatgrass, basin wildrye, and Idaho fescue. Higher elevations also commonly contain mountain mahogany, bitterbrush, quaking aspen, and black sagebrush (USBLM 1990a; USBLM, unpublished data). Riparian areas support willow, aspen, black cottonwood, narrowleaf cottonwood, alder, currant, wild rose, sedges, spikerushes, Baltic rush, bulrushes, cattails, greasewood, saltgrass, rabbitbrush, native meadow grasses, and forbs in varying degrees (Bradley and Nelson 1990; USBLM 1990a; USBLM, unpublished data; USFS, unpublished data).

Streams in the Marys River system support seven native fishes: Lahontan cutthroat trout, Lahontan redside shiner, speckled dace, mountain sucker, Tahoe sucker, tui chub, and Paiute sculpin (USBLM 1979; Johnson and Nokes 1995). As was common throughout the West, nonnative salmonids were introduced into the Marys River subbasin. Fortunately, the introductions were fairly unsuccessful and nonnatives—brook trout and rainbow trout—persist in only one stream in the subbasin. This stream, Currant Creek, does not flow into the Marys River because the channel has been diverted for irrigation.

All streams in the Marys River system exhibit high rehabilitation potential. Although streamside vegetation has been eliminated in some areas because of improper livestock management and, in the lower Marys River, from herbicide application and burning, remnant native vegetation exists throughout the subbasin. Monitoring has shown that reduced use of riparian areas by livestock has allowed vegetation to reestablish in many areas and to trap instream sediment. On the middle Marys River, noticeable bank building and stabilization occurred within three years of eliminating livestock use.

Livestock grazing is the major commodity use within the subbasin, which encompasses all or part of 12 USBLM or USFS grazing allotments. Commonly, ranchers turn livestock out to graze during the spring, summer, and fall on public or forest lands. The livestock are brought back to winter on private land after haying is completed. Hay production occurs at lower elevations on private lands adjacent to streams. Meadows are irrigated with water diverted from Marys River and T Creek.

Although mining is not being conducted in the Marys River subbasin, it does influence management. Because gold mining elsewhere within the Humboldt River basin is affecting or may affect Lahontan cutthroat trout and its habitat, several mining companies have been interested in restoration of stream systems for three primary reasons: (1) to mitigate impacts of mining operations by means of riparian restoration, (2) to improve the status of Lahontan cutthroat trout, and (3) to enhance community relations and involvement. Participation of the mining companies has facilitated project implementation. Mining companies have donated money, and company employees have volunteered time to the project. In addition, a land exchange between Independence Mining Company and the USBLM was completed in January 1996. This exchange resulted in the USBLM acquiring approximately 4,133 acres of land within the Marys River subbasin, including 20.8 miles of Lahontan cutthroat trout stream habitat (Figure 21.1).

Upper watershed.—This area consists of USFS and private lands north of the Humboldt National Forest boundary and includes the headwaters of the Marys River and several tributaries. Much of the upper Marys River lies within the Jarbidge Wilderness Area of the Humboldt National Forest. Riparian and aquatic habitat condition varies widely throughout the area (Table 21.2). Some good habitat exists in the upper tributaries, which are valuable spawning and nursery streams. The Marys River, however, exhibits a lack of streambank stability and vegetative cover and instream cover for trout (Johnson 1988).

Middle watershed.—This area consists of USBLM and private lands and includes several tributaries that contribute substantial flows to Marys River, as well as smaller tributaries that are valuable as spawning and nursery habitat. Surveys conducted by the USBLM from 1979 through 1993 (USBLM 1979, 1987b, 1988, 1990b, 1991, 1993) revealed streams in this area to be in poor to fair condition (Table 21.2). Although some improvement in riparian condition has occurred because of management changes initiated since 1991, streambank cover and stability are less than optimal, high-quality pools are scarce, and desirable stream bottom materials are lacking due to excessive sedimentation (USBLM 1993).

Lower watershed.—This area consists of USBLM and private lands and includes the fenced meadows acquired by the USBLM in the 1991 Marys River land exchange. These meadows are now entirely public land. They have never been adjudicated for livestock grazing and are not part of any grazing allotment. The Marys River in this area is in poor condition (Table 21.2). Bank cover and stability are lacking, resulting in channel incision and a lowered water table (USBLM 1992a). Woody vegetation had been sprayed, burned, and bulldozed by a previous landowner in a failed attempt to increase water yields. Portions of the channel in this reach have been straightened by bulldozer, but the decreased sinuosity and increased stream gradient in this reach is primarily in response to the removal of riparian vegetation and increased sediment load (Bradley and Nelson 1990). Lahontan cutthroat trout have occupied the upstream end of this reach as recently as the early 1980s (P. Coffin, USFWS, personal communication), but no trout were found here during the 1994 population sampling (Johnson and Nokes 1995).

ISSUES AND GOALS

The driving issue of the restoration effort was the deteriorated status of Lahontan cutthroat trout. Recovery efforts focused on public lands could substantially improve the trout's status while minimizing impacts to private landowners. Concerns varied from wanting Lahontan cutthroat trout to recover and be delisted to a desire to have a healthy, functional ecosystem supporting a diversity of native species. Because Lahontan cutthroat trout are dependent upon natural habitat conditions, they would be a good indicator for overall stream health. Related issues included sustaining viable livestock operations and using USBLM's acquired water rights.

Restoring Lahontan cutthroat trout habitat in the Marys River system to increase Lahontan cutthroat trout numbers and distribution is the primary objective for management of this area. The management goal is to contribute to the eventual delisting of the Lahontan cutthroat trout in the Humboldt River basin. A secondary goal is balancing use among various groups within a multiple-use framework. Although Lahontan cutthroat trout is the focus of the Marys River Project, from the beginning a watershed approach to management was considered necessary to achieve all goals.

The Marys River Master Plan (USBLM 1992b) was developed by the USBLM using an interdisciplinary approach, with input from the Elko District Multiple Use Advisory Committee, the USFS, the NDOW, the USFWS, a private consulting firm, and three of the private landowners within the subbasin. The master plan

TABLE 21.2.—Habitat condition[a] and Lahontan cutthroat trout population size (trout per mile) by stream and reach.

Stream and reach	Year	Habitat condition	Population size
Marys River			
Upper	1979	Poor	53
	1987	Fair	122
Middle	1979	Poor	47
	1987	Fair	
	1990	Poor	
	1993	Fair	0[b]
Lower	1992	Poor	0[b]
West Fork Marys River			
Upper	1979	Good	211
	1987	Good	294
East Fork Marys River			
Upper	1987	Fair	677
Marys River Basin Creek			
Upper	1979	Fair	211
	1987	Good	1,794
GAWS Creek			
Upper	1979	Excellent	211
Williams Creek			
Upper	1979	Fair	211
	1987	Good	2,693
Camp Draw Creek			
Upper	1979	Good	53
	1987	Good	422
Basin Creek			
Upper	1979	Poor	53
Cutt Creek			
Middle	1979	Poor	455
	1990	Poor	0[b]
Chimney Creek			
Middle	1979	Poor	764
	1987	Poor	
	1990	Poor	
	1993	Poor	26[b]
T Creek			
Upper	1979	Fair	1,386
Middle	1979	Poor	158
	1987	Poor	
	1990	Poor	26[b]
Draw Creek			
Upper	1979	Fair	
Middle	1979	Poor	915
	1987	Poor	53[b]
Wildcat Creek			
Upper	1979		1,947
Middle	1979	Poor	1,848
	1987	Poor	
	1993	Poor	739[b]

TABLE 21.2.—Continued.

Stream and reach	Year	Habitat condition	Population size
Currant Creek			
Middle	1979	Poor	370
	1994		0
Conners Creek			
Middle	1979	Good	238
	1987	Poor	
	1990	Poor	
	1993	Poor	0[b]
Hanks Creek			
Middle	1979	Poor	136
	1987	Poor	
	1993	Poor	0[b]

[a]Habitat condition is determined by comparing conditions at sampling to those optimal for Lahontan cutthroat trout. Optimal conditions are characterized by clear, cold water; a silt-free rocky substrate in riffle–run areas; an approximately 1:1 pool–riffle ratio with pools of slow, deep water; well-vegetated, stable streambanks; and abundant stream cover.

[b]Population sampling conducted in 1994 due to low water levels.

outlines the general actions necessary for the management and monitoring of the Marys River system and provides a framework for restoration of the subbasin.

RESTORATION EFFORTS

Officially, the Marys River Project began in 1991. However, various agencies had previously developed management plans and were implementing them to varying degrees. The Marys River Master Plan incorporated existing plans as much as possible. Previous monitoring and restoration efforts were valuable in providing information on existing conditions within the watershed, response potential of the habitats, and applicability of some restoration techniques.

Because improper livestock management was identified as the primary cause of habitat degradation in the watershed, emphasis has been placed on controlling livestock grazing (Table 21.3). Fence construction has been a major restoration activity in the watershed. Another approach designed to control livestock use by means of changing livestock distribution is the development of water sources away from streams. Little project work has been planned within riparian areas because substantial natural recovery has been documented following changes in livestock management (Figures 21.2, 21.3). For example, streambank stability and vegetative cover improved significantly on Chimney Creek following 2 years with no livestock use (Elko District, USBLM, unpublished data). Many examples exist of similar improvements following livestock management changes (Elmore and Beschta 1987; USGAO 1988; Kinch 1989; Myers 1989; Chaney et al. 1990, 1993). There are no plans to construct instream habitat improvement structures because streambanks are relatively unstable due to unconsolidated, noncohesive soils. Instream structures would likely lead to accelerated bank erosion.

Throughout the watershed, increased recreational values are anticipated from the watershed approach to management, which includes promoting the poten-

TABLE 21.3.—Restoration projects completed within the Marys River watershed.

Location	Year	Restoration project
Upper watershed		
Marys River	1990	U.S. Forest Service riparian pasture fencing
Middle watershed	1992	Prescribed burning
	1993	Prescribed burning
	1994	NW Carlson Spring development; Kelsey Spring development; Black Feather Spring development
	1995	Stud Creek fencing; Black Feather Spring fencing; NW Carlson Spring fencing; Kelsey Spring fencing; In-line Spring fencing; Pebble Bowl Spring fencing; East Hanks Spring fencing; West Hanks Spring fencing
	1996	Hay meadow irrigation (ongoing)
Marys River, Cutt Creek, and Chimney Creek	1991	U.S. BLM riparian pasture fencing
Wildcat Creek	1992	Riparian pasture fencing
Marys River	1991	Riparian exclosures 1 and 3
	1992	Riparian planting
	1993	Riparian exclosure 6; fence modification
	1994	Riparian exclosures 2, 4, and 7; riparian planting
Conners Creek	1991	Conners basin pasture fence reconstruction
Chimney Creek	1993	Riparian planting
Middle and lower watershed	1993	Road repair
Lower watershed		
Marys River	1992	South Cross Field fencing
	1993	Stockwater well construction; hay meadow irrigation (ongoing)
	1995	Meadow fence reconstruction; Cabin Field fencing
	1994	Allen Creek Reservoir development
	1995	Allen Creek Reservoir fencing

tial natural diversity of the area. Fishing, hunting, and nonconsumptive recreational opportunities are expected to increase. To facilitate public access to the area, extensive upgrading of existing roads was completed in 1993, and maintenance will be continued as necessary.

Upper Watershed

Fencing and riparian pasture.—Because much of the upper watershed is designated wilderness, less fence construction has occurred here than in other parts of the Marys River subbasin. Below the wilderness area boundary, the USFS constructed approximately 2.5 miles of fence to create a riparian pasture encompassing the Marys River. The riparian pasture was closed to livestock grazing in 1991 and will remain closed until woody riparian vegetation within the pasture is resistant to livestock impacts. The USFS has recommended that cottonwoods and willows, which became established following closure of the pasture, be at least 6-feet tall before grazing is resumed. The USFS has also constructed several segments of fence to eliminate unauthorized livestock use in other areas.

FIGURE 21.2.—Marys River upstream of Orange Bridge. Top photograph was taken August 1990, prior to restoration efforts. Note wide, shallow channel and lack of riparian vegetation. Bottom photograph was taken July 1996, following restoration efforts, and shows narrowing of channel and growth of willows and other riparian species. (Photographs courtesy of U.S. Bureau of Land Management.)

FIGURE 21.3.—Marys River at breached beaver dam. Top photograph was taken February 1992, prior to restoration efforts. Bottom photograph was taken July 1996, following restoration efforts, and shows deeper and narrower channel, stabilization of streambanks, and growth of riparian vegetation. (Photographs courtesy of U.S. Bureau of Land Management.)

TABLE 21.4.—Modifications to livestock management in the watershed to reduce and redistribute grazing impacts. Prescriptions are (1) manage sheep movement and grazing; (2) reduce sheep numbers; (3) rest area from grazing until desired condition achieved; (4) develop water sources away from streams; (5) restrict livestock to early-season grazing; and (6) reduce hot-season grazing.

Location	Prescription
Marys River basin sheep and goat allotment	1, 2
Upper Marys River sheep and goat allotment	1, 2
West Fork Marys River sheep and goat allotment	1, 2
U.S. Forest Service's Marys River riparian pasture	3
U.S. Bureau of Land Management's Marys River riparian pasture	3
Wildcat Creek riparian pasture	3
Hanks Creek field	4
Steer field	4
Winters Creek field	4
Fenced meadows along lower Marys River	3
Carlson field 1	3
Conners basin pasture	5
Anderson Creek allotment	6

Livestock management.—Within the wilderness area, the USFS has been revising livestock management practices to improve riparian and aquatic habitat conditions (Table 21.4). Domestic sheep are still allowed to graze in the subbasin, but sheep grazing and movements have been monitored more intensively to reduce impacts to Marys River, and in 1995 sheep numbers were reduced by 25%. The USFS has proposed converting areas where cattle grazing is allowed in the wilderness entirely to sheep use.

Middle Watershed

Fencing and riparian pasture.—In the middle watershed, many projects were funded by a US$45,000 contribution from Barrick Goldstrike Mines, Inc., which was matched with Federal Aid in Sport Fish Restoration monies for a total contribution of $180,000. The USBLM used some of these funds for the Marys River riparian pasture, which encompasses 20,000 acres and 19.5 miles of Lahontan cutthroat trout stream and was completed in 1991. It is closed to livestock grazing until riparian and aquatic habitat improves to at least good condition.

Other pastures in the middle watershed are also closed to grazing, including the Wildcat Creek riparian pasture, which encompasses 3,455 acres and includes 3.6 miles of Lahontan cutthroat trout stream and was completed in 1992. Livestock grazing will be allowed after riparian and aquatic habitat improves to at least good condition. A smaller pasture, Carlson field 1, was acquired by the USBLM in the Marys River land exchange and is closed to grazing until riparian and aquatic habitat improves to at least good condition. The pasture contains 1.5 miles of Conners Creek and a 77-acre wet meadow and spring complex that is tributary to Hanks Creek.

Where creating pastures has not been feasible, the riparian corridor has been fenced. Fences are constructed to include as much of the riparian area as is practical. Whenever possible, the minimum distance between fence and stream is 200 feet. Where streams flow through canyons, fences are located outside the canyon to reduce fence maintenance and allow for improvement of the riparian area in the canyon. The USBLM planned to construct approximately 12 miles of corridor fence to exclude livestock on 8 miles of stream on T Creek, Draw Creek, and Marys River. Most Marys River exclosures were completed in 1994; however, construction of the T Creek and Draw Creek exclosures is on hold because the Independence Mining Company land exchange should allow for different management opportunities. Instead of exclosures, riparian pastures may be constructed or the grazing system may be altered to enhance riparian habitat conditions.

Fences can create barriers to wildlife movement, so as miles of fence increase, potential barriers to wildlife increase. There was concern that big game populations may be affected by additional fence in the subbasin. To mitigate possible effects, new fences are constructed to USBLM standards that allow for wildlife movement. Fences near water, expected to receive heavy livestock pressure, are constructed with four wires; the bottom wire is barbless to facilitate wildlife access. The bottom wire is placed 16 inches from the ground and subsequent wires are at 6, 8, and 12 inch intervals. Line post spacing is 16.5 feet. Fences in areas not expected to receive heavy livestock pressure are constructed with three wires, the bottom wire being barbless and placed 16 inches from the ground. Subsequent wires are spaced at 10 and 12 inch intervals. Line post spacing is 22 feet. Several miles of existing fence in the middle watershed were modified to these standards.

Allotment management.—Proposed livestock grazing in the Conners Basin Pasture, which encompasses almost 6 miles of Conners Creek, is determined during annual meetings between the USBLM and the allotment permittee. The annual meetings allow grazing to be adjusted according to current conditions. Usually grazing occurs during the spring for one month, and livestock use ends prior to 20 June. In heavy snow years, somewhat later use is allowed because of the cooler, wetter conditions and slower plant growth. Streambank vegetative cover and stability have improved on many streams in the Elko District in response to similar grazing treatments. A definite trend of improvement in riparian condition on Conners Creek has been documented since initiating the annual meetings in 1992 (Elko District, USBLM, unpublished data).

When completing allotment management plans and multiple-use decisions, the USFS and USBLM have attempted to coordinate management and incorporate USBLM and USFS allotments into a single grazing system. The USBLM completed a multiple-use decision for the Anderson Creek allotment, a 22,038-acre grazing allotment entirely within the middle watershed. The decision altered livestock management practices to reduce hot-season use of riparian areas. Hot season is described as the period of diminishing growth of cool-season plant species and is typified by desiccation of upland forage, air temperatures sufficient to require livestock to use shade, and greater scarcity of water. Livestock grazing during this period usually results in greater use of riparian vegetation (Kinch 1989). The permittee on the Anderson Creek allotment held grazing permits on USFS lands as well. Reduction of hot-season use was possible by rotating livestock through all the USFS and USBLM pastures the permittee could use.

Vegetation management.—Vegetation is the key element in a properly functioning riparian system. Since 1992, over 20,000 aspen, alder, and choke-cherry seedlings have been planted on the Marys River and Chimney Creek within the Marys River riparian pasture by the USBLM and volunteers. Seedling planting may accelerate riparian vegetation establishment in these systems, but, perhaps just as importantly, these volunteer projects promote public participation and ownership in the restoration effort.

Prescribed burning has been used to improve upland forage production and create vegetative diversity for wildlife. In 1992 and 1993, 900 acres, consisting of small parcels from 50 to 200 acres each, were burned in the Marys River riparian pasture and an adjacent pasture to achieve a mosaic of burned and unburned areas of upland vegetation. Prescribed burning of mesic (requiring a moderate amount of moisture) vegetation is being considered as a means of maintaining productivity of the meadows.

Water resource management.—Alternate water sources for livestock are being developed throughout the middle watershed. Upland springs and seeps are being fenced and water piped outside the exclosure to protect the riparian areas and provide water for livestock. Construction of 11 water developments proposed by the USBLM or permittees was begun in 1994.

Native hay meadows acquired in the Independence Mining Company land exchange are being irrigated for hay production. A cooperative agreement involving Trout Unlimited, the USBLM, and a permittee outlines the haying strategy. In conjunction with irrigation, a minimum instream flow is being maintained for Lahontan cutthroat trout.

Additional efforts.—Cultural resource (historic and prehistoric artifacts) inventories associated with project construction have been completed. Information about the historical and cultural values of the area will be included in two interpretive sites planned for the middle watershed.

Lower Watershed

Fencing and riparian pasture.—The USBLM acquired about 8,000 acres of fenced meadows on the lower Marys River as a result of the 1991 Marys River land exchange. Fences were reconstructed to eliminate unauthorized livestock use, and about 6 miles of fence was built to create the South Cross field. Approximately 1,250 acres of fenced meadows near the Cross Ranch were made available to the permittee for livestock grazing to mitigate the effect of closing the Marys River riparian pasture to grazing. The South Cross field fence excludes the Marys River from grazing while allowing grazing in the meadows.

Water resource management.—Three stockwater wells have been developed in the lower watershed to eliminate the need for livestock access to the river. Because Newmont Gold Company donated equipment, the stockwater wells were completed much sooner than would have been possible had the work been contracted by USBLM. Design of a pipeline system to distribute water to several livestock watering troughs was begun in 1994 but has yet to be completed. A spring in the lower watershed was fenced in 1995; water will be piped outside the exclosure for livestock use.

Native hay meadows acquired in the 1991 Marys River land exchange are being irrigated to allow hay production at 50% of maximum capacity (USBLM 1990a). In addition, irrigation will maintain and enhance the meadows until riparian and aquatic recovery allows for natural subirrigation of the area. An irrigation and haying strategy is developed annually, but long-term management will be addressed in a management plan for the lower watershed. Among other things, the plan will evaluate the need to screen irrigation diversions to prevent fish movement into irrigation ditches. In conjunction with irrigation, a minimum instream flow of approximately 30 cubic feet per second is being maintained for Lahontan cutthroat trout. Some downstream water users were concerned that public ownership of water rights would somehow affect their rights. Maintaining the traditional irrigation use of the water and a minimum instream flow has ensured that at least as much water is available downstream under public ownership as would be under private ownership.

Additional efforts.—The Cabin field on the lower Marys River was designated as a Watchable Wildlife site because, as a functional wetland, the area supports large populations of waterfowl, shorebirds, and songbirds (Clark 1993). Construction of an overlook and interpretive site was begun in 1993 but has not yet been completed. This area provides an excellent location for environmental study by local schools.

Cultural resource inventories associated with project construction have been completed. The Marys River subbasin has numerous historical and cultural resource values, including prehistoric sites along the Marys River and the historic California Emigrant Trail. Information about the historical and cultural values will be incorporated into the two interpretive sites planned for the lower watershed.

Most management actions have been compatible with the primary goal of restoring Lahontan cutthroat trout habitat throughout the watershed. However, this primary goal has been compromised at times to accommodate the secondary goal of balancing the multiple uses of the area. Some adverse effects to the Lahontan cutthroat trout and its habitat were deemed acceptable in order to sustain viable livestock operations and improve recreational opportunities. For example, to avoid substantial costs to the livestock permittee by shortening the grazing season or reducing livestock numbers, since 1992 cattle have been allowed to graze in the Hanks Creek pasture throughout the hot season. Consequently, riparian and aquatic habitat condition of Hanks Creek has not improved. This situation will be addressed in an allotment evaluation that will be initiated in 1997. Options being considered include fence construction to establish a riparian pasture, stream exclosures, or both, along with changes in the duration and timing of livestock use. In 1995 three water sources away from the stream were developed to improve livestock distribution in this large pasture.

ADAPTIVE MANAGEMENT

Because of a lack of information on the components and relationships among components necessary for a healthy watershed, monitoring is essential. Monitoring should allow identification of the key indicators of watershed health and suggest appropriate management adjustments if downward trends are detected. In addition, plant and animal communities and hydrology of the system are being evaluated throughout the restoration process to determine the potential and value of various habitats in the subbasin and the impacts of land-use practices on watershed health.

Much of the monitoring proposed in the master plan (USBLM 1992b) is a continuation of previous efforts (Table 21.5). Monitoring focuses on riparian and aquatic habitats because the primary goal of the project is restoring Lahontan cutthroat trout habitat. In addition, management impacts are often concentrated in riparian areas.

Baseline information for the Marys River subbasin was available through previous inventory and monitoring efforts (USBLM 1979, 1987b, 1988, 1990b, 1991, 1993; Johnson 1988; USBLM, unpublished data; USFS, unpublished data). In a cooperative effort with the NDOW, intensive riparian and aquatic habitat and fish population inventories were completed in 1979 on USFS, USBLM, and much of the private lands in the Marys River system. Baseline information on macroinvertebrates and water quality was also collected during this inventory. Additional baseline information was collected from 1981 through 1990 on riparian vegetation and stream channel changes.

The USBLM conducted riparian and aquatic habitat inventory in 1992 on the lower Marys River and in 1993 throughout the middle watershed to assess current conditions and evaluate the effectiveness of management actions. Overall, riparian habitat has improved, but the quality of many other Lahontan cutthroat trout habitat components has declined. With the exception of 1993, northeastern Nevada has experienced below-average precipitation from 1987 through 1994. Extremely low streamflows have negatively affected habitat parameters such as pool quality and sedimentation.

The USBLM and the NDOW cooperatively conducted a fish population inventory on public and private lands in the lower and middle subbasin during the summer of 1994 to determine current numbers and distribution. The 1994 Lahontan cutthroat trout population in the middle and lower watershed represented a 96% decrease in numbers and an 84% decrease in occupied miles since 1979 (Table 21.2). The NDOW concluded that the drought of 1987 through 1994 has had a major effect on the Lahontan cutthroat trout population. Population declines for most other native species were also documented (Johnson and Nokes 1995).

The USFS conducted macroinvertebrate sampling on the Marys River in 1991

TABLE 21.5.—Monitoring conducted within the Marys River watershed.

Type of monitoring	Location in watershed			Frequency of monitoring
	Upper	Middle	Lower	
Riparian condition and trend				
Low-level aerial photos	X	X	X	Every 6 years
Riparian and stream habitat inventories	X	X	X	Every 3 years
Vegetation transects		X	X	Annually
Greenline[a]	X			Periodically
Springs inventory		X	X	As needed
Water temperatures measurements		X	X	Continuously
Water quality sampling	X	X	X	3 times per year
Macroinvertebrate sampling	X	X	X	3 times per year
Fish and wildlife populations				
Fish population composition, abundance, and structure	X	X	X	Every 3 years
Wildlife inventories		X	X	Annually
Hydrology and channel morphology				
Streamflow gauging stations measurements		X	X	Continuously
Valley bottom cross sections measurements		X	X	Every 2 years
Water table monitoring wells measurements		X	X	3 times per year
General watershed condition and trend				
Weather station measurements		X	X	As needed
Wildlife habitat inventories		X	X	Periodically
Livestock use assessments	X	X	X	Annually
Range condition assessments	X	X	X	Periodically
Recreational use				
Visitation measurements		X	X	Continuously

[a]The area in which one would encounter a more or less continuous vegetative cover when moving away from the center of a stream channel.

and on Wildcat Creek in 1992. The USBLM initiated macroinvertebrate sampling on the Marys River, and Cutt, Conners, Hanks, Chimney, T, and Wildcat creeks in 1993. Water quality sampling was conducted by both agencies in conjunction with macroinvertebrate sampling. Although sampling will help determine if water quality or food availability is limiting Lahontan cutthroat trout populations and will monitor changes in the riparian and aquatic system, the USFS has not conducted further sampling because it has been evaluating the effectiveness and need for this type of monitoring. The USBLM has conducted sampling three times per year annually, but in 1997 it will evaluate whether an adequate baseline exists and what future sampling frequency might be required.

Some monitoring is designed to identify long-term changes related to watershed restoration. The USBLM and USFS completed low-level aerial photography on the Marys River and its tributaries in 1992. These photographs were necessary for initial work on stream classification and will provide a baseline to assess long-term changes in riparian vegetation communities accurately.

In 1992, the USBLM installed two gauging stations on the Marys River, one in the middle reach and one in the lower reach. A gauging station was also installed on Hanks Creek in 1993. Because very few data exist documenting the relationship of streamflows and watershed improvement, these gauging stations can be used to assess the effects of improving grazing systems on streamflow patterns.

Cross sections of valley bottoms on Marys River were established at three sites in 1993. Cross sections will aid in determining the effects of land-use practices on riparian and aquatic habitats and will provide information on the recovery process by showing areas of sediment loss and trapping. Monitoring wells were installed along the valley bottom cross sections in 1993 and 1994. Monitoring wells monitor the depth of the water table and distance from the stream. These data could be helpful in promoting riparian management by quantifying the effects of riparian recovery on water regimes and floodplain water storage in a Great Basin stream system.

In 1993, five riparian vegetation monitoring sites were established on the middle and lower Marys River along the established valley cross sections and in conjunction with terrestrial wildlife inventories. These transects are being used to determine vegetation changes due to a rising water table and provide more information on the recovery process. To date, the vegetation transects have illustrated the variability of vegetation communities depending on precipitation.

Springs and seeps in the uplands are being inventoried to determine their condition and potential and to file for water rights on spring flows if necessary. This information will be useful in assessing watershed function and wildlife habitat conditions.

Because water temperature data throughout the subbasin is limited, thermographs will be installed at the gauging stations and in major tributaries. Temperature data will help in determining limiting factors for Lahontan cutthroat trout in particular stream reaches. Currently, the assumption is that summer water temperatures are limiting.

The NDOW established small-mammal and bird studies on the lower and middle Marys River in 1991. The USBLM and the NDOW are cooperatively monitoring species abundance and diversity annually. However, small-mammal trapping activities were suspended in August 1993 because of the Hantavirus outbreak. Information on the species occurring in riparian areas and the value of these areas to wildlife will be useful in making management decisions and in promoting riparian management elsewhere.

Most of the data collected in the subbasin are influenced by the weather, particularly precipitation. Monitoring has demonstrated that variability of conditions within the watershed depends on weather patterns. To interpret the other data collected accurately, precipitation data are being collected in the subbasin.

Parties involved in the Marys River project are attempting to determine sustainable levels of grazing and other commodity uses in the subbasin based on its potential natural condition. Potential of most of the uplands is known because an ecological site inventory was completed in 1987 (Elko District, USBLM, unpublished data). Because potential of riparian and aquatic habitats is unknown and will be a factor in determining sustainable-use levels, emphasis was placed on obtaining that information by conducting ecological site inventories of riparian habitats and stream system classification. The USFS and USBLM began the stream classification effort in 1992. The USBLM will complete classification in 1997. Sustainable-use levels will be determined through evaluation of monitoring data.

CRITIQUE

To date, significant progress has been made in restoring some riparian areas, much of the uplands are in good condition, and some solid data have been gathered to support future management decisions. The most significant challenge facing the Marys River Project is coordinating resource management by working as a team that involves the USBLM, the USFS, private landowners, and other entities interested in management of the watershed.

Resource management actions have been evaluated to determine strengths and weaknesses. For example, although the riparian planting projects are very successful in involving the public in watershed restoration, the resource benefit is limited. Two years after the first Marys River planting project, seedling survival was less than 8%. Other planting projects in the Elko District have had similar low seedling survival rates, but these projects continue to be very popular with the public.

A challenge for any restoration effort involving livestock management is the need to identify and remedy any unauthorized livestock use that may be occurring. Even small amounts of unauthorized use can negate recovery. Because of the cooperation of the grazing permittees, unauthorized livestock use has not been a significant problem in the Marys River watershed.

A substantial amount of monitoring data has been collected within the subbasin, but there is not a process to respond to new information in a timely manner. The data are analyzed in allotment evaluations and environmental analyses, but completing these documents is a time-consuming process. Both the USBLM and USFS have made adjustments in annual operating plans based on new information, but significant changes generally are reserved for the planning and decision process.

A great strength of the project is that data are being more widely shared among the USFS, the USBLM, and the NDOW. This has generally been accomplished informally. There has been more communication and coordination on monitoring and restoration efforts. For example, the USFS is comparing the USBLM Wildcat Creek riparian pasture, where there is no livestock grazing, to Wildcat Creek on the Humboldt National Forest, where a grazing system is in place, to determine varying rates of stream recovery. Private landowners within the subbasin also are being better informed of plans, accomplishments, and monitoring results through periodic meetings.

From its inception, the Marys River Project developed somewhat naturally into an interdisciplinary, interagency team process. Unfortunately, a formal team structure was never put in place to guide the process, and the lack of structure has been a major organizational problem. Depending on the issue, decisions have been made at various management levels, sometimes without input from other entities involved in the watershed. Because there has been no formal team structure or agreement on a decision-making process, decisions have often been based on who most strongly argued his or her position rather than on reaching consensus. This has sometimes resulted in adversarial interactions rather than teamwork. Although communication and cooperation have increased among entities involved in management of the watershed, there is no team that involves all these entities. Likewise, a comprehensive plan for management of the watershed is not expected to be developed. Depending on the entities involved there will continue to be some degree of coordination, but each federal agency, as well as private landowners, manage its lands independently.

PROSPECTS FOR THE FUTURE

The USFS and USBLM will continue established monitoring in the watershed, as funding and workforce allows. Because of budget reductions, funding to continue long-term monitoring is now in question. The USBLM is seeking universities or other government or private entities to continue monitoring and documenting the changes occurring in the watershed.

No new projects are planned in the upper watershed, due in part to workload constraints. In the middle and lower watershed the USBLM plans to continue implementing the actions outlined in the Marys River Master Plan, as funding and workforce allows. Through cooperation with conservation groups, the NDOW, and the livestock permittees, physical facilities such as a riparian pasture fence along Hanks Creek will be built to implement grazing strategies that enhance riparian and aquatic habitat and reduce sedimentation. Although more riparian planting projects were considered in the middle watershed, the USBLM is questioning proceeding with additional plantings because of the low survival rate of seedlings and the substantial natural reestablishment of riparian vegetation.

Based on monitoring, opportunities to expand or allow such uses as recreation, habitat management for sandhill cranes and other wildlife, and limited grazing compatible with riparian recovery are planned to be incorporated in two management plans, one for the middle and one for the lower Marys River system. In January 1996, the USBLM established an interdisciplinary team to develop the plan for the lower watershed.

To solidify or continue enhancement of the watershed resource base, conservation agreements, land exchanges, or both will be pursued as opportunities arise with willing adjacent private landowners.

In late 1995, Trout Unlimited increased its involvement in the Marys River Project and has facilitated a comprehensive watershed management approach by working with federal agencies and private landowners in the watershed. Although there is still no team or working group that includes all entities involved in management of the watershed, Trout Unlimited has enhanced communication among the entities involved and has ensured that attention remains focused on watershed management.

As the stream system recovers and responds to the management changes that have already been implemented, additional management changes that may be needed to further Lahontan cutthroat trout recovery will be explored as monitoring data are accumulated. The challenge will remain, with the decline of agency budgets, to involve already established partners and new partners to implement on-ground facilities to continue watershed recovery.

ACKNOWLEDGMENTS

We thank Sean Shea of the Biological Resources Division of the U.S. Geological Survey for creating the map; the Elko District, U.S. Bureau of Land Management staff for their review and assistance during preparation of the manuscript; and Glenn Clemmer and Osborne Casey for their reviews of this chapter.

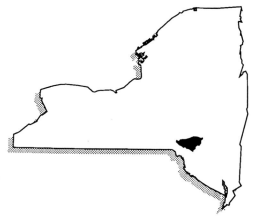

To a great many fly fishers, the Beaverkill is the standard by which all other trout streams are judged. It is first and oldest in reputation. . . . Fly fishers come from across the land, as if on a pilgrimage: to walk the well worn streamside paths; to fish and stand in the pools and shadows of the legendary anglers.
—Ed Van Put, 1996

CHAPTER 22

RESTORATION OF TROUT WATERS IN THE EAST: BEAVERKILL–WILLOWEMOC WATERSHED OF NEW YORK

Jock Conyngham and Joseph M. McGurrin

The Beaverkill–Willowemoc river system (hereafter referred to as the Beaverkill–Willowemoc; see Figure 22.1) in New York State's Catskill Mountains is widely acknowledged as the cradle of fly-fishing in America and a trout fishery of international caliber. From the nineteenth century to the present, this river system has formed the mainstay of a large tourist industry. Much of the Beaverkill–Willowemoc watershed appears relatively pristine, a remarkable quality given that it is only a 2-hour drive from New York City and other nearby metropolitan areas. Long stretches of the upper reaches and portions of the main stem offer excellent trout habitat in sparsely settled surroundings.

Whereas some headwater areas have never experienced significant disturbance, the pristine appearance of many others represents a healing from past conditions. Numerous environmentally destructive industries, such as tanneries, and wood alcohol distilleries, have shared the watershed with tourism and angling. The last of these industries ceased operation by the end of World War II, but its effects continue to reverberate.

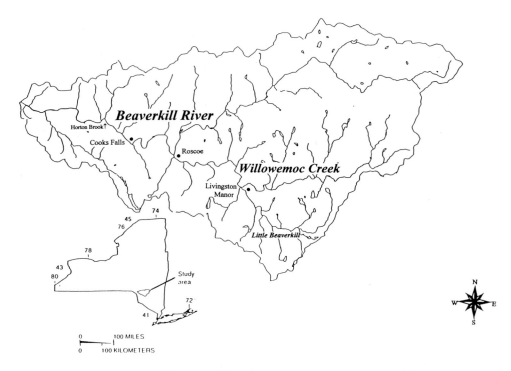

FIGURE 22.1.—The Beaverkill River–Willowemoc Creek drainage basin in New York State (map based on U.S. Geological Survey cartography). Map scale is 1 inch = 15.5 miles.

Gravel mining of the riverbed for road construction was a common practice in the past. Route 17, a four-lane divided expressway, was constructed through the watershed in the 1960s. For 25 miles it travels adjacent to reaches of the Little Beaverkill, the Willowemoc, and the Beaverkill and crosses them 16 times, which has entailed the construction of large drainage systems and bridges, deposition of fill in the floodplain, the placement of riprap, and the straightening of some channels. Following the completion of Route 17, the area experienced increased development for campgrounds and second homes. The U.S. Army Corps of Engineers performed flood control work on some reaches, most extensively after a destructive flood in 1969.

The watershed's flood of record occurred in January 1996. Culverts and bridges designed for a 50-year flood event backed up and caused severe flooding. County and town crews responded with channelization and bulldozing of streams, often at the mouths of tributaries critical for trout as thermal refugia (shelters from adverse instream temperatures) or spawning habitat. While these flood control projects have had or will have varying effects on the fishery, the fishery's vulnerability lies in local physiography. The watershed's steep topography and shallow overburden lead to recurring flow extremes and thermal stress on resident trout populations during summer droughts.

Longer-term impacts have been aggravated by the flow and thermal conditions that followed a series of droughts in 1988, 1991, 1993, and 1995. The severity of the 1991 drought combined with the high profile of the Beaverkill–Willowemoc fishery led to immediate but unsubstantiated press reports of a 90% decline in trout populations and American Rivers' 1992 listing of the system as America's ninth most endangered river. The angling and conservation press, including Trout Unlimited publications, further publicized the fishery's troubles. Angler visitation to the area dropped precipitously, leading to strong feelings in the local business community that the media had overstated the decline. Controversy quickly developed within and between the business and conservation communities about the degree and nature of declines in watershed condition and the fishery. Because of the importance of the Beaverkill–Willowemoc trout fishery, local Trout Unlimited members and the organization's New York State Council were actively involved in these debates and requested help from their national office.

Anglers have periodically predicted the death of the Beaverkill–Willowemoc fishery. New York State Department of Environmental Conservation (NYSDEC) files reveal public outcries in the early 1980s, the 1960s, the mid-1950s, and even back into the last century. The high emotion evident in these episodes and their concurrence with drought and changes in trout-stocking practices have masked the fact that the Beaverkill–Willowemoc is neither dying nor problem free. Local Trout Unlimited members believed that human activities had exacerbated natural variation in trout populations and decided that the watershed's history, qualities, and vulnerability demanded action. Most of Trout Unlimited's 95,000 members knew of the river and helped make it a priority on the organization's national agenda. With grants from the National Fish and Wildlife Foundation and the R. K. Mellon Foundation, the national office of Trout Unlimited initiated the Beaverkill–Willowemoc Watershed Initiative to involve government agencies (particularly the NYSDEC), academic institutions, conservation groups, the recreational fishing industry, and the local community in the development and implementation of watershed-based initiatives to conserve the Beaverkill–Willowemoc fishery.

THE BEAVERKILL–WILLOWEMOC WATERSHED

Geology, Geomorphology, Climate, and Vegetation

Bedrock in most of the watershed consists of gray sandstone, in association with shales and conglomerate, that is characterized by low groundwater yields except near fracture zones and bedding plates (Fisher et al. 1970; Cadwell 1989). Studies by local water districts found that the highest groundwater volumes appear in the coarser material of the valley bottoms, where unconsolidated glacial deposits of boulders, gravels, sands, and clays occur in depths of up to 90 feet (Leggette et al. 1994a, 1994b, 1995).

These valley bottoms are quite narrow and flanked by steep slopes. Particularly narrow valleys and steep slopes are prominent in the lower reaches of the system. Wetlands are present in only a few headwater areas. Though the system spreads over 300 square miles and contains considerable geomorphic variety, water storage is limited.

The watershed, with a mean annual temperature of about 44°F and mean annual precipitation of 45 inches (Ayers et al. 1994), supports an extensive forest. Most of the headwater stands are protected by inclusion in the Catskill State Park. An American beech–birch–maple association characterizes the forest, with oak becoming more common in lower reaches of the watershed. Poorly drained or moist streamside soils support local stands of eastern hemlock in the upper reaches, American sycamore in the lower reaches, and willow throughout. The stands are relatively even aged because of intensive harvesting to support the wood chemicals industry until the end of World War II. The recent maturation of the watershed's forest explains one of the lower river's missing habitat elements—large woody debris—which would provide structure and channel complexity. Large wood should reappear in the streams as the forest matures.

Including climate and flow figures from the Beaverkill, Ayers et al. (1994) analyzed computer projections on the implications of climate variability and the greenhouse effect in the Delaware River watershed. The specific implications for the Beaverkill-Willowemoc were that a discernible warming trend in the northern hemisphere of almost 1°F over the last century has led to increased winter flows, decreased spring floods (which form and maintain channels), and significantly decreased summer base flows (M. A. Ayers, U.S. Geological Survey, personal communication). The computer models suggest that if the warming trend continues, these effects will become more extreme.

Hydrologic Characteristics

The NYSDEC has classified 104 tributaries and main watercourses in the complex Beaverkill-Willowemoc drainage network (Figure 22.2). As one might expect in a system of this size (300 square miles), gradients vary dramatically and limit upstream movement of spawning fishes in upper reaches and habitat quality in the lower reaches. The Little Beaverkill falls about 63 feet per mile over its 10-mile course, and the Willowemoc falls approximately 43 feet per mile over 26 miles. In the 28 miles from the Beaverkill's source at 3,080 feet above mean sea level to its junction with the Willowemoc at Roscoe (elevation 1,280 feet above mean sea level), the river's gradient is 64 feet per mile. The gradient decreases to 24 feet per mile in the 16 miles from the junction at Roscoe down to Horton. This section contains the system's most famous and heavily fished pools. From Horton to its terminus with the East Branch of the Delaware River, the Beaverkill falls only 13 feet per mile.

Mean annual runoff at the U.S. Geological Survey (USGS) gauging station at Cooks Falls (drainage area 241 square miles) was 555 cubic feet per second for the period of record (1913–1994). The highly variable annual discharges (Figure 22.3) suggest both variable distributions of precipitation and a low component of regional groundwater in the river's base flow.

Trout populations lose condition (a measure of general health), become lethargic, and seek out refugia when water temperatures remain either below or above their preferred temperature range (Needham and Jones 1959; USEPA 1976;

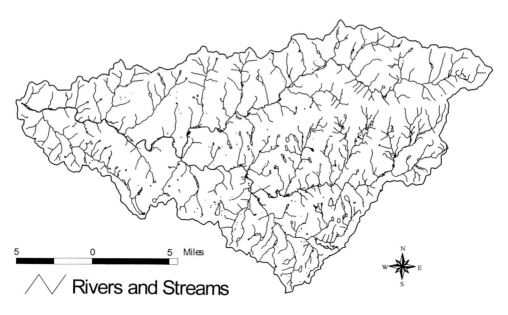

5   0   5   Miles

/\/ Rivers and Streams

FIGURE 22.2.—Drainage network of Beaverkill–Willowemoc watershed.

Hokanson et al. 1977; Spotila et al. 1979). Brown trout, for example, display maximum growth rates at water temperatures of 54 to 59°F; in one laboratory test juveniles lost weight regardless of feeding rates when water temperature was sustained at 71°F (Jobling 1981). If temperatures continue to increase, morbidity also climbs. In general, trout mortality occurs as water temperatures enter the high 70s. Trout in Catskill watersheds respond to such conditions by moving to thermal refugia at coldwater tributary mouths or groundwater outlets, but they are then vulnerable to predation, poaching, increased rates of disease and parasite transmission, and rapid loss of condition due to cessation of feeding (Sanford 1991; McBride 1996). The largest refuge in the lower river held approximately 400 fish for several weeks in the summer of 1995. W. Mincarelli (NYSDEC, personal communication) estimated the 1991 congregation at the same site to be more than 1,000 trout.

Beaverkill–Willowemoc trout populations in the lower part of the system experience severe thermal stress in many summers. Noting that water temperatures exceeded 80°F at the USGS station in Cooks Falls for 22 days in 1991 and 23 days in 1993 and that base flows were depressed relative to the mean base figures, the NYSDEC determined high summer temperatures and low summer flows to be factors limiting trout populations in the lower Beaverkill (McBride 1996). It is also possible that low winter temperatures and flows, combined with minimal groundwater inputs, limit overwintering and recruitment rates of tributary and headwater trout populations (Needham and Jones 1959; Latta 1965; Wesche 1985; Hunter 1991). Nonetheless, because of the system's size and its numerous tributaries and springs, considerable thermal variation exists.

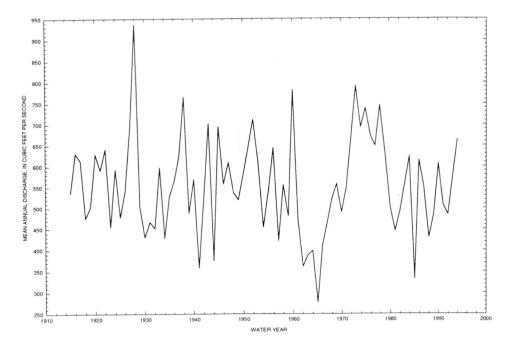

Figure 22.3.—Mean annual discharge at Cooks Falls, New York, 1913–1994 (data from U.S. Geological Survey gauging station).

The Beaverkill–Willowemoc's water chemistry has improved significantly over the last half century. Whereas fish kills occurred with some frequency before the closing of the acid factories in the 1940s, NYSDEC testing in the last two decades has shown good to excellent water quality in most of the watershed. Persistent heavy metals, however, are present in samples from Cooks Falls; copper levels in the water column are a concern, and sediment samples contain lead at levels significantly above background (NYSDEC 1990). Road runoff containing heavy metals from road salt and deposition of vehicle exhaust are the probable sources of these toxins. Both chemical analysis and a macroinvertebrate sampling program also suggest impacts from the Livingston Manor sewage treatment plant on the Willowemoc's water quality and biota (Bode et al. 1993a; 1995a, 1995b). The macroinvertebrate sampling program revealed no anthropogenic impacts in the Beaverkill (Bode et al. 1993b).

Acidity is a potential problem in some tributary and headwater reaches. A USGS precipitation chemistry station near the Beaverkill headwaters found rainfall pH averaging 4.3 to 4.5 (B. Baldigo, USGS, personal communication). A study currently underway on 38 Catskill streams suggests that peak and base flow nitrate concentrations are rising, symptomatic of mid- to late-stage nitrogen saturation by atmospheric deposition (G. Lovett, Institute of Ecosystem Studies, personal communication). Nitrate concentrations in water from a Beaverkill tributary and the upper Willowemoc are among the higher half of the samples.

Nonetheless, acidity in the lower Beaverkill tends toward neutral levels (Sanford 1991).

As one descends the Beaverkill–Willowemoc system, characteristics of the channel become cause for concern. Certain tributaries exhibit the disorganized appearance of out-of-phase, abnormally high rates of sediment load because of flood control activities or poor land use in their immediate watersheds. Eroding tributary banks, locally present where riparian vegetation has been disturbed, have led to aggradation (rising in elevation of the channel bottom by deposition of material eroded elsewhere) of the river bed in many places. Data from the Cooks Falls gauging station and stream morphology in other downstream reaches suggest that the channel is incising (cutting downward and outward into its bed and banks) because runoff systems, diking, and disposal of road construction fill in the floodplains have increased the erosiveness of flows in the lower river. The wide, shallow channel geometry characteristic of channels experiencing both incision and aggradation is particularly important in a river that experiences thermal stress.

Substrate in the Beaverkill–Willowemoc system tends to be large. Some larger boulders were dynamited during the early land-clearing era and more recently because of concerns about ice forming dams, but small boulders (10–40 inches), large cobble (5–10 inches), and small cobble (2.5–5 inches) substrates predominate. Bedrock ledges are common throughout the system's channels. Though they limit trout spawning largely to tributaries, these coarse bed materials create excellent nursery and adult habitat in most reaches.

Aquatic Populations

The aquatic and fishery resources of the Beaverkill–Willowemoc exemplify those of high-quality free-flowing trout waters in the Catskills (Figure 22.4). Aquatic vegetation is rare in the system with the exception of various algal species. Duckweed is locally present on tributaries and indicates sewage-based fertilization. Macroinvertebrate populations are robust in terms of both biomass (summed mass of individuals in a given area) and species diversity. A great deal of the Beaverkill–Willowemoc's fame with anglers is due to the heavy emergence episodes of aquatic insects, particularly the mayflies (Ephemeroptera), caddisflies (Trichoptera), and midges (Diptera). These populations, in combination with the shallow channels and large substrate common to the system, have brought the Beaverkill–Willowemoc international fame for its dry-fly fishing.

One of the objectives of the Beaverkill–Willowemoc Watershed Initiative is to understand, protect, and enhance populations of the system's indigenous brook trout as well as nonnative brown and rainbow trout. Other species in the system include the white sucker, longnose sucker, American eel, creek chub, cutlips minnow, slimy sculpin, mottled sculpin, common shiner, blacknose dace, longnose dace, and gizzard shad (D. K. Sanford, NYSDEC, personal communication). In the system's lower reaches, smallmouth bass, pumpkinseed, redbreast sunfish, fallfish, margined madtom, tessellated darter, shield darter, golden shiner, and American shad occur frequently (Sanford, personal communication).

FIGURE 22.4.—Photograph taken near Horton Pool, a famous fly-fishing area on the Big Beaverkill, in October 1995 by Jock Conyngham.

Human Populations and Industry

The human population in the watershed has long been stable and, more recently, even declining. The town of Rockland, which encompasses Roscoe and Livingston Manor, had 3,714 residents in 1905, 4,216 in 1960, and 4,096 in 1990 (Van Put 1996).

Settlement by European immigrants began in the late 1700s. Logs were often a family's first crop, and many of the piers and docks of Easton and Philadelphia, Pennsylvania, still contain timber rafted down the Beaverkill–Willowemoc and Delaware system from the time of first settlement to the 1880s. The physical damage that log transport likely inflicted on the channel was followed by the chemical impacts of the tanneries from the 1840s to the 1880s and that of 14 acid factories operating in the watershed from the 1880s to 1946. Fish kills from direct effluent flows and leachate from streamside sludge pits occurred many times until the end of World War II.

The acid factories' high fuelwood needs resulted in extensive forest clearing and severe alterations of the flow and thermal regimes as well as erosion impacts. One source calculated the factories' demand at 190,000 cords per year (E. Van Put, NYSDEC, personal communication). The standing order for fuel was wood 3 inches in diameter or larger, leading to total deforestation of large portions of the watershed. The upper Beaverkill was protected from cutting and log rafting through injunctions obtained by private fishing clubs prior to World War I (Van Put 1996).

Physical alteration of the river caused by construction of Route 17 in the early 1960s was extensive, but strong opposition led by Harry and Elsie Darbee, local residents and internationally famous fly tiers, reduced the extent of modifications. Original plans were to run virtually the entire highway down the plain immediately adjacent to the river and to construct 33 bridges, instead of today's 16. Although the concerns of the Darbees and other conservationists focused on physical instream disturbance and pollution runoff, Route 17's largest effect on the watershed may lie in its effects on basic hydrological processes. The road, median, shoulders, and an extensive culvert system have converted portions of the former valley infiltration zone and floodplain into areas of rapid and near-total runoff (Figures 22.5, 22.6).

The Fishery

Part of the Beaverkill–Willowemoc's attraction to anglers is the variety of its fishing opportunities. Some of the oldest and most exclusive fishing clubs in America are located in its headwaters, but it also offers miles of roadside public angling access. Public water is divided between three special-regulation areas (catch and release using only artificial lures) and water where less-restrictive regulations apply. The first special-regulation area was established in 1965; as such, it is one of the oldest in the country.

The public water, particularly the special-regulation areas, receives heavy pressure. By means of aerial surveys, the NYSDEC estimated an average annual number of 30,306 angler-trips to the lower sections from 1987 to 1993, then a dramatic decline to 16,523 trips in 1994 (NYSDEC, unpublished data). Many anglers apparently left the Beaverkill–Willowemoc to fish the more dependable tailwater flows of the rest of the Delaware system just to the west.

TROUT UNLIMITED'S INITIATIVE

The overall objectives of the Beaverkill–Willowemoc Watershed Initiative (hereafter, the project) are to (1) assess the current state of the watershed and fishery; (2) implement and evaluate restoration and management activities based on that assessment; and (3) provide a system of products and services that aid in the future conservation of the watershed and enhancement of the fishery. Trout Unlimited's national office had successfully used this type of strategy in funding local chapter and state council efforts to work with government agencies on small watersheds and with coalitions on larger systems, but not to the extent required by the Beaverkill–Willowemoc project.

As in most watersheds in the United States, a complex web of social and political boundaries divide the drainage with little correspondence to its physical properties. Given the Beaverkill–Willowemoc's size and tangle of boundaries, it is necessary that a variety of government and community interests work together to implement watershed level initiatives. In phase I of the project, a core study team consisting of Trout Unlimited and NYSDEC technical personnel collaborated to plan activities.

Building on the work of the study team, Trout Unlimited committed significant

FIGURE 22.5.—Channelization of Twadell Brook upstream of Route 17. (Photograph by Jock Conyngham.)

FIGURE 22.6.—Channelization and culverts combine to alter natural flow regime, infiltration, channel geometry, and biological qualities of many streams as they cross roads and highways. Photograph (by Jock Conyngham) of Route 17 crossing Twadell Brook.

resources to soliciting the support and participation of local businesses, community leaders, relevant nonprofit organizations, and individual citizens interested in the project. The long-term protection of the watershed lies in the hands of these various interest groups and individuals as well as those of the NYSDEC, but factionalism among these interests had hindered communication and cooperation. This contention has resulted in efforts by Trout Unlimited to ensure that technical information gathered in the project is both well disseminated and relevant to community and fishery needs. A communication system has been developed to accomplish this and includes (1) community meetings to encourage public input on the design and development of scientific studies; (2) a project mailing list to update all interested parties; and (3) a media campaign to inform the public about the importance of the trout fishery and its relationship to watershed health.

Watershed analysis can be expensive and difficult, especially if new data are needed at various spatial and temporal scales. For these reasons Trout Unlimited has gone to some lengths to aggregate existing data as well as collect primary data at several scales. The project is fortunate in that the USGS has had six different gauging stations in the system, four of them with a minimum of 33 years of concurrent measurements. Regrettably, only one continues operation. A local resident publishing a history of the river (Van Put 1996) has been another invaluable resource. Both of these sources have added temporal depth and perspective that would be impossible or unrealistically expensive to derive from other sources. In addition, many anglers and angling clubs have offered their logs, photographs, and reminiscences. This sort of folk knowledge is rich and often valuable in spite of common problems with recall and observer bias. Finally, Trout Unlimited has taken a problem-centered approach to the project. Those subjects and physical areas likely to yield results and that are amenable to cost-effective remediation or protection receive disproportionate attention. An example is the high priority the project gives to headwater and tributary areas because of their biological importance and potential to contribute cool water to the thermally stressed system.

Whereas phase I, initiated in 1994, focused on compilation of existing data, collection of new data, and analysis, phase II, begun in 1996, emphasizes pilot protection and enhancement efforts.

Phase I: Fishery and Watershed Diagnosis

Although much useful information had been gathered over the long history of the Beaverkill–Willowemoc fishery, phase I of the project marked the first time that a coordinated effort had been made to aggregate and coordinate analyses of the socioeconomic, biological, and physical characteristics of the entire watershed. Trout Unlimited's goal was to build a consensus on the existing state of the watershed and fishery and to establish priority conservation activities that could be implemented and evaluated in the second phase. A brief summary of each of the phase I components is provided below.

Socioeconomic assessment.—The purpose of the socioeconomic component was to provide information that government and business entities could use to evaluate the fishery and to manage development in ways that conserve or optimize the watershed, trout fishery, and associated local economy. The studies involved (1) measuring behavior and expenditures of different types of anglers and determining their preferences for various types of angling experiences and amenities; (2) measuring the economic impacts of the fishery on local businesses and the larger community; (3) measuring expenditures and tax contributions of private angling clubs; and (4) exploring the influence of the trout fishery on the value of real estate. The assessment also included recommendations for fishery development that could increase returns to angling-related businesses. Tyrrell et al. (1995) have presented most of the findings; information has also been provided to community leaders and the general public in the form of a short list of findings that are highlighted below.

1. Whereas some fishing destinations offer a variety of vacation activities to attract anglers, this does not appear to be the case for Beaverkill–Willowemoc anglers. The Beaverkill–Willowemoc trout fishery is the primary reason that anglers visit the town of Rockland, the largest population center in the watershed. Eighty-eight percent of anglers indicated they would not come to the area if there were no trout fishery.
2. The Beaverkill–Willowemoc fishery supported more than 90,200 angler-days in 1994 compared with 161,700 angler-days reported for 1988. This decline in fishing effort is consistent with data collected in aerial surveys of angler densities.
3. Total angler expenditures for 1994 were estimated at US$9.1 million based on an angler intercept survey. A survey of area businesses confirmed this estimate, with angler-related revenues estimated to be $10.1 million. Results of these two surveys were combined to estimate the economic impacts of these expenditures.
4. Revenues of greater local significance include wages, local taxes, local business profits, and local business supply purchasing. Based on a community model designed specifically for this project, angling's direct economic impact in Rockland was $4.8 million in 1994. Based on a multiplier of 1.87 to account for local respending, this method assessed annual community economic benefits from angling to be $8.9 million in 1994.
5. Of the anglers surveyed, 62% stated they prefer wild fish. Regarding catch-and-release, 50–65% of anglers surveyed release all fish caught, and a further 28% release more than 75% of fish caught. When asked what management changes would induce them to take more trips to the system, 46–57% of anglers chose "more wild fish, and less stocking"—more than any of seven other management options. The two figures represent different responses from anglers interviewed directly and those contacted by mail.

Although most residents had some awareness of the importance of the Beaverkill–Willowemoc fishery to the local community, the study data quantified the socioeconomic benefits of the fishery. Such information can be used to balance decisions about future economic growth relative to fishery protection.

Finally, the socioeconomic survey has set the stage for the biophysical assessment by engendering goodwill with local leaders and business people.

Biophysical assessment.—The goals of the project's biophysical component are to (1) develop a theoretical model to determine linkages among geomorphic, hydrologic, and biological attributes of the system, with specific reference to trout populations; (2) collect and analyze new and existing data on habitat characteristics and anthropogenic phenomena shaping them; (3) evaluate responses of trout populations to habitat and direct anthropogenic influences; and (4) develop an initial set of recommendations.

Trout Unlimited's biophysical assessment includes the identification of sources of biological vulnerability and resilience over time. Documenting, protecting, and optimizing these critical processes and areas are essential to the *Field of Dreams* strategy espoused by the project's technical advisors: that is, provide the habitat, and the trout will come. Trout population changes in a system with the Beaverkill–Willowemoc's flow characteristics are inevitable; the project hopes only to slow the rate of decline during periods of stress and to speed recovery after such periods. The 90% decline in trout populations that the press trumpeted after the 1991 drought turned out to be a figure of uncertain origin. Electrofishing by the NYSDEC in 1993 and 1994 suggested steep declines; at one site, the trout density of 93.1 fish per acre and biomass of 42.1 pounds per acre in spring 1993 dropped to 7.4 fish per acre and 6.1 pounds per acre in spring 1994 (McBride 1996). No determination of the relative roles of emigration versus mortality was possible.

Investigating sources of resilience in the fishery, river, and watershed (with specific reference to wild trout and hatchery trout that have overwintered), the project has enumerated (1) an extensive tributary network, (2) groundwater seeps, (3) thermal refugia created by the extensive tributaries and groundwater seeps, (4) trout movement between main stems and tributaries, (5) good to excellent water quality, and (6) abundant food resources. Social factors adding stability and resilience to the system include the protection of many headwater areas by Catskill State Park and private clubs, management expertise by the NYSDEC, and vigorous local support in Rockland.

Sources of potential or realized vulnerability include (1) a highly variable flow regime; (2) periodic thermal stress combined with limited thermal refugia; (3) tributary blockage at mouths; (4) localized areas of bed or bank erosion; (5) uniformly aged forest stands and a lack of large woody instream debris; (6) the thermal, chemical, physical, and flow regime impacts of road runoff; (7) decreased floodplain function brought about by diking, raised road and rail beds, and disposal of construction fill; and (8) localized inputs of toxins and fertilizers.

Sociopolitical factors that could exacerbate vulnerability or negatively affect wild and overwintering hatchery trout populations include inappropriate stocking practices (e.g., distributing many fish at a small number of sites), poor quality of stocked fish (based on visual assessment, angler opinion, and overwintering estimates), political and economic pressures on the NYSDEC, and factionalism among the various river constituencies.

Addressing the spatial and temporal scales deemed necessary has necessitated the construction of a geographic information system (GIS). Because of the substantial initial costs of assembling a GIS, Trout Unlimited's decision to go this route was not casual. The decision was made easier by an invitation to share data with the NYSDEC's Division of Fisheries GIS and a massive interagency GIS currently being compiled by the NYSDEC's Division of Waters, the U.S. Environmental Protection Agency, the U.S. National Park Service, and New York City's Bureau of Water Supply. Trout Unlimited's GIS is intended to serve both as a data storage, retrieval, and display system and as a tool to facilitate decision making at local and regional levels. Because Trout Unlimited is aggregating data at several scales and wants GIS products to be useful to local entities, the final products will be available on a desktop platform for users of both vector-based and raster-based software.

The GIS's core will combine the interagency datasets (on land use, soils, elevation, and hydrography), ground information (on flow characteristics, thermal patterns, habitat, and fisheries), and low-altitude, high-resolution remote-sensing data supplied by Cyberdyne, Inc. (Sisters, Oregon). The digital near-infrared data will efficiently capture information on channel geometry (including sinuosity and channel width to bank width ratios), substrate composition, pool to riffle ratios, shading, distribution of large woody debris, and vegetation characteristics in the riparian zone. A ground assessment of Rosgen stream classification at reference sites (Harrelson et al. 1994; Leopold 1994) will complement these datasets. The combination of these multiscale data on landforms, surficial geology, and hydrology will support the project's emphasis on applied fluvial geomorphology (how running waters shape channels and land formations) (Leopold et al. 1992; Ontario Ministry of Natural Resource 1994).

The USGS is conducting flow analyses (including high- and low-flow frequencies corrected for precipitation trends, duration curves, and base flow by subwatershed), a study of patterns of channel aggradation or incision, and groundwater mapping for the Beaverkill–Willowemoc system. A 1:48000 map resulting from the groundwater analysis and the study of the contribution of base flow in the subwatershed will be used as the foundation for protection efforts in phase II. The other analyses are designed to identify the presence and nature of any hydrologic change in the watershed. Of particular concern are the effects of road construction and runoff in the system as well as the effect of the siting of Route 17 on soils with the highest infiltration potential in the watershed (Figure 22.7). The New York State Department of Transportation has indicated a willingness to examine runoff control options (e.g., french drains [permeable, buried culverts], settling ponds, constructed wetlands, and removal of paved culverts) to increase precipitation residence time and filtration and toxin uptake by vegetation if the project or NYSDEC can provide quantitative data on impacts.

To construct a comprehensive thermal profile of the watershed, in 1995 the project placed 40 recording thermometers in main channels and important tributaries. This profile will quantify thermal restriction of trout habitat as well as identify reaches experiencing disproportionately rapid daytime warming or

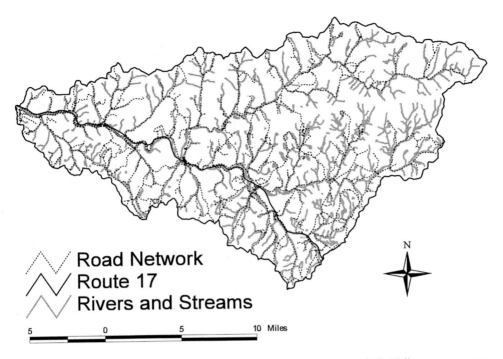

Road Network
Route 17
Rivers and Streams

N

5   0   5   10 Miles

FIGURE 22.7.—Stream drainage and road network in the Beaverkill–Willowemoc watershed.

evening cooling. These data are supplemented by on-the-ground surveys to determine groundwater inflow points. The NYSDEC is supplying past data from its water quality sampling. Trout Unlimited is also contracting with an NYSDEC team that specializes in deriving water quality analyses from macroinvertebrate population samples. The team's potentially most powerful index—impact source determination— compares community structure at the family and genus level to a database on the differential pollution tolerance of macroinvertebrates in order to group sites into general categories of pollution type and effect (Bode et al. 1995c). Impact source determination indices derived from 1993 and 1994 sampling raised questions on a range of pollution sources, particularly sewage, at several sites (Bode et al. 1993a, 1993b, 1995a, 1995b). In addition to increasing the extent of the main-channel survey, the macroinvertebrate team will also be asked to analyze samples from the larger and more significant tributaries. Another effort shared by Trout Unlimited and the NYSDEC is the mapping of all known hazardous waste sites and prioritizing their potential impacts based on proximity to groundwater flows and critical instream zones. No active mitigation is planned due to the high costs involved. Sources of potential or episodic pollution include abandoned waste and acid factory sites, the Livingston Manor sewage treatment plant, urban zones, Route 17 runoff points, and localized areas of dense riparian modification, such as campgrounds.

Trout Unlimited and NYSDEC personnel are working together on an inventory and assessment of thermal refugia, spawning areas, and tributary passage and habitat problems. In addition to the headwaters of the three main subwatersheds, 26 tributaries have been assessed as important for their roles as trout spawning and nursery habitat and refuge and for mitigation of thermal extremes of the main channel. These areas have received and will continue to receive primary attention.

A 1995 to 1996 radiotelemetry study funded by Trout Unlimited and NYSDEC assessed trout population movement within the Beaverkill and the larger Delaware River system. Movements related to thermal stress and spawning were the primary focus. Data on 12 fish tagged in 1995 showed movement between the Beaverkill, its tributaries, and the East Branch of the Delaware. The tagged trout were observed using shade and minimal groundwater seepage in their search for cool and oxygenated flows.

Trout population data are being supplied by the NYSDEC's records of electrofishing and creel surveys. Temporal trends, spatial distribution, and reproduction will be tracked. These data will also be used to assess efficacy of phase II mitigation and enhancement activities.

The phase I assessment has found no glaring sources of degradation comparable to those of the past. With the exceptions of effects from Route 17, physical alterations of floodplains, and sewage treatment systems, current and future impacts can be attributed to the cumulative effects of small, human perturbations, particularly in critical tributary and refuge areas. Small-scale restoration efforts focused on the roles of tributaries in thermal mitigation, refuge, spawning, and recruitment are likely to have the greatest short-term benefits for trout populations. Long-term benefits will most likely come from basinwide strategies incorporating ground and surface water conservation as well as protection of the processes that create and maintain desirable channel and flow characteristics.

Phase II: Implementation of Watershed Fisheries Initiatives

During phase II, data collection will continue for many of the types of biophysical information collected during phase I. Conditions in 1995 and 1996 have been unique; extreme low flows in summer 1995 were followed by bank-full discharges in October, the flood of record in January 1996 (Baldigo, personal communication), and several bank-full flows since. The four-year span of the entire project represents an inadequate time series for many types of data. Nevertheless, the project's emphasis will shift to trial mitigation and protection activities in phase II. It is thus not intended as a comprehensive watershed conservation plan but rather as a starting point for development of a long-term plan under the purview of the local community. Building on the technical information gathered in phase I and the input provided by the local community through the project's communications system, phase II work will implement priority activities in three areas: water conservation; watershed protection and restoration; and fishery and habitat management.

Water conservation.—Though Rockland's population levels have changed little, summer visitation and some water-intensive industries (including a poultry-processing facility) create significant demands on water resources. Residential water use peaks in summer and winter (winter because of practices designed to avoid frozen pipes), precisely when trout populations and habitat are most vulnerable to low flows (K. Benson and F. Getchell, Leggette, Brashears & Graham, Inc., in a letter to L. G. Biegel, New York State Department of Environmental Protection, 1995). Some of the town's water supply comes from surface sources with quality problems, and so, as part of ongoing infrastructural development, the town is sinking additional wells to convert the entire system to groundwater. Springwater withdrawals for export to the bottled water business represents another source of pressure on water resources. One such operation exists in Livingston Manor, and its effects on flow rates of an important tributary have been visible enough to stimulate public protest. In the course of phase I, two additional proposals for bottled water plants were submitted to Rockland's zoning board. These two proposals were voted down, but further proposals are in development.

Cognizant of the value of its water sources, the river, and the fishery, however, Rockland has proposed a new ordinance on extractive groundwater permits. The town of Colchester, which includes much of the lower basin, is considering a similar law. Phase II will provide some of the technical information required to make decisions on water use. It will also assist the towns in water conservation education through the preparation of a pamphlet to be distributed to residents and businesses.

Watershed protection and restoration.—The project's groundwater mapping will serve as a cornerstone for further legal protection at the local level. Rockland has indicated interest in protecting groundwater resources through zoning mechanisms. In phase II the project will bring in experts from the NYSDEC's Division of Water and other institutions to advise town leaders on protecting groundwater resources as well as the town leaders' desire to improve local handling of household toxic materials.

Rockland and Trout Unlimited have discussed voluntary upgrade alternatives for the Livingston Manor sewage treatment plant. These plans have been temporarily shelved by the announcement on 26 February 1996 that the Falls Poultry Plant would be closing. That processing facility's output represented 50 to 75% of the sewage treatment plant's volume (J. Sansalone, NYSDEC, personal communication).

A further set of phase II activities target base flow augmentation in addition to water quality and habitat improvement. The first endeavor is to renew or create wetlands where possible; the second involves increasing riparian vegetation, particularly with a planting program in spring and summer 1996 to address some of the effects of the January flood and subsequent bulldozing. The town of Rockland, local schools, and Trout Unlimited chapters from the region will assist in this effort. Trout Unlimited and the NYSDEC are planning a retreat for town, county, and state public works and road personnel to discuss post-flood repair work, flood prevention, road runoff handling, and trout habitat needs. A third element involves breaching dikes and installing equalization culverts in raised road and rail beds to reconnect floodplains to the river. A fourth initiative, to be tried on a pilot basis, is the conversion of one or two of the dozens of small dams in the watershed to release water from the deeper, cooler part of the water column. Finally, and most extensively, Trout Unlimited, the USGS, and New York State Department of Transportation are discussing pilot conversions of Route 17's runoff system to one designed to maximize water infiltration and toxin uptake by vegetation.

Fishery and habitat management.—Trout Unlimited and the NYSDEC are currently examining various protection alternatives for critical instream zones. In addition, a set of trial physical enhancement installations for thermal refugia are under discussion. These installations would be designed to concentrate coldwater flow paths and protect the congregated fish from poaching, predation, and exposure to sun. Anticipated trials may consist of temporarily anchoring natural objects, such as large snags, between the refugia and main current and monitoring the response of the trout population.

More extensive instream work is anticipated in or at the mouths of those headwater and tributary reaches that have experienced degradation. Trial projects designed to improve fish passage and optimize summer and winter habitat will be conducted during phase II.

The list of anticipated phase II activities has included several designed to help local leaders and communities manage the Beaverkill–Willowemoc's resources according to their priorities. Trout Unlimited also wants to support the NYSDEC in its longstanding and critical stewardship of the watershed, river, and fishery. Many conservationists remain concerned about the NYSDEC's issuance of more than 1,200 emergency permits to landowners in one nine-county region alone; hundreds of violations (by NYSDEC's estimates) of these permits and of the memoranda of understanding issued to local government entities, as well as unpermitted actions, have resulted in only one summons issued in 1996. The apparent suspension of stewardship activities in response to floods and other "emergencies" remains a long-term problem.

Recurring budget cuts, and prodevelopment and antiregulatory political pressures, may limit the NYSDEC's enforcement, monitoring, and management capabilities. Phase II will include lobbying efforts both to increase NYSDEC funding and to close loopholes in New York's Stream Protection Act. The project has engaged the NYSDEC in dialogue regarding the closure of thermal refugia to angling during critical periods, assessment of stocking practices, improvement of the quality of state hatchery products, and post-flood permitting and enforcement procedures. Should the NYSDEC decide to go ahead with refugia closures or other management changes, Trout Unlimited will urge other angling constituencies to support these decisions.

Phase II contains an educational component with three major directives. The first objective deals with youth education about angling, the environment, and the Beaverkill–Willowemoc. The primary methods to achieve this objective are

(1) financial support for the Catskill Fly Fishing Center's ongoing youth education programs and (2) rehabilitation of a tributary encased in a culvert to an open channel flowing through the Roscoe Central School property, enabling monitoring and education on the school grounds. The second objective is public education regarding economic values of the river and fishery. Angler education about responsible practices, natural variation in populations, and the Beaverkill–Willowemoc ecosystem constitutes the third objective. The primary means to realizing the second and third objectives are through the project's communications campaign, which consists of periodic community meetings, dissemination of report summaries through a mailing list of interested parties, and regular media contact.

The public meetings and contact have been and continue to be especially productive. Although conducting the scientific and technical aspects of the project without extensive public involvement would have been more expedient, such an approach ignores community support, the key element in initiating and executing a watershed conservation plan. In the past, Trout Unlimited technical assessments were generally fully developed by professional staff and then distributed for public review. In the Beaverkill–Willowemoc Watershed Initiative, the public was invited to comment on all technical developments from preliminary design to data collection to evaluation of preliminary results. This created a significant administrative burden but resulted in greater regional and local ownership in the project. Trout Unlimited considers that sense of ownership to be the key to long-term conservation in the Beaverkill–Willowemoc watershed, particularly if it leads to the formation of an effective watershed coalition.

CRITIQUE

The project's lack of success thus far in building that coalition is probably its greatest disappointment. Great rivers lead to high emotions from diverse constituencies, and in this project the national office of Trout Unlimited entered a hornet's nest. It encountered three competing local fishery conservation groups (Trout Unlimited's BeaMoc Chapter, the BeaMoc Coalition, and Headwaters) and five regional groups (Trout Unlimited's New York State Council, Trout Unlimited's New York City Chapter, Theodore Gordon Flyfishers, and the federations of sportsmen's clubs of Sullivan County and Delaware County), all with large stakes in the local fishery. The presence and activities of the national office of Trout Unlimited initially escalated tensions instead of easing them. Effort has been expended in building bridges to these local groups as well as to the NYSDEC and other project counterparts; the situation has improved significantly. The building of consensus remains one of the project's top goals for the future.

An offsite presence and staff may have initially helped in dealing with these highly charged personality and conservation politics, but at some cost to efficiency, profile, and communications. The project manager moved to the area in June 1996 to eliminate those particular shortcomings.

The project has not carried out a formal limiting factor analysis (quantitative documentation of a single factor that limits a system or population from reaching its highest potential). This decision was based partly on the system's size and heterogeneity—what limits trout populations in one reach often may not apply to another—and partly on the project's orientation to addressing obvious stressors. Nonetheless, limiting factor analysis is always desirable; it will enter the geomorphology and habitat classification work described earlier.

Another possible weakness of the project was its strategy to embed itself as deeply as possible in the long-term operating procedures of the NYSDEC and the town of Rockland. If these long-term capabilities had been ignored, results would have come earlier and faster. This long-term approach also leads to certain gambles. The NYSDEC may not be able to conduct project monitoring if funding cuts limit the NYSDEC's resources. Trout Unlimited, however, stands by its decision to emphasize long-term capabilities and support existing management institutions.

The project has enjoyed some success in the difficult act of balancing good technical work with good conservation. An example is the process by which consideration of watershed processes has helped determine cost-effective protection and mitigation strategies. Trout Unlimited also believes that its slow, long-term approach will be to the ultimate benefit of the watershed and its residents as well as the river and its fish. The diverse skills and perspectives of the Trout Unlimited–NYSDEC technical team, combined with public input and careful planning, are leading to an effective systems approach to the social ecology, biological realities, and political–economic contexts of the Beaverkill–Willowemoc and its many actors. Finally, and importantly, public opinion toward the project itself has swung from polite suspicion and alienation to enthusiasm, hospitality, and support.

ACKNOWLEDGMENTS

The authors would like to thank the Beaverkill–Willowemoc Watershed Initiative's primary funders, the R. K. Mellon Foundation and the National Fish and Wildlife Foundation, and its technical advisors, John Fitzgibbon and Jack Imhof. It is also indebted to the project's counterpart and cooperating institutions: the New York State's Department of Environmental Conservation (Bureau of Fisheries, Regions 3 and 4; Division of Water); the towns of Rockland and Colchester; the U.S. Geological Survey; the New York State Department of Transportation; the New York City Department of Environmental Protection, Bureau of Water Supply; the Catskill Center for Environment and Development; the Delaware River Basin Commission; the U.S. National Park Service; the College of Environmental Science and Forestry, State University of New York at Syracuse; and the U.S. Military Academy, West Point. The current and former NYSDEC personnel who have given most to the project and this report are Phil Hulbert, Wayne Elliot, Bob Angyal, Jack Isaacs, Ed Van Put, Bill Kelly, Russ Fieldhouse, Walt Keller, Kay Sanford, Bob Bode, Larry Abele, Ricardo Lopez, and Howard Pike. Ed Eckel, a local resident, and Martin Redcay, president of Trout Unlimited's BeaMoc Chapter, have helped generously both in the field and in conversation.

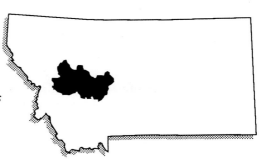

It isn't the biggest river we fished, but it is the most powerful. It runs straight and hard. . . . It runs hard all the way. It is a tough place for a trout to live.
—Norman Maclean, 1976

CHAPTER 23

RESTORATION OF TROUT WATERS IN THE WEST: BLACKFOOT RIVER OF MONTANA

Gary Aitken

The Blackfoot River, known locally as the Big Blackfoot, is a legendary trout stream, immortalized in Norman Maclean's (1976) best-selling book *A River Runs Through It* and the Robert Redford movie of the same name. By 1987, a number of people in widely separated sections of the Blackfoot River watershed didn't see the river in quite the way either of these popular-press vehicles portray it. Concerned with the current and future condition of the river, these people believed that the river and its watershed were in serious trouble and on a downhill slide.

In the lower reaches of the river, outfitters and guides, local residents, and returning clients of guest ranches had noticed the quality of the fishery was steadily declining. In the headwaters area near the town of Lincoln, Montana, a proposed gold mine using cyanide extraction technology threatened to contaminate the river and surrounding area. Independently, parties from different parts of the watershed approached the Montana Department of Fish, Wildlife and Parks (MDFWP), the state agency responsible for fisheries, but with little success. There were few data available on the condition of the fishery and none available to reveal trends. The regional MDFWP office was short on funds and had only two fisheries biologists, who were split between two offices, responsible for the region. Cleanup work at the largest Superfund site in the nation, involving another Montana watershed and a large lawsuit, loomed on the horizon. The regional office was overextended—in short, the Blackfoot River was "too far down the list."

In the fall of 1987, a meeting was held between people interested in the Blackfoot River and the regional manager of the MDFWP. River restoration and habitat work were not at the forefront of the discussions; many of those at the meeting believed overharvest to be the major problem of the river, and that stricter angling regulations could solve it. However, local politics and public opinion simply would not tolerate radical changes in regulations without solid evidence supporting those changes. Unfortunately, MDFWP funding was not available to perform the needed studies. Finally someone asked if private parties came up with the funds, would the MDFWP do the work? The MDFWP agreed to this arrangement and drew up a plan that would start with a fishery inventory. The plan indicated a need for seasonal employees to help with the groundwork at a cost of US$13,000 to $15,000 for one year.

At this point, those concerned with the plight of the river had a clear short-term goal. They formed a local chapter of Trout Unlimited because its mission, "conserving, protecting, and restoring coldwater fisheries and their watersheds," was appropriate. The group composed a fund-raising letter, developed a list of potentially sympathetic people, sent the letter out, and began knocking on doors. In four months, the Big Blackfoot Chapter of Trout Unlimited (Big Blackfoot Chapter) raised $16,000, and the restoration effort on the Blackfoot River was begun.

THE BLACKFOOT RIVER WATERSHED

The Blackfoot River rises in the high mountains of the Scapegoat Wilderness near the Continental Divide in western Montana (Figure 23.1). The river flows 132 miles in a westerly direction to its confluence with the Clark Fork River. Primary tributaries, from the headwaters to the mouth, are Landers Fork, North Fork of the Blackfoot, Monture Creek, and the Clearwater River. The watershed's physical geography may be viewed in three sections: headwaters dominated by coniferous forest and partially located within designated wilderness areas; a middle section consisting primarily of a broad agricultural valley with a conifer-ous forest perimeter; and a lower section dominated by steeper canyons and coniferous forest. From the headwaters to the lower elevations much of the watershed is used for timber production. Nonwilderness portions of the upper section contain abandoned mining operations. Agricultural enterprises consist mostly of livestock production; land holdings are large and are devoted to native pastures and irrigated hay production. In the forested canyons of the lower reaches, timber production and recreation dominate. Lands administered by the U.S. Forest Service constitute 45% of the basin; 43% is in private lands, about

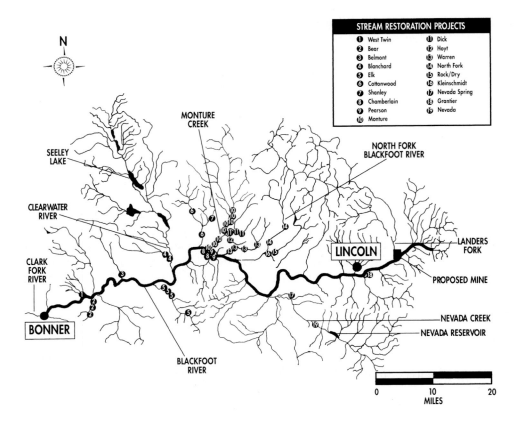

FIGURE 23.1.—The Blackfoot River watershed in west-central Montana. Stream project locations are shown: (1) West Twin, (2) Bear, (3) Belmont, (4) Blanchard, (5) Elk, (6) Cottonwood, (7) Shanley, (8) Chamberlain, (9) Pearson, (10) Monture, (11) Dick, (12) Hoyt, (13) Warren, (14) North Fork, (15) Rock and Dry, (16) Kleinschmidt, (17) Nevada Spring, (18) Grantier, and (19) Nevada Creek.

one-half of which is large timber company holdings; 7% is state owned; and 5% is administered by the U.S. Bureau of Land Management.

The upper and middle sections of the watershed lie in an area characterized by severe winter conditions. The coldest temperature ever recorded in the lower 48 states was in the headwaters area, and winter temperatures in the −30 to −50°F range occur almost every year. Many tributary streams and the main stem of the river experience severe anchor ice formation as well as surface freezing. Anchor ice forms in all types of water—pools, runs, and riffles. In essence, portions of the river and its tributaries freeze from the bottom up, the top down, and in from the shallows.

During its course through the chain of lakes through which it flows, water in the Clearwater River is warmed. In addition, some of this water is diverted for irrigation. However, Clearwater River water entering the Blackfoot River at their confluence is significantly cooler due to groundwater recharge; the Clearwater River does not significantly raise temperatures in the main stem.

The Blackfoot is essentially a free-flowing stream, although Milltown Dam, a small dam just below the river's confluence with the Clark Fork River, has no fish passage facilities and prevents upstream migration of all species of fish. Large agricultural operations in the watershed make significant withdrawals of surface water during the growing season, typically from May through September. With the exception of the subbasin drained by the Clearwater River in the middle section, there are few lakes in the watershed; almost all water originates as snowmelt or stream runoff or emerges as springs and seeps. Discharge data are available from 1898 onward (U.S. Geological Survey, unpublished data); the average discharge at the mouth is 1,578 cubic feet per second (cfs), with wide variation on both an annual and a seasonal basis. Seasonal lows occur between late fall and midwinter; the record low is 200 cfs. Seasonal highs occur during spring runoff; the record high is 19,200 cfs. Seasonally based instream flow appropriations to the state (known as Murphy Rights) are in place on the lower 34.7 miles of the river and vary from a low of 650 cfs to a high of 2,000 cfs. Higher priority rights for substantial amounts of water exist for agricultural diversions and other purposes.

In 1975, an abandoned tailings pond at Mike Horse Mine in the headwaters washed out, sending contaminated water and sediments directly into the river. In addition, acidic drainage from surrounding mines and mining waste piles still causes problems today. Sections of the headwaters have depressed pH values and elevated levels of metals in the sediments, water, and the food chain. These pollution problems have resulted in depressed and stunted benthic (bottom-dwelling) communities and fish populations (Moore et al. 1991). A state effort began addressing these problems in 1993 and is still in progress.

In 1988, a 247,600-acre forest fire (Canyon Creek fire) burned a substantial portion of the headwaters area of the North Fork of the Blackfoot. The burned area has a steep terrain with thin soils. Subsequent storms and snowmelt resulted in increased sediment loads from this portion of the watershed; however, sediment loads are now decreasing. Prior to the fire, portions of the lower reaches of the North Fork were intermittent in very dry years. Since the fire, this has not occurred, perhaps because the increased sedimentation reduced ground infiltration.

Much of the watershed has been logged during the past 100 years. Large areas have been clear-cut, and extensive road building has occurred. Historically, logging has extended all the way to the river and tributary streambanks, resulting in increased erosion and sediment loads. The removal of old-growth timber from riparian areas has resulted in a lack of woody material available for channel complexity and fish habitat. Agricultural effects on the watershed include bank degradation due to livestock trampling; increased sediment loads from agricultural runoff and bank degradation; water temperature increases due to degraded tributary streams that have little shading and are excessively slow and shallow because of bank degradation; loss of fish into irrigation ditches; barriers to fish migration caused by culverts and irrigation diversion structures; and loss of woody streambank vegetation.

Recreational use has also adversely affected the watershed. During the fall hunting season, deer and elk are under constant pressure, which causes frequent movement of large herds of animals or concentration of large herds on private bottomlands. This unnatural concentration of native animals in sensitive bottomlands results in bank degradation, sometimes worse than that caused by well-managed domestic animals. Recreational usage of the river itself is steadily increasing. Angling pressure has resulted in overharvest of trout and depressed populations in some areas. Overcrowding of the river has become an issue among both recreational and nonrecreational river users.

Dominant native fishes in the Blackfoot watershed include westslope cutthroat trout, bull trout, and mountain whitefish. Rainbow trout, brown trout, and brook trout are the predominant nonnative fishes. Because of varying environmental conditions, the dominant species changes greatly as one travels from the headwaters to the mouth. Westslope cutthroat and brook trout dominate the headwaters, brown trout dominate some reaches in the middle section, and rainbow trout dominate the remaining reaches and the lower section. Wild populations of all species are supported in the river. The headwaters of the Blackfoot River is one of the few systems in which fluvial bull trout (those that spend their entire life in the river system) still exist in viable numbers, although their range has been severely reduced because of barriers that prevent migration and degradation of spawning habitat (D. Peters, MDFWP, personal communication). The river has been managed primarily as a wild, native fishery since regular stocking of rainbow trout ceased in 1979. Since then, there have been occasional introductions of westslope cutthroat trout using wild stocks of adults and fry to reestablish spawning runs. Almost all spawning and rearing habitat in the basin lies in tributary streams, many of which are surrounded by privately owned ranches or timberland. The aquatic growing season is limited by coldwater temperatures, which extend throughout much of the year, and growth rates are correspondingly slower than in warmer waters.

The bull trout has repeatedly been considered for listing pursuant to the U.S. Endangered Species Act, and the westslope cutthroat trout is considered to be a sensitive species. Throughout its range, the bull trout is threatened by degradation and fragmentation of its habitat and by interactions with introduced brook trout (Rieman and McIntyre 1993, 1995). Like the bull trout, the westslope cutthroat trout depends upon maintenance of cold, high-quality water and is easily threatened by habitat degradation and the introduction of nonnative salmonids (Behnke 1992; McIntyre and Rieman 1995).

EARLY SURVEYS AND INITIATIVES

Little formal scientific baseline data are available for the Blackfoot River predating the significant mining, agricultural, timber harvest, and recreational activities that have occurred in the basin. However, eyewitness accounts, angler's logs, and photographs provide substantial documentation of the river's past productivity. Mining activity in the headwaters area clearly has degraded the watershed, contributing heavy metals and acids to both ground and surface water. Agricultural and timber harvest practices generally result in nonpoint pollution sources and are more difficult to assess.

Few studies existed for the river at the time the Big Blackfoot Chapter and the MDFWP initiated their cooperative fishery inventory. Those which did exist did not cover all important reaches but did provide some basis for comparison. Coffin and Kilke (1971) provided information on sites in the headwaters area, Spence (1975) provided reference fisheries data at five sites in the upper section, and Walker (1977) provided historical data on recreational usage in the lower section.

That the Big Blackfoot Chapter raised $16,000 in a short period to fund the initial fishery inventory indicated substantial support for restoration. The Big Blackfoot Chapter hired a contractor to inventory all reaches of the main stem as specified by the MDFWP, which reassigned a fisheries biologist to work on the river full time. The general objectives of the inventory were to

- develop baseline information on the main-stem fishery,
- develop baseline information on tributary fisheries and habitat,
- determine the status of species of special concern (westslope cutthroat trout and bull trout),
- develop fisheries management options for public review, and
- develop a management plan to deal with public concerns and resource issues in the drainage.

The fishery inventory (Peters and Spoon 1989) focused on the main stem of the river and consisted of a five-year inventory from the headwaters to the mouth. It emphasized sampling of juvenile trout, repeated earlier fish population surveys in the headwaters area for comparison purposes, and included some creel census work. It also included a bull trout redd (nest) survey in two tributaries known to be important spawning areas for that species. Peters and Spoon (1989) established that fish populations in the Blackfoot varied greatly, both in terms of densities and species composition. Reasons for this variation in fish populations included variation in habitat characteristics, recruitment sources, and human influences (e.g., habitat degradation and fishery exploitation). The survey recommended that future work include surveying major tributaries to identify current and potential use for spawning by trout. The survey also found that bull trout and westslope cutthroat trout were in need of additional protection to insure the viability of their populations.

In early 1989, the Big Blackfoot Chapter began looking for sources of funding to continue fishery studies and conduct restoration work. Grant writing was considered to be critically important, so a professional contractor was hired for this purpose. Funding for the ongoing study is now provided by the MDFWP, and restoration work is cooperatively funded by the MDFWP, other public agencies, private foundations, and the Big Blackfoot Chapter.

Peters and Spoon (1989) indicated that some main-stem problems were caused by a lack of trout population recruitment. The following year, studies (Peters 1990) were directed at 19 tributary streams in the basin, to identify fishery characteristics, habitat, and problems. Toxic metals accumulation in trout was also studied in the headwaters area, and population monitoring was continued in the main stem. As in the 1989 study, a creel census was conducted, and a brochure was developed to distribute with creel census cards to explain some of the problems of the river and its fishery.

In addition, the Big Blackfoot Chapter pushed for and received priority funding from the state for cleaning up some of the major mine tailings in the headwaters. Unfortunately, little real cleanup resulted. The effort was brought to a standstill, initially by concerns about how the state awarded the contract to do the work and subsequently by Superfund liability concerns.

The initial tributary survey results (Peters 1990) showed that almost all of the trout spawning and rearing habitat had significant problems—primarily due to agricultural activities associated with livestock grazing but aggravated by other practices such as timber harvest. Peters (1990) concluded that potential management tools for improving the fishery included stock enhancement, changes to fish harvest regulations, and habitat restoration. In 1990 a third study (Pierce 1991), funded by the U.S. Bureau of Land Management and the MDFWP, was completed. Its purpose was to develop detailed habitat inventories on U.S. Bureau of Land Management-managed streams and on some streams with future restoration potential. Uncontrolled grazing on agricultural lands and poor logging practices in higher-elevation riparian areas were contributing to stream habitat degradation and depressed fisheries (Pierce 1991). In addition, fish passage barriers were observed that limited access to tributaries for spawning. Pierce (1991) did not recommend elimination of traditional land-use practices. Rather, he suggested cooperative efforts be used to develop and implement management practices that recognized other uses of riparian areas and that would be beneficial to fish and wildlife.

ISSUES AND GOALS OF RESTORATION

After both the main-stem and tributary studies were completed, sufficient data were available to establish preliminary long-term goals. These findings, objectives, and strategies (Peters 1990) were species specific and are summarized below.

Westslope Cutthroat Trout

The Blackfoot and its tributaries have the potential to produce greatly enhanced populations of westslope cutthroat trout, with fish up to 20 inches. There are two life history forms of westslope cutthroat trout in the watershed: a headwaters form that lives its entire life in smaller tributary streams and a larger riverine form that spawns and rears in the tributary streams and then migrates to the main stem for growth and maturation. Slow growth rates and environmental limiting factors coupled with overharvest result in failure to reach full growth and recruitment potential. High background sediment loads and increased sedimentation from commercial timber harvest and the 1988 Canyon Creek fire make the population even more sensitive to overharvest. Upstream from Lincoln, populations have crashed since the Mike Horse Mine tailings pond failure in 1975. At present no harvest of westslope cutthroat trout can be sustained in this reach. Westslope cutthroat trout occupy reaches farther upstream than do rainbow and brown trout. Young-of-the-year sizes indicate a temporal isolation as well, the westslope cutthroat spawning after rainbow trout. From Lincoln to the mouth of the Blackfoot the average density of westslope cutthroat trout six inches or longer is 1.8 per 1,000 feet.

The management objective for westslope cutthroat trout is to increase population density of fish six inches or longer to 100 per 1,000 feet by the year 2000. To achieve this objective, strategies have been developed that consist of (1) implementing catch-and-release regulations to protect adult spawners and juvenile fish—all available wild fish are needed for recovery purposes and any reintroductions will need protection from harvest; (2) improving spawning and rearing habitat where needed; (3) reestablishing spawning runs in unoccupied habitat (suitability of habitat should be tested with a trial introduction of westslope cutthroat trout); and (4) sampling westslope cutthroat and brook trout, stream bottom sediments, and aquatic insects for toxic metals concentrations in suspect areas.

Bull Trout

Bull trout spawn and rear in tributary streams and then migrate to the main stem for growth and maturation, where they have historically reached 15 pounds. Angler pressure is increasing, and a 59% decline in the number of redds has occurred in the two major remaining spawning runs over a five-year period. For the first time in 1989 no fish over 7 pounds were sampled. The species cannot afford any harvest at this time: the average density of adult fish is only 0.05 per 1,000 feet.

The management objective for bull trout is to increase the standing crop (the summed mass of individuals in a given volume or area) of adults larger than 5 pounds to 1 per 1,000 feet by the year 2000. To achieve this, strategies consist of (1) implementing catch-and-release regulations for the main stem and all tributaries; (2) improving spawning and rearing habitat where needed; and (3) reestablishing spawning runs in unoccupied habitat. Hopefully there are still remnant populations in the system to repopulate such habitats.

Rainbow Trout

Rainbow trout is the dominant trout species downstream from Monture Creek, and lower reaches of tributary streams provide most of the spawning opportunity for rainbow trout. Young of the year appear to leave the tributaries shortly after hatching. Below Belmont Creek (river mile 21.9), the population of rainbow trout 12 inches and longer is 12 per 1,000 feet and is limited by overharvest. Upstream of Belmont Creek, the population density of adult fish is 0.5 per 1,000 feet; juvenile density is also depressed. This low density appears to be attributable to an overwintering loss of juveniles, making the population extremely sensitive to overharvest. Environmental variables, which favor native species, may be the cause of juvenile mortality in winter.

The management objective from the mouth of the Blackfoot to Belmont Creek is to increase the population of rainbow trout 12 inches and longer to 100 per 1,000 feet. Above Belmont, the objective is to increase the population of fish 12 inches and longer to 50 per 1,000 feet. Strategies to achieve these objectives consist of (1) reducing harvest on the main stem and in key spawning streams by establishing a three-fish limit for fish under 12 inches and catch and release for fish over 12 inches; and (2) improving key spawning habitat where possible.

Brown Trout

Brown trout is the dominant species in the main stem of the Blackfoot from Lincoln to Monture Creek. A lack of tributary spawning access for other trout species appears to favor brown trout, which spawn in the main river. Stream habitat from Nevada Creek to Monture is more characteristic of that for westslope cutthroat and rainbow trout, but a lack of recruitment sources for rainbow and westslope cutthroat trout probably favors brown trout. From Lincoln to Nevada Creek, the meandering channel and abundance of woody debris and overhead cover provide habitat favoring brown trout. However, large sediment deposits limit production of fish and prey organisms. The numbers of yearling brown trout have declined significantly since the 1970s.

The management objective from Lincoln to Monture is to maintain the density of brown trout larger than eight inches to at least 20 fish per 1,000 feet. From Monture to the mouth, the objective is to maintain the density of fish larger than eight inches to at least 20 per 1,000 feet. Strategies consist of reducing harvest on the main stem and all tributaries by establishing a three-fish limit for fish under 12 inches and catch and release for fish over 12 inches. These regulations match those for rainbow trout for simplicity.

RESTORATION EFFORTS

The first project was relatively minor in terms of funding requirements, although this and subsequent projects have demonstrated that removal of a simple migration barrier can have dramatic effects. The project involved replacing a culvert that was located on private land and which blocked fish passage on Rock Creek with a bridge. More importantly, the project established an operating pattern. This project also demonstrated to the landowner that a reasonable working relationship with government agencies was possible. Subsequently, this project grew into one of the more ambitious projects undertaken to date and has resulted in restoration of over a mile of critical spawning and rearing habitat for rainbow and brown trout on two tributary streams.

The pattern established by this first project, that is, of a small pilot project followed by a larger project, has been repeated several times and has proven to be a good model for both the agencies involved and the landowners. It gives all parties a chance to establish working relationships and determine whether or not they wish to continue with more ambitious projects.

Restoration efforts have varied from simple to complex. A summary of the types of work completed or underway in the Blackfoot River basin follows. Figure 23.2 summarizes the projects undertaken and the types of practices and the organizations involved in each project. Figure 23.1 shows the project sites and the stream reaches affected by each project.

Replacement of fish passage barriers.—These barriers are typically either culverts with an outflow pipe perched above the water level or having excessively high water velocities during spring runoff, or irrigation diversion structures that have the same effects. The culverts have been replaced either with bridges or with much larger culverts that contain periodic baffles to slow water velocities and provide temporary resting havens. Irrigation diversion structures have been redesigned to allow for fish passage through or past structures in the stream channel (Figure 23.3).

Prevention of fish loss in ditches.—These projects involve installing self-cleaning trash and fish screens at the diversion structures for irrigation ditches. The screens are designed to prevent fishes of all age-classes from leaving the natural river system and entering artificial irrigation systems, where they are washed onto fields or die when the ditch is drained at the end of the irrigation season. Screen cleaning can be water powered, electrically driven, or manual (Figure 23.4).

Spawning habitat restoration.—Spawning habitat restoration is an indirect result of other, general habitat improvement practices. Spawning habitat in poor condition has been restored through removing passage barriers, improving livestock management, increasing stream structure by adding boulders and large woody debris, stabilizing banks, and enhancing streamflows. Spawning areas suffering from silt and sediment deposition have been restored by narrowing and deepening the channel and increasing sinuosity. In restored channels, spawning gravels maintain themselves through periodic, natural scouring flows. Roadside erosion control has involved relocating roads away from streams to reduce sedimentation during runoff, as well as hardening road surfaces in areas with soils prone to severe erosion.

Several extensive channel restoration projects have been completed that involve restoring sinuosity to artificially straightened streams, narrowing and deepening channels, and rebuilding streambanks. Because of the major land disturbances involved in these projects, they always involve instream structural habitat work, riparian vegetation enhancement, and sometimes protection measures such as riparian fencing. These efforts have returned barren stream reaches to highly productive spawning and rearing habitat (Figure 23.5).

Riparian vegetation enhancement.—Vegetation enhancement has been applied primarily on streambanks rebuilt during channel restoration work and has involved planting native grasses (rough fescue, bluebunch wheatgrass, slender

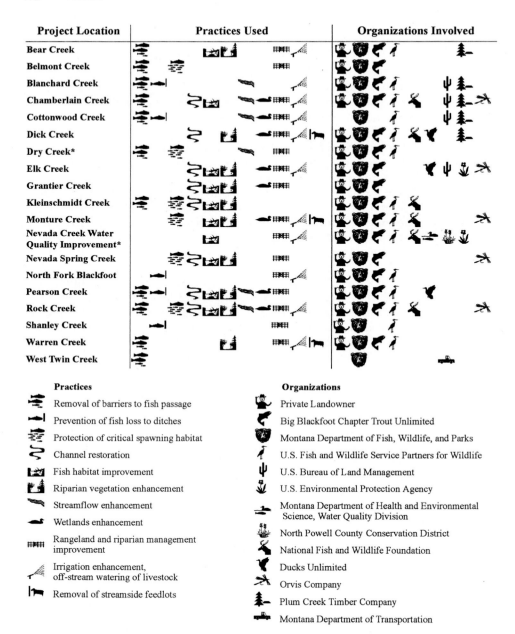

Project Location	Practices Used	Organizations Involved

Practices

- Removal of barriers to fish passage
- Prevention of fish loss to ditches
- Protection of critical spawning habitat
- Channel restoration
- Fish habitat improvement
- Riparian vegetation enhancement
- Streamflow enhancement
- Wetlands enhancement
- Rangeland and riparian management improvement
- Irrigation enhancement, off-stream watering of livestock
- Removal of streamside feedlots

Organizations

- Private Landowner
- Big Blackfoot Chapter Trout Unlimited
- Montana Department of Fish, Wildlife, and Parks
- U.S. Fish and Wildlife Service Partners for Wildlife
- U.S. Bureau of Land Management
- U.S. Environmental Protection Agency
- Montana Department of Health and Environmental Science, Water Quality Division
- North Powell County Conservation District
- National Fish and Wildlife Foundation
- Ducks Unlimited
- Orvis Company
- Plum Creek Timber Company
- Montana Department of Transportation

FIGURE 23.2.—Summary of the projects completed to date and in progress (asterisk). The types of practices used and the organizations involved are shown. The significance of a project should not be gauged by the number of practices involved; some of the most important projects may involve only one or two practices.

FIGURE 23.3.—Denil fish ladder installed on Cottonwood Creek.

wheatgrass, thickspike wheatgrass, and sodar streambank wheatgrass), forbes (purple prairie clover), and shrubs and trees (common chokecherry, woods rose, bebb willow, sandbar willow, Geyer willow, black hawthorne, Saskatoon serviceberry, bog birch, Rocky Mountain juniper, red osier dogwood, Ponderosa pine, Engelmann spruce, quaking aspen, and black cottonwood).

Streamflow enhancement.—Several instream flow leases have been arranged in conjunction with other restoration work, and more are under consideration. Montana water law requires that water appropriated in a water right be put to beneficial use; failure to do so can result in the loss of some or all of the water right. When a project results in increased efficiency of water use, instream flow leases allow a water right holder to leave the saved water in a stream while still protecting the holder's rights to that saved water.

Wetlands restoration.—Restoration efforts in the Blackfoot valley have involved several wetlands projects, ranging in size from one acre to several hundred. These have often required small dams, generally earthen, to replace or restore natural barriers removed when the wetlands were originally drained.

Riparian management and rangeland and irrigation practices.—Rangeland practices have involved cross fencing of pastures (dividing large pastures into smaller ones to concentrate grazing) and implementation of grazing systems that reduce constant pressure on sensitive streamside areas. Management strategies

FIGURE 23.4.—Fish screen installed on Lund-Jorgenson irrigation ditch.

have included restricting grazing to only certain periods, such as late summer or fall, when banks are drier and more durable, or allowing grazing any time during the growing season but only for short periods. In some cases, fences have been installed to restrict livestock access to only a small section of stream instead of allowing unlimited access to the entire riparian area. Control plots have been established to compare differences between well-managed grazing systems with fenced riparian zones and well-managed grazing systems in unfenced riparian zones; results are not yet available.

Irrigation enhancement practices have involved helping land managers to irrigate their ground more efficiently, thereby allowing more water to be left in the stream. These practices have included ditch consolidation and ditch lining to prevent ground infiltration losses and conversion to more efficient sprinkler systems. In almost any agricultural area, there are situations in which feedlots and corrals are built alongside streams to provide animals easy access to water. By providing off-stream watering, the entire feedlot–corral system can be moved to upland areas so the riparian area can be restored. Off-stream watering projects have involved, in some cases, installing watering devices hundreds of feet away from the stream and subsequent fencing to establish a buffer zone to allow banks and streamside vegetation to regenerate.

Weed control.—As in most parts of the arid western United States, noxious weeds are a serious problem in the Blackfoot watershed. Although normally considered an agricultural problem, noxious weeds have a severe negative effect

FIGURE 23.5.—Stream rehabilitation work on Rock Creek. Top photograph, taken in September 1991, shows work in progress to reduce wide channel. Bottom photograph taken in August 1992.

on native streamside vegetation, and indiscriminate use of herbicides can have devastating effects on aquatic resources. The most damaging species are spotted knapweed, leafy spurge, and yellow toadflax. Spotted knapweed, first reported in Montana in 1920, has now spread to 4 million acres in the state (USBLM 1996). Restoration efforts have included coordinated weed control programs and introduction of biological controls at numerous sites. For spotted knapweed, a root-boring moth *Agapeta zoegana*, a root-boring weevil *Cyphocleonus achates*, and two seed head gall flies *Urophora affinis* and *U. quadrifasciata* have been introduced. For leafy spurge, flea beetles *Aphthona abdominalis*, *A. flava*, and *A. nigriscutis* and a stem-boring beetle *Oberea erythrocephala* have been introduced. For yellow toadflax, a defoliating moth *Calaphea lunula* and a seed-capsule-feeding weevil *Gymnetron antirrhini* have been used. The design of restoration projects attempts to reduce weed problems associated with disturbed land by minimizing land disturbance.

Landowner Participation

Funding for these projects has come from a variety of sources. Typically, the landowner contributes material or labor, and grants from various foundations, private companies, federal fish and wildlife programs, state fisheries programs, or mitigation funds make up the remainder. Although funding is obviously important, the key to all of these restoration efforts is landowner participation. It has been our experience that the old saying, "Where there is a will, there is a way," applies directly to river restoration. If all involved parties want a project to happen, it will get done.

Restoration efforts are by definition long-term projects. It is imperative that the landowners involved in a project embrace it and genuinely assume ownership. Their pride of accomplishment, pride of ownership, and pride of stewardship are virtually the only factors that can insure long-term success of a project. Without landowner support, the money spent often is wasted, for the resource will almost certainly degrade over time.

Some members of the agricultural community in the Blackfoot valley were associated with people involved in the early formation of the Big Blackfoot Chapter and the fishery inventories. Consequently, they became aware of the problems in the watershed and the work being done. This familiarity provided a valuable bridge to the agricultural community and helped the Big Blackfoot Chapter gain the acceptance needed to complete the initial projects.

Current Status

Restoration efforts to date have involved numerous different activities at widely separated sites. Projects addressing the overall goal of restoring and improving the condition of the watershed as a whole, not just riparian corridors or streams themselves, are encouraged. Consequently, some projects not normally considered river restoration work, such as weed control, have been embraced. More indirect efforts will probably be done in the future, such as assisting landowners in designing, implementing, and monitoring grazing management in upland areas. This broadscale approach not only helps the restoration effort directly but also pulls the community together by helping bridge the gap between residents with conservation and recreational interests and those with agricultural and timber interests. As of 1996, specific accomplishments include

- establishment of 23 acres of wetlands;
- restoration of 407 acres of wetlands;
- enhancement of 1,080 acres of wetlands;
- restoration of 13.5 miles of riparian area;
- restoration of 17.2 miles of stream channel;
- reopening of 185 miles of tributary streams to fish passage through barrier removal;
- control of erosion along 17 miles of roads;
- reseeding of 430 acres of native prairie;
- coordinated weed control on 32,000 acres;
- protection of 3,700 acres through acquisition by fee title; and
- protection of 40,750 acres through conservation easements.

ADAPTIVE MANAGEMENT

Many people involved in the initial push to restore the Blackfoot River originally believed that overharvest of fishes was the major problem in the middle and lower sections of the watershed. Indeed, one of the reasons for funding the original fishery study was to provide data to justify more restrictive angling regulations. While the study did justify more restrictive regulations, it and subsequent tributary studies also brought to light numerous other more significant problems that have become the focus of restoration efforts.

Things have not always gone as planned. In the early stages, when studies to provide baseline data were critically needed, some potential sources of funding could not be used because they were restricted to on-the-ground projects. Rather than simply ignore these potential funding sources, projects were proposed for some of the obvious problem areas.

Because many critical reaches of the Blackfoot River system lie on private lands, restoration efforts have, of necessity, been focused on projects involving landowners willing and desiring to be part of the restoration effort. Fortunately, many of the critical areas have conscientious landowners more than willing to participate. However, there are a few sections where projects are needed but landowner cooperation has been lacking. Hopefully the examples set through other projects will lead to cooperative efforts to restore these remaining areas.

Today, the Big Blackfoot Chapter is in the envious position of having more projects proposed than it can fund. To help make the selection process as objective as possible, the Chapter has developed a 200-point evaluation (Appendix 23.1) that is applied to all potential projects. The top-ranking projects are selected, and the Big Blackfoot Chapter attempts to secure funding for them. If other funding sources become available that do not fit the top-ranking projects but which may be used for other projects, the Big Blackfoot Chapter attempts to obtain funding for those projects as well.

Some landowners have fundamentally distrusted the Big Blackfoot Chapter of Trout Unlimited simply because of its name. Due to its origins and focus, the board of directors of the Big Blackfoot Chapter is unconventional by normal Trout Unlimited standards (besides anglers, it includes landowners, biologists, foresters, water quality specialists, and outfitters). However, Trout Unlimited is viewed as an environmental organization, and many ranchers and timber companies do not like to join, contribute money to, or work with environmental organizations.

In order to get projects with tangible benefits completed, and because of the sensitive nature of landowner relationships with government agencies and conservation organizations, the initial projects undertaken were conducted "with no strings attached." In some cases, these projects involved convincing landowners to change management practices that have been handed down from generation to generation. In pursuing such efforts, one must recognize that change sometimes comes slowly and that people respond better to encouragement than criticism. Every step forward is of benefit to the resource, even if imperfect.

Restoration efforts are moving toward a model for which project approval is contingent upon involved landowners signing documents that outline conditions which must be maintained for a period of several years after project completion. The purpose of this formality is not to provide any basis for subsequent action but to help insure all parties agree on what is expected of them. The Wildlife Extension Agreement of the U.S. Fish and Wildlife Service is serving as a model.

There have been some issues that the Big Blackfoot Chapter board members have felt needed to be addressed but which they also felt were outside the organization's sphere of influence and responsibility. One of these was a lack of communication among some of the agencies working on the river. In the course of his work on the initial fishery inventory, the biologist conducting the study discovered that different state and federal agencies were planning or conducting studies and projects at various sites on the river. Some proposed projects conflicted with one another, and some projects were not in the best interest of watershed restoration. The Big Blackfoot Chapter believed that better progress could be made if everyone worked together in a cooperative manner rather than individually—or at least if everyone were more aware of each other's activities and concerns. At the same time, some of the agency personnel were looking for an introduction to work with landowners.

As a first step, the Big Blackfoot Chapter sponsored a conference to bring together various parties working in the watershed to talk about what they were doing so that everyone would have some idea of what others were doing and why. The Big Blackfoot Chapter contacted representatives from the agencies involved, and all agreed to participate. Attendance at the first conference was good and included members of the U.S. Fish and Wildlife Service, the U.S. Bureau of Land Management, the U.S. Forest Service, the MDFWP, the Montana

Department of State Lands, the Montana Department of Transportation, and Champion International, a timber company with large holdings in the basin. The format of the conference was informal; representatives from each agency gave a brief overview of what they were doing in the basin, why they were doing it, and where their interests lay. There was little interaction during the actual presentations, but during breaks representatives began to mix and discuss common interests with those from other agencies.

The initial conference was considered a success, and another was held the next spring. Subsequently, a suggestion was made that the Big Blackfoot Chapter sponsor an organization to initiate technology transfer between the various agencies, to work on cooperative funding of projects so there was less duplication of effort, to promote maintaining rural values in the valley, and perhaps ultimately to serve as a mediator for disputes involving resource issues. The general idea was that people disagree on lots of things, but they tend to agree on many things as well, such as the need and desirability of restoration work. The Big Blackfoot Chapter was not well suited to this task; an organization was needed that could be presented as neutral on all issues and whose focus was discussion, not restoration.

The Blackfoot Challenge was formed in 1992 to fulfill this need. Its mission is to provide a forum that promotes cooperative resource management of the Blackfoot River, its tributaries, and adjacent lands, and to coordinate efforts that will enhance, conserve, and protect the natural resources and rural lifestyle of the Blackfoot valley for present and future generations. The Blackfoot Challenge supports environmentally responsible resource stewardship through the cooperation of public and private interests. The Blackfoot Challenge's goals are broad in nature, reflecting its mission.

1. Provide a forum for the timely distribution of technical and topical information from public and private sources.
2. Foster communication between public and private interests to avoid duplication of efforts and capitalize on opportunities.
3. Recognize and work with the diverse interests in the Blackfoot Valley to avoid confrontation.
4. Examine the cumulative effects of land management decisions and promote actions that will lessen the adverse impacts of land management in the Blackfoot valley.
5. Provide a forum on public and private resources to resolve issues.

CRITIQUE

Perhaps the biggest problem for restoration work is not spending enough time and effort on landowner education and relations. The key to restoration efforts lies in good relationships with the stewards of the Blackfoot River and its tributaries. More time should be devoted to helping landowners understand why restoration efforts are important and listening to their concerns so that those personnel implementing restoration efforts have a better understanding of the problems faced in the daily management of farm and ranch operations. Efforts expended in this arena will pay off in terms of improved landowner trust and cooperation. It is an extremely difficult issue on which to make progress, and there are few tangible milestones or goals.

Restoration project planning and execution are handled primarily by the state, federal, and local agency professionals involved in a particular project, individual landowners, and private contractors specializing in restoration work. Communication among all parties is sometimes not as good as it should be, and miscommunication or lack of communication can result in a less-than-ideal long-term relationship or project outcome. Facilitating communication should receive more attention.

The restoration effort continues to forge ahead with new projects, but there is also a need to examine the results thus far in aggregate to determine if original goals are actually being accomplished and, perhaps, to reevaluate those goals. Individual project monitoring is taking place, and the progress of individual projects needs to be evaluated in relation to the desired overall condition of the watershed. The first in-depth progress evaluation was made available in 1997 (Pierce et al. 1997).

A management plan for the river is in progress but has not yet been completed. It is a cooperative effort involving many agencies, landowners, outfitters, business owners, recreationists, and the general public. The MDFWP serves as the lead in developing the management plan.

The restoration effort has been a very active program, targeted at restoring specific streams. Major improvements have resulted from projects completed on over a dozen streams, and more projects are in progress. Results have been excellent: there have been improvements in fish densities, species composition and distribution, and average size. Removal of barriers to fish passage and streamflow improvements have opened entire watersheds to repopulation by species that depend on fish populations and the riparian corridor for their survival. The cooperative, nonconfrontational efforts made have demonstrated that we can improve the health of a riparian system with benefits to all concerned. Some project areas are now used by species of special concern, including the olive-sided flycatcher, bald eagle, northern goshawk, black tern, and Howell's gumweed. Other species of special concern known to be in the area but not specifically sighted on project areas include the white-faced ibis, harlequin duck, ferruginous hawk, Columbian sharp-tailed grouse, loggerhead shrike, grizzly bear, fisher, wolverine, Canadian lynx, and gray wolf.

Large increases in populations of bull trout and westslope cutthroat trout have yet to occur. Early reports indicated that these species will take longer to recover than nonnative species because of the initial lack of mature adult fish and these species' naturally lower reproductive rates. Their numbers appear to be increasing slowly, as was expected by the biologists involved in the restoration effort (D. Peters and R. Pierce, MDFWP, personal communication).

PROSPECTS FOR THE FUTURE

The fishery in the river has, in general, improved, although it has not rebounded to the levels described in Maclean's novel, Redford's movie, or many old-time residents' memories. Restoration efforts have demonstrated that river restoration does not necessarily have to be done at the expense of business enterprises such as agriculture and timber. They also demonstrate that if everyone involved will deal with problems in a forthright and cooperative manner, solutions can be developed in which everyone benefits.

The landowners who have participated in these restoration efforts deserve a hearty thanks. All of the public relations, grant writing, conflict resolution, funding efforts, and actual project work are of little use without the continued support and involvement of the landowners. It is their long-term stewardship and pride of ownership that provides the quality habitat necessary for fish and wildlife resources to survive and thrive.

As more is learned about watersheds, it becomes clear that in order to maintain a healthy river system, or restore an unhealthy one, management of upland areas as well as the riparian zone must improve. Uplands have a large influence on tributary condition and overall water quality. Projects on the uplands are now underway involving multiple landowners and rangeland management. These projects involve difficult problems and will take longer to show results but also have the potential for large gains. These efforts involve yet another agency, the local districts of the U.S. Natural Resources Conservation Service. Additional projects are being undertaken that will hopefully reconnect tributary systems that have been severed from the overall Blackfoot drainage for many years because of water diversions and severe habitat degradation.

Most restoration efforts undertaken to date have concentrated on the middle and lower sections of the river. These areas, primarily affected by agriculture, recreation, and timber production, were identified in the fishery inventory as high-priority sections for recovery of depressed populations of fishes and offered the best chance for genuine cooperative efforts. The problems caused by agriculture, recreation, and timber production are generally reparable damage, in the sense that they are relatively easily corrected via restoration of natural habitat, replacement of artificial structures with more ecologically sound structures, or simply enlightened management and natural succession. Unfortunately, problems in the upper section of the river are not so easily remedied. Past mining activity has had far-reaching effects in the upper basin, distributing toxic metals over wide areas. Cleanup of some known problem sites has begun by the principal parties responsible for the sites. Adding to existing problems, Phelps Dodge Corporation and Canyon Resources Corporation, operating jointly as the Seven-Up Pete Joint Venture, are proceeding with a proposal to open a huge open pit gold mine just east of the town of Lincoln. The mine would use cyanide heap leach extraction technology, and would be situated on top of a sole-source aquifer that feeds a major tributary of the Blackfoot. The project is surrounded on two sides by tributaries and on a third side by the main stem of the river. The mining operation involves massive manipulation of the aquifer and includes leach pads sited within 3,000 feet of the river and tailings piles within 1,500 feet of the river. Mining operations in Montana have consistently underestimated the severity of problems that will be encountered and the difficulty of solving them and have a history of abandoning operations and leaving the public to deal with the mess left behind. Every cyanide heap leach mine ever built in the state has leaked, and the impacts from such operations are long lasting and extremely expensive to clean up.

The proposed mine has the potential to reverse the gains already achieved in restoration efforts. It has caused a huge public outcry on a local, state, and national level. However, at least some segments of our society seem to have forgotten that it is much cheaper in the long run to protect our resources than to resurrect and attempt to restore them. Over the long term, the monetary value of the gold extracted will be incomparable to the monetary, psychological, and social value of the Blackfoot River as a free-flowing, unpolluted, vibrant natural resource to the people and the economy of the Blackfoot watershed, the state of Montana, and the entire country. The mine is being justified on a short-term economic basis, with operation peaking in seven years and then declining. In contrast, the river ecosystem and the communities in the Blackfoot Valley operate on a long-term basis.

The mining issue is a dark cloud hanging over the Blackfoot River and its valley. How this issue is dealt with will probably determine the character of the region for years to come. The residents of the watershed have exhibited a tremendous desire to maintain their rural and sustainable-natural-resources-based lifestyles. Many of them are exceptional land stewards. We look toward the future with some apprehension but also with a tenacity, energy, and dedication that will hopefully see the river peacefully recovering and steadily improving in the year 2000.

ACKNOWLEDGMENTS

Most of the data for this chapter were gleaned from Peters and Spoon (1989), Peters (1990), and Pierce (1991), the reports produced by the MDFWP as part of the restoration effort. Special thanks to Don Peters and Ron Pierce for their patience in explaining to me some of the problems they have encountered as well as the results of their studies and the implications of those studies. The watershed map was generated by Ron Pierce and Meggan Laxalt and modified to show the location of the proposed mine. The restoration efforts described in this chapter are the result of cooperative efforts and funding from many organizations and individuals. My thanks to board members of the Big Blackfoot Chapter of Trout Unlimited for their time and effort in educating me about past activities and problems; a special thanks to Betty duPont, Becky Garland, Mark Gerlach, Hank Goetz, Land Lindbergh, and Jim Masar. Thanks also to Jack Thomas of Hydrotech Water Resource Consultants for information concerning grant applications and past activities and procedures. Finally, a heartfelt thanks to the landowners, organizations, corporations, and agencies who have put their best foot forward and cooperated to make this restoration effort possible.

Appendix 23.1: Prioritization Criteria for Improvement Projects

Stream Name (or river reach): _____

Is the stream or river reach listed as fishery or habitat impaired in an existing report?

Summary Scores	**Maximum points**	**Score**
Severity and extent of the impact to fishery or habitat	100	_____
Anticipated benefit	40	_____
Economic and technical feasibility	60	_____

I. Severity and Extent of the Fishery or Habitat Impairment

Type of water body:	Perennial	6	_____
	Intermediate or ephemeral	2	_____
Impaired uses:	Fisheries (general condition)	5	_____
	Spawning habitat	5	_____
	Rearing habitat	5	_____
	Cover for refuge	5	_____
	Value to species of special concern	5	_____

Degree of impairment: Presence x Severity = Degree
                                          of problem   (0–3)

Toxics (e.g., metals)	5	x _____ = _____
Dewatering	5	x _____ = _____
Sediment	4	x _____ = _____
Habitat alteration	4	x _____ = _____
Temperature	3	x _____ = _____
Nutrients	2	x _____ = _____

II. Anticipated Public Benefit

Habitat or fishery problem in the watershed is a common or widespread problem, and results of this project are likely to have a high demonstration, technology transfer, or education value to other parts of the watershed

(0–10): _____

Project improves particularly sensitive and ecologically significant stream in watershed with potential for exceptional fishery or habitat (0–10): _____

The general public has use of and access to the proposed project area

(0–10): _____

Project addresses the problem by proposing a streamwide planning approach (to the maximum extent possible), incorporating off-stream land uses and management in the overall project area (0–10): _____

III. Economic and Technical Feasibility

Problems in the stream can be addressed and are likely to be corrected with proposed remedies (likelihood of producing demonstrable fishery or habitat improvement) (0–15): _____

Local sponsors, including the Big Blackfoot Chapter, have the ability and interest to fund and administer an improvement project effectively, including the capability to plan and coordinate project activities and to monitor success in meeting project objectives (0–15): _____

Local landowners are willing and financially able to proceed with the project. There is good commitment on the landowners' part to invest their own resources in the project (0–15): _____

Necessary coordination with local, state, and federal agencies and other entities has been done during the initial assessment and will likely continue through the life of the project (0–15): _____

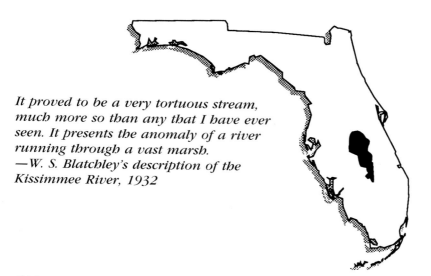

It proved to be a very tortuous stream, much more so than any that I have ever seen. It presents the anomaly of a river running through a vast marsh.
—W. S. Blatchley's description of the Kissimmee River, 1932

CHAPTER 24

HEADWATER RESTORATION AND REESTABLISHMENT OF NATURAL FLOW REGIMES: KISSIMMEE RIVER OF FLORIDA

Louis A. Toth, D. Albrey Arrington, and Glenn Begue

The Kissimmee River basin was once part of a vast, contiguous wetland system that extended from central Florida south to Florida Bay (Figure 24.1). The northern portion of the Kissimmee River basin formed the headwaters of this wetland landscape and included a 1,633 square-mile watershed with 26 lakes that ranged in size from a few acres to 55.5 square miles. The headwater lakes were linked by interconnecting sloughs (marsh that commonly transports overland water flow) and flanking wetlands and, during wet-season months, by high water stages (Parker 1955). The Kissimmee River originated at the southern end of Lake Kissimmee and provided the principal outlet for the upper basin watershed. The river meandered approximately 103 miles through a 1- to 2-mile-wide floodplain before emptying into Lake Okeechobee, the second largest (668 square miles) lake in the conterminous United States. The historical Everglades originated as overflow from the southern end of Lake Okeechobee and formed a 60-mile-wide, 3,800-square-mile "river of grass" that dominated the south Florida landscape. The river, lakes, and wetlands of the Kissimmee–Okeechobee–Everglades basin were a haven for fish and wildlife, including large flocks of wading birds, overwintering waterfowl, and a nationally recognized centrarchid-based sport fishery. (USFWS 1959; Myers and Ewel 1990; Toth 1993; Davis and Ogden 1994b; Florida Game and Fresh Water Fish Commission, unpublished data).

During the last half century the physical configuration, hydrology, and ecology of the Kissimmee–Okeechobee–Everglades system were greatly altered (Betz 1984; Kushlan 1990) by the Central and Southern Florida Flood Control Project (U.S. House 1949). The system is now compartmentalized with a network of canals, levees, and water control structures used to manage water levels and flow within and among each basin (Light and Dineen 1994; Toth and Aumen 1994). Creation and operation of this flood control system altered hydrologic regimes and drained a large portion of the area's historical wetlands. The flood control project resulted in a wide range of localized and regional impacts on fish and wildlife resources and ecosystem structure and function (Kushlan 1990; Toth 1990a, 1993; Davis and Ogden 1994b; Aumen 1995).

The Kissimmee basin portion of the flood control project (Figure 24.2) was constructed between 1962 and 1971. The headwater lakes were connected by canals and divided into a series of water storage reservoirs with dam-like water control structures used to regulate lake stages (levels) according to flood control schedules and operation rules. Lake levels are maintained at or below the prescribed regulation schedule elevations (Figure 24.3), which are held low during the high-flood-risk tropical storm season, allowed to rise to a maximum of 52.5 feet during late fall and winter, and lowered during spring. The spring decline approximates historic (preregulation), dry-season drawdowns and lowers lake levels in preparation for the rainy season. Water is discharged from the headwater lakes only during this spring drawdown period and when rainfall causes lake stages to exceed the regulation schedule.

Although flood control schedules for the upper basin lakes were formulated with some provisions for conserving fish and wildlife resources (Florida Game and Fresh Water Fish Commission, unpublished), wetland habitats and their associated values diminished as a result of a reduction in the range of water level fluctuation that had occurred prior to regulation (Figure 24.4). Lower water levels eliminated the outer fringe of littoral (shoreline) wetlands and facilitated drainage of adjacent upland marshes (Williams 1990). Reduced water level fluctuations adversely affected productivity and quality of remaining littoral wetlands by providing favorable conditions for rapid accumulation of unconsolidated organic sediments (Wegener and Williams 1975). In addition, littoral habitat availability for fish was blocked by organic sediment berms that developed around the lakes at elevations corresponding to the consistent low annual stages prescribed by the regulation schedules (Williams 1990; Moyer et al. 1995).

The Kissimmee River and floodplain were channelized to provide an outlet canal (C-38) for draining floodwaters from the headwater basin (Figure 24.5). Water control structures divide the channelized river into a series of pools with stepped water surface profiles (Figure 24.2). Excavation of the canal and deposition of spoil obliterated 35 miles of river channel and 7,000 acres of

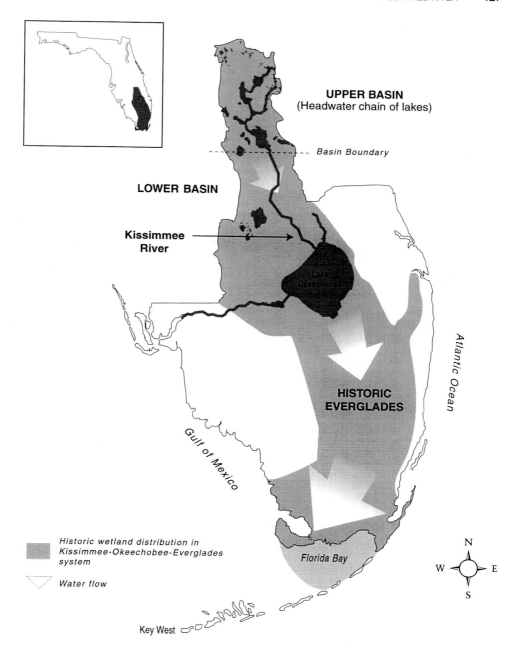

FIGURE 24.1.—Historical (approximately 1900) distribution of wetlands in the Kissimmee–Okeechobee–Everglades system in south-central Florida.

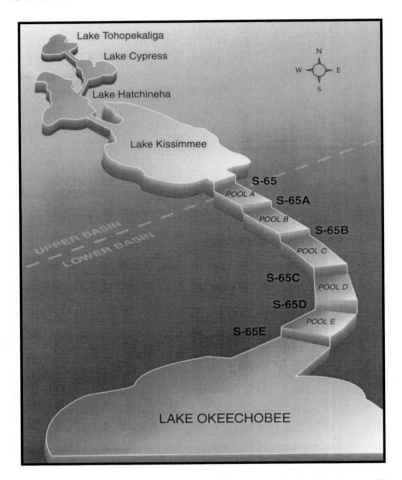

FIGURE 24.2.—Schematic of flood control modifications of the Kissimmee River basin, including canals connecting the lower group of headwater lakes, water control structures (S-65, S-65-A–E), and steplike impoundments (Pools A–E) along the channelized Kissimmee River.

floodplain wetlands. Transformation of the river–floodplain ecosystem into a series of impoundments lowered and stabilized water levels, which, in turn, drained approximately 30,000 acres, or two-thirds, of the historic floodplain wetlands (Pruitt and Gatewood 1976). Most physical, chemical, biological, and functional attributes of the river ecosystem were altered, if not destroyed (Toth 1990a, 1993). Waterfowl populations declined by 92%, wading birds were largely eliminated, and game fish resources continue to decline (Perrin et al. 1982; Miller 1990; FGFWFC 1994). Channelization of the Kissimmee River is particularly significant because the river's historical hydrologic characteristics, including frequency of overbank flows and prolonged floodplain inundation, were unique among North American riverine systems (Toth 1990b, 1994; Toth et al. 1995).

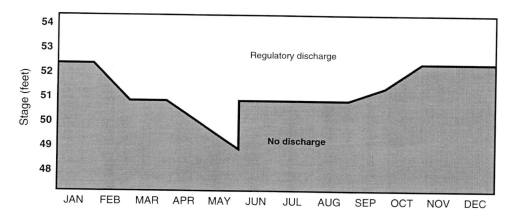

FIGURE 24.3.—Existing flood control regulation schedule for the lower group of headwater lakes (Kissimmee, Hatchineha, and Cypress) in the Kissimmee River basin. Water levels (stages) in these lakes are maintained at or below prescribed regulation schedule elevations through discharges to the channelized river. No releases are made when lake stages are less than the flood control schedule. Stages are measured relative to the National Geodetic Vertical Datum (established in 1929).

The extensive spatial and temporal hydrologic connectivity between the river and floodplain was key to the functional integrity of the ecosystem and was particularly important in affording fish and wildlife with continuously available feeding, reproduction, and nursery habitat.

The broad array of environmental impacts resulting from channelization of the river provided impetus for a restoration initiative, which eventually led to a comprehensive restoration plan for the entire river–floodplain ecosystem (Toth 1994; Toth and Aumen 1994; Koebel 1995). Restoration of flows from the Kissimmee's headwater basin lakes is a critical element of this plan. This chapter describes why a watershed perspective is needed to restore the channelized Kissimmee river–floodplain ecosystem and how restoration will be accomplished.

FOUNDATIONS OF THE RESTORATION PLAN

The Kissimmee River restoration movement began as a grassroots initiative during the latter stages of channelization. Murky, sediment-laden water, barren spoil banks deposited on former wetlands, an obliterated river channel, and a wide canal (C-38) through the valley provided highly visible evidence of environmental devastation of an unanticipated magnitude. As the backlash over aesthetic impacts subsided, concern for and documentation of losses of ecological resources mounted, and the restoration movement became increasingly organized and politically active (Woody 1993). Environmental conservation organizations such as the National Wildlife Federation, Audubon Society, and Sierra Club championed the restoration cause, which received consistent support from the state's governors, legislature, and congressional delegations.

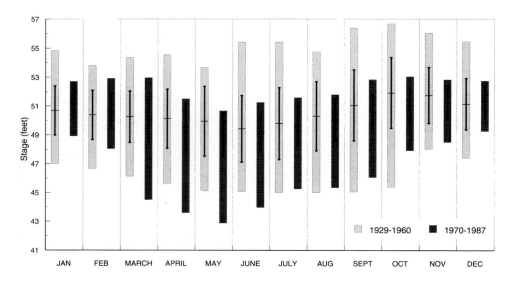

FIGURE 24.4.—Mean monthly stages (horizontal line on bars) in Lake Kissimmee during preregulation (1929 to 1960) and postregulation (1970 to 1987) periods. Shaded bars represent maximum and minimum range of monthly stages. Dark vertical bars show means plus or minus one standard deviation.

Although analyses of restoration alternatives began in the early 1970s, the restoration planning and evaluation process became impeded by a history of somewhat vague and subjective environmental objectives, which were viewed and assessed independently. An initial focus was perceived downstream impacts on water quality in Lake Okeechobee and measures to reestablish nutrient filtration and absorption processes that were lost with drainage of the river's floodplain wetlands (Florida Department of Administration 1976; McCaffrey et al. 1977; Davis 1981). Subsequently, the scope of evaluations of potential restoration measures ranged from enhancing the largemouth bass fishery and endangered species populations to increasing wading bird and waterfowl use of the floodplain, reestablishing floodplain wetlands, and "returning the river to its natural state" (USACOE 1985). By the late 1970s, dechannelization was the favored, albeit generic, restoration objective and "filling in the ditch" became a rallying cry of river restoration activists.

At this point in the restoration initiative, the focus on dechannelization did not consider how continued regulation of the headwater lakes might affect the potential for restoring ecological resources of the river and floodplain. For example, the lowering of lake stages and associated drainage of the upper basin's littoral wetlands may have affected regional populations and recovery potential of wading birds and waterfowl throughout the landscape, including the Kissimmee River floodplain. Moreover, in the early 1980s evidence began to accumulate suggesting that postregulation inflows from the headwater lakes to the channel-

FIGURE 24.5.—Aerial photograph of channelized Kissimmee River (C-38) at its origin at the outlet of Lake Kissimmee. The water control structure (S-65) is used to regulate stages in Lake Kissimmee and discharges to C-38. Remnants of the meandering river channel occur on both sides of the canal (foreground).

ized river would limit restoration of the river's floodplain wetlands (USACOE 1985).

Although the total volume of water discharged from the headwater lakes (approximately 58% of average annual river inflows) has not changed since channelization, regulation of lake stages for flood control has greatly altered discharge regimes, particularly timing and regularity of outflows. Prior to regulation, the river received continuous inflows from the headwater lakes. Lowest discharges typically occurred during the dry season (December to May) and steadily increased to a peak at the end of the wet season (November; Figure 24.6a). Since regulation, natural seasonality of high- and low-flow periods has been reversed, discharges are intermittent, and there are extended periods during each year without inflows from the headwater lakes. Low or no discharge is common during most wet-season months (June to November), whereas highest annual flows often occur during dry-season months (particularly February to May), when stages in the lakes are lowered to provide storage capacity for flood control during the hurricane season (Figure 24.6b).

These changes to historical flow regimes greatly limit potential for restoring biological resources and functions in the river and its floodplain. Extended postregulation periods of no inflow from the headwater lakes would keep most of the river floodplain dry during the wet-season months even if the river were not channelized. Hypothetical stage hydrographs (charts of water levels over time) and associated floodplain inundation frequencies were generated based upon an 18-year period of postregulation daily discharge records from the headwater lakes, historic relationships between Lake Kissimmee discharges and river stages at the Fort Kissimmee gauging station (river mile 32) (Toth et al. 1995), and elevation data for the floodplain adjacent to Fort Kissimmee. These hypothetical hydrographs indicated postregulation inflows would have kept the entire floodplain dry more than 70% of the time during the wet season, whereas approximately 80% of the floodplain was inundated more than 70% of the time during these months prior to implementation of flood control regulation (Figure 24.7). Even during particularly wet years, discharges would be predominantly intermittent and lead to only short-term floodplain inundation followed by rapid drainage, which generally would preclude utilization of the floodplain by fishes, wading birds, and waterfowl. Even without channelization, postregulation discharges would have reduced hydrologic connectivity between the river channel and floodplain, thereby largely eliminating nutrient and sediment filtration processes once provided by floodplain wetlands. The seasonal shift of high-flow periods from summer–fall to late winter–spring months likely would have interfered with fish reproduction and recruitment, because the spawning period and reproductive habits of Kissimmee River game fish species (McLane 1955) make their eggs and young vulnerable to high flows during spring months.

Perhaps the most significant step in the development of the Kissimmee River restoration plan occurred in 1988 with adoption of an explicit goal of "reestablishing a river–floodplain ecosystem that is capable of supporting and maintaining a balanced, integrated, adaptive community of organisms having a species composition, diversity, and functional organization comparable to that of the natural habitat of the region" (Toth 1990b). This goal of ecological integrity (e.g., Karr and Dudley 1981) changed the perspective of the restoration planning process from a focus on independent environmental objectives to a comprehensive view of all structural and functional attributes of the ecosystem. The scope

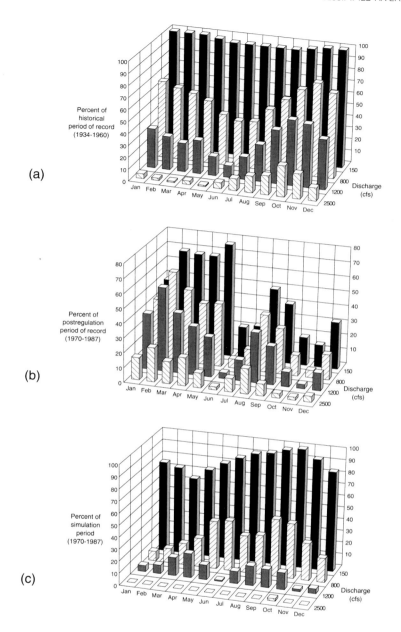

FIGURE 24.6.—Lake Kissimmee discharge characteristics during (**a**) preregulation (1934 to 1960) and (**b**) postregulation (1970 to 1987) periods and (**c**) for a simulated postregulation period based upon a new regulation schedule and operation rules. Graphs show percent of each period that monthly discharges equaled or exceeded the volumes given along z-axis. For example, during the preregulation period (**a**), December discharges exceeded 150 cubic feet per second (cfs) 100% of the time but were greater than 2,500 cfs only 11% of the time.

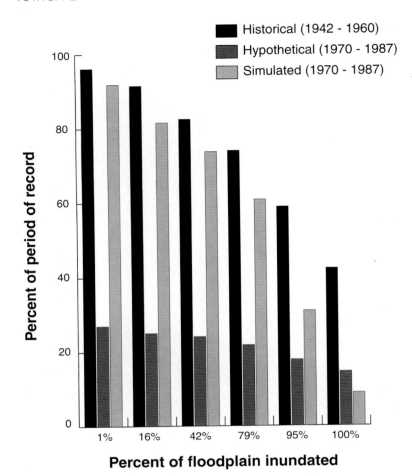

FIGURE 24.7.—Prechannelization floodplain inundation frequencies compared with hypothetical floodplain inundation that would have resulted from postregulation inflows through an unchannelized river and simulated floodplain inundation resulting from a new regulation schedule. Graphs show percent of each period that river stages exceeded the range of floodplain elevations adjacent to Fort Kissimmee (river mile 32) during the wet season (June–November). Floodplain elevations are represented as proportions of the floodplain inundated.

of the ecological integrity goal and the limitations imposed by regulated discharge regimes from the headwater lakes established that dechannelization alone was insufficient to accomplish ecological restoration of the river–floodplain ecosystem. Modifications of the flood control schedule and operation rules were necessary to address impacts of lake regulation. Although the primary objective of this component of restoration planning was to reestablish discharge regimes necessary to restore the ecological integrity of the river, the goal was balanced by a desire to alleviate simultaneously environmental impacts of the flood control project within the headwater lakes.

MODIFICATION OF MANAGEMENT OF THE HEADWATER LAKES

Analyses of potential modifications to the regulation schedule and operation rules for the headwater lakes were based on the assumption that environmental restoration objectives for the river, floodplain, and lakes will require reestablishment of historical hydrologic characteristics. Restoration criteria for the river-floodplain ecosystem include reestablishment of continuous inflows from the headwater lakes, particularly during July to October, natural seasonality of high- and low-inflow regimes, and a wide range of discharge variability (Toth 1994). These characteristics are necessary to restore floodplain wetland structure and function, minimize disruptive flows during fish reproductive periods, and reestablish spatial and temporal components of river habitat heterogeneity, including diverse channel morphology and substrate characteristics. The wide range of discharge variability will provide for a continuously shifting range of hydrologic conditions needed to maintain biodiversity (persistent populations of natural complement of flora and fauna) in wetland ecosystems (Zedler 1988). The primary criterion for enhancing environmental resources in the headwater lakes is increasing the range and temporal dynamics of water level fluctuations, particularly maximum extent and duration of high stages. Achievement of this criterion would lead to an expansion of flanking wetlands and increased quality and productivity of littoral habitat within the lakes.

However, potential for modifying the regulation schedule and operation rules to replicate historical stage and discharge characteristics fully is limited by flood control and navigation constraints. Although the restoration initiative provided a mechanism for at least partially accommodating flood control needs through land acquisition (Jones and Malone 1990; Horton and Meierer 1992), some level of flood protection is required within the headwater drainage basin. Need for flood control limits maximum lake stages and is particularly constraining during the high-risk tropical storm season from June to September. Navigation constraints stem from heavy recreational boating in the lakes, primarily for sportfishing. Recreational boating has increased substantially since the flood control project was completed, and most supporting infrastructure, such as boat ramps, docks and marinas, and particularly associated access channels to the lakes, have been constructed based upon the constricted postregulation range of water level fluctuations. Consequently, the ability to provide continuous inflows to the river by allowing lake stages to fall to historic low levels is limited by socioeconomic concerns associated with the potential for reducing recreational boating. This navigation constraint is compounded by a postregulation invasion of the nuisance exotic plant *Hydrilla verticillata*, which severely affects navigation in the headwater lakes (Williams 1990) and is an increasing impediment to boating as lake stages decline.

Within constraints imposed by flood control and navigation requirements, which establish upper and lower limits for lake stages, alternative operation rules were devised to allow lake stages to fluctuate more naturally with rainfall and associated inflows from the watershed and to reestablish outflow regimes that approximate historic (preregulation) stage–discharge relationships for the head-water lakes. Fluctuations caused by rainfall would lead to higher lake stages and increased water storage and thereby accommodate maintenance of continuous discharge from the headwater lakes. Preliminary analyses (Loftin et al. 1990a; Toth 1993) demonstrated that the flood control regulation schedule and opera-tion rules had to be modified for only the lower group of headwater lakes (i.e., Kissimmee, Hatchineha, and Cypress) to provide discharge regimes necessary to accomplish the river restoration goal.

Twenty-one alternative regulation schedules and operation schemes for lakes Kissimmee, Hatchineha, and Cypress were evaluated using simulation modeling and input data from an 18-year, postchannelization (1970 to 1987) period. The model (Fan 1986) used measured rainfall, average monthly temperature, and average solar radiation data to generate daily lake stages and flow for the simulation period. The model was calibrated for stage but appeared to underes-timate actual discharge during the simulation period by approximately 20%. During the 18-year period used for model simulations, the headwater basin received approximately 10% less rainfall, and, as a result, contributed 40% less average annual runoff, than during the preregulation years (Obeysekera and Loftin 1990).

Environmental analyses of alternatives focused on comparisons of their simulated hydrologic performance with established restoration criteria. Potential environmental effects in the headwater lakes were based on comparisons of observed and simulated stage duration data for the postchannelization period. Although maximum hydrologic restoration would be achieved by reestablishing lake stage frequencies similar to the preregulation period, flood control and navigation constraints limited the range of fluctuation in lakes Kissimmee, Hatchineha, and Cypress between 48.5 and 54 feet. During the 1929 to 1960 period of record, Lake Kissimmee stages exceeded 54 feet approximately 7% of the time and were less than 48.5 feet approximately 19% of the time. Because frequencies of lake stages between 49.6 and 50.8 feet were similar for postregu-lation and historical periods, and the primary environmental objective for the headwater lakes is to reestablish littoral wetlands that were drained by lowered stages, analyses focused on the degree to which alternatives increased frequen-cies of lake stages between 50.8 and 54 feet.

Potential environmental effects in the river–floodplain ecosystem were based on comparisons of simulated discharge regimes with prechannelization flow characteristics. Because the simulation period was drier than the prechanneliza-tion period and model outputs underestimated total volumes of flow, analyses of alternatives focused on reestablishment of continuous outflow and natural seasonal distributions of high- and low-discharge periods.

Initial screening of hydrologic performance of each alternative indicated that 13 options produced hydrologic regimes that met minimum requirements of restoration criteria for the headwater lakes and river–floodplain ecosystem. One alternative outperformed others by producing lower frequencies of no-flow periods, redistributing average- and high-flow volumes in a more natural seasonal pattern, and providing the greatest increase in frequencies of lake stages greater than 50.8 feet. This alternative allows lake stages to rise 1.5 feet higher than the existing schedule and has four discharge zones, that will provide for continuous outflows which vary with lake stages within the flood control and navigation constraints (Figure 24.8). Simulations indicated that the new schedule will lead to higher lake stages approximately 30% of the time and will increase inundation frequencies of lake elevations greater than 50.8 feet by 5% (Figure 24.9). Even during the relatively dry simulation period, this schedule provided continuous flows at least 94% of the time from May to February (100% of the time during October to December), 81% of the time during March, and 87% of the time in April. During at least 75% of the simulation period, discharges were highest during May to December and declined to 150 to 400 cubic feet per second (cfs) during dry-season months (Figure 24.6c). During the 10% of the simulation period when winter–spring rainfall was above normal, the new schedule provided January to April discharges (1,000–1,400 cfs) comparable to flows during the wet season. Although the new schedule appears to provide lower frequencies of high flows (e.g., discharges greater than 2,500 cfs) compared with historical discharge characteristics, a return of wetter climatological conditions would likely lead to higher discharges.

RESTORATION OF ECOLOGICAL INTEGRITY

Simulated stage frequencies indicate that implementation of this new regulation schedule will lead to reestablishment of approximately 5,900 acres of littoral wetlands around Lakes Kissimmee, Hatchineha, and Cypress. Since stages have been regulated for flood control, peripheral elevation boundaries of littoral wetlands around these lakes have shifted from 54 or 55 feet (Florida Game and Fresh Water Fish Commission, unpublished data) to 52 feet (USFWS 1994), a loss of approximately 36,500 acres of adjacent marsh. Based upon multiyear inundation frequencies associated with both historic and existing wetland boundaries, we expect littoral wetlands to reestablish at elevations exposed to inundation frequencies greater than 10%, which simulations indicate will occur between 52 and 52.5 feet (Figure 24.9). Habitat suitability modeling (USFWS 1994) indicates these changes in the flood control schedule and operation rules will lead to significant increases in habitat availability for mottled duck, ring-necked duck, great egret, snowy egret, the endangered snail kite, and the endangered wood stork.

Because lake stages will be more closely linked to climatic conditions, water level fluctuations will be more variable. Simulated variance of average monthly Lake Kissimmee levels resulting from the new regulation schedule was significantly higher than the variability that occurred during 1970 to 1987 with the existing schedule. Increased stage variability should reduce the rate at which

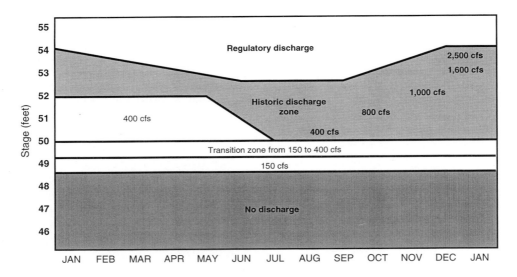

FIGURE 24.8.—New regulation schedule and operation rules for lakes Kissimmee, Hatch-ineha, and Cypress. This "zoned" schedule provides discharges to the lower river basin that vary according to lake stages. Within the historic discharge zone, flows will vary linearly between given values (e.g., between 400 and 800 cfs when lake stages are between 50 and 51 feet). Stages are measured relative to the National Geodetic Vertical Datum (established in 1929).

sediment berms develop and thereby should increase accessibility of the outer fringe of littoral habitats to fishes.

Although restoration of ecological integrity of the river–floodplain ecosystem also requires dechannelization (Table 24.1), reestablishment of several key structural and functional attributes is dependent upon recreating historical inflow characteristics from the headwater lakes. Reestablishment of continuous flow through 43 miles of reconnected river channel will lead to improved dissolved oxygen regimes, which are currently a major limiting factor and cause of fish mortality during summer and fall months (Wullschleger et al. 1990; FGFWFC 1994). In addition to providing seasonal flows conducive to successful fish reproduction, restored discharge regimes will provide hydrodynamic pro-cesses that will lead to restoration and maintenance of spatial and temporal aspects of river channel habitat heterogeneity, including diverse substrate and channel morphology (Toth 1991; Toth et al. 1993).

Restored discharge regimes also will reestablish lateral connectivity and associated physical, chemical, and biological interactions between the restored river channel and floodplain. These interactions include key functional attributes of ecosystem integrity, such as nutrient and sediment filtration processes, energy exchange (e.g., export of small fish, invertebrates, and organic matter from the floodplain to river), and use of the floodplain for feeding, reproductive, nursery, and refuge habitat by fishes, wading birds, and waterfowl. Restoration of structure and function of the river's floodplain wetlands requires reestablishment

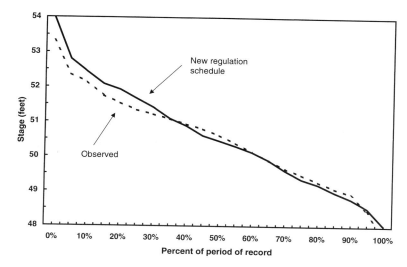

FIGURE 24.9.—Stage duration curves for existing (observed) and new (simulated) regulation schedule for Lake Kissimmee. Curves show percent of the period of record (1970 to 1987) for which recorded or simulated lake stages equaled or exceeded given elevations (stages). Stages are measured relative to the National Geodetic Vertical Datum (established in 1929).

of historical floodplain inundation characteristics (Toth 1994; Dahm et al. 1995). Assuming the river will be dechannelized, we used historic Lake Kissimmee discharge–Fort Kissimmee stage relationships to generate expected stage hydrographs and associated floodplain inundation from simulated daily discharges from the headwater lakes. Even during the relatively dry simulation period, the new regulation schedule and operation rules produced inundation frequencies comparable with the historical period over 80% of the floodplain (Figure 24.7).

TABLE 24.1.—Key features and expected ecological benefits of the Kissimmee River restoration plan.

Construction and operational components
Backfilling of 22 miles of canal
Recarving of 9 miles of obliterated river channel
Removal of two water control structures
Modifications of the flood control regulation schedule and operation rules for the Kissimmee's headwater lakes
Expected ecological benefits
Restoration of 40 square miles of river and floodplain ecosystem
Restoration of 43 miles of contiguous river channel
Restoration of 27,500 acres of floodplain wetlands and 5,900 acres of littoral wetlands
Restoration of habitat for approximately 320 fish and wildlife species including the endangered wood stork, bald eagle, and snail kite

FINE-TUNING THE RESTORATION PLAN:
AN EXPANDED VISION

In addition to restoring hydrologic regimes that will effectively accomplish integrated environmental restoration objectives for both the headwater lakes and river–floodplain ecosystem, the new regulation schedule will provide operational flexibility for fine-tuning the restoration plan and adaptive management of the recovering and restored ecosystem. Adaptive restoration planning and implementation requires a commitment to a well-organized evaluation program for monitoring system responses (Walters et al. 1992). Following implementation, the four discharge zones and stage–discharge rules for the new regulation schedule will be continuously reevaluated. Potential modifications will be based upon future hydrologic performance and results of the project's comprehensive restoration evaluation program (Toth 1993; Koebel 1995). Given uncertainties of the simulations, which result from underestimated discharge and the relatively dry simulation period, it may be possible to modify the schedule to maintain higher lakes stages for longer durations and still provide necessary river inflows to achieve restoration of the river–floodplain ecosystem. Alternatively, adjustments may be required to refine river discharge regimes, particularly high flows. At least periodic high flows are necessary to restore and maintain geomorphic (landform) structure and associated biological diversity in riverine systems (Brookes 1995).

Although the new regulation schedule is expected to lead to significant increases in wetland resources, flood control and navigation constraints preclude full restoration of ecological processes and fish and wildlife habitat values associated with the historical range of water level fluctuation in the headwater lakes. Because the revised schedule will not significantly affect annual low stages, management programs for the lakes may require continued use of periodic extreme drawdowns of lake water levels to simulate droughts. Periodic lake drawdowns revitalize littoral wetlands by consolidating accumulated sediments and providing for germination of desirable plant species (Holcomb and Wegener 1972) and facilitate control of exotic and nuisance plants, which compete with native species and adversely affect navigation. Although extreme drawdowns could potentially interfere with river restoration needs (discharge regimes), conflicts can be minimized by executing drawdowns during dry climatic periods. This will require flexibility in planning and implementation of these management measures, and means for mobilization will be needed to take advantage of natural, but unpredictable, climatic cycles.

With incorporation of the headwaters component, the scope of the Kissimmee River restoration plan embraces a watershed perspective that will be the paradigm for future planning and fine-tuning of the project. This broader restoration vision will focus next on the river's lateral tributary systems. Since the river was channelized, many tributary watersheds have been extensively ditched to provide drainage for more intensive agricultural uses, particularly improved pasture (Obeysekera and Loftin 1990). Increased drainage density (Gatewood and Bedient 1975), including new connections between previously isolated wetlands, combined with increased outlet capacity provided by construction of C-38 has changed the hydrologic response within lateral watersheds from upland retention with slow runoff to upland drainage and rapid runoff. As a result, inflow regimes to the channelized river have shifted from predominantly base flow runoff to surface (direct) runoff, with a corresponding increase in both the rate and magnitude of inputs from lateral tributary watersheds (Huber et al. 1976). Secondary drainage networks also have eliminated over 300,000 acres of wetlands within the lower (below the headwater lakes) river basin (USACOE 1985).

The Kissimmee River restoration project's land acquisition program provides opportunity for rehabilitation of at least some of these watersheds. Where project acquisitions extend beyond the floodplain, reestablishment of the basin's wetlands is a foremost goal. Although complete elimination of agricultural drainage networks is not always feasible, more natural hydroperiods can be provided artificially with, for example, impoundments. Such measures would be particularly beneficial in tributary slough systems, where detention of surface water runoff would extend wetland corridors beyond the restored floodplain and contribute to reestablishment of more natural flow regimes within the restored river system. Improved water management and conservation systems also could benefit existing agricultural land uses by maintaining adequate water supply (e.g., groundwater levels) during drought and seasonal dry periods. Strategies integrating a mix of agricultural, protective, and compromise (sensu Odum 1969) environments, or watersheds, are required to manage water resources effectively on a landscape scale (Karr and Dudley 1981; Karr 1994).

The Kissimmee River restoration project is expected to have regional, landscape, national, and even international benefits. Reestablishment of nutrient filtration processes on 27,500 acres of restored floodplain will help alleviate effects of degraded water quality on fish and wildlife habitats in Lake Okeechobee (Aumen 1995). Positive effects on bird communities, particularly wading birds and waterfowl, should occur throughout the south-central Florida landscape as restoration of littoral and floodplain wetlands will provide additional breeding and feeding habitat. Because refuge will be available within the restored river ecosystem when habitat conditions in other portions of the Kissimmee–Okeechobee–Everglades system are not favorable, waterfowl and wading bird populations will be more buffered against natural and anthropogenic impacts, including those related to management of these regional ecosystems for conflicting water resource needs such as water supply and flood control. South Florida wetlands provide critical habitat for meeting goals of the international North American Waterfowl Management Plan (USFWS 1986). Similar benefits will accrue for other birds, including migratory species, that range throughout this landscape and use wetland habitats.

Dechannelization of the Kissimmee River began with a pilot fill project that was completed in August 1994 (Toth 1996). The reconstruction required for the headwater lakes component of the restoration project will be initiated in 1997. The restoration project is scheduled to be completed in 2010 (Koebel 1995).

Restoration of the Kissimmee River is an integral component of plans for rehabilitating the entire Kissimmee-Okeechobee-Everglades system. In addition to contributing to recovery of regional waterfowl and wading bird populations and the endangered wood stork and snail kite, the Kissimmee River restoration project will reestablish more natural inflow regimes to the downstream lake, wetlands, and estuaries. As in the Kissimmee, reestablishment of historical hydrologic characteristics is a key requisite for restoration of wetland habitats throughout the south Florida landscape.

ACKNOWLEDGMENTS

Paul Angermeier, Chris Carlson, Walt Courtenay, Jr., Bill Helferrich, Steve Lin, Ed Moyer, Sue Newman, Lawson Snyder, and Jack Williams provided many useful comments and suggestions on an earlier draft of this chapter. Lilia Ana Bazo, Jane Walters, and Barb Welch prepared most figures.

PART 5

A Vision for the Future

This final section attempts to review the watershed restoration practices that have been most successful in restoring watershed structure and function and provides options for future management. The first two chapters concentrate on describing the lessons that we have learned thus far from our experiences with restoration efforts. In Chapter 25, "Learning to Live Within the Limits of the Land: Lessons Learned from the Watershed Restoration Case Studies," the restoration efforts described earlier in this volume are reviewed to describe how successful collaborative relationships are established, how scientific principles are built into restoration programs, and how these efforts pay off in long-term changes in watershed health. It is this final measure, actual restoration of healthy, productive, and properly functioning watersheds in the long term, that success or failure of our restoration efforts will be determined. Chapter 26, "What Works, What Doesn't, and Why," combines the experiences of the resource conservation director of Trout Unlimited with the national fisheries program manager of the U.S. Forest Service in revealing why certain restoration efforts are successful and why others are not.

Environmental protection and restoration efforts are incorrectly seen by many people as costing jobs and economic vitality. As described in Chapter 27, "Saving and Creating Jobs Through Watershed Restoration," just the opposite is true. Strong rural economies often depend on a healthy, aesthetically appealing environment. Whether attracting new businesses or decommissioning forest road networks, watershed restoration programs can create jobs and benefit local economies. As described in this chapter, experiences from the Jobs in the Woods Program that created watershed restoration employment for displaced timber workers in the Pacific Northwest can serve as a model for rural areas throughout the country.

443

In the final chapter, "Eco-Societal Restoration: Creating a Harmonious Future Between Human Society and Natural Systems," one of the most concerned practitioners and prolific authors in the field of environmental restoration writes about society's options for the future. Whether we recognize it or not, by continually increasing human pressures on the environment through our burgeoning human populations and resource consumption, we are eliminating future options for our society. The author describes possible future options and what we have to do—very soon—to have a positive effect on the quality of life for members of the next generation.

We must begin to lessen the stress on our finite resources. Effective watershed restoration programs can play a major role. The choice is ours, but as we ponder our future, activities continue that forestall or even eliminate our options.

Watershed councils lay the cultural foundations for a lasting way of life. They establish the traditions of responsible speech, of civil democracy, and of making decisions based on factual information and well-articulated values. They embody the long-term perspective of sustainability, seeking not quick fixes but deeper understanding and new alternatives.
—Alan T. Durning, 1996

CHAPTER 25

LEARNING TO LIVE WITHIN THE LIMITS OF THE LAND: LESSONS FROM THE WATERSHED RESTORATION CASE STUDIES

Christopher A. Wood, Jack E. Williams, and Michael P. Dombeck

The relationship between human society and the ecological systems upon which we depend often seems combative. We extract food, fiber, and water from natural habitats in increasing amounts while seeking technological fixes to watershed damage, endangered species crises, and invasions of exotic species. Homes and businesses are built in floodplains only to be hit by 100-year flood events that seem to occur at increasing frequencies and with ever more devastating consequences: witness recent floods in Idaho and North Dakota and along the Mississippi River. Forests are dissected with roads, logged, and replanted until they become even-aged monocultural stands of timber that are, in turn, susceptible to catastrophic wildfire, erosion, and disease.

Our failure to consider the long-term consequences of how societal decisions can diminish the health, diversity, and productivity of the land has profound effects. A more harmonious relationship is needed for human society to prosper into the future (Cairns 1997, this volume). Watershed restoration provides a means to achieve this harmony. Effective restoration entails restoring structure and function to our natural surroundings by integrating ecological principles with a deeper understanding of ways that social and cultural mores influence how people use the land (Preister and Kent 1997, this volume). Restoration ecology is a broadening discipline that includes ecology, politics, economics, and sociology (Cairns and Heckman 1996). Our knowledge of the ecological principles that support restoration has advanced to the point that the Society for Ecological Restoration, formed in 1992, publishes a journal, *Restoration Ecology*, dedicated to the ecological science of restoration.

How, or whether, society will implement this ecological foundation is less clear. A colleague who reviewed one of the watershed restoration case studies for this book was surprised to learn that citizens were working together to restore their local watersheds. His thinking mirrors that of many national environmental groups and resource management agencies who often look askance at local coalitions, thinking members of these coalitions lack either the knowledge, willingness, or foresight to restore the health of their lands and waters. Conversely, local citizens often criticize federal and state agencies and national environmental groups for not involving and working with local communities on federal projects.

As the case studies in this book demonstrate, however, conservation and restoration are undergoing profound changes from an emphasis on top-down approaches inspired by federal legislation to watershed-based community coalitions motivated by residents' desires to improve their surroundings. These local efforts do not diminish the importance of federal mandates to clean our air and waters, preserve endangered species, or protect public lands. Rather, they amplify the effectiveness of mandates by making stakeholders of civic organizations, communities, and local conservation groups.

The community coalitions described in this volume are geographically diverse and include a broad array of people, perspectives, and problems. Some of the citizen coalitions, such as the Mattole Restoration Council (Zuckerman 1997, this volume) have been working together since the early 1980s; others were organized over the past few years. Several of the restoration efforts, such as that on Fish Creek in north-central Oregon (Reeves et al. 1997, this volume), were initiated by preeminent aquatic ecologists and hydrologists, others by local citizens and members of organizations such as Trout Unlimited. Each of the case studies depicts places where people realized, and acted on, the need to reconnect their social and economic cultures to the land that sustains them.

The case studies are emblematic of the social and scientific issues, problems, and opportunities that organizations and people face when undertaking watershed restorations. They provide a window through which to view the types of restoration work occurring in communities across America. By no means are they an exhaustive list, nor are they necessarily replicable. Each watershed presents its own set of unique challenges, obstacles, and opportunities. Despite their unique characteristics, existing restoration efforts provide important lessons to expand

and improve such work beyond the relatively small number of watersheds where restoration now is underway. Our purpose in this chapter is to aid in this expansion and improvement by summarizing the lessons learned from the case studies presented in this volume.

EDUCATION AND COMMUNICATION

Given adequate popular support and political will, society could probably "fix" what ails most watersheds. According to national surveys, 74% of polled Americans consider themselves either sympathetic to environmental concerns or active environmentalists (NEETF 1996). Unfortunately, this support has not yet translated itself into widespread action. Less than 1% of the federal budget is spent on natural resources management (Tilt and Williams 1997, this volume).

All of the case studies, be they on public, private, or lands of mixed ownership, emphasize the need for education and communication. For example, in the 85% privately owned Mattole River watershed of California, technical experts from Redwood National Park taught mapping techniques and aerial photo interpretation to local citizens (Zuckerman 1997). The collaboration resulted in a catalog of watershed erosion sources and appropriate prescriptions. Most importantly, the Mattole Restoration Council made the catalog available to additional private landowners in the watershed—prompting many to take corrective action before upslope problems percolated to downslope neighbors and fish and wildlife species.

Nolte (1997, this volume) describes how two high school teachers from a small town in Oregon worked with fisheries biologists from the U.S. Forest Service and U.S. Bureau of Land Management to develop the Crooked River Ecosystem Education Council. This council's actions have served the dual purpose of helping the watershed recover from years of degradation while increasing students' awareness through classroom lessons, field studies, work experience, and annual restoration projects. Students have also participated by monitoring results of previous projects. Lessons are reinforced through annual conservation fairs and other community events.

As the children of the Crooked River schools grow and take the places of their parents in operation of farms, ranches, and community businesses, with hope, they will maintain a more sustainable lifestyle and have a lighter touch on the landscape. As Cairns (1997) notes, if those who are affected do not support watershed restoration itself and the necessary maintenance thereafter, chances for long-term success greatly diminish.

The same message of communication, education, and the need to work with a broader spectrum of society holds for resource professionals, whether they are biologists, geologists, land managers, or technical specialists. None of us can afford to work in isolation. Most natural resource professionals selected their fields because of a deep appreciation for the outdoors, hunting, fishing, or wildlife viewing. Yet society's most pressing resource issues—population growth, overconsumption, endangered species, and pollution—are less technical than they are social.

It is often easier to rally popular support behind cleanup of an isolated environmental catastrophe (witness the billion-dollar cleanup of Prince William Sound after the *Exxon Valdez* oil spill) than it is to motivate people to address chronic, long-term degradation of a watershed. There is no more important task of the scientist or biologist than to ensure that average citizens understand the consequences of resource degradation and the benefits of watershed restoration. All of the money and technical expertise in the world cannot overcome disinterest in or, worse, distrust of restoration actions.

COLLABORATIVE STEWARDSHIP

Most successful watershed restoration projects share one critically important common feature: they bring together diverse groups of people—environmentalists, industry representatives, community leaders—under the common goal of restoring the health and productivity of their watersheds. These coalitions can be formal and structured or informal and loosely organized. What is most important is that they are diverse and balanced and create clear lines of communication among watershed users and restorers. For example, damage from a sediment spill that dumped 50,000 tons of sediment from Island Park Dam into the Henrys Fork ultimately provided the impetus for the Henry's Fork Foundation and the Fremont-Madison Irrigation District to create the Henry's Fork Watershed Council (Van Kirk and Griffin 1997, this volume). Both parties, long at odds over water use in the basin, believed the spill to be the result of poor communication among government agencies. The two former adversaries created a consensus-based council that (1) promotes cooperation in watershed planning, (2) critiques and prioritizes proposed watershed projects, (3) identifies funding for research and long-term monitoring, and (4) educates people about the importance of a healthy watershed.

Watershed coalitions and other stakeholder coalitions increase the effectiveness and promote citizen support of restoration efforts. As part of the Tennessee Valley Authority's Clean Water Initiative, 12 small teams of resource specialists were formed for each of the Tennessee River's large watersheds. The first of these River Action Teams was formed for the Hiwassee watershed in North Carolina, Georgia, and Tennessee (Bowling et al. 1997, this volume). The Hiwassee team coordinates with community leaders to promote a holistic approach to restoration by (1) balancing human use of the watershed with the need to restore resource integrity, (2) factoring in the needs of stakeholders in watershed management, and (3) prioritizing protection of high-quality or rare aquatic resources. The Hiwassee team agreed to a methodology that helps ensure that scarce funds are allocated where they will be most effective. The team's criteria are (1) protecting "high-value" resources, (2) solving problems that most threaten the sustainability of the watershed, and (3) targeting and emphasizing collaborative partnerships.

As Tilt and Williams (1997) point out, in an era of government downsizing, it is critical that less government spending does not translate to less conservation. Collaborative partnerships are essential to financing, promoting, and sustaining successful restoration efforts. As stated earlier, watershed restoration depends as much on social factors such as revenue, public interest, and stakeholder concern as it does on knowledge of disciplines such as hydrology, geomorphology, or ecology. Partnerships among state and federal agencies, conservationists, and industry help leverage funds and scarce technical skills, build ownership, and promote understanding among partners.

Other tools and techniques that help ensure participation include expanding restoration efforts from public lands onto the lands of willing and interested private landowners. In the Coquille watershed of western Oregon, 70% of the lands in the watershed are private, and approximately 1,000 landowners live within the basin (Hudson and Heikkila 1997, this volume). Private lands are the priority for restoration work in the watershed because they most often occur in valley bottoms where the best fish habitat historically occurred. The Coquille Watershed Association identifies opportunities on private lands that can build upon projects initiated on public lands and tries to secure matching funding for the Association's work from organizations such as the National Fish and Wildlife Foundation.

Perhaps of all the case studies described in this volume, restoration of the Anacostia watershed of the District of Columbia and Maryland is the most challenging. With a population density exceeding 4,700 people per square mile and some of the most polluted water in the country, the Anacostia deserves its ranking as one of the most degraded watersheds in the United States. Achieving broad community and governmental support for restoration efforts requires a good outreach effort, coordinated planning, and persistence. More than 60 citizen and government parties now participate in the Anacostia restoration effort. In the 10 years of active restoration on the Anacostia, more than 125 projects have been completed (Shepp and Cummins 1997, this volume). Development of an ecologically based six-point action plan to restore the Anacostia River watershed (MWCOG 1991) provided the vision to keep the coalition together and bring additional support and funding.

Regardless of whether the collaboration occurs through watershed coalitions, resource advisory councils (Dombeck et al. 1997, this volume), or habitat conservation agreements between regulatory and land management agencies, bringing together stakeholders early in the restoration process greatly increases the probability of success.

WATERSHED-SCALE ANALYSIS

One of the principle problems facing restoration practitioners is the matter of geographic scale. How can observed problems such as declining fisheries, diminished water quality, degraded riparian vegetation, or soil erosion be placed within the context of the entire watershed? Is the observed problem a primary causal factor in the overall declining health of a watershed, or is it merely a symptom of a larger difficulty? How does one go about measuring the health of a watershed?

The desire to restore a watershed can be prompted by declining fisheries, polluted streams, or threats to human health and safety. For example, in the fall of 1987, restoration of the Blackfoot River watershed in Montana was initiated by a group of anglers who were concerned with the decline of the legendary trout fishery (Aitken 1997, this volume). After a public debate over how best to proceed, several anglers proposed that they come up with the funds to conduct a fishery inventory, and the Montana Department of Fish, Wildlife, and Parks do the work. The anglers formed the Big Blackfoot Chapter of Trout Unlimited, and the restoration began. The fishery inventory led to the realization that more comprehensive studies on how to restore the watershed were needed. Those concerned with fishery declines began to look at broader land management problems throughout the watershed. The importance of taking a broader perspective was one of the principle lessons learned by the Mattole Restoration Council (MRC). At first, the MRC was concerned about declining runs of anadromous salmon and steelhead. As the MRC began to see the cumulative problems of roads and timber harvests that degraded streams which, in turn, supported the fish, the MRC broadened its thinking from fish to streams and finally to entire watersheds in order to address the problem of declining fish runs successfully (Zuckerman 1997).

Successful restoration requires a thorough knowledge of the observed problem and how this problem interacts with other processes in the watershed. Watershed analysis is one tool to understand problems on a watershed scale. Ziemer (1997, this volume) describes six steps for conducting such an analysis. This procedure is being followed by federal agencies in the Pacific Northwest in dealing with problems of endangered species and declining salmon runs, but it also is a logical thought process for anyone trying to understand problems observed in local watersheds. The process of watershed analysis is intended to assist land managers in viewing the seemingly confusing array of multiple projects and disturbances that operate at any given period of time. The Henry's Fork Watershed Council reviews each proposed project by its Watershed Integrity Review and Evaluation Criteria to help insure that the project is viewed within the context of watershed-scale processes and interactions (Van Kirk and Griffin 1997). These criteria could be useful to many watershed coalitions across the country.

For many restoration efforts it is important to place the observed problem in the context of the overall health or condition of the watershed. To understand the health of a watershed, it is important to identify some benchmark or relatively healthy watershed for comparison. Historical data on habitat condition can be used to determine how much degradation has occurred and to help define desired future conditions (Wissmar 1997, this volume). Using old maps, the MRC (Zuckerman 1997) was able to compare the extent of old-growth coniferous forests in the Mattole watershed in 1947 and 1988. Ninety-one percent of the softwood forest had been harvested since World War II. Restoration efforts are aimed at cutover slopes most likely to fail and send debris torrents into the Mattole and its principal tributaries. If historical data from the watershed are not available, comparisons can be made to relatively undisturbed habitats in the same geological and ecological region. Watersheds in wilderness areas, national park lands, areas isolated by extreme geography, or other relatively undisturbed areas can be useful for comparisons.

Although fish populations are often excellent indicators of riverine health, all watersheds possess specific characteristics that can be measured to determine the health of the entire watershed. As part of their Clean Water Initiative, the Tennessee Valley Authority has employed small teams of scientists to monitor the health of specific watersheds (Bowling et al. 1997). These River Action Teams assess watershed health by examining fish and macroinvertebrate community composition through the index of biotic integrity (Karr et al. 1986; Plafkin et al. 1989) and by examining an index based on the number of pollution-sensitive mayflies, stoneflies, and caddisflies (the EPT index of Bowling et al. 1997). Both techniques examine the diversity and abundance of pollution-sensitive aquatic species versus pollution-tolerant species. By comparing such attributes of aquatic communities from more degraded streams with the same attributes of aquatic communities from streams in relatively pristine areas, the relative health of streams can be determined. Further details and examples of these techniques are discussed in Angermeier (1997, this volume) and Lyons et al. (1996).

Once the need has been articulated and the problems have been understood in their broader watershed context, a socioeconomic assessment of the watershed can help build public support for restoration priorities while encouraging people to understand the economic values associated with healthy watersheds. For example, prior to initiating the Beaverkill–Willowemoc Watershed Initiative, Trout Unlimited conducted a socioeconomic assessment of the Beaverkill River and Willowemoc Creek watershed (Conyngham and McGurrin 1997, this volume). Among other things, the assessment (1) measured the economic impacts of the trout fishery on local businesses and the larger community; (2) measured expenditures and tax contributions of anglers; and (3) evaluated the influence of the trout fishery on the value of real estate. The resulting information instilled in residents a better understanding of the value of the fishery and helped community leaders to balance other forms of economic growth against the need to protect the river and its fishery.

RESTORATION AND ECOLOGICAL INTEGRITY

One of the critical first steps in any restoration effort is to decide on the overall goal of the project. What is the desired future condition of the habitats and watershed in question? As discussed by Angermeier (1997), watersheds can assume a wide variety of future conditions. Knowledge of historical conditions can provide a benchmark to determine the desired future status of a watershed's physical and biological elements. Historical flow regimes provided the information to determine restorative actions and desired future conditions for the headwaters of the Kissimmee River (Toth et al. 1997, this volume). Conditions in wilderness areas or other relatively undisturbed areas from similar ecological settings also can provide reference or benchmark conditions for restoration efforts (Flebbe and Dolloff 1995).

The most effective restoration efforts begin with all parties agreeing to one or more restoration goals. These should include a broad geographic-scale goal of reestablishing the ecological integrity of habitats (Karr and Dudley 1981). Toth (1990b) described such a goal for the Kissimmee River restoration: "reestablishing a river–floodplain ecosystem that is capable of supporting and maintaining a balanced, integrated, adaptive community of organisms having a species composition, diversity, and functional organization comparable to that of the natural habitat of the region." Even in highly degraded watersheds such as the Anacostia, substantial improvements in the ecological integrity of habitats can be achieved. Several goals of the six-point action plan to restore the Anacostia River are focused on restoring ecological integrity of wetlands, tributary streams, and forested uplands (MWCOG 1991; Shepp and Cummins 1997).

A common problem in urban and agricultural watersheds is accelerated runoff caused by an increase in the surface area of impermeable structures, such as asphalt and buildings, and loss of natural habitats. Wetlands and other natural habitats slow runoff and allow water to percolate slowly underground and flow downstream. In many urban environments, including the Anacostia watershed, wetlands have been filled and small tributaries have been channelized or encased in storm sewers. Roads, parking lots, roof gutters, and storm drains cover much of the urban watershed and contribute to accelerated runoff. As a result, rivers experience floods of greater-than-normal intensity, whereas summer flows can be a fraction of what they were under natural conditions. This effect was documented in a long-term study of Watts Branch, a headwater tributary near Rockville, Maryland (Leopold 1994). Urbanization of this small watershed began increasing substantially in the early 1950s. From 1958 to 1972 the average annual flood discharge was 781 cubic feet per second (cfs). From 1973 to 1987, increased building and road construction had increased the average annual flood discharge to 959 cfs, an increase of 23% (Leopold 1994). Average annual floods in Seneca Creek, a larger stream in Maryland, increased more than 100% between 1931 to 1960 (2,973 cfs) and 1961 to 1991 (6,014 cfs) as a result of increased urbanization (Leopold 1994). Restoring wetlands, recreating natural sinuosity in river channels, and reconnecting rivers to their floodplains are among the methods to restore the natural hydrological regime of a watershed and reduce average flood intensities.

Another critical prerequisite for successful restoration efforts is to recognize the dynamic, interconnected character of natural systems. Failure to recognize these linkages, whether between soil and plants or between streams and riparian areas, can undermine efforts to restore ecological integrity and watershed health. For example, efforts to reforest areas of the Klamath Mountains of southern Oregon and northern California have been hampered by a loss of the soil microflora (Perry et al. 1989). These fungi, and other soil organisms, are critical to many plants in promoting the uptake of nutrients through roots and in maintaining plant defenses against disease. Soils in clear-cut stands lose porosity, microbial communities, and microflora; the loss of these principal factors has been determined to cause failure of reforestation efforts in the Pacific Northwest

(Perry et al. 1989; Amaranthus and Perry 1994). On the other hand, restoring and maintaining a high diversity of tree species within forested lands has been shown to increase forest resiliency to disturbances such as wildfire, insects, and diseases (Perry and Maghembe 1989).

Restoring and maintaining connectivity within aquatic and riparian systems are equally important. Erosion and deposition of sediments is a fundamental characteristic of river channels. High river flows that occur during floods alternately scour channels and redeposit sediments. This creates a complex mosaic of channel and riparian condition that is critical in supporting fish and wildlife populations. In watersheds in which rivers have been channelized or dammed, this dynamic is lost and habitats become simplified and less able to support a wide range of species. Although returning to predisturbance conditions may not be a viable option in many watersheds, there remain many ways to restore at least partial biological function to rivers, such as allowing reconnection of floods to floodplains, which increases habitat complexity and productivity (Stanford et al. 1996).

LAND MANAGEMENT PRACTICES

One of the basic principles of effective restoration efforts is that such efforts treat and correct the fundamental causes of loss of ecosystem structure and function rather than merely the symptoms of these problems. For example, site-specific rehabilitation efforts to decrease stream erosion are not likely to succeed in the long term if factors degrading riparian vegetation are not corrected. Treating the fundamental causes of degradation, whether road construction, timber harvest, mining, or overgrazing by livestock, will facilitate natural healing and repair of eroding slopes and streambanks.

Frissell (1997, this volume) describes the problems that can be encountered when technological "quick fixes" are proposed rather than long-term ecological solutions. When the U.S. Forest Service attempted to restore anadromous salmon and steelhead runs in Fish Creek in north-central Oregon, they identified a lack of pools and an overabundance of riffle habitat as part of the problem (Reeves et al. 1997). For several years, instream structures were installed in Fish Creek in an attempt to increase instream habitat complexity. Although the structures produced some desired benefits in the short term, flooding eventually destroyed many of the structures, and numerous landslides and debris torrents brought large quantities of sediment into the stream from adjacent uplands. Restoration efforts that focused on closing and recontouring roads, revegetating harvested slopes, and restoring natural hydrographic regimes would have had more long-term benefits to steelhead and salmon populations than did the instream efforts. As noted by Reeves et al. (1997), more attention should have been paid to upslope conditions rather than concentrating solely on stream habitat projects. Stream conditions largely are a consequence of the overall health of the watershed and should not be managed as if they are separate from this context (Bisson et al. 1992; Doppelt et al. 1993). Attempts to manage habitats in isolation from natural processes that occur on broader geographic and temporal scales will likely increase degradation despite the best intentions of land managers.

Understanding the principal causes of ecological problems has been a cornerstone of success in efforts to restore fish populations in the Mattole River. Early efforts to restore chinook salmon and steelhead runs in the Mattole River consisted of hatchbox programs designed to increase fish numbers artificially through propagation. Over the years, citizens of the Mattole watershed have learned that the solutions to declining fish runs lie not only in the stream but in the watershed as a whole (Zuckerman 1997). Poor timber harvest practices and too many roads with poorly constructed stream crossings have contributed substantially to deteriorating stream conditions. Efforts to increase fish runs through artificial propagation are poor substitutes for solving the problems that degrade spawning and rearing habitat and often cause unexpected problems for remaining wild fish, which must compete with artificially produced fish for what little quality habitat remains (Meffe 1992; Williams and Williams 1995).

On a broad geographic scale, efforts to restore salmon in the Columbia River basin illustrate the tendency to treat the symptoms rather than the fundamental problems when attempting to resolve major natural resource issues. Considerably more than US$1 billion has been spent attempting to restore salmon in the Columbia River, but native stocks continue to decline because restoration efforts have focused on hatchery production to mitigate continuous declines in habitat condition rather than on correcting hydroelectric operations, timber harvest, road construction, and other activities that degrade freshwater habitats (Nehlsen et al. 1991; Stanford et al. 1996). As wisely noted by Orr (1994), many of our ecological problems will be solved only when people understand that we cannot engineer our way out of these problems through improved technology.

The cause of degradation in some watersheds, at least in general terms, seems to be obvious. In the Marys River watershed of Nevada the problem was overgrazing by livestock (Gutzwiller et al. 1997, this volume). Long-standing effects of grazing by cattle had reduced riparian vegetation, eroded streambanks, and reduced water quality in streams that supported the threatened Lahontan cutthroat trout. The U.S. Bureau of Land Management, which controlled livestock grazing on allotments of public land but lacked authority on privately owned lands along the river, was able to address grazing concerns throughout most of the watershed through creative use of land exchanges that acquired critical habitat for the threatened Lahontan cutthroat trout by trading desirable development lands. The U.S. Bureau of Land Management acquired 47,000 acres in the watershed, including 55 stream miles along the Marys River, in exchange for 660 acres within the Las Vegas metropolitan area. They also acquired 4,133 acres with 21 miles of stream by means of a land exchange with Independence Mining Company. This enabled the agency to reduce and redistribute livestock grazing impacts to the overall benefit of aquatic habitats.

Many watersheds owe their problems to a suite of complicating factors. Health of the Blackfoot watershed in Montana has been degraded by historical mining operations, overgrazing by livestock, clear-cut logging operations, construction of numerous roads along rivers and creeks, high recreational use, and introduction and spread of nonnative species (Aitken 1997). Similarly, the Beaverkill–Willowemoc system of New York has been adversely affected by gravel mining of riverbeds, channelization, point source and nonpoint source pollution, removal of riparian vegetation, high recreational use, and road construction (Conyngham and McGurrin 1997). In such complex situations, there are several potential courses of action. Watershed analysis and other broad-scale assessments can be used to determine where land management changes can be most effective at the watershed scale. Ziemer (1997) and Frissell (1997) provide guidance to prioritize restoration efforts.

If properly designed and implemented, restoration efforts do not have to encompass all or a major portion of the watershed to be effective. Pilot restoration projects in Iowa, which consist of planting and protecting riparian buffer strips and establishing small wetlands adjacent to private agricultural lands, have demonstrated techniques that could resolve a large component of nonpoint source pollution problems throughout agricultural lands (Isenhart et al. 1997, this volume). In this instance, small but significant changes in agricultural practices eliminated the primary causal factors of stream degradation. Smaller pilot efforts also have been used extensively on the Blackfoot. The importance of these smaller projects exceeds what might be expected because of their effect in establishing good working relationships between private landowners and government agencies (Aitken 1997). Pilot projects provide the opportunity to demonstrate success in changing land management practices that can be viewed by other potential partners; seeing such success can improve potential partners' willingness to participate in restoration efforts.

Riparian areas often are an appropriate starting point in efforts to improve stream conditions. Despite their often meager acreage and geographic extent relative to other habitats, riparian areas critically influence stream conditions by buffering upslope impacts, such as erosion, providing shade, reducing flood intensity, and providing woody structure for pool formation (Gregory et al. 1991; Naiman et al. 1992). Improved management of riparian areas yielded significant improvements to aquatic habitats in many of the case studies in this volume. Some of the increased emphasis on riparian management discussed herein included changing livestock grazing regimes and constructing fencing for livestock control (Gutzwiller et al. 1997), revegetating riparian areas with native species (Isenhart et al. 1997; Nolte et al. 1997), and relocating water sources away from streams to discourage livestock use (Gutzwiller et al. 1997; Hudson and Heikkila 1997). One of the principle functions of riparian zones is to serve as a source of large wood that can be recruited into streams as trees fall and decay. Rather than installing large pieces of wood or logs in streams to create complex artificial structures that may alter hydrologic regimes or otherwise quickly fail (Frissell and Nawa 1992; Reeves et al. 1997), it would be more ecologically sound and cost-effective to increase instream habitat complexity by stocking streambanks and riparian zones with large woody debris that can be naturally relocated to the stream channel during floods. This could provide a near-term solution to increasing stream channel complexity while riparian zones are protected and vegetation is allowed to mature.

Restoring flow and water temperature regimes that resemble historical flow conditions is a critical component of restoration of rivers affected by dams and hydroelectric operations. High flows may be critical for reshaping sand bars, flushing silt and other fine sediments from a system, stimulating natural plant succession in riparian areas, and removing nonnative fishes from a river. Low flows also may be important to ecosystem structure and function, as are the pattern of flow changes and the rate at which river levels fluctuate. The entire flow regime, including the timing, duration, and intensity of flows, often is critical to ecosystem function (Stanford et al. 1996; Toth et al. 1997). Perhaps one of the best opportunities to restore altered flow regimes is through the Federal Energy Regulatory Commission dam relicensing process. This relicensing process was used successfully to restore, at least partially, flow regimes and ecosystem structure to the Deerfield River watershed in New England (Kimball 1997, this volume). Similar opportunities present themselves across the country as hundreds of federally permitted dams come due for relicensing.

ADAPTIVE MANAGEMENT

Restoration is an expensive and long-term process. Watersheds, whether the Blackfoot River in Montana or Fish Creek in Oregon, are highly complex. The effects of natural disturbances such as floods or wildfire blend with human-induced disturbances such as road building and livestock grazing in a synergistic or cumulative fashion. These multiple agents of change can baffle restoration practitioners and mask the results of their efforts (Stanford and Ward 1992). Monitoring and adaptive management are critical ingredients to unraveling the mysteries of cause and effect.

Monitoring efforts should determine whether (1) the restoration plan has been designed and implemented properly, (2) the expected results are being achieved, and (3) modifications to the plan, if any, are needed in future efforts (Kershner 1997, this volume). Despite all the funds, staff time, and energy invested in restoration, very few examples of well-designed monitoring efforts exist. A lack of monitoring data is a common weakness to restoration efforts, including most of the case studies examined in this volume. Long-term data sets of 5 to 20 years, which are the most valuable at unraveling the complexities of watersheds, are exceedingly rare.

The Fish Creek restoration effort (Reeves et al. 1997) is one of the best examples of a long-term monitoring effort properly integrated into a restoration program. Several years of data on Fish Creek habitat conditions were collected before restoration efforts began. But just as importantly, data collection continued after instream work was completed. This data provided a quantitative evaluation of the effectiveness of work to improve fish habitat and fish populations. Monitoring also is an important component of restoration of the Henrys Fork (Van Kirk and Griffin 1997). It is interesting to note that early restoration efforts on the Henrys Fork were geared to site-specific projects without regard to understanding or prioritizing needs throughout the watershed. Only later did the practitioners on the Henrys Fork step back from isolated projects and begin to examine restoration needs on a broader scale. Although early restoration projects were valuable to foster trust and working relationships among interest groups, the long-term success of restoration on the Henrys Fork and in other watersheds is dependent on this broader geographic view.

All too often, watershed coalitions are discouraged from designing or implementing restoration efforts because of what they perceive as their inadequate research or statistical capabilities. As described by Kershner (1997), however, most elements of a well-designed monitoring effort are straightforward and easy to understand by scientist and layperson alike. Monitoring efforts must be able to detect change in the condition of a watershed and to determine the causal factors for this change. Monitoring efforts that deal with only the implementation and effectiveness phases of restoration can be exceedingly valuable. All interests in a watershed coalition are concerned about whether their project has been implemented according to plan and whether they were effective in achieving the desired conditions.

Monitoring provides a valuable opportunity for agency scientists to enlist the support of local citizens in gathering ecological data. With guidance from agencies, citizens can build valuable long-term data sets and become more knowledgeable about how natural and artificial disturbances affect the integrity of the watershed in which they live.

Monitoring efforts can focus on small geographic scales or pilot-project areas and still yield valuable information that may then be applied at the watershed scale. Monitoring efforts on Bear Creek in Iowa (Isenhart et al. 1997) examined changes in water quality as a result of creating riparian vegetation buffer zones between agricultural fields and streams. Isenhart et al. (1997) were able to demonstrate the effectiveness of their plantings and buffer zones by monitoring effluent from small pilot areas. Their results demonstrate the potential value of widespread implementation of such a program on water quality and aquatic biological diversity in Midwest agricultural lands.

Monitoring and adapting management based on the results are iterative and long-term processes. Initial restoration plans are unlikely to be completely adequate or able to resolve all problems. Monitoring results should reveal ways to modify management prescriptions in order to achieve desired structure and function of watersheds. In this respect, well-designed monitoring and adaptive management efforts provide some insurance against failure. At the very least, we should allow ourselves to learn from our mistakes.

CONCLUSIONS

At first glance, it is hard to imagine what seed potato growers and fly fishers of eastern Idaho might have in common. Both have fought over increasingly scarce water supplies in Idaho's Henrys Fork. One group, the seed potato growers, have been residents of the watershed for many generations, whereas most fly fishers were viewed as outsiders. Both groups found a common cause in the degradation of water quality in the Henrys Fork, which adversely affected both pursuits. Improving land-use practices, reservoir operations, and interagency coordination provided common ground for action to improve water quality. Once the two parties were brought together in the nonadversarial setting provided by a community-based watershed coalition, they found they had more in common and more issues they could agree upon than either had imagined.

Even diverse interests can see the advantages of healthy and productive watersheds. Ranchers that manage riparian areas to maintain adequate vegetative cover, which traps sediments during floods and prevents streambanks from eroding, also benefit from reduced soil erosion and improved water quality. The reward for anglers who spend weekends with state and federal land managers pulling noxious weeds or replanting streambanks is healthier fish populations. Teachers who educate their students about the importance of restoration are helping their students become better, more responsible citizens. All of us benefit from healthy, diverse, and productive watersheds. Communicating such benefits is a critical function of the watershed restoration process.

Each watershed is distinct. Finding common ground to remedy watershed degradation, whether the problems are water quality, endangered species, declining salmon runs, or forest health, is a fundamental first step in developing consensus about how to restore watersheds. In this way, we develop the long-term perspective of sustainability by seeking, in the words of Alan Durning (1996), not quick fixes but deeper understanding, which is reflected in new alternatives to land management.

A person ought, I suppose, to support his home; and this planet is the only home we have. And, too, there ought to be some time set aside for mountains and mountain streams, and fat, truculent trout. After all, where there is such water and trout, there is, I choose to believe, hope—hope not only for what is but for what is possible.
—Harry Middleton, 1991

CHAPTER 26

WHAT WORKS, WHAT DOESN'T, AND WHY

Joseph McGurrin and Harv Forsgren

For centuries, humans have altered the physical character and water quality of streams and rivers throughout the world. Major transformations of North American rivers that changed the spawning, incubation, rearing, feeding, and migratory habitats necessary for fish survival began in the early nineteenth century (Mrowka 1974). Because rivers are formed by their surrounding drainage basins or watersheds, the health of rivers and their fisheries can be directly affected when humans use and alter the landscapes within a watershed (Meehan 1991). Although our knowledge of the basic relationships among fish populations, fish habitat, and watersheds is improving rapidly, there is only partial documentation of broadscale trends of fish and habitat loss.

An inventory of the nation's rivers was initiated in the late 1970s by the U.S. Heritage Conservation and Recreation Service. After elimination of the agency in 1980, in 1982 the U.S. National Park Service completed the study, known as the Nationwide Rivers Inventory. The inventory's purpose was to identify streams or stream segments of high quality that had the potential for designation as wild, scenic, and recreational rivers. The Nationwide Rivers Inventory represents the only available comprehensive survey of U.S. streams in the 1980s. The survey indicated that of approximately 3.1 million miles of streams in the nation, exclusive of Alaska and Hawaii, only 2% (<62,000 miles) had sufficient high-quality habitat features to be worthy of federal protection status (Benke 1990).

One indicator of the declining health of our rivers is the dramatic increase in the number of threatened and endangered freshwater fish species. An American Fisheries Society study (Williams et al. 1989) documented a 45% increase in the number of endangered North American freshwater fishes from 1979 to 1989. The study found nearly one-third of all North American fishes to be endangered, threatened, or of special concern. Recovery of these fishes, if possible, will require restoration not only of aquatic habitats but of entire river systems (Williams and Rinne 1992).

Given the status of rivers and native fish populations, the task of restoration is daunting. The issues involved in watershed management and fishery conservation are complex. The solutions to many of the problems facing resource managers have not been determined, and some problems have just been identified. Although we have substantial information on the impacts of altering watersheds for human uses, a similar database of successful watershed restoration efforts is sorely lacking. Despite the lack of a comprehensive record of restoration efforts, there has been enough success to offer useful information to those interested in restoring watersheds. The purpose of our chapter is to outline fundamental principles and offer practical guidelines for developing and implementing community-based watershed restorations. We present a "fish-centric" view of watershed restoration but suggest the principles and guidelines are more broadly applicable.

FUNDAMENTAL PRINCIPLES OF WATERSHED RESTORATION

Watersheds are ecosystems composed of different lands that are connected and drained by a network of streams and rivers (Doppelt et al. 1993). Watershed boundaries are defined in terms of the land basins that define the flow of water to different river systems. Although Aldo Leopold identified the importance of the connections between land and water almost a century ago, the loss of fisheries and other watershed benefits, such as clean water, have been the stimulus for the recent rediscovery of the importance of taking a watershed approach to fishery conservation and management efforts. The "new" watershed approaches have helped to define better the dependence of fish populations on their surrounding watersheds by emphasizing the natural means through which watersheds and their rivers form aquatic and riparian habitats.

Integrating this new science with a socioeconomic system not based on watershed boundaries has proven difficult. Even though all social and economic benefits are based on the biological and physical properties of watersheds, the political debate often centers on "economics versus the environment." This conflict-generating paradigm is beginning to change as watersheds become further degraded and less economically productive and as more people understand the socioeconomic benefits of healthy watersheds. A more harmonious integration of environmental and economic principles, however, has yet to fully occur.

Despite the complex and ever changing nature of the biophysical and socioeconomic processes that affect watersheds, a number of basic principles have emerged and proven essential to developing successful watershed restoration programs. These fundamental principles have been stated in a number of different ways by a variety of authors in this book and elsewhere (e.g., Angermeier 1997; Frissell 1997; both this volume). Following are some of the most important biophysical and socioeconomic principles. Although presented in two distinct sections, they cannot be viewed, understood, or applied effectively in isolation of each other.

Biophysical Principles

Watershed protection is first priority.—Although the great majority of U.S. watersheds are already degraded to some degree, most also still have some healthy components. Protecting the remaining healthy components of a watershed provides a strong base to which future restoration work can be anchored. Over the long run, watershed protection costs less than restoration, and the outcome is more certain. Some have recognized the importance of both passive and active protection (Doppelt et al. 1993). Passive protection refers to placing restrictions on human activities that may adversely affect watershed conditions or processes. This has been effected through federal land management planning processes, the National Environmental Policy Act process at the project level, and, in some cases, litigation. Active protection includes on-the-ground actions taken to reduce the risk of watershed damage associated with management activities. Using appropriately sized and designed culverts to reduce the risk of landslides associated with forest roads is one example of active watershed protection. In this example, use of larger culverts may initially cost more than use of smaller culverts, but that additional cost is insignificant when compared with the maintenance, replacement, and restoration costs associated with smaller culverts that wash out from periodic storm events and debris flows.

Sustainability must be the basis for watershed restoration and conservation.—Conservation must be based on the sustainable use of watersheds and their resources. Sustainable management can be likened to the Hippocratic oath of the medical profession—"Do no harm." For land managers, this amounts to taking no actions that limit future management options. For example, activities that reduce soil productivity, isolate streams from their floodplains, or contribute to the extirpation (local extinction) or extinction of species fail the test of sustainability. Although most are quick to recognize land management activities that fail the sustainability test, in our efforts to restore fisheries we must recognize the potential for similar adverse consequences if our actions are too narrowly focused on single species without consideration of that species's relationship with other members of the biotic community. The objective should be to reestablish stream and watershed processes, not to create habitats specific to single species. Sustainability must be one of the cornerstones of responsible land management and fishery conservation.

Streams are inextricably connected to the lands within their watershed.—Stream system health, diversity, and productivity are largely the products of the land conditions within the stream's watershed. Riparian and upland conditions pronouncedly influence the magnitude, frequency, timing, and duration of drought and flood events that shape rivers and the base flows that often limit aquatic production. Watershed conditions help determine the delivery of coarse structural material such as boulders and large wood that contribute to the physical characteristics of the stream channel and water column which fish recognize, by use, as suitable habitat. Riparian and upland conditions also significantly influence the delivery of fine sediments, leaves, twigs, and terrestrial insects that contribute to the productive capacity of a stream. Stream restoration, therefore, generally has more to do with taking appropriate management actions outside the stream channel than within it. As Beschta et al. (1994) observed, "Pouring time and money into a degraded stream that is continuously perturbed by human land use activities is not only futile, but it raises false public expectation that aquatic conditions will be improving."

Healthy streams and their watersheds exist within a state of dynamic equilibrium.—The term *stable* has little relevance in describing stream and watershed conditions, unless a long enough temporal scale is considered. Mackin (1948) noted the result of trying to control rivers by altering stream channels

> [A] safe general implication is that the engineer who alters natural equilibrium relations . . . will often find that he has a bull by the tail and is unable to let go. As he continues to correct or suppress undesirable phases of the chain reaction of the stream to the initial "stress" he will necessarily place increasing emphasis on study of the genetic aspects of the equilibrium in order that he may work with rivers, rather than merely on them.

Streams tend toward a state of dynamic equilibrium, in which systems are in balance—products of flowing water and the surrounding land forms. Interactions produce small but virtually continuous changes in fish habitat conditions as the watershed moves toward a new state of equilibrium (Swanston 1991). The actual effect of disrupting events on habitat quality and productivity depends largely on an event's intensity and timing. Some events are regular and cyclical in occurrence and are the product of climate or geography (e.g., seasonal and annual precipitation, freezing and ice formation, and low flows). Others are of a more random nature, triggered by extreme storms, earthquakes, volcanic eruptions, fires, or disease. Events may be natural or human induced.

As the name suggests, the dynamic equilibrium condition is not static but constantly changing. Daily, seasonal, and annual differences in physical, chemical, and biological attributes are the norm rather than the exception for river systems. Natural differences can be of a profound magnitude—sometimes dwarfing the smaller, more chronic changes induced by land management activities and sometimes inappropriately used to justify, or minimize, the significance of management-induced perturbations. In recent years there has been much discussion about managing ecosystems within their natural range of variation. This concept embraces the principle of dynamic equilibrium but may fail to protect the resiliency of stream systems if management activities result in an increased frequency or duration of conditions near the "edges" of the range. Episodic conditions near the edges must be infrequent or of short enough duration to permit recovery to conditions well within the range of variation. It is also folly to assume that we are skilled enough to manage close to the edge without regularly falling from the cliff.

Socioeconomic Principles

Beyond biophysical principles, those considering watershed restoration should understand the political, social, and economic systems that drive human management of watersheds. These systems involve potential challenges that are at least as formidable as the biophysical ones. In fact, an August 1995 survey of watershed groups in Washington, Oregon, and California suggested the most common problems watershed groups need help solving are of a socioeconomic nature (B. Bradbury, For the Sake of the Salmon, personal communication). Common problems identified by the survey include securing long-term, dependable funding; obtaining landowner interest; receiving understanding and support of efforts from city and county governments; motivating agencies to work with citizens; and influencing government planning decisions that have the potential of undermining watershed health. Five of the most significant socioeconomic principles follow.

Legal and regulatory systems create the social framework for restoration efforts.—As increasing demands are placed on finite natural resources, conflicts become more prominent. Environmental legislation to address such conflicts began, possibly, with early English riparian water use regulations. The United States formally recognized the importance of environmental protection with passage of the National Environmental Policy Act. As Sager (1977) pointed out, the National Environmental Policy Act

> contained concepts fundamentally different from any previous legislation. It gave the federal government the primary responsibility for the nation's environment; it provided a legal forum in which citizens could bring suit against either the government at any level or any private person or group which, in the citizen's opinion, was guilty of abusing any segment of the environment, biotic or abiotic.

The National Environmental Policy Act and other federal (e.g., U.S. Endangered Species Act, Clean Water Act, Clean Air Act, National Forest Management Act, and Federal Land Policy and Management Act) and state environmental legislation, regulations, and policies cumulatively establish a political framework for watershed protection and restoration. Because they establish specific standards or restrictions, and the means for citizen involvement in identifying protection and restoration needs, those interested in watershed conservation and restoration should not only be aware of legislative and regulatory requirements, but understand how they are applied in an overall watershed conservation program.

Multiple jurisdictions are the rule for most watershed restorations.— Whereas wildlife and fish are usually considered public resources, the lands on either side of the channel may be in a variety of ownerships. Land ownership patterns and a division of regulatory responsibilities among federal, state, tribal, and local governments in the United States more often than not necessitate collaborative efforts among a variety of government agencies and private landowners to effect successful watershed restorations. Matching the social, economic, and cultural objectives of multiple organizations to design and implement watershed restoration is often a significant complicating factor. However, because of the essence of the water–land interaction of watersheds, failure to secure the participation of all affected jurisdictions may negate the beneficial actions taken on the lands of those who cooperate. Recent community-based efforts, including those documented in the case studies of this book, demonstrate the importance of gaining the ownership and support of affected interests.

Perceived values of restoration must exceed perceived restoration costs.—Watershed restoration has often proven to be a lengthy, difficult, and costly process. To establish a broad base of support for watershed restoration the perceived values of restoration must exceed the perceived costs of restoration. That is, the benefits of healthy watersheds (e.g., sustainable levels of goods and services such as fishing, hunting, swimming, and boating; wildlife and fish conservation; improved water quality and quantity; reduced flood protection and mitigation costs; and enhanced reservoir life span) must exceed the short-term costs related to direct investment of limited capital resources. Restoration and protection values are often hard to quantify in terms of current economic systems and often conflict with short-term resource use opportunities— often opportunities with more readily quantifiable economic benefits. It is important to recognize that perceived values and costs are often galvanized around some focal issue rather than all the potential benefits and values that are derived from healthy watersheds.

Success is dependent on a catalyst and effective organization.—The most important factor in successful watershed restoration is people. We previously recognized this principle in the collective political sense but wish here to distinguish the critical role of individuals and their ability to help engage others. A common attribute of the successful watershed restoration case histories cited in this volume, and elsewhere, is the presence of an individual or small group of individuals with the vision, energy, persistence, and communication skills to serve as a catalyst, or sparkplug, to help shape community attitudes and generate support. To capitalize on these attributes, someone with organizational skills has to step forward to help direct that vision and energy to produce results.

Table 26.1.—Ten practical guidelines for developing community-based restoration efforts.

1. Select an appropriate scale
2. Define a purpose and goal
3. Develop and select project leaders
4. Build community ownership through partnerships
5. Use and integrate the best available science to develop a restoration plan
6. Design a business plan to implement individual projects
7. Follow the cardinal rules of environmental restoration
8. Monitor and evaluate the results of restoration efforts
9. Communicate results and reward accomplishment
10. Maintain a long-term perspective for success

Effective communication *is critical to restoration success.*—Even the most technically sound and ecologically significant restorations are unlikely to succeed if people do not understand and support them. The importance of communication can not be overstated. Effective communication is crucial to help affected publics understand the need for restoration, its relative costs and benefits, and their opportunities to take an active role.

PRACTICAL GUIDELINES FOR A WATERSHED RESTORATION PROGRAM

The previous discussions of fundamental principles were general in nature, but having some understanding of the existing strengths and weaknesses of watershed science and policy when initiating local restoration efforts will increase chances for success. Successful fish habitat restoration projects begin with the premise that a stream and its watershed function as an ecological unit. Projects that do not account for the interactions between the waterway and adjacent lands yield only short-term results or, often, fail altogether (Frissell and Nawa 1992; Frissell 1997). Although many fish habitat projects focus on outcomes for individual species or stream segments, such projects should be developed with an understanding of where the project fits into the big picture—the health of the entire watershed.

The following section outlines ten practical guidelines for developing community-based restoration efforts by means of a big-picture watershed approach (Table 26.1). Some of these ideas were developed as part of a guide for grassroots fishery projects (Nickum et al. 1995), whereas others are rooted in scientific and technical approaches to environmental restoration.

Unfortunately, the path to success is not as simple as following a set of rules or formulas. Individual watersheds and their communities are unique. The guidelines herein are core strategies that have contributed to the success of a wide variety of restoration efforts. They are presented from the perspective of a private citizen or interest group that is just starting. How and which of these guidelines are applicable depend on the starting point. Do relationships exist with other watershed users and those involved in management activities? Have fishery and watershed assessments already been prepared? Is there an existing plan for watershed restoration? Just as the starting point may differ, so may the end point envisioned for watershed restoration efforts.

Select an appropriate scale.—The first issue in developing a restoration project is to select a candidate watershed. The selection of an appropriate watershed highlights a variety of practical concerns about a community's own interests and capabilities, as well as the desires of other users within the larger community. Restoration efforts can be initiated by a wide variety of organizations ranging from coalitions of governments to small volunteer groups. The interests and capabilities of different organizations are likely to determine the appropriate scope and physical scale of watershed restoration activities. Watershed restoration efforts may focus on a first-order mountain stream less than 2 miles in length to the restoration of a major river system such as the Columbia basin. Some watersheds are so large and complex that restoration efforts are beyond the capabilities of any single institution or group. In such situations, a group or individual can still contribute to restoration efforts by working as part of a larger coalition. Larger watersheds might also be broken into more manageable pieces (e.g., subwatersheds) that can be tackled in turn.

Define a goal.—Selecting a watershed that is appropriate to an organization's interests and capabilities involves defining the purpose of the watershed restoration. Defining the purpose for watershed restoration may be fairly straightforward when working on a small spring creek watershed managed by a single property owner. It may be a much more complex task if it involves a large multistate, public–private cooperative effort, such as that occurring on the Chesapeake Bay watershed. Regardless of the size and complexity of the effort, it is essential to understand the needs of the watershed and the needs of its users. A scoping process that identifies these needs is an essential step in defining the overall purpose or goal of watershed restoration.

A clearly stated goal helps keep efforts focused throughout the duration of the restoration effort. At the scoping stage, ask individual user groups for their goals for watershed restoration. Look at the commonalities and differences among various groups. If a consensus of interests is readily apparent, develop a mission statement outlining the general purpose of the restoration. For many fisheries groups, this might be something such as: "Improve habitat to increase natural reproduction of native fishes in the stream." As discussed earlier, watershed restoration should be broader than improving just fish reproduction. If so, the mission statement may read "Restore watershed functions to improve water quality, fish and wildlife habitat, and scenic value." At this early stage, the goal should be stated qualitatively. More specific, quantitative objectives can be established as detailed information about the watershed is collected (Ziemer 1997, this volume) and as greater consensus emerges among various interests.

Develop project leadership.—Once restoration goals are defined, begin building relationships with potential partners and finding experts to help plan and conduct the actual restoration work. Before recruiting outside partners, however, it is important to have a small leadership group with specific responsibilities, including one person designated as the overall project coordinator. Without a coordinator, participants may end up duplicating work or, worse, inadvertently undermining each other. The coordinator serves as a central source whom potential partners or volunteers can contact for information. Depending on the scope of a project, the coordinator will need other participants to take responsibility for specific aspects of the restoration, such as conducting surveys or community outreach and publicity, raising funds, or organizing work days. Clearly defined sharing of leadership responsibilities can ensure that individual talents are maximized. Sharing leadership responsibilities also is important in terms of sustaining a long-term restoration effort. It allows newer participants an opportunity to gain experience by serving an "apprenticeship" under a veteran restoration leader, thus laying the groundwork for future leadership.

Build community ownership.—Without support from the local community, watershed restoration will fail. Opportunities for community involvement should be available at all stages, from preliminary scoping through restoration evaluation. Partnerships are the cornerstone of cooperative restoration efforts. Depending on the scope and scale of the restoration, partnerships may take the form of a few people in an informal network to a large watershed coalition involving thousands of people and having formal bylaws and membership requirements. At a minimum, watershed coalitions are well served to gain the support of state fish and wildlife agencies, private landowners, and public land managers.

Understanding the backgrounds of partners, their potential contributions, and what benefits healthy watersheds offer them is essential. Keep expectations reasonable. Restoration involves significant fiscal resources and time commitments. Know how restoration efforts may affect partners' workloads and be realistic in asking for assistance. Get an early start. Building good working relationships with partners takes time!

Use the best available science.—Restoration efforts vary widely, each depending on the characteristics of the ecosystem, public understanding of the issues, the general geographic area, the kinds of fishes and other wildlife involved, and the nature of degradation or alteration that has occurred. Restoration details are dictated almost entirely by local circumstances and must be worked out on the ground with experts familiar with each watershed's unique characteristics. Assessing these characteristics will help ensure clear, achievable objectives for restoration (Wissmar 1997, this volume; Ziemer 1997).

Technical assistance from experts familiar with local conditions is needed to assess the biological health of watershed streams, determine the factors limiting production, and develop appropriate restoration strategies and objectives. Recruiting expert advisors is a vital step in planning the technical details of restoration projects. There are abundant sources of information on stream and watershed restoration. People with knowledge and experience in restoration are often willing to help local communities with restoration projects. Good sources of assistance include government officials, agency employees, civic associations, local colleges and universities, private landowners, conservation organizations, and professional societies.

Design a business plan.—Successful watershed restoration efforts depend on strong leadership, community involvement, and careful planning. As the watershed restoration proceeds consider all of the details—from permits to funding, from equipment to volunteer labor. A general business plan that includes specific objectives, tasks, and strategies is necessary for just about every type of watershed and restoration activity.

Specific objectives will help keep individual restoration efforts focused and offer milestones for measuring progress. Objectives should highlight the limiting factors acting on the watershed, fish habitat, and fish population issues identified in technical assessments. They should readily fit under the framework of the initial watershed restoration goals or mission statements. However, initial goals may need to be modified based on the preliminary scoping process, assessment work, and changing conditions.

Develop a list of tasks to accomplish restoration objectives. Tasks could range from education campaigns that address incompatible watershed uses to community awareness projects on water conservation in the home to riparian planting projects. Depending on the scope of the overall watershed restoration effort, the length of individual tasks will vary from days to several years.

Based on the objectives and tasks, formulate strategies for project completion. Develop a timetable and budget for individual projects and tasks and ensure that someone takes responsibility for each item. Timetables and budgets make useful guides in planning for fund-raising needs. Build flexibility into plans, timetables, and budgets. Things rarely go exactly as planned when dealing with the complexities of watersheds and their communities, so backup plans or alternative strategies may be needed.

Follow the cardinal rules of environmental restoration.—The fundamental principles and practical guidelines discussed herein suggest a few hard and fast rules that should guide all types of environmental restoration efforts. The first rule is that the best strategy is to avoid degradation in the first place. Although the great majority of large watersheds in North America are already degraded to some degree, most also still have some healthy components. Protecting healthy components of a watershed—in cooperation with local communities, government agencies, private landowners, and conservation groups— offers a strong base from which future restoration work can expand. There is nothing more frustrating than starting a watershed toward recovery only to see new human actions reverse progress. Community involvement in protecting a watershed from bad zoning decisions or plans for water use will usually provide more long-term benefits for watershed health than restoring or mitigating watershed damage after it occurs.

The second rule is to focus on the causes of the problems, not just the symptoms. Watershed management often involves a determination of factors that limit production of various outputs. For example, habitat quantity and quality are the basis of fishery productivity. When habitat is degraded, restoration efforts sometimes focus on only physical and biological symptoms of the problem. Repairing excessive erosion through bank stabilization projects may be totally ineffective if the source of the problem remains untreated. If existing land management practices are the cause of the problem, changing management strategies may be enough to allow natural recovery processes to proceed.

The third rule is to work with rather than against natural watershed restoration processes. Long-term restoration success is associated with complementing and accelerating natural recovery processes. Incorporating basic principles of fluvial geomorphology (how running waters shape land formations) will help in several ways, doing so will (1) provide a technical basis for addressing watershed impacts on streams, (2) incorporate the reality that streams are dynamic and not static entities, (3) focus attention on the natural processes that restoration should imitate, and (4) highlight the causes of habitat change and not just the symptoms. Fluvial or hydrological processes will determine what can and cannot work within a particular watershed. If a project works against the natural forces that bring a stream toward equilibrium instead of with them, then restoration will almost certainly fail. As stream and watershed conditions are assessed, it is important to consult someone knowledgeable about fluvial geomorphology (such as hydrologists or geomorphologists) to understand the physical character of streams and the processes at work in the watershed.

Monitor and evaluate results.—Monitoring and evaluation are crucial to an adaptive management approach to restoration. Although there is always some sense of closure when a milestone of a restoration effort is achieved, restoring and protecting aquatic systems requires an ongoing commitment. Stream and watershed conditions must be monitored to see if projects are having the expected results and whether modifications or additional projects will be needed. Postproject monitoring allows identification of improvements in the watershed's condition. By comparing this information to preproject assessments, one can characterize changed conditions, document successes, and garner support for additional restoration work. Equally important, monitoring will show where projects have not met expectations or have resulted in unexpected side effects. The restoration may be modified in response to this new information. Often modification will require only fine-tuning, but at times a new limiting factor or harmful side effect may require a major change in approach.

As with preliminary watershed assessments, expert technical assistance is desirable in monitoring. Arrange for cooperative monitoring before, not after, conducting restoration activities. If an initial commitment to design and implement monitoring is not made, there is a good chance that postrestoration assessments will not be made. Remember, monitoring is not just important to assess the results of restoration and make any adjustments that may be appropriate, it also has great educational value. What is learned from successes and failures will make future restoration efforts more effective.

Communicate results and reward accomplishments.—Communication is a key to maintaining enthusiasm and promoting cooperation and understanding within a watershed community. A whole chapter could be developed just on the process of forming watershed coalitions or associations through environmental education and communication campaigns. For active project participants, communicating results and rewarding accomplishments is particularly important. Most people thrive on recognition, especially when it is given in public. When people are publicly recognized, it conveys the idea that they are special and their contributions are truly appreciated. It generates a sense of well-deserved pride in contributing to something worthwhile. As a result, recognition is a valuable tool in maintaining support for restoration initiatives. Appropriate recognition can fuel enthusiasm in ways nothing else can. It is also common courtesy due those who help in restoration efforts.

Many kinds of recognition are available. Some examples are letters of appreciation, mementos given to participants, awards, plaques, coverage by various media, and recognition at a formal event such as a project dedication.

Maintain a long-term perspective.—Watershed restoration is an ongoing process. Rivers have changed and been degraded over a period of decades. For most large watersheds, restoration of healthy watershed processes and functions will take equally as long. Ultimately, conservation of watersheds is akin to managing a trust fund. A trust fund provides income for the beneficiary, but the assets of the trust are to be protected and enhanced. Similarly, watersheds can provide some of the values that people want—clean water, opportunities for fishing, hunting, and other outdoor recreation, and scenic views, as well as livestock forage, timber, and minerals. Enjoying these values is possible, however, only if communities maintain the assets, such as the diversity and productivity of the watershed. Maintaining these assets requires a long-term investment perspective.

CONCLUSIONS

Watershed restoration is much more complicated than understanding the fundamental principles and practical guidelines outlined in this chapter. Perhaps the greatest shortcoming of a watershed approach to restoration is the difficulty of translating watershed-based concepts into real-world actions. Whereas restoration using watershed boundaries makes perfect sense from a biological and ecological perspective, very few political, social, and economic systems are based on watershed delineations. Given this chasm between the biological and social frameworks that affect watersheds, restoration is inherently an uphill battle. Despite this basic problem, we have offered practical strategies in an attempt to bridge some of the gaps between existing theory and practice.

For communities and individuals just becoming interested in their watersheds, the prospect of restoration can be daunting. There have been plenty of reasons why watershed protection and restoration hasn't worked: from political and economic systems that fail to protect watershed functions and values to well-intentioned local efforts that have ended in piece-meal and ineffective outcomes. The resulting degradation of U.S. rivers and the loss of native fish populations and biodiversity (i.e., variety of life) have reached crisis levels. Against all odds, some fish populations have kept footholds in the healthier portions of even some highly degraded watersheds. For most species, however, these footholds are becoming tenuous. There are fewer and fewer streams with the high-quality habitats needed to maintain native fishes. Native species, from the coho salmon to the razorback sucker, are at increasing risk, and the unique adaptations to local environments carried by populations of these fishes are being irretrievably lost. The most immediate socioeconomic outcome has been the loss of important fisheries and the industries they support. Entirely new strategies and policies must be established based on principles of watershed dynamics and ecosystem health, but it will take time for society to make these changes. In the meantime, communities need to move forward with watershed restoration in the context of political and socioeconomic systems that are not oriented to river and watershed conservation.

No matter what sort of scientific or political watershed approach is developed in the future, it will fail without the support of local communities. A multitude of local watershed restoration projects are being conducted throughout the United States and demonstrate the interest of local citizens in changing the way we look at our rivers. Generally, these efforts are conducted by watershed coalitions focused on fish or wildlife and the associated recreational and commercial activities they support. Although it is obvious that the nation needs a new comprehensive political and economic approach to watershed conservation, local efforts do not need to wait until such programs are developed. A grassroots approach to watershed restoration, where actions start within local communities, will inspire or prompt governments, institutions, or states to action.

Small projects, whether they be a stream cleanup day or a highly technical fishery restoration effort, can be worthwhile if they are developed within the watershed context. Some may consider these efforts to be only piecemeal investments, involving work on single, small watersheds or on subwatersheds of a larger system. This criticism is particularly interesting because it was through a piecemeal process that most of our watersheds became degraded in the first place. River systems have unraveled because they were taken apart for different human uses on a piece-by-piece basis. The term *simplification* is often used to describe this process whereby watershed complexity is lost through the cumulative effects of many small actions. The only way to rebuild "simplified" river systems is to involve communities in a logical piece-by-piece process guided within the context of a watershed framework.

Although we offer no one solution or complex national strategy for watershed restoration, our belief in the basic principles behind successful grassroots watershed initiatives is confirmed by the results of many of the case studies described in this book. Whether these initiatives will ultimately translate into the systematic restoration of U.S. watersheds remains to be seen.

Owls versus jobs was just plain false. What we've got here is quality of life. And as long as we don't screw that up, we'll always be able to attract people and business.
—Bill Morrisette, mayor of Springfield, Oregon (New York Times, 11 October 1994.)

CHAPTER 27

SAVING AND CREATING JOBS THROUGH WATERSHED RESTORATION

Bob Doppelt

The debates over land, water, and fish and wildlife conservation in rural communities are typically framed as choices between jobs or the environment. This impression of "Sophie's Choice" occurs primarily because the public lacks an understanding of how environmental degradation leads both to direct job losses and sometimes to profound economic consequences for many sectors of society. Similarly, the public is generally unaware of the important role that protecting environmental amenities plays in the ability of many rural communities to attract the new people and investments that can establish businesses, create jobs, and help secure a rural community's economic future.

Massive global and national economic forces are profoundly changing our economies and communities. Employment is continuing to decline in the traditional natural resource extraction and agricultural industries that once supported rural economies. To survive and prosper, rural communities must attract new people and investments that can establish businesses and create jobs. The environmental quality-of-life amenities that are so attractive in many rural areas are an advantage that can be instrumental in building new strategies to attract businesses and jobs.

This chapter examines the role that environmental protection, in general, and watershed restoration, in particular, can play in maintaining and enhancing economic viability for many rural communities. It concludes with a set of recommendations for establishing income- and job-producing programs for watershed and native fish protection and restoration. Although written with the rural western United States in mind, many of the principles discussed herein also would apply to other rural areas of the country where watersheds are healthy or are being restored.

RURAL ECONOMIES AND THE ENVIRONMENT

Few people would dispute that rural communities in the western United States have suffered in the past few decades. Key indicators, such as per capita income, job availability, real earnings per job, and the rate of college completion remain far lower in rural counties than in metropolitan areas. Although the temptation is strong for some rural residents to blame their woes on environmental policies, the reality is that the problems are, in part, the result of past environmental abuses and due to profound changes in the regional, national, and international economies.

Environmental Degradation and Jobs

Environmental degradation has played a major role in the loss of some rural jobs. There is perhaps no better example than the job losses caused by the watershed degradation that led to closure of nearly the entire West Coast ocean salmon fishery by the Pacific Fisheries Management Council in 1993. In 1988, the Pacific Northwest commercial and recreational salmon fishing industries generated about US$1.2 billion and supported about 60,000 staff-years of employment annually (PRC 1992). With the closure of the ocean fishery, those numbers plummeted by over 90%, and many rural coastal communities that were heavily dependent on fishing jobs are now imperiled. This caused the Clinton administration in 1994 to declare the coasts of Washington, Oregon, and northern California official disaster areas and distribute over $15 million in disaster relief.

Many factors have caused the decline of Pacific salmon, including impacts from hatcheries, dams, and overfishing. However, most scientists agree that the universal problem affecting almost all declining stocks is the loss and degradation of freshwater habitat (Williams et al. 1989; Nehlsen et al. 1991; Reeves and Sedell 1992). The places where salmon spawn and rear their young have been fragmented, degraded, or lost across the western United States. Some of the greatest impacts to salmon habitat have occurred in the forested watersheds of the Northwest because of decades of unsustainable road building and timber harvest (FEMAT 1993). So, upon closer inspection, it turns out that watershed degradation actually pitted jobs in certain sectors of the economy (such as timber, mining, and grazing) against jobs in other sectors (such as commercial and recreational fishing industries).

Having economic interests pitted against each other is actually not a new phenomenon in the rural western United States. For example, late-arriving miners and farmers often moved upstream from the pioneers who had arrived earlier and developed the lower floodplains. As newcomers moved upstream, they took water from streams, leaving those downstream dry. This created intense conflicts and led to the water rights doctrine of prior appropriations, also known as "first in time, first in right."

The controversy continues today. Unsustainable timber harvesting, grazing, farming, mining, and other land-use practices threaten the thousands of jobs once associated with the fishing industry, urban needs for clean water, the booming tourism industry, and new information-based technology businesses that rely on clean water, as well as many other economic sectors. The real issues regarding environmental and watershed degradation and the loss of native fishes are not about jobs versus the environment—they are about jobs versus jobs (PRC 1992; Niemi and Whitelaw 1995; Power 1995a, 1995b).

Environmental Degradation and the Economy

The consequences of environmental degradation go well beyond jobs losses. The far-reaching and profound consequences of watershed degradation are rarely assessed or explained to the public. For example, growing evidence indicates that jobs and incomes generated by the continuation of unsustainable logging, mining, road building, and grazing in the rural western United States come at the expense of jobs, incomes, and economic vitality elsewhere. A good example of this was outlined in an analysis, based on a new comprehensive economic analysis format, of the economic consequences of protecting watersheds and salmon habitat in six national forests in Idaho. The report came to the following five major conclusions (Niemi and Whitelaw 1995).

1. The primary economic issue is jobs versus jobs, not jobs versus salmon, because the watersheds do not have to be unsustainably grazed, logged, or mined to increase jobs and standards of living for Idahoans and others in the Pacific Northwest. Thus, this is a choice between economic futures. There are economic trade-offs with any watershed management policy in the six forests. Curtailing unsustainable grazing, logging, and mining or not, either decision will have both positive and negative effects on the economy. To weigh the economic trade-offs of protecting salmon habitat and watersheds, one must examine the mechanisms by which forest resources contribute to the economy and determine which holds greater promise.

2. Watershed and salmon habitat protections would likely strengthen, not weaken, the national, regional, and statewide economies because (a) unsustainable grazing, logging, and mining in the watersheds in the six national forests diminishes the productivity of fish habitat and other resources; (b) workers and business owners in other industries and United States taxpayers heavily subsidize these activities, which therefore transfers funds from society as a whole to a small sector; and (c) curtailing these activities and arresting the

subsidies will stimulate jobs and incomes by reinforcing the state's and region's singular economic strength: the natural environment and the quality of life.

3. Acting now to stop further degradation of the watersheds and salmon habitat will only accelerate economic transitions that are occurring anyway. Idaho's cattle, timber, and mining industries are declining. Idaho's economic future lies elsewhere. The cattle, timber, and mining industries impose substantial costs on other industries and households. Thus, they impede jobs in other sectors and lower the standards of living of residents throughout Idaho, the Pacific Northwest, and the United States.

4. Protecting watersheds and salmon habitat can reinforce efforts to increase the economic opportunities for workers and families, and help diversify local economies. The economic forces causing the fall in relative wages of Idaho's rural residents and the economic dislocations in Idaho's household communities are powerful, pervasive, and national, even global. Allowing the habitat-degrading grazing, logging, and mining in the national forests to continue, or repealing the Endangered Species Act, will do absolutely nothing to arrest these forces. Other sectors and groups increasingly will challenge the preferential access to forest resources historically afforded the cattle, timber, and mining industries. As more information becomes available on the subsidies accorded grazing, logging, and mining on the national forests, business owners and workers in other industries, as well as taxpayers, will act to terminate them. As investors, banks, and other sources of capital become more aware of the damage that unsustainable grazing, logging, and mining activities inflict on the productivity of forest resources, they will reduce the supply of financial capital to the firms and households dependent on these activities.

5. Curtailing unsustainable activities will have minimal adverse consequences for Idaho's economy, and even these will be offset, especially in the long run, by positive impacts on other elements of the state's economy. Grazing, logging, and mining activities on the six national forests play a small and declining role in the state's economy. The levels of jobs and incomes, per acre of grazing, logging and mining, have fallen dramatically and will fall farther in the future. The sectors of the economy sensitive to the quality of Idaho's forest environment are larger than cattle, timber, and mining industries, and they are growing. The environmental legacy of conventional grazing, logging, and mining hinders growth in these sectors. Ending the subsidies to conventional logging, grazing, and mining on the six national forests will benefit those who pay the subsidies and stimulate investment, jobs, and incomes in other sectors.

These conclusions, that environmental degradation can lower incomes and depress economic conditions, have been supported by additional studies (Been 1994; Templet and Farber 1994).

Macroeconomic Forces Affecting Rural Communities

While environmental degradation has led to direct job losses and profound economic consequences for many sectors of the rural western United States, a number of other changes are occurring in regional, national, and global economies. Although there are many forces at play, five trends stand out.

The economy is becoming increasingly globalized.—Global corporations have broken down national boundaries. New technologies allow vast sums of money to flow instantaneously across the world, virtually uncontrolled. National governments are increasingly incapable of influencing the activities of these global corporations or the new global market. Corporations operating in the highly competitive worldwide marketplace are often ignorant of, or not particularly concerned about, the needs and desires of individual communities (Rifkin 1995). As a result, local communities are often left powerless to affect the consequences of corporate activities on the community or the local environment. A growing trail of fragmented local communities and environments have been left in the wake of the new global economy, not just in the rural western United States but worldwide.

Blue-collar jobs will continue to disappear because production is being increasingly separated from labor inputs and technology is continuously displacing workers (Rifkin 1995).—For example, in 1880 it took more than 20 hours of labor to harvest an acre of wheat. By 1916, due to mechanization, the labor was reduced to 12.7 hours, and just 21 years later only 6.1 hours were needed. In 1850, 60% of the working population was employed in agriculture. By 1900 the number had been reduced to 48%, and by 1990 the proportion of the population was just 3%. In the 1960s alone, long before the nation's environmental laws took effect, nearly 40% of the remaining agricultural workforce was replaced by machinery (Rifkin 1995).

The same long-term trends have applied to the mining industry and to manufacturing in general. In 1992, 45,000 mining jobs were eliminated in the United States. In 1900, 19% of Americans worked in manufacturing jobs, and by 1950, the heyday of mass production line manufacturing, the percentage rose to 34%. By 1990, the substitution of software and machines for humans, as well as other factors, had shrunk the manufacturing workforce back to 17%. The downward trend continues today (Rifkin 1995).

Labor needs have continued to decline even as productivity has increased. Despite the common belief that the United States has lost its manufacturing base, manufacturing productivity has actually risen by roughly half over the last 15 years (more than 2% annually since 1980) while manufacturing employment has fallen (Drucker 1993). International competition will continue to force steady increases in agricultural, natural resource extraction, and manufacturing productivity, and U.S. agricultural productivity will continue to outpace demand. Hence, there will be a continuous downward pressure on agricultural employment.

Further, the most important factor in productivity has shifted from physical work to strategic thinking (Reich 1992; Drucker 1993). These trends make the long-term prospects for blue-collar manufacturing, agricultural, and natural resource industry workers bleak: "The ratio of blue-collar workers in the total labor force was one in three in the 1920s, one in four in the 1950s, less than one in six today, and likely to be, at most, one in 10 by the year 2010" (Galston and Baehler 1995). The inevitable processes of decoupling production from employment and replacing people with technology remains the greatest cause of rural job loss and community fragmentation. Without innovative approaches to attract new investments that can establish new businesses and create jobs, many rural communities in the West are certain to face increased dislocation and pain.

There is decreasing demand for U.S. raw materials.—As production and technological efficiencies have improved, raw materials have become increasingly less important as inputs for production (Drucker 1989; Power 1995a). With the exception of wartime, the amount of raw materials needed per unit of output has been dropping throughout this century. "There is less 'stuff' in the stuff we buy today. Wood products, for instance, used to mean plywood and dimensional lumber . . . [but today] less of the total value of the product comes from wood and more of the value from engineering" (Conerly 1995). In short, although there may be a few exceptions, rural development strategies built upon sustaining or increasing demand for primary products or raw materials will fail (Galston and Baehler 1995).

Public investments in rural development and worker retraining have waned, and the trend will not reverse soon.—Financial pressures on all levels of government suggest that public funds will remain limited. Public programs are certain to employ fewer people and to focus on fostering private efforts, instead of bearing long-term financial burdens. The nearly one-fifth of rural jobs that depend on government activities will continually be threatened by government downsizing (Galston and Baehler 1995). Private investment will therefore be key, although private investment may remain limited as consumers struggle to get out from under debt at a time when real wages are stagnating in many regions. Lack of public and private funds suggests there will be less money available to fund investments in the rural western United States. The net result will be increased pressures and competition over existing resources. Again, new strategies will be required for communities to beat the competition to lure private and public investments.

Most regions of the rural western United States are going through major economic restructuring.—The process has created a confusing picture of apparently contradictory trends; most of the traditional, mainstay rural industries are declining just when the overall regional economy is booming. A recent major economic report (Power 1995a), endorsed by 65 independent economists from Montana, Oregon, Washington, Idaho, and California, summed up the restructuring occurring in the Pacific Northwest as follows:

Since the mid-1980s the economies of the Pacific Northwest states have outperformed the national economy according to almost any measure of economic vitality that one might choose to use. . . . Idaho's performance has been the most impressive, outpacing the national economy by almost a factor of four between 1988 and 1994. . . . The "worst" performances (Oregon and Montana) had employment growing at a rate "only" 2.2 times the national average and real earnings climbing at a rate "only" 2.5 times the national average. . . . The fundamental strength of this economic performance is all the more apparent given that it took place just as two industries, timber and aerospace, that historically dominated the region's economy were being dramatically down-sized. . . .

As the massive job losses in aerospace and timber were taking place, this region led the nation in job creation, income generation, and success in attracting new businesses and residents. . . . There are important lessons here regarding the resiliency and diversity of the regional economy. In particular, it is apparent that the economic fate of the entire region no longer rests solely, or even predominantly, on these two industries.

The Future of the Rural West

The implications of these trends for the rural western United States should be apparent enough: jobs and communities will continue to be affected by macroeconomic forces that local communities cannot easily influence. The traditional rural economic development strategies that rely on natural resource extraction and blue-collar manufacturing will not provide the jobs and incomes needed to maintain or enhance future economic vitality. Innovative new solutions must be developed.

As with most new ideas, creative solutions arise only after a problem is viewed in new ways. This can occur only if rural communities seek to understand the large-scale economic forces affecting them. Once this shift in perspective has occurred, they may begin to understand the importance of protecting, and where necessary restoring, their "new" comparative advantages: quality-of-life amenities, including environmental assets.

Drucker (1989) summarized the economic trend well by stating that the decoupling of output from employment (the changing relationship between the number of people employed and the amount of production) in the traditional economic sectors is likely to continue. New sources of rural jobs must be found, and to enhance long-term job prospects and community stability these jobs must be more responsive to the comparative advantages of the rural western United States and less exposed to global economic pressures.

Much of the rural West's major advantages are social and environmental quality-of-life factors. These attributes are the key to attracting new residents and investments that will establish new businesses and jobs. Environmental amenities are one of the most important factors in attracting new residents and investments (Galston and Baehler 1995; Niemi and Whitelaw 1995; Power 1995a).

Clean water, for example, provides a competitive advantage for many companies in the Pacific Northwest. Siltec, the largest manufacturing employer in Salem, Oregon, uses large quantities of water to wash silicon wafers and clearly sees abundant supplies of clean water as a major advantage and reason for locating high-technology firms in that state (Seideman 1996). According to Siltec's environmental engineer (Seideman 1996) "The bottom line is, water quality is a competitive advantage."

The 65 endorsers of the economic report discussed earlier summarized the economic importance of environmental amenities in the Pacific Northwest as follows (Power 1995a).

> Anyone working with the new residents and businesses coming to the Pacific Northwest is familiar with one factor that is driving the economy; the region is perceived as providing a superior, attractive environment in which to live, work, and do business. . . . The natural landscapes of the Pacific Northwest are its signature. They also appear to be a major drawing card. . . . These environmental features are not just aesthetic qualities that are nice but of little economic importance. [Rather, they are central to] the economic vitality of the region because of these two basic economic facts: (1) many people move to the region, and remain here, because they want to enjoy its high-quality living environment and (2) that growth in population stimulates the development of new businesses and the expansion of existing ones. . . . In short, one of the reasons jobs and incomes in the Pacific Northwest are growing faster than elsewhere in the United States is because the region's population is growing faster, and the population is growing faster because many people believe that this is a wonderful place to live, work, and raise a family.

WATERSHED PROTECTION AND RESTORATION INITIATIVES

One of the most important steps any rural community can take today to secure its economic future is to protect, and, where necessary, to restore, its environmental amenities. Healthy watersheds that produce ample, consistent supplies of clean water, productive soils, scenic and recreational opportunities, and self-sustaining native fish and wildlife populations provide some of the key environmental amenities.

Every community is located within a watershed. Each community either directly or indirectly depends on the "ecological services" provided by these watersheds for its well-being. Watersheds are highly integrated ecological systems. The water flowing through a watershed connects a mosaic of habitats in which energy, organisms, and materials move in complex yet highly integrated ways (Doppelt et al. 1993). This means that instream conditions are largely determined by processes that occur throughout the watershed. The decline and extinction of native fishes are often directly or indirectly associated with the degradation of watersheds themselves. The loss of native fishes, therefore, not only creates economic hardships for sectors such as the fishing and tourism industries, but is also often a key indicator of serious broadscale watershed health problems.

Many rural communities fear that conservation programs will harm local economic conditions. However, environmental regulations enacted in the United States over the past several decades have not created significant job losses (Meyer 1992, 1993; Goodstein 1995). In fact, U.S. Department of Labor figures show that environmental and safety regulations have led to less than 0.01% of all large-scale layoffs in the United States (Goodstein 1995). This should not be a surprise. Environmental protection involves investments in additional equipment and more personnel to operate and monitor the equipment; the restoration of environmental quality in a degraded setting requires the employment of economic resources just as any other productive activity does (Power 1995a, 1995b).

This is not to suggest that environmental policies have a totally benign effect on rural communities. There are localized impacts, especially on communities dependent on single industries. However, environmental protections generally cause economic dislocations only when a business is already nearing obsolescence, has declining sales, or has more efficient competitors elsewhere.

Data from across the nation consistently show that, in the aggregate, environmental policies have saved and created more jobs than they have eliminated. Studies indicate that at the national level the net cost of environmental regulation is quite low (Meyer 1992, 1993). To the extent that the rural backlash against environmental protection is aimed at ineffective environmental policies and poor delivery mechanisms, the outcry is valid. However, to blame the massive social and economic problems rural communities face today on environmental policies is simply factually wrong. Healthy watersheds, clean water, abundant fisheries, and other environmental amenities are the most important comparative advantages enjoyed by many rural communities in today's marketplace.

All of the factors discussed above converge to form at least one conclusion: protecting and restoring watersheds will be critical to the future economic health of many rural communities. Restoration initiatives cannot only help save some existing jobs, such as those in the fishing industry; they can also create jobs in habitat restoration and help protect or restore key environmental amenities that are vital to attracting new residents and investments. Each of these benefits will help rural communities gain more control over their futures and buffer themselves from global economic forces.

Watershed restoration initiatives can mean jobs and lots of them. For example, employment can be provided for those displaced by the changes in traditional farming and resource extraction economies. Restoration can allow many local workers to remain employed in the outdoors, using their existing or similar skills, such as heavy equipment operation and restoration forestry.

The creation of jobs in watershed restoration can parallel the jobs and businesses being created in other environmental clean-up industries. Revenues and sales of the products from the environmental industry directly supported about 1.1 million jobs in 1993 and are projected to grow to 1.3 million jobs by 1998 (Goodstein 1995). Interestingly, studies have found that most of these jobs have been created in the private sector. However, the businesses and jobs were created only after the government shifted policy toward the restoration of natural resources.

The Question of Sustainability

One of the most important issues that communities throughout the rural West must address is the level of protection and restoration needed to maintain environmental amenities, and therefore, the communities' long-term comparative economic advantages. The level of protection and restoration required to achieve a comparatively better environment than that found elsewhere may be considerably less than that needed to maintain properly functioning watersheds and sustainable fisheries.

The larger question of how to make the needed transition to an ecologically sustainable society is not the focus of this chapter. Instead, the primary point made here is that there are clear economic benefits to protecting and restoring watersheds. It is important to understand that economics alone, however, will not sufficiently define the levels of protection and restoration needed to maintain sustainable biological systems. Sustainability must be the ultimate goal, which means that many other issues, such as ethics and responsibilities to future generations, must also play central roles in decision making.

A Spectrum of Restoration Programs

Two general types of watershed restoration initiatives are needed: those primarily supported by government agencies on public lands and those focused on private-land watersheds. These are often intermixed.

Federal efforts.—One of the best examples of an attempt to establish a large-scale, public-land watershed restoration and jobs program recently unfolded in the Pacific Northwest. The lessons learned from this program can help guide similar efforts elsewhere.

On 2 April 1993, President Clinton convened his Northwest Forest Conference in Portland, Oregon, to learn firsthand about the environmental, social, and economic problems revolving around federal forest management in the Pacific Northwest. At the conference, scientific evidence about the degraded conditions of the region's forests and watersheds was shared, along with the information that 60,000 fishing-related jobs were dependent upon habitat in watersheds where ecological functions remain healthy. Out of this conference arose two new interrelated initiatives: the Forest Ecosystem Management Assessment Team (FEMAT) report (FEMAT 1993) and the associated Record of Decision for lands administered by the U.S. Forest Service and U.S. Bureau of Land Management. These became known as the Northwest Forest Plan and the Northwest Economic Adjustment Initiative, which included the Jobs in the Woods restoration employment program. The Northwest Forest Plan laid out a new approach to federal forest management in western Washington, western Oregon, and northwestern California. The entire plan was built upon a new watershed protection and restoration program called the Aquatic Conservation Strategy, which included a regional emphasis for watershed restoration to protect and restore endangered salmon habitat and stream ecosystems.

The watershed restoration program as first described in FEMAT (1993) is arguably the most far-reaching, ecologically based watershed restoration and job creation program ever attempted in this nation. Although not a hard and fast rule, restoration treatments initially emphasized activities designed to prevent further harm to the most biologically intact remaining watersheds. Activities focused on forest road improvements, such as road obliteration and decommissioning, and sediment reduction projects, such as culvert repairs and replacement. After the best remaining areas were secured, the strategy called for restoration of the more degraded areas through riparian reforestation and some instream projects.

The watershed approach was new and faced many obstacles. Consequently, although projects were implemented, nearly all the funds expended, and some dislocated workers found employment, the program suffered many bumps and bruises in the first 2 years. For example, the U.S. Congress wanted the entire $27 million appropriated for the project expended and wanted projects to begin implementation within the initial fiscal year. This left insufficient time to undertake National Environmental Policy Act reviews and still have ecologically effective projects underway in 1994. The goal of establishing broad peer review and public participation in the process was thwarted by the Federal Advisory Committee Act, which makes it difficult for the federal government to relinquish decision-making authority to local citizens. In addition, U.S. Forest Service and U.S. Bureau of Land Management staff were overburdened, some managers resisted the program, and there was a lack of sufficiently trained staff to implement the program. Yet, many agency personnel and others went the extra mile to see that much was accomplished (Tarnow 1995).

No procedures were established to monitor or assess either the short- or long-term appropriateness or effectiveness of specific restoration projects. Given the push to spend the funds and get projects on the ground, most projects were taken off the shelf, that is, they were planned prior to the new restoration goals of FEMAT (Tarnow 1995).

The technical aspects of the program have improved every year since its inception. However, Congress fundamentally altered the program in 1995 by cutting the watershed restoration budget, insisting that timber harvest once again become the agencies' top priority, and passing a timber salvage rider that forced many damaging timber sales to proceed in key watersheds and riparian areas. Nevertheless, the watershed restoration program created by the Northwest Forest Plan is sound in principle and holds great promise. It can and should be used as an example for others to follow to establish restoration programs.

The Northwest Economic Adjustment Initiative was the correlated program intended to help workers, businesses, communities, and Native American tribes that have relied on a forest-based economy to adjust to reduced timber harvest levels. A significant part of the initial funding, $27.8 million in fiscal year 1995, was aimed at employing dislocated workers and others in the watershed restoration component of the Northwest Forest Plan. This employment initiative became known as Jobs in the Woods. Its objective was to combine priority watershed restoration work with longer-term, good-wage jobs for dislocated workers. During fiscal year 1995, this program employed more than 2,200

TABLE 27.1.—Workforce by state for the Jobs in the Woods Program during fiscal year 1995. These numbers reflect only those jobs completed by 1 March 1996, or about 62% of total dollars allocated (data from U.S. Department of the Interior and U.S. Department Agriculture 1996).

Workforce measure	State			
	Washington	Oregon	California	Total
Number of worker-days generated	16,146	31,178	11,525	58,849
Number of workers employed	435	1,350	441	2,226
Number of displaced workers hired	163	621	226	1,010
Average wage and benefit (per hour)	$19.08	$15.33	$20.01	$18.14

workers in the Pacific Northwest at an average wage of $18.14 per hour for the purposes of obliterating roads, replacing poorly designed culverts, and restoring fish habitat (Table 27.1).

The economic initiative's watershed restoration contracting program was aided by a federal waiver that allowed contracts to be awarded only to contractors from the target counties (those including timber-dependent communities) in the three states. Although helpful, this restriction did not ensure that local, dislocated timber workers gained employment in watershed restoration. More support was needed to assist local contractors to compete for contracts and to help connect local, dislocated workers with the contractors. In addition, multiyear stewardship contracts were needed, because such contracts would have offered benefits to contractors, workers (job predictability), and watersheds (Tarnow 1995). Furthermore, agency decision makers were not equipped to make the hard decisions of steering money toward higher-priority restoration projects and away from some needy communities (Tarnow 1995). In the end, almost all of the $27.8 million appropriated for Jobs in the Woods in fiscal year 1995 was spent, and over 600 contracts were awarded. However, because of the obstacles discussed above, the job effectiveness of the program was limited.

Nevertheless, the Jobs in the Woods program has tremendous potential. In late 1994, a demonstration project developed in Sweet Home, Oregon, met with a high level of success. Ten dislocated workers were retrained in watershed restoration and contracting and were employed over 6 months. In 1995, building on the success of the Sweet Home pilot project, numerous other pilot projects began in the three states. Two exciting projects have taken Jobs in the Woods to a whole new level. First, the Ecosystem Workforce Curriculum Committee has begun work on a universal curriculum that would be used to train all ecosystem workforce participants in the Pacific Northwest. Second, the states of Oregon, Washington, and California jointly proposed a natural resource partnership proposal, a cost-effective public–private cooperation. This program uses the Oregon Options Model (developing a series of performance standards and benchmarks for success) to develop a partnership between federal and state agencies to revolutionize the ecosystem restoration field. The governors of all three states endorsed this proposal, although it seemed to stall as it made its way toward the White House.

On private lands a number of watershed restoration and jobs creation programs have evolved in the western United States. Some were created by government, and others have evolved at the community level. For example, in fiscal year 1995, the National Oceanic and Atmospheric Administration established a program to employ in restoration efforts commercial fishers affected by the loss of Pacific salmon. Fifteen million dollars were appropriated in 1995, and $15 million more were targeted in 1996. Some commercial fishers were hired by local watershed councils to do restoration work, and some became involved with the pilot projects that evolved out of the Jobs in the Woods program. However, anecdotal evidence suggests that the program has achieved limited employment success so far.

State efforts.—In 1993, the state of Washington established a program called Jobs and the Environment, the goal of which was to hire dislocated timber workers in watershed restoration. In total, $6.5 million was appropriated through the 1993 to 1995 budgets and dispersed through a competitive grants process to 13 watershed restoration projects that trained and employed displaced forest products workers. A portion of the funds was used to support 22 Washington Conservation Corps restoration projects that employed displaced forest products workers and disadvantaged youths (Office of Financial Management, Washington State, unpublished data).

The state of Washington also established a watershed restoration partnership program and appropriated $10 million in 1994. The program provided some funding for local stream enhancement groups and other efforts (Office of Financial Management, Washington State, unpublished data). Despite these efforts, most restoration and jobs programs in Washington State remain limited in scope, outlook, and effectiveness.

The state of Oregon established a watershed health program in 1993 to restore watersheds and native fishes statewide. The program is set up to empower local citizen groups, called watershed councils, to develop and apply restoration strategies. By 1996 over 60 local watershed councils existed throughout Oregon. These councils have focused on public education and stewardship building, watershed assessments, and implementing restoration plans. Although local employment is not a primary goal of the program, many local jobs have been created. Some initiatives, such as that of the McKenzie River Watershed Council, have proven very effective. Others have evolved slowly, and still others have become captured by special corporate interests that have used the programs mostly as public relations "greenwash," while thwarting meaningful restoration. Nevertheless, Oregon is one of the states farthest in the lead on establishing a large number of community-based watershed restoration initiatives. Numerous other public and community-based watershed restoration and jobs programs have begun in states across the nation (see case studies in this volume).

A Note of Caution

It should be noted that a good deal of controversy exists about the types of activities that can best help restore watersheds, and therefore the types and numbers of restoration jobs that can be created. Certain restoration activities seem to have clear ecological and economic benefits. For example, few would argue about the ecological benefits of reducing the threats of catastrophic landslides from forest roads through activities such as replacing culverts and obliterating inappropriate roads. The job creation potential for those with heavy equipment skills is also high. However, the benefits from many other activities, such as some instream channel modification projects, are much less certain.

Often, the best form of restoration is to protect an area from further degradation, eliminate the factors causing further disturbance, and give the system time to recover naturally (Frissell 1997, this volume). Although some careful experimentation is needed, given the limited experience and knowledge we have with ecological restoration, active restoration and job creation should focus primarily on those activities that are known to produce ecological and economic benefits and avoid those that may have risks.

SUGGESTIONS FOR RESTORATION PROGRAMS

Despite the tremendous economic and ecological benefits that watershed restoration initiatives can provide for rural communities, many obstacles exist. A bold new vision is needed to remove the obstacles and foster effective programs throughout the nation. The following recommendations are offered to help foster watershed restoration programs.

1. Rural community leaders, elected officials, and public agencies alike need information, technical assistance, and financial support to shift from looking at their economies through their "rearview mirrors" (believing that the future will look like the past) to taking a forward-looking approach that embraces the critical role that quality-of-life and environmental amenities, such as healthy watersheds and clean water, will play in their economic futures. There must be a special effort to correct the inadequacies of traditional economic analysis because, as long as these shortcomings persist, there is little hope for overcoming the inertia that sustains the rearview mirror approach to rural economic development.
2. All levels of government must make long-term policy commitments to restoration. At the least, government must not undercut restoration efforts once they begin. The undermining of the Northwest Forest Plan due to congressional action is a good example of how restoration can be thwarted when policies and funding levels constantly change.
3. Independent funding mechanisms that are separate from state or federal appropriations processes are needed for restoration programs. Both public and private restoration programs consistently struggle for stable, long-term funding. Long-term stability is needed, and political demands for immediate action must be avoided. Long-term funding can take the form of (a) regional restoration trust funds, such as was established in 1993 through the California Central Valley Project; (b) community restoration revolving loan funds that can provide small grants and revolving loans, such as has been proposed by

the Pacific Rivers Council; and (c) community development banks, such as is being created in Willapa Bay, Washington, through Ecotrust and Shore Bank, as well as other options.

4. The institutional barriers to linking restoration with long-term jobs in federal land programs must be removed. Such institutional adjustments would include (a) continuing and expanding contractor assistance programs to empower local contractors to compete effectively for agency contracts and ensuring that contractors are committed to both high-quality work and good-wage jobs; (b) securing agency cooperation to remove barriers to innovative partnerships; (c) reforming project design and packaging to ensure the development of larger multiple-objective, multiyear contracts; (d) improving local contractor access to bonding (and increasing bonding capacity for multiple-project application) through programs such as the Small Business Administration's Surety Bond Guarantee; (e) providing business development support to expand local contractor's business capacity for doing restoration work; (f) providing and coordinating wage and hour enforcement for compliance with the goal of high wages for high-quality work; and (g) reducing and streamlining the bureaucratic process required to implement restoration programs (Tarnow 1995).

5. Public and private investments must be sought to foster and support innovation and entrepreneurial private business ventures that can enhance local job opportunities and protect and restore watersheds. This investment can help communities in the rural western United States create jobs while maintaining and restoring vital environmental amenities.

6. Finally, rural communities must move quickly to protect their options. To have a chance to restore watersheds successfully, the best remaining habitats must be protected immediately. Restoration must be built around those protected habitats. Restoration treatments without the initial protections is meaningless and usually a great waste of money and time.

CONCLUSION

As the twenty-first century speeds toward us, it is clear that swift and profound economic forces will continue to change the rural landscape. To navigate these turbulent waters, we need to heed Cato the Elder's sage advice to a farmer contemplating the purchase of land: "Go in and keep your eyes open, so that you may be able to find your way out." Similarly, rural communities need to keep their eyes open to understand what is really occurring in order to determine how to proceed. Looking in the rearview mirror, or flying blind, will not help rural communities secure their economic futures. Only by maintaining and restoring quality-of-life amenities, including environmental amenities such as healthy watersheds, can many rural communities survive and prosper.

*We have probed the Earth, excavated it, burned it, ripped
things from it, buried things in it, chopped down its forests,
leveled its hills, muddied its waters and dirtied its air. That
does not fit my definition of a good tenant. If we were here on
a month-to-month basis, we would have been evicted long ago.
—Rose Bird, Chief Justice of the California Supreme Court,
1977 to 1987 (quoted in Ohnstad 1985)*

CHAPTER 28

ECO-SOCIETAL RESTORATION: CREATING A HARMONIOUS FUTURE BETWEEN HUMAN SOCIETY AND NATURAL SYSTEMS

John Cairns, Jr.

Sustainable use of the planet will not be possible unless a new relationship develops between human society and natural systems. The first step in establishing this new relationship will be to establish a balance between ecological destruction and repair. This balance can be achieved by simultaneously reducing the rate of destruction and increasing the rate of repair until a steady state is reached that does not deprive future generations of either the ecological services (defined as those ecological functions perceived as beneficial by human society) or amenities that present generations now enjoy.

Eco-societal restoration is defined herein as ecological restoration with the human component of the ecosystem actively participating in the process, which usually requires a willingness to alter social behaviors to enhance the integrity of natural systems. The concept is based on the assumption that successful ecological restoration is most likely to occur at the landscape level (a scale large enough to include the heterogeneity in ecosystems) and both large temporal and spatial scales are routinely to be involved (e.g., NRC 1992a). Undertakings at a landscape level are either unlikely to be initiated or, once accomplished, unlikely to endure if human society affecting these ecosystems does not support both the restoration itself and the maintenance thereafter (e.g., Cairns 1994).

Ecological restoration at the landscape level under such conditions clearly transcends the boundaries of classical ecology as it has been practiced since the field developed. Eco-societal restoration should involve disciplines such as engineering, economics, sociology, ecology, political science, and, of course, ethics and philosophy. Too often, specialized disciplines are separated by highly specialized disciplinary jargon and by substantive perceptual gulfs. For years I have struggled with frighteningly large interdisciplinary teams (the largest one included 52 specialists from 14 departments) required to generate a database perceived as necessary to resolve complex environmental problems. Specialists often gather information according to the dictates of their peers rather than the needs of the decision makers. About 15 years ago, I came to the conclusion that a few multidimensional people were more amenable than teams to resolving issues that transcended the capabilities of individual disciplines. I have recently been encouraged to see that this process is beginning to occur in industry as evidenced by the bestseller, *Reinventing the Corporation* (Hammer and Champy 1994). Frequently, data are lacking on crucial issues (e.g., how many species now occupy the planet and how fast are they being destroyed?), and even when robust data are available (e.g., rate of destruction of old-growth forests), conceptual and bureaucratic frameworks are often lacking for addressing the issues effectively.

This chapter examines the relationship between human society and natural systems that may permit sustainable use of the planet more effectively than practices presently used. There are six options with regard to eco-societal restoration (modified from Cairns 1995):

1. Cease ecological restoration or limit it to research. If, during the remainder of this century and the first half of next century, human society does not reach a new relationship with natural systems, attempting to repair these natural systems makes little sense because they will not persist unless society's attitudes change. It would be a pity if this option were selected, either through ignorance or resistance to change, because it would be giving up hope of having sustainable use of the planet and would show little or no compassion for future generations.

2. Continue present practices, in which ecological destruction far exceeds ecological repair and population growth continues unabated, until there is more certainty about the consequences of these strategies. This option assumes that, once consequences have become apparent, society will not have precluded other options. Because there is uncertainty about how the natural portion of the life support system will function under new conditions, some might argue that acting on uncertain evidence is not wise. Society may be able to destroy even more of the Earth's natural systems without incurring major penalties, but the Earth may also have already exceeded some critical ecological threshold. Regrettably, this is a global experiment with no prece-

dent, and the degree to which it is reversible is equally uncertain. Even if the chance of unfortunate consequences is 1 in 20, society should act more prudently because the outcome of the one chance might be catastrophic. There is no robust evidence that present practices constitute anything close to those necessary for sustainable use of the planet.

3. Establish a balance between destruction and repair of natural systems—a no-net-loss policy. Damage to an ecosystem can occur within a matter of minutes or hours (as was the case with the *Exxon Valdez* oil spill in Prince William Sound, Alaska) but may take years, decades, or even centuries to be repaired. Therefore, a true no-net-loss policy would mean that some ecosystems now impaired would be restored in anticipation of ecological accidents that would damage other ecosystems. As a consequence, the no-net-loss strategy must necessarily involve regions, landscapes, or entire countries. Achieving a no-net-loss balance in a small political area, such as a county, might well not be possible. The National Research Council (NRC 1992a) recommended establishing a National Restoration Trust Fund, which would be extremely helpful in implementing a no-net-loss policy. Exercising the no-net-loss option with continued human population growth would mean that the acreage of natural systems per capita would continue declining.

4. Restore damaged ecosystems at a greater rate than they are being damaged, that is, exceed a no-net-loss policy. Exceeding no net loss of natural systems by restoring more systems than are destroyed would require a dramatic departure from present practices in virtually every country in the world. Yet, if the human population continues to grow at its present rate, the per capita ratio of natural systems to each individual in the human population will continue to decline, although less precipitously than in option 3.

5. Adopt a policy of no net loss of natural systems coupled with a stabilized human population. This option would maintain the status quo in the amount of natural systems per capita. It would, of course, represent a dramatic departure from present practices. Furthermore, this is the first of the six options that fits the strategy of sustainable use of the planet, making it reasonably probable that future generations will have the same relationship with natural systems that exists when both the population and the acreage of natural systems remain in a constant relationship. Establishing a policy of no net loss of natural systems is obviously not going to be effective for sustainable use of the planet if the natural systems are all but eradicated by the time that the acreage of natural systems stabilizes. Also, stabilizing the human population at, for example, 9 to 15 billion, does not ensure that sustainable use for many generations is possible. Even if it did, one wonders whether the quality of life would be acceptable.

6. Restore ecosystems at a greater rate than they are being damaged and stabilize the human population. If this option can be accomplished sometime early in the next century, it has a very reasonable chance of ensuring sustainable use of the planet and an improved quality of life. Before rejecting this option, policy makers should consider whether it is best to have a huge number of

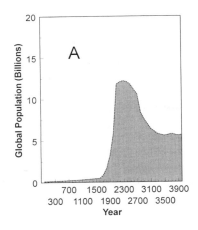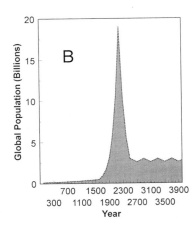

FIGURE 28.1.—Alternate population growth scenarios. (A) Population growth is controlled, ecosystem services are protected, and a larger number of people live over a longer time. (B) Population is not controlled, ecosystem services are damaged, and the population stabilizes at a lower level.

humans on the planet in the very near future or a much larger number of humans over a longer period of time (Figure 28.1). The latter would almost certainly provide a better quality of life for each individual, although adopting this policy would require much more sacrifice from present generations than any of the other options.

Obviously, it would not be prudent to postpone selecting an option until the world population has reached 15 billion and the remaining old-growth forests and other quality ecosystems disappear. What could be used for restoration models and from where would species for recolonization come?

The United Nations Population Fund (UNPF 1991) felt "optimistic" that the planet's human population might stabilize at 9 billion by 2050. Because most of these more than 3-billion additional persons will be added to developing countries, where environmental destruction is frequently rampant, the only cause for optimism seems to be that things could get worse. In fact, they well could get worse, as Garrett (1994) abundantly documented. She believes that human society has grossly underestimated the challenge to *Homo sapiens* by microscopic competitors for domination of the planet and notes that the debate is not about whether the threat exists but the whys, hows, and whens of its occurrence. One example cited by Garrett is Siddique et al. (1991), who found that the El Tor strain of cholera was capable of shrinking itself 300-fold when suddenly plunged into cold salt water. Obviously, it was then even more difficult to detect, particularly when the search was made by relatively untrained people, which would be necessary for monitoring its presence or absence on a large scale. Rogers and Packer (1993) are convinced that global warming would greatly expand the territory and infectivity ratio of the East African tsetse fly, which is responsible for transferring the trypanosomes that cause sleeping sickness.

The message in Garrett's (1994) book is loud and clear—large predators such as the tiger, lion, and wolf may have been either eliminated or virtually eliminated—but increases in human population densities have lowered resistance in starving people. In addition, rapid transport of people and diseases by airplanes and the like has increased human society's vulnerability to organisms that cannot be seen without microscopes or even sometimes cannot be found with microscopes when they are in low densities. Garrett believes not only that the factors just mentioned are important, as they definitely are, but also that by altering habitats worldwide, society has driven some species, such as the rodent carriers of hanta virus, into closer proximity to humans and, as in the case of Lyme disease, has increased the abundance of the carriers (white-tailed deer) so dramatically that their overpopulation forces them into ever closer proximity to humans. Garrett's book opens another vista in the relationship between human society and natural systems, namely that destroying them and polluting the environment facilitates plagues.

OUR RELATIONSHIP WITH FUTURE GENERATIONS

Thomas Jefferson believed no generation should pass on any debt to succeeding generations. Jefferson probably included in the term debt the legacy of nature, although it was expressed in the context of his particular time and the lexicon of that day was not the same as that of today. However, it is abundantly clear that Jefferson had a strong empathy for natural systems. Converting his philosophy to current terminology means not destroying ecological capital (e.g., old-growth forests, groundwater aquifers, the integrity of natural systems, and the like) at a rate faster than it can be restored or replaced. I am well aware that Jefferson, the individual, was often deeply in debt, especially toward the end of his life. But Jefferson, the president, managed the country's resources frugally and saw to it that the country did not pass on any large debt to succeeding generations. Almost certainly, Jefferson would support in principle the concept of sustainable use of the planet, although he would almost certainly have difficulty putting such beliefs into practice. Passing resources from one generation to another requires enormous restraint, especially when there are many persons less affluent, even in the United States, than society would like. At the same time, all developed countries have been responsible for activities in developing countries that would not be permitted in their own.

The basic question facing global human society is whether it wishes to optimize the number of humans that can exist on the planet during the next century or during the next million years. Each choice requires a quite different strategy. It seems highly improbable that the level of use of natural resources characteristic of Europe, North America, and a few other industrialized areas of the world could possibly be reached by the present global human population of 5.6 billion and even more unlikely for the projections of 9 to 15 billion somewhere before the first half of the next century. Ecological capital, such as old-growth forests, simply will not sustain both an expanding population and increased per capita affluence. The major problem is how to shift from a growth to a maintenance society because perpetual material growth is not compatible with sustainable use. Many believe that cessation of growth is equivalent to stagnation. This is not true for humans for which increase in size stops relatively early in life; one hopes that intellectual growth and other types of experiential growth continue.

HUMAN SOCIETY'S LIFE SUPPORT SYSTEM

For most of the time that humans have existed on the planet, their life support system has been entirely ecological. Natural systems maintained the atmospheric gas balance, provided food and fiber, maintained water quality, and reabsorbed the harvested materials once they had been used. In other words, not only were human life support systems entirely ecological but both the extraction and use of natural resources was such that the waste products and discarded products were generally readily reabsorbed by natural systems. Then came the agricultural and, subsequently, the industrial revolutions, which permitted humans to increase in number and become concentrated in urban areas. Consequently, for the first time in history, human society is now dependent upon a life support system that is both ecological and technological.

If the idea of a two-component life support system is accepted, with a concomitant acknowledgment that the technological component operating without restraint can damage the ecological component, then a balance must be achieved so that the amenities and the necessities provided by both components are optimized (Cairns 1996). However, simultaneously maintaining the integrity of both systems requires that development of the technological component of the life support system be restrained whenever it threatens the integrity of the ecological component. Restraining the activities and development of the technological component of the life support system inevitably requires changes in human societal behavior.

There is persuasive evidence that human behavior can be altered not only to protect but to restore damaged ecosystems (NRC 1992a; Cairns and Pratt 1995). Geller (1991) developed some schematics identifying three basic types of actively caring. He espouses intervention, attempting to influence another individual's behavior in desired directions through giving, rewarding, or correcting feedback; demonstrating or teaching desirable behavior; and developing or implementing a behavioral change intervention program. Clive Seligman was quoted by Geller (1994) as saying "Unless business can make money from environmental products or politicians can get elected on environmental issues, or individuals can get personal satisfaction from experiencing environmental concern, then individuals and organizations will simply do whatever competes with environmentalism if they see the payoff as greater." In one of the classic publications on human behavior, Skinner (1938) argued persuasively that human behavior is selected for or determined by its consequences and that substantial numbers of people cannot be expected to change their behavior as a result of information or advice alone, especially when the information is about a distant future. Skinner further stated that people might follow advice when the information from the advice giver has led to beneficial consequences in the past; however, the situation requires that people experience the reinforcing consequences of prior compliance with similar advice givers or similar rules. Such learning or response selection by reinforcing consequences is difficult or impossible when reinforcement lies in the distant future or punishing consequences are unclear, uncertain, or remote. Lack of clarity, uncertainty, and remoteness are all common characteristics in making judgments about complex, multivariate systems such as ecosystems.

It is highly probable that the average person gives little, if any, daily thought to the ecosystem services benefitting him or her. Ecosystem services are those ecosystem functions deemed useful to human society. An illustrative list of ecosystem services (Westman 1978) includes:

1. capture of solar energy and conversion into biomass, which is used for food, building materials, and fuel;
2. decomposition of wastes such as sewage;
3. regeneration of nutrients in forms essential to plant growth (e.g., nitrogen fixation);
4. storage, purification, and distribution of water (e.g., flood control, drinking water purification, and transportation);
5. generation and maintenance of soils;
6. control of pests by insectivorous animals, such as birds, bats, and insects;
7. provision of a genetic library for development of new foods and drugs by means of both Mendelian genetics and bioengineering;
8. maintenance of breathable air;
9. control of both micro- and macroclimate;
10. provision of buffering capacity for adapting to changes to recover from natural stresses such as flood, fire, and pestilence;
11. pollination of plants, including agricultural crops, by insects, bats, and other pollinators; and
12. aesthetic enrichment from vistas, recreation, and inspiration.

THE COEVOLUTION OF HUMAN SOCIETY AND NATURAL SYSTEMS

Coevolution has been defined as the simultaneous development of adaptations of two or more populations, species, *or other categories* (emphasis by author) that interact so closely that each is a strong, selective force on the other (Raven and Johnson 1986). Arguably, human society and natural systems have such a relationship. Failure of the ecological life support system to deliver services upon which human society depends is a real possibility. Change in such a way that there is a hostile relationship between human society and natural systems dominated by "pests" is also possible. There is no question that human society can affect natural systems. The question that most people would debate is whether natural systems can have a major effect on human society. Books such as Garrett's (1994); episodic events such as floods, earthquakes, volcanos, and hurricanes; and ozone holes show that human society is indeed vulnerable. However, some members of human society still think environmentalists are wrong (e.g., Chase 1995) or that human society's problems should be solved before spending money on the environment (e.g., Preston 1995). The real question is whether human society can ignore effects on natural systems without penalty while it pursues short-term agendas.

There appear to be two types of coevolution possible between human society and natural systems (Cairns 1994): (1) a benign, mutually supportive relationship—that is, the model of those who visualize sustainable use of the planet (e.g., The World Commission on Environment and Development 1987), or (2) a hostile coevolution—that is, human society eliminates all species not resistant to anthropogenic effects and is left with those species impossible to control, such as pests. Arguably, it is impossible for human society to eliminate all species other than domesticated animals and plants without itself suffering major deleterious effects. By eliminating the less-resistant species, the advantage goes to pests, which not only cause the most trouble but are also the most likely to persist. This hostile coevolution might lead to an ecological arena in which human society is battling uncontrollable species for scarce resources.

HUMAN SOCIETY AND PESTS

The Four Seas are rising, clouds and waters raging,
The Five Continents are rocking, wind and thunder roaring.
Away with all pests!
Our force is irresistible.

(Mao Zedong as quoted in Horn 1969)

Despite the fervor with which Mao's exhortations were followed many years ago, pests were not eliminated from the People's Republic of China. In the world's most populous country, where enormous numbers of people could be thrown at a problem by the words of one man in a way rarely achieved by human society, not only were pests not eliminated but arguably not significantly damaged.

If other species possessed the cognitive abilities of *Homo sapiens*, undoubtedly humankind would be regarded as a pest. Humans have co-opted 40% of the Earth's photosynthetic energy (Vitousek et al. 1986), have destroyed a substantial amount of the habitat for other species, have eliminated a huge number of other species (E. O. Wilson 1992), and are still expanding quite rapidly worldwide into habitats formerly occupied by other species. Pests are those species that compete with humans for grain, foodstuffs, and the like; grow where society does not wish them to grow; carry diseases or are causes of disease; or irritate in a variety of other ways. They are frequently species quite capable of coexisting with humans, such as rats, cockroaches, houseflies, sparrows, and coyotes. Society has not been able to eliminate pests without danger to itself, those domesticated species humans cherish, or the wildlife that society does not wish to destroy. Garrett (1994) argued persuasively that plagues are becoming more prevalent precisely because habitats of species have been destroyed and the displaced species are now invading societal habitats (as in the case of the carriers of hanta virus) or society has created habitats beneficial to a disease-carrying species (the white-tailed deer and Lyme disease). She argues that humans are at risk precisely because they have felt themselves immune from the laws of nature and have believed that they could control nature (in this case, disease-causing organisms) to a much greater degree than they actually can. Garrett's book provides one highly probable way in which nature can strike back at human society, through disease and pests, and one for which, particularly in impoverished countries, control will be extraordinarily difficult. There are other ways nature can strike back as well, such as the loss of some or many of the ecosystem services mentioned earlier. It is worth remembering that more than one of these catastrophes can occur simultaneously and that, as is often the case under these circumstances, the negative effects may be more than additive (i.e., synergistic).

THE EVOLUTION OF A NEW RELATIONSHIP

Clearly, if each individual acts in such a way that sustainable use of the planet is possible, human society will have progressed far beyond its present exploitive relationship. The new relationship would go beyond destroying ecological capital for short-term gain, thus permitting future generations at least the possibility of enjoying the natural amenities now enjoyed. Reexamining some of the established dogma and platitudes that impair society's relationship with natural systems may help diminish the barriers that impede establishing a sustainable relationship with natural systems. Some illustrative questions that may aid this process follow.

1. Is there anything worth pursuing apart from money and self-interest? If the answer is no, it is probably not worth reading further, but it is worth considering whether sustainable use of the planet is possible if sufficient numbers of people feel this way.
2. If each of us were to detach ourselves from immediate preoccupations, such as private property rights and the like, and examine the planet as a system with both temporal and spatial dimensions, what would be our place in this system? If we view ourselves and our property as part of an ecological landscape with a temporal dimension, can we expect society to continue supporting private property rights if exercising these rights is more destructive than constructive to societal goals?
3. Is it possible to have a meaningful life if one's goals are narrow and self-centered? Is there a deeper satisfaction from destroying natural systems to achieve these narrow, short-term goals or would life have more meaning if members of society acted in such a way that sustainable use of the planet would be possible if other members also did so?
4. Following hurricanes, earthquakes, floods, and terrorist bombings, many good qualities of human nature are exhibited (e.g., consideration for others, risking one's life for the greater good of society, foregoing individual prerogatives to assist others, and so on), but it is curious that these qualities seem to be most

evident when counterbalanced by the worst in human nature, such as sacrificing others to prove an ideological point. It is ironic that comparable violence to other species is not only tolerated by much of human society but is applauded as progress. Does this mean that humankind's sense of equity and fairness is limited to its own species, and then only when there has been a flagrant violation of the guiding beliefs? Or, is it possible that the ethos, or set of guiding beliefs, together with equity and fairness extends to all species on the planet—with whom the fate of human society is inextricably linked?

5. Tullock (1994) contrasted human and nonhuman societies in a very interesting fashion. He noted that he can put devices on his car to reduce the amount of air pollution released and thus reduce the amount of pollution in the city of Tucson at a considerable personal expense. He noted, however, that assuming his system reduces the total amount of pollution by 1/100,000 (and thus a comparable reduction for the contaminants he breathes), installing pollution-reducing devices would not be worth the cost to many citizens. Tullock rightly concluded that while some people might be sufficiently altruistic to install such devices, most humans are not as charitable. Humans, in this case, turn to coercion and the use of government by voting in favor of compelling all the citizens of Tucson to install pollution control devices on their cars. This coercion is only accepted if there is some formal system for producing it, and society may then be satisfied, or perhaps happy, about it. Tullock further noted that there is no comparable system of coercion in insect societies and generally, with some possible exceptions, not in any nonhuman societies. Ants and other social insects do not use coercion on their own members; they must have some other method of handling the problem of free riders (those who benefit from society but do not necessarily contribute to it), and this method has something to do with their genetic structure. If human society does indeed lack such genetic structure, is it possible that social evolution will remedy this in a fashion (i.e., social evolution) described by Ornstein and Ehrlich (1991)?

6. What will the world look like in the year 2050 if present practices are continued?

7. What will the world be like if human society breaks present barriers to establish a good relationship between itself and natural systems?

8. What might the planet be like in 2050 if eco-societal restoration becomes widespread?

Vice President Gore (1992) stated "Those who have a vested interest in the status quo will probably continue to stifle any meaningful change until enough citizens who are concerned about the ecological system are willing to speak out and urge their leaders to bring the earth back into balance." However, can humanity wait for leaders to act? How much environmental destruction can the planet tolerate (E. O. Wilson 1992)?

Damaging disturbances of ecosystems with robust or even modest ecological integrity should be accompanied by restoration of similar ecosystems in the same region or repair of the damaged ecosystem itself. All altered ecosystems (e.g., surface mines) should be bonded for restoration once the temporary use has ceased. If the use is intended to be permanent, that is, 20 years or more, then replacement in kind elsewhere should be immediately undertaken. Merely protecting existing ecosystems, however laudable, will not change the already established ratio of destruction and repair.

NATURAL RECOVERY PROCESSES

One of my first major experiences with the damage of roughly 100 miles of river (Cairns et al. 1971) encouraged the belief that natural recovery would make restoration efforts unnecessary. The Clinch River in Virginia was damaged by an enormous fly ash spill when a power plant dam broke and let the fly ash slurry into the river (appropriately by way of Dumps Creek). This spill eliminated practically every species in the river, but, in just a few years, nearly every taxonomic group had returned, with the exception of the mollusks. This first spill was followed by a much less severe acid spill, and again the river recovered. For both spills, no residual toxicants of any significance remained, and therefore the habitat was essentially undamaged. The water quality, of course, improved nearly immediately as the pulse of contaminants moved downstream. There were approximately 17 tributaries and the undamaged headwaters upstream of the spill. In fact, one could initially identify the contribution of each tributary to the recovery process, although this quickly blurred as the entire river recovery got underway. It quickly became apparent that this was a unique situation—a badly disturbed ecological patch, even with so large a size, was surrounded by healthy ecosystems that could and did furnish recolonizing organisms. Moreover, the problem of invasion of damaged areas by exotics, while always a factor, was not nearly as important in this natural recovery process as it would have been if the damaged area were surrounded by other damaged areas already colonized by nonnative species. If organisms capable of recolonization are nearby and propagules (any easily relocated life history stage) can reach the damaged ecosystem without human assistance in a relatively short period of time, then, of course, natural recolonization is a very desirable phenomenon for a number of reasons.

1. Because the recolonizing organisms are from the ecological landscape in which the damaged patch occurs, they will merge and integrate with the landscape quite nicely. This is advantageous when a damaged patch is surrounded by a healthy ecosystem. However, it is not advantageous when the damaged patch is the only remaining healthy ecosystem in the landscape and may be colonized by pioneering or nonnative species, often in the pest category.

2. Because natural epicenters (sources of colonizing species) will furnish invading species at appropriate times of the year and over long time spans, there is no concern about whether the introductions are made at the optimal time or even at an appropriate time. Natural selection will take care of that if invasion pressure is maintained.

3. If the recolonizing species are from natural sources and arrive unassisted by humans, they are likely to have interacted with each other before; therefore, resources will be partitioned in a way similar to the larger ecological landscape in which the damaged patch occurs.

4. Natural recovery is more likely to produce a self-maintaining (i.e., self-perpetuating) system than is assisted recolonization. for above reasons

5. Clearly, natural recolonization and ecological recovery are much cheaper than assisted colonization and recovery—in most cases, at least 10 times less costly. It is also possible, however, that the ecological landscape in which the damage occurred was in a climax or old-growth (or very late) successional stage. The pioneering species representing the earlier stages of succession may no longer be nearby; therefore, species suitable for preparing the "ecological stage" for later arrivals will not be present to do so. Furthermore, although these species may represent earlier successional stages that had occurred in the larger ecological landscape in which the damage occurred, the new damaged patch may not be integrated with the landscape very well. Small damaged patches (through falling of old trees, fire, and similar events) are quite common in natural systems. The important question is, are the early stages of colonization under those circumstances common to the system or relatively rare?

Vogl (1980) postulated two types of ecosystems: (1) perturbation-dependent systems—in which long-term stability causes a reduction in species richness, and, thus, diversity is maintained by periodic, partial collapse of the system, and (2) perturbation-independent systems—which maintain their ecological integrity by continual maintenance of ecological diversity and habitat patchiness.

The prospects for a natural recovery following a severe perturbation was dramatically illustrated by the recolonization of Mount St. Helen's in Washington State following the eruption in 1980. Franklin et al. (1988) recounted this event. It had been quite some time since the previous eruption, and the devastation was substantial. Nevertheless, recolonization from propagules and other materials that somehow escaped destruction by the eruption, from seeds brought in by birds, and from a variety of other factors astonished ecologists. One could speculate that the biota in the area surrounding a natural volcano is perturbation dependent and needs the sort of stress that an eruption provides. In any case, natural recolonization worked quite well in this situation. Sparks (1996) noted that the Mississippi River floods restructured the morphology of the river and provided many other stimuli that increased the complexity of the biota inhabiting the river. Neither of these rather dramatic and large-scale recolonizations were based on human-assisted introduction of species, at least not to any great extent.

ETHOS, EQUITY, AND FAIRNESS

If ecological restoration is carried out primarily to ensure delivery of ecosystem services to society or to furnish amenities that human society considers desirable, then ecosystems damaged during the period when ecosystem protection was not exemplary may have to be repaired by the society that permitted their degradation and destruction. Of course, if the ecological damage was caused by an organization or an individual, by carelessness or deliberate action, they should be responsible for the ecological repair. However, in many cases, ecosystems were damaged by cumulative effects, any one of which might be tolerated by the ecosystem but which could not be tolerated in the aggregate.

There should be equity in both intergenerational distribution of ecosystem restoration costs and benefits as well as geographical equity. It might also be possible to establish demographic equity, especially on a global scale, but this would be exceedingly difficult to implement. As an example of geographical equity in wetland loss, the National Research Council (NRC 1992a) noted that California has lost 91% of its wetlands, whereas the state of Virginia has lost only 42%. Additionally, the National Research Council (NRC 1992a) recommended restoring 10 million wetland acres by the year 2010, which represents less than 10% of the total number of acres of wetlands lost in the last 200 years. One viewpoint is that the loss of wetlands in California is ecologically disastrous, and perhaps other areas should forego their share of the national restoration effort to help California. Another viewpoint is for each area to take care of its own problems. A third alternative is to use money where the greatest ecological benefits can be obtained for the lowest cost. This last alternative might be good practice in both economics and ecology, but may not exemplify a sense of equity and fairness. For example, the money spent on the *Exxon Valdez* oil spill in Prince William Sound in Alaska might have produced more ecological benefits elsewhere in the United States, but possibly none of the activities would have directly benefitted Alaskans. It is quite possible that a very serious problem might be neglected in Alaska because the money can be spent more efficiently and with better benefits per dollar elsewhere. Again, while this distribution of resources might be justified on an ecological basis, the equity and fairness issues are not so easily resolved. Cairns and Dickson (1995) discussed equity and fairness from the perspective of intergenerational distribution of ecological benefits and costs; there are other discussions of this as well. Brower and Chapple (1995) noted that society is, in a very real sense, "stealing the earth from our children." That is, ecological capital is being taken away that society has neither the ability nor the desire to replace before the next generation will be in a position to use it.

ACKNOWLEDGMENTS

I am indebted to Darla Donald, editorial assistant, for preparing this manuscript for publication. My colleagues B. R. Niederlehner and Bruce Wallace provided useful comments on the second draft.

Literature Cited

Aitken, G. 1997. Restoration of trout waters in the West: Blackfoot River of Montana. Pages 402–424 *in* Williams et al. (1997b).

Amaranthus, M. P., and D. A. Perry. 1994. The functioning of ectomycorrhizal fungi in the field: linkages in space and time. Plant and Soil 159:133–140.

American Rivers. 1995. North America's most endangered and threatened rivers of 1995. American Rivers Inc., Washington, DC.

Amoros, C., J. C. Rostan, G. Patou, and J. P. Bravard. 1987. The reversible process concept applied to the environmental management of large river systems. Environmental Management 11:607–617.

Anderson, J. W., and nine coauthors. 1992. Upper Grande Ronde River anadromous fish habitat protection, restoration, and monitoring plan. Upper Grande Ronde Technical Working Group. Wallowa-Whitman National Forest. Baker City, Oregon. (Available from Wallowa-Whitman National Forest, Post Office Box 907, Baker City, Oregon 97814.)

Angermeier, P. L. 1995. Ecological attributes of extinction-prone species: loss of freshwater fishes of Virginia. Conservation Biology 9:143–158.

Angermeier, P. L. 1997. Conceptual roles of biological integrity and diversity. Pages 49–65 *in* Williams et al. (1997b).

Angermeier, P. L., and J. R. Karr. 1994. Biological integrity versus biological diversity as policy directives. BioScience 44:690–697.

Applegate River Watershed Council. 1994. Applegate watershed assessment. Prepared for State of Oregon Watershed Health Program and Strategic Water Management Group. Rogue River National Forest, Medford, Oregon. (Available from Rogue River National Forest, Post Office Box 520, Medford, Oregon 97501.)

Arbuckle, C. H. 1900. First biennial report of the State Fish and Game Warden. Idaho Department of Fish and Game, Boise.

Armour, C. L., K. P. Burnham, and W. S. Platts. 1983. Field methods and statistical analyses for monitoring small salmonid streams. U.S. Fish and Wildlife Service Biological Service Program, FWS/OBS-83-33.

Atzet, T., and D. L. Wheeler. 1984. Preliminary plant associations of the Siskiyou Mountain province. U.S. Forest Service Pacific Northwest Region, Portland, Oregon.

Auble, G. T., J. M. Friedman, and M. L. Scott. 1994. Relating riparian vegetation to present and future streamflows. Ecological Applications 4:544–554.

Aumen, N. G. 1995. The history of human impacts, lake management, and limnological research on Lake Okeechobee, Florida (USA). Archiv fur Hydrobiologie, Ergebnisse der Limnologie 45:1–16.

Ayers, M. A., D. M. Wolock, G. J. McCabe, L. E. Hay, and G. D. Tucker. 1994. Sensitivity of water resources in the Delaware River basin to climate variability and change. U.S. Geological Survey, Water Supply Paper 2422, Reston, Virginia.

Babcock, W. H. 1986. Tenmile Creek: a study of stream relocation. Water Resources Bulletin 22:405–415.

Baharav, D. 1991. Pitkin County, Colorado, wildlife study. Final report. Pitkin County Task Force, Aspen, Colorado. (Available from Pitkin County Manager's Office, 530 East Main, Aspen, Colorado 81612, USA.)

Bain, M., J. T. Finn, and H. E. Brooke. 1988. Streamflow regulation and fish community structure. Ecology 69:382–392.

Ball, K., V. Moore, and J. Curran. 1982. Regional fishery management investigations. Idaho Department of Fish and Game, Federal Aid in Fish Restoration Project F-71-R-6, Job VI-b, Idaho Falls.

Bandler, B. 1988. Anacostia: the other river. Interstate Commission on the Potomac River Basin, Publication 88-1, Rockville, Maryland.

Bandler, B. 1993. Restoring the Anacostia's fisheries. In the Anacostia Watershed (Interstate Commission on the Potomac River Basin, Rockville, Maryland) 6(1):3.

Barnum, J. B. 1984. Ecological structure and function of small agricultural streams in central Iowa. Doctoral dissertation. Iowa State University, Ames.

Beechie, T., E. Beamer, and L. Wasserman. 1994. Estimating coho salmon rearing habitat and smolt production losses in a large river basin, and implications for habitat restoration. North American Journal of Fisheries Management 14:797–811.

Been, V. 1994. Locally desirable land use in minority neighborhoods: disproportionate siting or market dynamics. Yale Law Journal 103:1383.

Behnke, R. J. 1992. Native trout of western North America. American Fisheries Society Monograph 6, Bethesda, Maryland.

Bella, D. A., and W. S. Overton. 1972. Environmental planning and ecological possibilities. Journal of the Sanitary Engineering Division, American Society of Civil Engineers 98:579–592.

Benda, L., T. Beechie, R. C. Wissmar, and A. C. Johnson. 1992. Morphology and evolution of salmonid habitats in a recently deglaciated river basin, Washington State, U.S.A. Canadian Journal of Fisheries and Aquatic Sciences 49:1246–1256.

Benke, A. C. 1990. A perspective on America's vanishing streams. Journal of the North American Benthological Society 9:77–88.

Benner, P. 1989. Historical resurrection of the Coquille River and the surrounding area. Doctoral dissertation. Oregon State University, Corvallis.

Beschta, R. L., and W. S. Platts. 1986. Morphological features of small streams: significance and function. Water Resources Bulletin 22:369–379.

Beschta, R. L., W. S. Platts, J. B. Kaufman, and M. T. Hill. 1994. Artificial stream restoration: money well spent or an expensive failure? UCWR (Universities Council on Water Resources) Annual Conference. UCWR, University of Illinois, Carbondale.

Best, D. W. 1984. Land use of the Redwood Creek basin. Redwood National Park Research and Development Technical Report 9. National Park Service, Crescent City, California.

Betz, J. V. 1984. The human impact on water. Pages 122–128 in E. A. Fernald and D. J. Patton, editors. Water resources atlas of Florida. Florida State University Press, Orlando.

Bickford, W. E., and M. S. Tisa. 1992. Flawless fisheries through watershed management. Pages 95–103 in Stroud (1992).

Bilby, R. E., B. R. Fransen, and P. A. Bisson. 1996. Incorporation of nitrogen and carbon from spawning coho salmon into the trophic system of small streams: evidence from stable isotopes. Canadian Journal of Fisheries and Aquatic Sciences 53:164–173.

Bilby, R. E., and J. W. Ward. 1991. Large woody debris characteristics and function in streams draining old-growth, clear-cut, and second-growth forests in southwestern Washington. Canadian Journal of Fisheries and Aquatic Sciences 48:2499–2508.

Bisson, P. A., J. L. Neilson, R. A. Palmason, and L. E. Grove. 1982. A system of naming habitat types in small streams, with examples of habitat utilization by salmonids during low stream flow. Pages 62–73 in N. B. Armantrout, editor. Acquisition and utilization of aquatic habitat inventory information. American Fisheries Society, Western Division, Bethesda, Maryland.

Bisson, P. A., T. P. Quinn, G. H. Reeves, and S. V. Gregory. 1992. Best management practices, cumulative effects, and long-term trends in fish abundance in Pacific Northwest river systems. Pages 189–232 in Naiman (1992a).

Bisson, P. A., and J. R. Sedell. 1984. Salmonid populations in streams in clearcut vs. old-growth forests of western Washington. Pages 121–129 in W. R. Meehan, T. R. Merrell, Jr., and T. A. Henley, editors. Fish and wildlife relationships in old-growth forests. American Institute of Fishery Research Biologists, Morehead City, North Carolina.

Bivens, R. D. 1985. History and distribution of brook trout in the Appalachian region of Tennessee. Tennessee Wildlife Resources Association Technical Report 85-1, Nashville.

Blatchley, W. S. 1932. In days agone: notes on the fauna and flora of subtropical Florida in the days when most of its area was a primeval wilderness. The Nature Publishing Company, Indianapolis, Indiana.

Bode, R. W., M. A. Novak, and L. E. Abele. 1993a. Biological stream assessment: Willowemoc Creek, Sullivan County. 1993 survey. New York State Department of Environmental Conservation, Division of Water, Albany.

Bode, R. W., M. A. Novak, and L. E. Abele. 1993b. Biological stream assessment: Beaverkill, Sullivan and Delaware counties. 1993 survey. New York State Department of Environmental Conservation, Division of Water, Albany.

Bode, R. W., M. A. Novak, and L. E. Abele. 1995a. Biological stream assessment: Little Beaverkill, Sullivan County. 1994 survey. New York State Department of Environmental Conservation, Division of Water, Albany.

Bode, R. W., M. A. Novak, and L. E. Abele. 1995c. Biological stream assessment: upper Esopus Creek, Ulster County. 1995 survey. New York State Department of Environmental Conservation, Division of Water, Albany.

Bode, R. W., M. A. Novak, and L. E. Abele. 1995b. Biological stream assessment: Willowemoc Creek, Sullivan County. 1994 survey. New York State Department of Environmental Conservation, Division of Water, Albany.

Bolling, D. M. 1994. How to save a river: a handbook for citizen action. Island Press, Washington, DC.

Bookchin, M. 1990. The philosophy of social ecology: essays on dialectical naturalism. Black Rose Books, Montreal.

Bowling, D. L., Jr., and five coauthors. 1997. Using river action teams to restore water quality: Hiwassee River of North Carolina, Georgia, and Tennessee. Pages 277–293 *in* Williams et al. (1997b).

Bowman, M., editor. 1996. River renewal: restoring rivers through hydropower relicensing. National Park Service and American Rivers, Washington, DC.

Bradley, P., and K. Nelson. 1990. Wildlife and wildlife habitats associated with the Humboldt River and its major tributaries—Mary's and Humboldt rivers, Rafter Diamond Ranch Olympic Management Incorporated. Nevada Department of Wildlife, Federal Aid to Wildlife Restoration Project Report W-53-R, Study II, Job 1, Reno.

Bradshaw, A. D. 1983. The reconstruction of ecosystems. Journal of Applied Ecology 10:1–17.

Bradshaw, A. D. 1984. Ecological principles and land reclamation practice. Landscape Planning 11:35–48.

Bravard, J. P., C. Amoros, and G. Patou. 1986. Impacts of civil engineering works on the succession of communities in a fluvial system: a methodological and predictive approach applied to a section of the Upper Rhone River. Oikos 47:92–111.

Brock, W. A. 1986. Enhancement of rearing habitat for juvenile steelhead trout (*Salmo gairdneri*) by boulder placement in a tributary to the Klamath River. Master's thesis. Humboldt State University, Arcata, California.

Brodie, G. A., C. R. Britt, T. M. Tomaszewski, and H. N. Taylor. 1993. Anoxic limestone drains to enhance performance of aerobic acid drainage treatment wetlands: experiences of the Tennessee Valley Authority. Pages 129–138 *in* G. A. Moshiri, editor. Constructed wetlands for water quality improvement. CRC Press, Boca Raton, Florida.

Brookes, A. 1988. Channelized rivers: perspectives for environmental management. Wiley, New York.

Brookes, A. 1995. The importance of high flows for riverine environments. Pages 33–49 *in* D. M. Harper and A. J. D. Ferguson, editors. The ecological basis for river management. Wiley, Sussex, UK.

Brooks, C. E. 1986. The Henry's Fork. Winchester Press, Piscataway, New Jersey.

Brostrom, J. K., and L. Watson. 1988. Evaluation of Henrys Lake trout stocking program. Idaho Department of Fish and Game, Federal Aid in Fish Restoration Project F-71-R-11, Subproject II, Job 3, Idaho Falls.

Brouha, P. 1987. Agency policy and practices: wildlife and fisheries management in the USDA Forest Service. Fisheries 12(3):8–10.

Brower, D. R., and S. Chapple. 1995. Let the mountains talk, let the rivers run: a call to those who would save the earth. HarperCollinsWest, New York.

Bullard, L. 1994. Watershed committee of the Ozarks: ten year portfolio. Watershed Committee of the Ozarks, Springfield, Missouri. (Available from Watershed Committee of the Ozarks, 300 West Brower Street, Springfield, Missouri 65802-3817, USA.)

Burkhart, M. R., S. L. Oberle, M. J. Hewitt, and J. Pickus. 1994. A framework for regional agroecosystems characterization using the National Resources Inventory. Journal of Environmental Quality 23:866–874.

Busby, M. S., R. A. Barnhart, and P. P. Petros. 1988. Natural resources of the Mattole River estuary, California: natural resources and habitat inventory summary report. Prepared for the U.S. Bureau of Land Management, Arcata Resource Area, and Cooperative Fishery Research Unit, Humboldt State University, Arcata, California.

Butler, J. 1996. Angler's environment 1996. Fly Rod and Reel 18(2):27.

Cadwell, D. H. 1989. Surficial geologic map of New York, lower Hudson sheet. New York State Geological Survey, Albany.

Caine, N. 1980. The rainfall intensity-duration control of shallow landslides and debris flows. Geografiska Annaler A: Physical Geography 62(1–2):23–27.

Cairns, J., Jr. 1988a. Increasing diversity by restoring damaged ecosystems. Pages 333–343 *in* E. O. Wilson, editor. Biodiversity. National Academy Press, Washington, DC.

Cairns, J., Jr. 1988b. Restoration ecology: the new frontier. Pages 1–11 *in* J. Cairns, Jr., editor. Rehabilitating damaged ecosystems, volume I. CRC Press, Boca Raton, Florida.

Cairns, J., Jr. 1993. The current state of watersheds in the United States: ecological and institutional concerns. Pages 11–17 *in* Watershed '93: a national conference on watershed management. U.S. Environmental Protection Agency, Report 1994-300-781/12415. U.S. Government Printing Office, Washington, DC.

Cairns, J., Jr. 1994. Abel Wolman distinguished lecture. National Academy of Sciences, Washington, DC. (Reprinted as "Ecosocietal restoration: reestablishing humanity's relationship with natural systems" *in* Environment 37(5):4–9,30–33.)

Cairns, J., Jr. 1995. Restoration ecology. Pages 223–235 *in* W. A. Nierenberg, editor. Encyclopedia of environmental biology, volume 3. Academic Press, San Diego, California.

Cairns, J., Jr. 1996. Determining the balance between technological and ecosystem services. Pages 13–30 *in* P. C. Schulze, editor. Engineering within ecological constraints. National Academy Press, Washington, DC.

Cairns, J., Jr. 1997. Eco-societal restoration: creating a harmonious future between human society and natural systems. Pages 487–499 *in* Williams et al. (1997b).

Cairns, J., Jr., J. S. Crossman, K. L. Dickson, and E. E. Herricks. 1971. The recovery of damaged streams. Association of Southeastern Biologists Bulletin 118(3):79–106.

Cairns, J., Jr., and K. L. Dickson. 1995. Individual rights versus the rights of future generations: ecological resource distribution over large temporal and spatial scales. Pages 175–187 *in* S. R. Ingman, X. Pei, C. D. Ekstrom, H. J. Freidsam, and K. R. Bartlett, editors. An aging population, an aging planet, and a sustainable future. Texas Institute for Research and Education on Aging, Denton.

Cairns, J., Jr., and J. R. Heckman. 1996. Restoration ecology: the state of an emerging field. Annual Review of Energy and the Environment 21:167–189.

Cairns, J., Jr., and J. R. Pratt. 1995. Ecological restoration through behavioral change. Restoration Ecology 3(1):51–53.

Callicott, J. B. 1989a. The conceptual foundations of the land ethic. Pages 75–99 *in* J. B. Callicott, editor. In defense of the land ethic. State University of New York Press, Albany.

Callicott, J. B. 1989b. On the intrinsic value of nonhuman species. Pages 129–155 *in* J. B.

Callicott, editor. In defense of the land ethic. State University of New York Press, Albany.

Callicott, J. B. 1991. Conservation ethics and fishery management. Fisheries 16(2):22–28.

Carpenter, S. L., and W. J. D. Kennedy. 1988. Managing public disputes: a practical guide to handling conflict and reaching agreements. Jossey-Bass Inc., San Francisco.

Carson, R. 1962. Silent spring. Houghton Mifflin Company, Boston.

Carter, K. B. 1955. Pioneer irrigation: upper Snake River valley in 1955. Daughters of Utah Pioneers, Utah Printing Company, Salt Lake City.

Carter, W. R., III. 1992. Problems of incorporating fish habitat management into watershed development. Pages 167–174 *in* Stroud (1992).

Castelle, A. J., A. W. Johnson, and C. Conolly. 1994. Wetland and stream buffer size requirements—a review. Journal of Environmental Quality 23:878–882.

Chaney, E., W. Elmore, and W. Platts. 1990. Livestock grazing on western riparian areas. Produced for the U.S. Environmental Protection Agency by the Northwest Resource Information Center, Inc., Eagle, Idaho. (Available from Northwest Resource Information Center, Highway 44, Eagle, Idaho 83616.)

Chaney, E., W. Elmore, and W. S. Platts. 1993. Managing change. Produced for the U.S. Environmental Protection Agency by the Northwest Resource Information Center, Inc., Eagle, Idaho. (Available from Northwest Resource Information Center, Highway 44, Eagle, Idaho 83616.)

Chase, A. 1995. What if environmentalism is a big mistake? Environmental Solutions (Gelman Sciences) Spring issue:1, 7.

Chatters, J. C., D. A. Neitzel, M. J. Scott, and S. A. Shankle. 1991. Potential impacts of global climate change on Pacific Northwest spring chinook salmon (*Oncorhynchus tshawytscha*): an exploratory case study. Northwest Environmental Journal 7:71–92.

Clark, J. L. 1993. Nevada wildlife viewing guide. Falcon Press, Helena, Montana.

Clayton, J. L., and D. A. Kennedy. 1985. Nutrient losses from timber harvest in the Idaho Batholith. Soil Science Society of America Journal 49:1041–1049.

Coats, R. N., and T. O. Miller. 1981. Cumulative silvicultural impacts on watersheds: a hydrologic and regulatory dilemma. Environmental Management 5:147–160.

Coffin, D. L., and K. R. Kilki. 1971. Water resources of the upper Blackfoot River valley, west-central Montana. Montana Department Natural Resources Technical Report 1, Helena.

Coffin, P. D. 1983. Lahontan cutthroat trout fishery management plan for the Humboldt River basin. Nevada Department of Wildlife, Federal Aid to Fish Restoration Project Report F-20-17, Study IX, Job 1-P-1, Reno.

Colletti, J. P., and six coauthors. 1994. A socio-economic assessment of the Bear Creek watershed. Pages 295–302 *in* R. C. Schultz and J. P. Colletti, editors. Opportunities for agroforestry in the temperate zone worldwide: proceedings of the third North American agroforestry conference. Iowa State University, Ames.

Conerly, W. 1995. The death of regional economics. Pacific Northwest Economic Indicators 4(3):1–9.

Conquest, L. C., S. C. Ralph, and R. J. Naiman. 1994. Implementation of large-scale stream monitoring efforts: sampling design and data analysis issues. Pages 69–89 *in* S. Loeb and A. Spacie, editors. Biological monitoring of aquatic systems. CRC Press, Boca Raton, Florida.

Conyngham, J., and J. M. McGurrin. 1997. Restoration of trout waters in the East: Beaverkill–Willowemoc watershed of New York. Pages 381–401 *in* Williams et al. (1997b).

Cooke, G. D., and W. R. Jordan, III. 1995. Ecosystem rehabilitation. Pages 1–4 *in* Using ecological restoration to meet Clean Water Act goals. U.S. Environmental Protection Agency, Chicago, Illinois.

Cooke, G. D., E. B. Welch, S. A. Peterson, and P. R. Networth. 1993. Restoration and management of lakes and reservoirs. Lewis Publishers, Boca Raton, Florida.

Coon, J. C. 1978. Angler opinions, attitudes and preferences at Henrys Lake. Idaho

Department of Fish and Game, Federal Aid in Fish Restoration Project F-53-R-12, Job XIV-d, Idaho Falls.

Cooper, J. R., J. W. Gilliam, R. B. Daniels, and W. P. Robarge. 1987. Riparian areas as filters for agricultural sediment. Soil Science Society of America Journal 51:416–420.

Coppin, N. J., and I. G. Richards. 1990. Use of vegetation in civil engineering. Butterworth, London.

Costanza, R. 1992. Toward an operational definition of ecosystem health. Pages 239–256 *in* R. Costanza, B. G. Norton, and B. D. Haskell, editors. Ecosystem health: new goals for environmental management. Island Press, Washington, DC.

Courtenay, W. R., Jr., and P. B. Moyle. 1992. Crimes against biodiversity: the lasting legacy of fish introductions. Transactions of the North American Wildlife and Natural Resources Conference 57:365–372.

Covington, W. W., and five coauthors. 1994. Historical and anticipated changes in forest ecosystems of the inland west of the United States. Journal of Sustainable Forestry 2:13–63.

Coy, L. 1985. A livestock industry perspective on nonpoint source pollution control. Pages 19–20 *in* M. L. Moore, editor. Perspectives on nonpoint source pollution. Reprinted by Terrene Institute, Washington, DC (1994).

Coyle, K. J. 1995. Watershed conservation in America: the Swift River principles. Pages 21–22 *in* P. M. Lavigne, editor. Proceedings of the watershed innovators workshop. River Network, Portland, Oregon.

Croonquist, M. J., and R. P. Brooks. 1991. Use of avian and mammalian guilds as indicators of cumulative impacts in riparian–wetland areas. Environmental Management 15:701–714.

Crumpton, W. G., T. M. Isenhart, and S. W. Fisher. 1993. Transformation and fate of nitrate in wetlands receiving nonpoint source agricultural inputs. Pages 283–291 *in* G. A. Moshiri, editor. Constructed wetlands for water quality improvement. CRC Press, Boca Raton, Florida.

Currens, K. P. 1994. Genetic variation in Crooked River rainbow trout. Oregon Cooperative Fishery Research Unit Genetic Laboratory Progress Report 94 (2a), Oregon State University, Corvallis.

Dabney, S. M., K. C. McGregor, L. D. Meyer, E. H. Grissinger, and G. R. Foster. 1993. Vegetative barriers for runoff and sediment control. Pages 60–70 *in* J. K. Mitchell, editor. Integrated resource management and landscape modification for environmental protection. American Society of Agricultural Engineers, St. Joseph, Michigan.

Dahl, T. E. 1990. Wetland losses in the United States, 1780s to 1980s. U.S. Fish and Wildlife Service, Washington, DC.

Dahm, C. N., K. W. Cummins, H. M. Valett, and R. L. Coleman. 1995. An ecosystem view of the restoration of the Kissimmee River. Restoration Ecology 3:225–238.

Daniels, R. B., and J. W. Gilliam. 1996. Sediment and chemical load reduction by grass and riparian filters. Soil Science Society of America Journal 60:246–251.

Davis, S. M. 1981. Mineral flux in the Boney Marsh, Kissimmee River: mineral retention in relation to overland flow during the three-year period following reflooding. South Florida Water Management District, Technical Publication 81-1, West Palm Beach.

Davis, S. M., L. H. Gunderson, W. A. Park, J. R. Richardson, and J. E. Mattson. 1994. Landscape dimension, composition, and function in a changing Everglades ecosystem. Pages 419–444 *in* Davis and Ogden (1994b).

Davis, S. M., and J. C. Ogden. 1994a. Toward ecosystem restoration. Pages 769–796 *in* Davis and Ogden (1994b).

Davis, S. M., and J. C. Ogden, editors. 1994b. Everglades: the ecosystem and its restoration. St. Lucie Press, Delray Beach, Florida.

Deacon, J. E. 1988. The endangered woundfin and water management in the Virgin River, Utah, Arizona, Nevada. Fisheries 13(1):18–24.

Dengler, L., G. Carver, and R. McPherson. 1992. Sources of north coast seismicity. California Geology 45(2):40-53.

Diamond, J. 1987. Reflections on goals and on the relationship between theory and practice. Pages 329-336 *in* W. R. Jordan, III, M. E. Gilpin, and J. D. Aber, editors. Restoration ecology: a synthetic approach to ecological research. Cambridge University Press, Cambridge, UK.

Dombeck, M. P., J. W. Thomas, and C. A. Wood. 1997. Changing roles and responsibilities for federal land management agencies. Pages 135-144 *in* Williams et al. (1997b).

Dombeck, M. P., and J. E. Williams. 1995. Roles, responsibilities, and opportunities for the Bureau of Land Management in aquatic conservation. Pages 430-433 *in* J. L. Nielsen, editor. Evolution and the aquatic ecosystem: defining unique units in population conservation. American Fisheries Society Symposium 17, Bethesda, Maryland.

Doppelt, B. 1997. Saving and creating jobs through watershed restoration. Pages 472-486 *in* Williams et al. (1997b).

Doppelt, B., M. Scurlock, C. Frissell, and J. Karr. 1993. Entering the watershed: a new approach to save America's river ecosystems. Island Press, Washington, DC.

Dowie, M. 1995. Losing ground, American environmentalism at the close of the twentieth century. Massachusetts Institute of Technology Press, Cambridge.

Drake, J. A. 1991. Community-assembly mechanisms and the structure of an experimental species ensemble. American Naturalist 132:1-26.

Drost, C. A., and G. M. Fellers. 1996. Collapse of a regional frog fauna in the Yosemite area of the California Sierra Nevada, USA. Conservation Biology 10:414-425.

Drucker, P. 1989. The new realities. Harper and Row, New York.

Drucker, P. 1993. Post-capitalist society. Harper Collins, New York.

Durning, A. T. 1996. This place on earth: home and the practice of permanence. Sasquatch Books, Seattle.

Dynesius, M., and C. Nilsson. 1994. Fragmentation and flow regulation of river systems in the northern third of the world. Science 266:753-762.

Ebersole, J. L., W. J. Liss, and C. A. Frissell. 1997. Restoration of stream habitats in the western United States: restoration as re-expression of habitat capacity. Environmental Management 21:1-14.

Echeverria, J., P. Barrow, and R. Roos-Collins. 1989. Rivers at risk—the concerned citizen's guide to hydropower. Island Press, Washington, DC.

Ehlers, R. 1956. An evaluation of stream improvement devices constructed eighteen years ago. California Fish and Game 42:203-217.

Ehrenfeld, D. W. 1976. The conservation of non-resources. American Scientist 64:648-656.

Ehrenfeld, D. W. 1978. The arrogance of humanism. Oxford University Press, New York.

Ehrlich, P. R., and H. A. Mooney. 1983. Extinction, substitution, and ecosystem services. BioScience 33:248-254.

Ehrlich, P. R., and E. O. Wilson. 1991. Biodiversity studies: science and policy. Science 253:758-762.

Elliot, R. 1982. Faking nature. Inquiry 25:81-93.

Elmore, W., and R. L. Beschta. 1987. Riparian areas: perceptions in management. Rangelands 9(6):260-265.

ERI (Ecosystems Research Institute). 1995. Environmental report for the proposed spillway modification. Prepared for FERC (Federal Energy Regulatory Commission) Island Park Hydroelectric Project 2973 by ERI, Logan, Utah.

Everest, F. H., and five coauthors. 1986b. Abundance, behavior, and habitat utilization by coho salmon and steelhead trout in Fish Creek, Oregon as influenced by habitat enhancement. Bonneville Power Administration, Division of Fish and Wildlife, Annual Report, Project 84-11, Portland, Oregon.

Everest, F. H., G. H. Reeves, and J. R. Sedell. 1986a. Salmonid habitat: new beginnings through enhancement, but not without uncertainty, risk and failure. Pages 9-19 *in*

D. Guthrie, editor. Wild trout, steelhead and salmon in the 21st century. Oregon Sea Grant Publication ORESU-W-86-002, Oregon State University, Corvallis.

Everest, F. H., G. H. Reeves, J. R. Sedell, D. Hohler, and T. Cain. 1987. The effects of habitat enhancement on steelhead trout and coho salmon production, habitat utilization, and habitat availability in Fish Creek, Oregon, 1983–86. Bonneville Power Administration, Division of Fish and Wildlife, Annual Report, Project 84-11, Portland, Oregon.

Everest, F. H., G. H. Reeves, J. R. Sedell, D. Hohler, and T. Cain. 1988. Changes in habitat and populations of steelhead trout, coho salmon, and chinook salmon in Fish Creek, Oregon, 1983–87, as related to habitat improvement. Bonneville Power Administration, Division of Fish and Wildlife, Annual Report, Project 84-11, Portland, Oregon.

Everest, F. H., and J. R. Sedell. 1984. Evaluation of fisheries enhancement projects on Fish Creek and Wash Creek, 1982–1983. Bonneville Power Administration, Division of Fish and Wildlife, Annual Report, Project 84-11, Portland, Oregon.

Everest, F. H., J. R. Sedell, G. H. Reeves, and J. Wolfe. 1985. Fisheries enhancement in the Fish Creek basin—an evaluation of in-channel and off-channel projects. Bonneville Power Administration, Division of Fish and Wildlife, Annual Report, Project 84-11, Portland, Oregon.

Fan, A. 1986. A routing model for the upper Kissimmee chain of lakes. South Florida Water Management District, Technical Publication 86-5, West Palm Beach.

Fausch, K. D., J. Lyons, J. R. Karr, and P. L. Angermeier. 1990. Fish communities as indicators of environmental degradation. Pages 123–144 in S. M. Adams, editor. Biological indicators of stress in fish. American Fisheries Society Symposium 8, Bethesda, Maryland.

Fedler, T. 1996. The 1994 economic impacts of fishing, hunting, and wildlife related recreation on national forest lands. American Sportfishing Association, Alexandria, Virginia.

FEMAT (Forest Ecosystem Management Assessment Team). 1993. Forest ecosystem management: an ecological, economic, and social assessment. U.S. Forest Service, Portland, Oregon.

Fennema, R. J., C. J. Neidrauer, R. A. Johnson, T. K. MacVicar, and W. A. Perkins. 1994. A computer model to simulate natural Everglades hydrology. Pages 249–289 in Davis and Ogden (1994b).

FERC (Federal Energy Regulatory Commission). 1996. Final environmental impact statement for Deerfield River projects, Vermont–Massachusetts (FERC projects 2323-012 and 2334-001). FERC/DEIS-0105 FERC, Office of Hydropower Licensing, Washington, DC.

Firth, P., and S. G. Fisher. 1992. Global climate change and freshwater ecosystems. Springer-Verlag, New York.

Fisher, W. F., Y. W. Isachsen, and L. V. Richard. 1970. Geologic map of New York. Map and chart series 15. New York State Museum and Science Service, Albany.

Flader, S. L., and J. B. Callicott, editors. 1991. The river of the mother of God and other essays by Aldo Leopold. University of Wisconsin Press, Madison.

Flebbe, P. A., and C. A. Dolloff. 1995. Trout use of woody debris and habitat in Appalachian wilderness streams of North Carolina. North American Journal of Fisheries Management 15:579–590.

Florida Department of Administration. 1976. Final report on the special project to prevent eutrophication of Lake Okeechobee. Florida Department of Administration, Division of State Planning, Tallahassee, Florida.

FGFWFC (Florida Game and Fresh Water Fish Commission). 1994. Lake Okeechobee-Kissimmee River–Everglades resource evaluation project. Federal Wallop-Breaux completion report (F-52-5). Florida Game and Fresh Water Fish Commission, Tallahassee.

Fontaine, B. L. 1988. An evaluation of the effectiveness of instream structures for steelhead trout rearing habitat in the Steamboat Creek basin. Master's thesis. Oregon State University, Corvallis.

Fox, S. 1981. John Muir and his legacy, the American conservation movement. Little, Brown and Company, Boston.

Franklin, J. F., P. M. Frenzen, and F. J. Swanson. 1988. Re-creation of ecosystems at Mount St. Helens: contrasts in artificial and natural approaches. Pages 1–38 *in* J. Cairns, Jr., editor. Rehabilitating damaged ecosystems, volume II. CRC Press, Boca Raton, Florida.

Frazee, R. W., and D. P. Roseboom. 1993. Stabilizing eroding streambanks with the willow-post method. Pages 208–213 *in* J. K. Mitchell, editor. Integrated resource management and landscape modification for environmental protection. American Society of Agricultural Engineers, St. Joseph, Michigan.

Fremont, J. C. 1845. Report of the exploring expeditions to the Rocky Mountains in the year 1842, and to Oregon and northern California in the years 1843-1844. Report of the U.S. Senate, Washington, DC.

Frissell, C. A. 1991. Water quality, fisheries, and aquatic biodiversity under two alternative forest management scenarios for the west-side federal forests of Washington, Oregon, and California. Consultant's report prepared for The Wilderness Society, Washington, DC.

Frissell, C. A. 1993. Topology of extinction and endangerment of native fishes in the Pacific Northwest and California (U.S.A.). Conservation Biology 7:342-354.

Frissell, C. A. 1997. Ecological principles. Pages 96–115 *in* Williams et al. (1997b).

Frissell, C. A., and D. Bayles. 1996. Ecosystem management and the conservation of aquatic biodiversity and ecological integrity. Water Resources Bulletin 32:229-240.

Frissell, C. A., D. Bayles, and W. J. Liss. 1993. An integrated, biophysical strategy for ecological restoration of large watersheds. Pages 449–456 *in* D. F. Potts, editor. Changing roles in water resources management and policy. American Water Resources Association Technical Publication Series TPS-93-2, Bethesda, Maryland.

Frissell, C. A., W. L. Liss, R. E. Gresswell, R. K. Nawa, and J. L. Nebersole. 1997. A resource in crisis: changing the measure of salmon management. Pages 411–444 *in* D. J. Stouder, P. A. Bisson, and R. J. Naiman, editors. Pacific salmon and their ecosystems: status and future options. Chapman and Hall, New York.

Frissell, C. A., W. J. Liss, C. E. Warren, and M. D. Hurley. 1986. A hierarchical framework for stream habitat classification: viewing streams in a watershed context. Environmental Management 10:199-214.

Frissell, C. A., and R. K. Nawa. 1992. Incidence and causes of physical failure of artificial habitat structures in streams of western Oregon and Washington. North American Journal of Fisheries Management 12:182-197.

Fuller, D. D., and A. J. Lind. 1992. Implications of fish habitat improvement structures for other stream vertebrates. Pages 96–104 *in* H. M. Kerner, editor. Proceedings of the symposium on biodiversity of northwestern California. University of California, Division of Agriculture, Wildland Resources Center and Natural Resources, Report 29, Berkeley.

FUND (Foundation for Urban and Neighborhood Development) Pacific Associates. 1981. A social impact management system for Honolulu. Final phase 2 report. Pacific Associates Division of Foundation for Urban and Neighborhood Development, City and County of Honolulu, Hawaii. (Available from Post Office Box 3165, Aspen, Colorado 81612, USA.)

Galston, W., and K. Baehler. 1995. Rural development in the United States. Island Press, Washington, DC.

Gamblin, M., S. Elle, and J. Tharp. 1993a. Regional fisheries management investigations for 1989. Idaho Department of Fish and Game, Federal Aid in Fish Restoration Project F-71-R-14, Idaho Falls.

Gamblin, M., S. Elle, and J. Tharp. 1993b. Regional fisheries management investigations for 1991. Idaho Department of Fish and Game, Federal Aid in Fish Restoration Project F-71-R-16, Idaho Falls.

Gard, R. 1972. Persistence of headwater check dams in a trout stream. Journal of Wildlife Management 36:1363–1367.

Garreau, J. 1982. Nine nations of North America. Avon Books, New York.

Garrett, L. 1994. The coming plague. Farrar, Straus and Giroux, New York.

Gatewood, S. E., and P. B. Bedient. 1975. Drainage density in the Lake Okeechobee drainage area. Florida Department of Administration, Division of State Planning, Tallahassee.

Geller, E. S. 1991. If only more would actively care. Journal of Applied Behavior Analysis 24:607–612.

Geller, E. S. 1994. The human element in integrated environmental management. Pages 5–26 *in* J. Cairns, Jr., T. V. Crawford, and H. Salwasser, editors. Implementing integrated environmental management. Virginia Polytechnic Institute and State University, Blacksburg.

Gibson, K. D., J. B. Zedler, and R. Langis. 1994. Limited response of cordgrass (*Spartina foliosa*) to soil amendments in a constructed marsh. Ecological Applications 4:757–767.

Gierach, J. 1996. Cutthroats: a passion for going native. Fly Rod and Reel 18(3):94–96.

Gigliotti, L. M. 1992. Environmental attitudes: 20 years of change? Journal of Environmental Education 24(1):15–26.

Gillis, A. M. 1991. Bringing back the land. BioScience 41:68–71.

Goodstein, E. 1995. Jobs and the environment: myth of the national trade-off. Report of the Economic Policy Institute, Washington, DC.

Gordon, M. K. 1987. The origins of the Anacostia River improvement project. Soundings (Potomac River Basin Consortium) 4(Spring):10–18.

Gore, A. 1992. Earth in the balance. Houghton Mifflin, New York.

Gore, J. A., editor. 1985. The restoration of rivers and streams: theories and experience. Butterworth, Boston.

Gore, J. A., and F. L. Bryant. 1988. River and stream restoration. Pages 23–38 *in* J. Cairns, Jr., editor. Rehabilitating damaged ecosystem, volume I. CRC Press, Boca Raton, Florida.

Gore, J. A., and F. D. Shields, Jr. 1995. Can large rivers be restored? BioScience 45:142–152.

Gough, S., B. Turner, and M. Petersen. 1990. Building and using the stream table. Missouri Department of Conservation, Jefferson City.

Gowan, C., and K. D. Fausch. 1996. Long-term responses of trout populations to habitat manipulations in six Colorado streams. Ecological Applications 6:931–946.

Graham, E. U. 1994. The Anacostia restoration. In the Anacostia Watershed (The Interstate Commission on the Potomac River Basin, Rockville, Maryland) 7(3):2.

Green, D. H. 1990. History of Island Park. Gateway Publishing, Ashton, Idaho.

Green, R. H. 1979. Sampling design and statistical methods for environmental biologists. Wiley, New York.

Gregory, S. V., and L. Ashkenas. 1990. Riparian management guide. U.S. Forest Service, Willamette National Forest, Eugene, Oregon.

Gregory, S. V., F. J. Swanson, W. A. McKee, and K. W. Cummins. 1991. An ecosystem perspective of riparian zones. BioScience 41:540–551.

Greiwe, R. J. 1980. An introduction to social resource management (Training Handbook Number One), Procedures for characterizing and delineating a human resource unit using cultural descriptors (Training Handbook Number Two), Procedures for identifying and evaluating public issues, management concerns and management opportunities (Training Handbook Number Three), and Social analysis procedures for land management and planning (Training Handbook Number Four). Foundation for Urban and Neighborhood Development, Denver, Colorado.

Griffith, G. E, J. M. Omernik, T. F. Wilton, and S. M. Pierson. 1994. Ecoregions and subecoregions of Iowa: a framework for water quality assessment and management. Journal of the Iowa Academy of Science 101:5–13.

Griffith, J. S., and R. W. Smith. 1995. Failure of submerged macrophytes to provide cover for rainbow trout throughout their first winter in the Henrys Fork of the Snake River, Idaho. North American Journal of Fisheries Management 15:42–48.

Griggs, G. B., and J. R. Hein. 1980. Sources, dispersal and clay mineral composition of fine-grained sediment off central and northern California. Journal of Geology 88:541–566.

Gruessner, B. 1996. A regional action plan for managing contaminated sediments in the Anacostia River. The Interstate Commission on the Potomac River Basin, Rockville, Maryland.

Grunbaum, J. B. 1996. Geographical and seasonal variation in diel habitat use by juvenile (age 1+) steelhead trout (*Oncorhynchus mykiss*) in Oregon coastal and inland stream. Master's thesis. Oregon State University, Corvallis.

Gunn, A. S. 1980. Why should we care about rare species? Environmental Ethics 2:17–37.

Gutzwiller, L. A., R. M. McNatt, and R. D. Price. 1997. Watershed restoration and grazing practices in the Great Basin: Marys River of Nevada. Pages 360–380 *in* Williams et al. (1997b).

Guy, C. S., and S. Denson-Guy. 1995. An introduction to learning and teaching styles: making the match. Fisheries 20(2):18–20.

Hackett, B., and B. Bonnichsen. 1994. Volcanic crescent. Pages 24–61 *in* T. Shallat, editor. Snake: the plain and its people. Boise State University, Boise, Idaho.

Hagans, D. K., and W. E. Weaver. 1987. Magnitude, cause and basin response to fluvial erosion, Redwood Creek basin, northern California. Pages 419–428 *in* R. L. Beschta, T. Blinn, G. E. Grant, F. J. Swanson, and G. G. Ice, editors. Erosion and sedimentation in the Pacific Rim. International Association of Hydrologic Sciences Publication 165.

Hagans, D. K., W. E. Weaver, and M. A. Madej. 1986. Long term and off-site effects of logging and erosion in the Redwood Creek basin, northern California. Pages 38–66 *in* Papers presented at the American Geophysical Union meeting on cumulative effects. National Council for Air and Stream Improvement Technical Bulletin 490, New York.

Hambrock, M. J., and P. Murto. 1993. The St. Louis River: a grass-roots approach to protection. Pages 51–56 *in* Watershed '93: a national conference on watershed management. U.S. Environmental Protection Agency, Report 1994-300-781/12415, U.S. Government Printing Office, Washington, DC.

Hamilton, J. B. 1989. Response of juvenile steelhead to instream deflectors in a high-gradient stream. Pages 149–157 *in* R. E. Gresswell, B. A. Barton, and J. L. Kershner, editors. Practical approaches to riparian resource management: an educational workshop. American Fisheries Society, Montana Chapter, Billings.

Hammer, M., and J. Champy. 1994. Reinventing the corporation. Harper Business, a Division of Harper Collins Publishers, New York.

Hankin, D. G., and G. H. Reeves. 1988. Estimating total fish abundance and total habitat area in small streams based on visual estimation methods. Canadian Journal of Fisheries and Aquatic Sciences 45:834–844.

Hanson, M. L. 1987. Riparian zones in eastern Oregon. Oregon Environmental Council, Portland.

Harden, D. R., R. J. Janda, and K. M. Nolan. 1978. Mass movement and storms in the drainage basin of Redwood Creek, Humboldt County, California—a progress report. U.S. Geological Survey Open-File Report 78-486, Menlo Park, California.

Hargrove, E. C. 1989. Foundations of environmental ethics. Prentice Hall, Englewood Cliffs, New Jersey.

Harmon, M. E., and 12 coauthors. 1986. Ecology of coarse woody debris in temperate ecosystems. Advances in Ecological Research 15:133–302.

Harrelson, C. C., C. L. Rawlins, and J. P. Potyondy. 1994. Stream channel reference sites:

an illustrated guide to field technique. U.S. Forest Service Rocky Mountain Forest and Range Experiment Station, General Technical Report GTR-RM-245, Fort Collins, Colorado.

Harris, S. L. 1980. Fire and ice: the Cascade volcanoes. Pacific Search Press, Seattle.

Hartig, J. H., and R. L. Thomas. 1988. Development of plans to restore degraded areas in the Great Lakes. Environmental Management 12:327–347.

Hartman, G. F. 1965. The role of behavior in the ecology and interaction of underyearling coho salmon (*Oncorhynchus kisutch*) and steelhead trout (*Salmo gairdneri*). Journal of the Fisheries Research Board of Canada 22:1035–1081.

Hartman, G. F., and J. C. Scrivener. 1990. Impacts of forestry practices on a coastal stream ecosystem, Carnation Creek, British Columbia. Canadian Bulletin of Fisheries and Aquatic Sciences 223:1–148.

Hartman, G. F., J. C. Scrivener, and M. J. Mills. 1996. Impacts of logging Carnation Creek, a high-energy coastal stream in British Columbia, and their implication for restoring fish habitat. Canadian Journal of Fisheries and Aquatic Sciences 53(Supplement 1):237–251.

Hausbeck, K. W., L. W. Milbrath, and S. M. Enright. 1992. Environmental knowledge, awareness and concern among 11th-grade students: New York State. Journal of Environmental Education 24(1):27–34.

Heady, H. F., and five coauthors. 1988. Coastal prairie and northern coastal scrub. Pages 733–760 *in* M. G. Barbour and J. Major, editors. Terrestrial vegetation of California. California Native Plant Society Special Publication 9, second edition, Sacramento.

Heede, B. H., and J. N. Rinne. 1990. Hydrodynamic and fluvial morphologic processes: implications for fisheries management and research. North American Journal of Fisheries Management 10:249–268.

Helms, D. 1993. Watershed management in historical perspective: the Soil Conservation Service's experience. Pages 89–93 *in* Watershed '93: a national conference on watershed management. U.S. Environmental Protection Agency, Report 1994-300-781/12415, U.S. Government Printing Office, Washington, DC.

Hemstrom, M. A., and S. E. Logan. 1984. Preliminary plant associations and management guide. Siuslaw National Forest, Corvallis, Oregon. (Available from Siuslaw National Forest, Post Office Box 1148, Corvallis, Oregon 97339.)

Henjum, M. G., and seven coauthors. 1994. Interim protection for late-successional forests, fisheries, and watersheds, national forests east of the Cascade crest, Oregon and Washington. Eastside Forests Scientific Society Panel. The Wildlife Society Technical Review 94-2, Bethesda, Maryland.

Hesse, L. W., and G. E. Mestl. 1993. An alternative hydrograph for the Missouri River based on the precontrol condition. North American Journal of Fisheries Management 13:360–366.

Hicks, B. J., R. L. Beschta, and R. D. Harr. 1991a. Long-term changes in streamflow following logging in western Oregon and associated fisheries implications. Water Resources Bulletin 27:217–226.

Hicks, B. J., J. D. Hall, P. A. Bisson, and J. R. Sedell. 1991b. Responses of salmonid populations to habitat changes caused by timber harvest. Pages 483–518 *in* Meehan (1991).

Hight, J. 1996. For love and money: restoring salmon habitat. North Coast Journal 7(8):10–14, Eureka, California.

Hill, A. R. 1996. Nitrate removal in stream riparian zones. Journal of Environmental Quality 25:743–755.

Hokanson, K. E. F., F. Kleiner, and T. W. Thorshund. 1977. Effects of constant temperature and daily temperature fluctuations on specific growth and mortality and yield of juvenile rainbow trout (*Salmo gairdneri*). Journal of the Fisheries Research Board of Canada 34:639–648.

Holcomb, D. and W. Wegener. 1972. Hydrophytic changes related to lake fluctuation as measured by point transects. Proceedings of the Annual Conference Southeastern Association Fish and Wildlife Agencies 25(1971):570–582.

Holling, C. S. 1978. Adaptive environmental assessment and management. Wiley, London.

Holling, C. S. 1992. Cross-scale morphology, geometry, and dynamics of ecosystems. Ecological Monographs 62:447–502.

Holling, C. S., L. H. Gunderson, and C. J. Walters. 1994. The structure and dynamics of the Everglades system: guidelines for ecosystem restoration. Pages 741–756 *in* Davis and Ogden (1994b).

Holtby, L. B., and J. C. Scrivener. 1989. Observed and simulated effects of climatic variability, clear-cut logging and fishing on the numbers of chum salmon (*Oncorhynchus keta*) and coho salmon (*O. kisutch*) returning to Carnation Creek, British Columbia. Canadian Special Publication of Fisheries and Aquatic Sciences 105:62–81.

Hooks, G. M., T. L. Napier, and M. V. Carter. 1983. Correlates of adoption behaviors: the case of farm technologies. Rural Sociologist 48(2):308–323.

Horn, J. S. 1969. Away with all pests: an English surgeon in People's China. Modern Reader, London.

Horton, C. A., and D. J. Meierer. 1992. Flood plain identification using elevation data: Kissimmee River, Florida. Pages 47–52 *in* Proceedings of the twelfth annual Environmental Systems Research Institute user conference. Environmental Systems Research Institute, Redlands, California.

House, F. 1990. To learn the things we need to know. Whole Earth Review 66:36–47.

House, R. 1996. An evaluation of stream restoration structures in a coastal Oregon stream, 1981–1993. North American Journal of Fisheries Management 16:272–281.

House, R. A., and P. L. Boehne. 1985. Evaluation of instream enhancement structures for salmonid spawning and rearing in a coastal Oregon stream. North American Journal of Fisheries Management 5:283–295.

House, R., V. Crispin, and R. Monthey. 1989. Evaluation of stream rehabilitation projects—Salem District. U.S. Department of the Interior, Bureau of Land Management, Oregon State Office, Technical Note T/N R-6, Portland.

Howell, P. J., and D. V. Buchanan, editors. 1992. Proceedings of the Gearhart Mountain bull trout workshop. American Fisheries Society, Oregon Chapter, Corvallis.

Huber, W. C., J. P. Heaney, P. B. Bedient, and J. P. Bowden. 1976. Environmental resources management studies in the Kissimmee River basin. University of Florida, Department of Environmental Engineering Sciences, Final Report (ENV-05-76-2), Gainesville.

Hudson, W. F., and P. A. Heikkila. 1997. Integrating public and private restoration strategies: Coquille River of Oregon. Pages 235–252 *in* Williams et al. (1997b).

Hunt, R. L. 1993. Trout stream therapy. University of Wisconsin Press, Madison.

Hunter, C. J. 1991. Better trout habitat: a guide to stream restoration and management. Island Press, Washington, DC.

Hynes, H. B. N. 1975. The stream and its valley. Internationale Vereinigung fuer Theoretische und Angewandte Limnologie Verhandlungen 19:1–15.

IDC (Idaho Department of Commerce). 1994. County profiles of Idaho: Fremont County. IDC, Boise.

IDFG (Idaho Department of Fish and Game). 1940. Eighteenth biennial report, 1939–40. IDFG, Boise.

IDFG (Idaho Department of Fish and Game). 1958. Twenty-seventh biennial report, 1956–58. IDFG, Boise.

IDFG (Idaho Department of Fish and Game). 1973. Annual report for fiscal year 1972–73. IDFG, Boise.

IDFG (Idaho Department of Fish and Game). 1975. Annual report for fiscal year 1974–75. IDFG, Boise.

Interagency Ecosystem Management Task Force. 1995. The ecosystem approach:

healthy ecosystems and sustainable economics. Volume 1, overview (PB95-265583). Report of the Interagency Ecosystem Management Task Force. (Available from the National Technical Information Service, U.S. Department of Commerce, 5285 Port Royal Road, Springfield, Virginia 22161.)

Interrain Pacific. 1996. Road/stream crossing density. Pages 29–32 *in* Coquille subbasin working atlas. Interrain Pacific, Portland, Oregon.

Iowa Agricultural Statistics Service. 1994. 1994 Iowa agricultural statistics. Iowa Department of Agriculture and Land Stewardship and U.S. Department of Agriculture National Agricultural Statistics Service, Des Moines.

Isenhart, T. M. 1992. Transformation and fate of nitrate in northern prairie wetlands. Doctoral dissertation. Iowa State University, Ames.

Isenhart, T. M., R. C. Schultz, and J. P. Colletti. 1997. Watershed restoration and agricultural practices in the Midwest: Bear Creek of Iowa. Pages 318–334 *in* Williams et al. (1997b).

IWRB (Idaho Water Resources Board). 1992. Comprehensive state water plan: Henrys Fork basin. IWRB, Boise.

Jacobs, T. C., and J. W. Gilliam. 1985. Riparian losses of nitrate from agricultural drainage waters. Journal of Environmental Quality 22:467–473.

Jeppson, P. W. 1967. Tests for increasing the returns of hatchery trout. Idaho Department of Fish and Game, Federal Aid in Fish Restoration Project F-32-R-10, Job 11, Boise.

Jeppson, P. W. 1969. Tests for increasing the returns of hatchery trout. Idaho Department of Fish and Game, Federal Aid in Fish Restoration Project F-32-R-11, Job 8, Boise.

Jeppson, P. W. 1973. Snake River fisheries investigations. Idaho Department of Fish and Game, Federal Aid in Fish Restoration Project Project F-63-R-3, Job III-a, Idaho Falls.

Jobling, M. 1981. Temperature tolerance and the final preferendum—rapid methods for the assessment of optimum growth temperatures. Journal of Fish Biology 19:439–455.

John, D. 1994. Civic environmentalism: alternatives to regulation in states and communities. Congressional Quarterly Press, Washington, DC.

Johnson, B. L., W. B. Richardson, and T. J. Naimo. 1995. Past, present, and future concepts in large river ecology. BioScience 45:134–141.

Johnson, G. L. 1988. Stream fisheries management. Nevada Department of Wildlife, Federal Aid in Fish Restoration Progress Report F-20-23, Job 206, Elko.

Johnson, G. L., and B. Nokes. 1995. Fisheries survey of the Bureau of Land Management portion of the Marys River drainage system. Nevada Division of Wildlife, Elko.

Johnson, W. E. 1965. On mechanisms of self-regulation of population abundance in *Oncorhynchus nerka*. Mitteilungen der Internationalen Vereinigung für Theoretische und Angewandte Limnologie 13:66–87.

Jones, J. A., and G. E. Grant. 1996. Peak flow responses to clear-cut and roads in small and large basins, western Cascades, Oregon. Water Resources 32:959–974.

Jones, M. L., and six coauthors. 1996. Assessing the ecological effects of habitat change: moving beyond productive capacity. Canadian Journal of Fisheries and Aquatic Sciences 53(Supplement 1):446–457.

Jones, R., and B. Malone. 1990. Mission improbable, Kissimmee River land acquisition. Pages 249–251 *in* Loftin et al. (1990b).

Judy, R. D., Jr., T. M. Murray, S. C. Svirsky, M. R. Whitworth, and L. S. Ischinger. 1984. 1982 national fisheries survey, volume 1. U.S. Fish and Wildlife Service, FWS/OBS-84/06. Washington, DC.

Junk, W. J., P. B. Bayley, and R. E. Sparks. 1989. The flood pulse concept in river-floodplain systems. Pages 110–127 *in* D. P. Dodge, editor. Proceeding of the international large river symposium. Canadian Special Publication in Fisheries and Aquatic Sciences 106.

Kaeding, L. R., G. D. Boltz, and D. G. Carty. 1996. Lake trout discovered in Yellowstone Lake threaten native cutthroat trout. Fisheries 21(3):16–20.

Kapala, C. 1995. Deerfield River settlement agreement: a settlement for the future? Hydro Review 14(4):30–39.

Karr, J. R. 1981. Assessment of biotic integrity using fish communities. Fisheries 6(6):21–27.

Karr, J. R. 1987. Biological monitoring and environmental assessment: a conceptual framework. Environmental Management 11:249–256.

Karr, J. R. 1991. Biological integrity: a long neglected aspect of water resource management. Ecological Applications 1:66–84.

Karr, J. R. 1993. Measuring biological integrity: lessons from streams. Pages 83–104 *in* S. Woodley, J. Kay, and G. Francis, editors. Ecological integrity and the management of ecosystems. St. Lucie Press, Delray Beach, Florida.

Karr, J. R. 1994. Landscapes and management for ecological integrity. Pages 229–251 *in* K. C. Kim and R. D. Weaver, editors. Biodiversity and landscapes: a paradox of humanity. Cambridge University Press, New York.

Karr, J. R., and D. R. Dudley. 1981. Ecological perspective on water quality goals. Environmental Management 5:55–68.

Karr, J. R., K. D. Fausch, P. L Angermeier, P. R. Yant, and I. J. Schlosser. 1986. Assessing biological integrity in running waters: a method and its rationale. Illinois Natural History Survey Special Publication 5.

Karr, J. R., and I. J. Schlosser. 1978. Water resources and the land–water interface. Science 201:229–234.

Karr, J. R., L. A. Toth, and D. R. Dudley. 1985. Fish communities of midwestern rivers: a history of degradation. BioScience 35:90–95.

Karr, J. R., L. Toth, and G. Garman. 1983. Habitat preservation for midwest stream fishes: principles and guidelines. U.S. Environmental Protection Agency Project Summary, EPA-600/S3-83-006, Corvallis, Oregon.

Katz, E. 1992. The big lie: human restoration of nature. Research in Philosophy and Technology 12:231–241.

Keddy, P. A., H. T. Lee, and I. C. Wisheu. 1993. Choosing indicators of ecosystem integrity: wetlands as a model system. Pages 61–79 *in* S. Woodley, J. Kay, and G. Francis, editors. Ecological integrity and the management of ecosystems. St. Lucie Press, Ottawa.

Keenleyside, M. H. A., and F. T. Yamamoto. 1962. Territorial behavior of juvenile Atlantic salmon (*Salmo salar* L.). Behavior 19:139–169.

Kelsey, H., M. A. Madej, J. Pitlick, P. Stroud, and M. Coghlan. 1981. Major sediment sources and limits to the effectiveness of erosion control techniques in the highly erosive watersheds of north coastal California. Pages 493–509 *in* T. R. H. Davies and A. J. Pearce, editors. Erosion and sediment transport in Pacific rim steeplands. International Association of Hydrological Sciences Publication 132.

Kennedy, V. C., and R. L. Malcolm. 1977. Geochemistry of the Mattole River of northern California. U.S. Geological Survey, Open-File Report 78-205, Menlo Park, California.

Kent, J. A. 1992. Issue management handbook. Washoe County Planning Department, Reno, Nevada.

Kent, J. A., D. Drigot, and D. Baharav. 1994. Thinking beyond our borders: a bio-social ecosystem approach to resource management on public land. James Kent Associates, Aspen, Colorado.

Kepler, D. A., and E. C. McCleary. 1994. Successive alkaline-producing systems (SAPS) for treatment of acidic mine drainage. Pages 195–204 *in* Volume 1, Proceedings International Land Reclamation and Mine Drainage Conference. U.S. Department of the Interior, Bureau of Mines (SPO6A-94), Pittsburgh, Pennsylvania.

Kershner, J. L. 1997. Monitoring and adaptive management. Pages 116–131 *in* Williams et al. (1997b).

Kershner, J. L. In press. Setting riparian/aquatic restoration objectives within a watershed context: the value of watershed analysis. Restoration Ecology.

Kimball, K. D. 1997. Using hydroelectric relicensing in watershed restoration: Deerfield River watershed of Vermont and Massachusetts. Pages 179–197 *in* Williams et al. (1997b).

Kinch, G. 1989. Riparian area management: grazing management in riparian areas. U.S. Bureau of Land Management, Technical Reference 1737-4, Denver, Colorado.

King County. 1993. Cedar River current and future conditions report. King County, Department Public Works, Surface Water Management Division, Seattle, Washington.

King County. 1996. Proposed lower Cedar River basin and nonpoint pollution action plan. King County, Department of Public Works, Surface Water Management Division, Seattle, Washington.

Klingeman, P. C., and J. B. Bradley. 1976. Willamette River Basin streambank stabilization by natural means. U.S. Army Corps of Engineers, Portland, Oregon.

Koebel, J. W., Jr. 1995. An historical perspective on the Kissimmee River restoration project. Restoration Ecology 3:149–159.

Krueger, C. C., J. E. Marsden, H. L. Kincaid, and B. May. 1989. Genetic differentiation among lake trout strains stocked into Lake Ontario. Transactions of the American Fisheries Society 118:317–330.

Kushlan, J. A. 1990. Freshwater marshes. Pages 324–363 *in* R. L. Myers and J. J. Ewel, editors. Ecosystems of Florida. University of Central Florida Press, Orlando.

Kusler, J. A. 1995a. What is wetlands and watershed management? Pages 13–25 *in* Wetlands and watershed management: meeting landowner and resource conservation needs through partnership approaches. Association of State Wetland Managers, Berne, New York.

Kusler, J. A. 1995b. Key issues, steps, and procedures, in wetlands and watershed management. Pages 1–12 *in* Wetlands and watershed management: meeting landowner and resource conservation needs through partnership approaches. Association of State Wetland Managers, Berne, New York.

La Rivers, I. 1962. Fishes and fisheries of Nevada. Nevada State Fish and Game Commission, Reno.

LaLande, J. 1995. An environmental history of the Little Applegate River watershed, Jackson County, Oregon. U.S. Forest Service, Rogue River National Forest, Medford, Oregon.

Lamberti, G. A., S. V. Gregory, L. R. Ashkenas, R. C. Wildman, and K. M. S. More. 1991. Stream ecosystem recovery following a catastrophic debris flow. Canadian Journal of Fisheries and Aquatic Sciences 48:196–208.

Larsh, E. B. 1995. Mack and the boys as consultants. Pages 57–76 *in* Doc's lab: myth and legends of Cannery Row. PBL Press, Monterey, California.

Latta, W. 1965. Relationship of young-of-the-year trout to mature trout and groundwater. Transactions of the American Fisheries Society 94:32–39.

Lawson, P. 1993. Cycles in ocean productivity, trends in habitat quality, and the restoration of salmon runs in Oregon. Fisheries 18(8):6–10.

LeCren, E. D. 1965. Some factors regulating the size of populations of freshwater fish. Mitteilungen der Internationalen Vereinigung für Theoretische und Angewandte Limnologie 13:88–105.

Lee, R. G. 1992. Ecologically effective social organization as a requirement for sustaining watershed ecosystems. Pages 73–90 *in* Naiman (1992a).

Leggette, Brashears & Graham, Inc. 1994a. Hydrogeologic assessment of test wells #5 and #6, Part 1. Livingston Manor Water District, Rockland, New York. (Available from Leggette, Brashears & Graham, Inc., 500 B Lake Street, Ramsey, New Jersey 07446, USA.)

Leggette, Brashears & Graham, Inc. 1994b. Hydrogeologic assessment of test wells #5 and #6, Part 2. Livingston Manor Water District, Rockland, New York. (Available from

Leggette, Brashears & Graham, Inc., 500 B Lake Street, Ramsey, New Jersey 07446, USA.)

Leggette, Brashears & Graham, Inc. 1995. Ground-water supply development. Rockland-Roscoe Water District, New York. (Available from Leggette, Brashears & Graham, Inc., 500 B Lake Street, Ramsey, New Jersey 07446, USA.)

Lein, L. 1994. FACA holds the boat for Alabama sturgeon. Wildlife Law News Quarterly. Fall 1994:8.

Leonard, P. M., and D. J. Orth. 1986. Application and testing of an index of biotic integrity in small coolwater streams. Transactions of the American Fisheries Society 115:401–414.

Leopold, A. 1947. The ecological conscience. Bulletin of the Garden Club of America. September 1947:45–53, Minneapolis, Minnesota.

Leopold, A. 1949. A Sand County almanac. Oxford University Press, New York.

Leopold, L. B., editor. 1953. Round River: from the journals of Aldo Leopold. Oxford University Press, New York.

Leopold, L. B. 1980. The topology of impacts. Pages 1–21 in R. B. Standiford and S. I. Ramacher, editors. Cumulative effects of forest management on California watersheds: an assessment of status and need for information. Proceedings of the Edgebrook conference. University of California, Berkeley.

Leopold, L. B. 1994. A view of the river. Harvard University Press, Cambridge, Massachusetts.

Leopold, L. B., M. G. Wolman, and J. P. Miller. 1992. Fluvial processes in geomorphology. Dover Publications, New York.

Li, H. W., and 12 coauthors. 1995. Safe havens: refuges and evolutionarily significant units. Pages 371–380 in J. L. Nielsen, editor. Evolution and the aquatic ecosystem: defining unique units in population conservation. American Fisheries Society Symposium 17, Bethesda, Maryland.

Liang, S.-H. 1995. Fish communities in small agricultural streams in Iowa: relationships with environmental factors. Doctoral dissertation. Iowa State University, Ames.

Lichatowich, J. A. 1989. Habitat alterations and changes in abundance of coho (*Oncorhynchus kisutch*) and chinook (*O. tshawytscha*) salmon in Oregon's coastal streams. Pages 92–99 in Proceedings on national workshop on effects of habitat alteration on salmonid stocks. Canadian Special Publication of Fisheries and Aquatic Sciences 105.

Lichatowich, J., L. Mobrand, L. Lestelle, and T. Vogel. 1995. An approach to the diagnosis and treatment of depleted salmon populations in Pacific Northwest watershed. Fisheries 20(1):10–18.

Light, S. S., and J. W. Dineen. 1994. Water control in the Everglades: a historical perspective. Pages 47–84 in Davis and Ogden (1994b).

Limerick, P. N. 1987. Legacy of conquest: the unbroken past of the American West. Norton, New York.

Lisle, T. E. 1981. Channel recovery from recent large floods in north coastal California: rates and processes. Pages 153–160 in R. L. Coats, editor. Watershed rehabilitation in Redwood National Park and other Pacific coastal areas. Center for Natural Resource Studies, JMI Inc., National Park Service, Arcata, California.

Loftin, M. K., L. A. Toth, and J. T. B. Obeysekera. 1990a. Kissimmee River restoration: alternative plan evaluation and preliminary design report. South Florida Water Management District, West Palm Beach.

Loftin, M. K., L. A. Toth, and J. T. B. Obeysekera, editors. 1990b. Proceedings of the Kissimmee River restoration symposium. South Florida Water Management District, West Palm Beach.

Lopez, B. 1991. Foreword. Page v in R. Nilsen, editor. An introduction to environmental restoration. A Whole Earth Catalog/Ten Speed Press Publication, Berkeley, California.

Loucks, S., and K. Preister. 1995. Stewardship in the Applegate Valley: issues and opportunities in watershed restoration. A report of the Outreach and Education

Project of the Applegate River Watershed Council through the State Watershed Health Program. The Rogue Institute for Ecology and Economy, Ashland, Oregon.

Lowrance, R. R. 1992. Groundwater nitrate and denitrification in a coastal plain riparian forest. Journal of Environmental Quality 21:401–405.

Lowrance, R. R., and five coauthors. 1984. Riparian forests as nutrient filters in agricultural watersheds. BioScience 34:374–377.

Lowrance, R. R., R. Leonard, and J. Sheridan. 1985. Managing riparian ecosystems to control nonpoint source pollution. Journal of Soil and Water Conservation 40:87–91.

Ludwig, D., R. Hilborn, and C. Walters. 1993. Uncertainty, resource exploitation, and conservation: lessons from history. Science 260:17, 36.

Lyons, J., L. Wang, and T. D. Simonson. 1996. Development and validation of an index of biotic integrity for coldwater streams in Wisconsin. North American Journal of Fisheries Management 16:241-256.

MacDonald, L. H., A. W. Smart, and R. W. Wissmar. 1991. Monitoring guidelines to evaluate effects of forestry activities on streams in the Pacific Northwest and Alaska. U.S. Environmental Protection Agency. EPA 910/9-91-001, Seattle.

Mackin, J. H. 1948. Concept of the graded river. Bulletin of the Geological Society of America 59:463–511.

Maclean, N. 1976. A river runs through it and other stories. University of Chicago Press, Chicago.

MacMahon, J. A., and W. R. Jordan, III. 1994. Ecological restoration. Pages 409–438 *in* G. K. Meffe and C. R. Carroll, editors. Principles of conservation biology. Sinauer, Sunderland, Massachusetts.

Madej, M. A. 1984. Recent changes in channel-stored sediment, Redwood Creek, California. National Park Service, Redwood National Park Research and Development Technical Report 11, Crescent City, California.

Mahood, J. D. 1985. Vermont's LaPlatte River watershed project: lessons learned. Pages 408–411 *in* M. L. Moore, editor. Perspectives on nonpoint source pollution. Reprinted by Terrene Institute, Washington, DC (1994).

Management Institute for Environment and Business. 1993. Conservation partnerships: a field guide to public–private partnering for natural resource conservation. National Fish and Wildlife Foundation, Washington, DC.

Marston, R. A., and J. E. Anderson. 1991. Watersheds and vegetation of the greater Yellowstone ecosystem. Conservation Biology 5(3):338–346.

Maser, C., and J. Sedell. 1994. From the forest to the sea: the ecology of wood in streams, rivers, estuaries, and oceans. St. Lucie Press, Delray Beach, Florida.

Master, L. 1990. The imperiled status of North American aquatic animals. Biodiversity Network News 3:1–8, Arlington, Virginia.

May, R. M. 1994. The effects of spatial scale on ecological questions and answers. Pages 1-17 *in* P. J. Edwards et al., editors. Large-scale ecology and conservation biology. 35th Symposium on the British Ecological Society with the Society for Conservation Biology. Blackwell Scientific Publications, Oxford, England.

McAtee, W. L. 1918. Natural history of the District of Columbia. Biological Society of Washington Bulletin 1:1–44.

McBride, N. D. 1996. Summary of lower Beaverkill trout studies in 1993 and 1994. Delaware River Watershed File 407. New York State Department of Environmental Conservation Region 4 Fisheries Office, Stamford.

McCaffrey, P. M., W. H. Hinkley, J. M. Ruddell, and S. E. Gatewood. 1977. First annual report to the Florida legislature. Coordinating Council on the Restoration of the Kissimmee River Valley and Taylor Creek–Nubbin Slough Basin, Tallahassee, Florida.

McIntosh, B. A., and six coauthors. 1994. Historical changes in fish habitat for select river basins of eastern Oregon and Washington. Northwest Science (Special Issue) 68:268-285.

McIntyre, J. D., and B. E. Rieman. 1995. Westslope cutthroat trout. Pages 1-15 *in* M. K. Young, editor. Conservation assessment for inland cutthroat trout. U.S. Forest

Service Rocky Mountain Forest and Range Experiment Station General Technical Report RM-GTR-256.

McLane, W. M. 1955. The fishes of the St. John's River system. Doctoral dissertation. University of Florida, Gainesville.

Mechenich, G. 1990. Manual for use of the sand-tank groundwater flow model, 2nd edition. Central Wisconsin Groundwater Center, University of Wisconsin–Extension and University of Wisconsin–Stevens Point, College of Natural Resources, Steven Point.

Meehan, W. R., editor. 1991. Influences of forest and rangeland management on salmonid fishes and their habitats. American Fisheries Society Special Publication 19, Bethesda, Maryland.

Meffe, G. K. 1992. Techno-arrogance and halfway technologies: salmon hatcheries on the Pacific coast of North America. Conservation Biology 6:350–354.

Megahan, W. F., J. P. Potyondy, and K. A. Seyedbagheri. 1992. Best management practices and cumulative effects from sedimentation in the South Fork Salmon River: an Idaho case study. Pages 401–414 *in* Naiman (1992a).

Menzel, B. W. 1983. Agricultural management practices and the integrity of instream biological habitat. Pages 305–329 *in* F. W. Schaller and G. W. Bailey, editors. Agricultural management and water quality. Iowa State University Press, Ames.

Menzel, B. W., J. B. Barnum, and L. M. Antosch. 1984. Ecological alterations of Iowa prairie–agricultural streams. Iowa State Journal of Research 59:5–30.

Meyer, S. M. 1992. Environmentalist and economic prosperity: testing the environmental impact hypothesis. Project on Environmental Politics and Policy, Massachusetts Institute of Technology, Cambridge.

Meyer, S. M. 1993. Environmentalism and economic prosperity: an update. Department of Political Science, Massachusetts Institute of Technology, Cambridge.

Mickelson, M. R. 1994. A not-so-brief, somewhat adulterated, unabridged and biased history of the Henry's Fork Foundation, 1984–1993. Henry's Fork Foundation Newsletter, Spring 1994. Island Park, Idaho.

Middleton, H. 1991. On the spine of time: an angler's love of the Smokies. Simon and Schuster, New York.

Milbrath, L. W. 1989. Envisioning a sustainable society. State University of New York Press, Albany.

Miller, G. T., Jr. 1988. Living in the environment. Wadsworth, Belmont, California.

Miller, S. J. 1990. Kissimmee River fisheries: a historical perspective. Pages 31–42 *in* Loftin et al. (1990b).

Milton, S. J., W. R. J. Dean, M. A. duPlessis, and W. R. Siegfried. 1994. A conceptual model of arid rangeland degradation: the escalating cost of declining productivity. BioScience 44:70–76.

Minckley, W. L., and J. E. Deacon, editors. 1991. Battle against extinction: native fish management in the American West. University of Arizona Press, Tucson.

Minckley, W. L., and G. K. Meffe. 1987. Differential selection by flooding in stream fish communities of the arid American Southwest. Pages 93–104 *in* W. J. Matthews and D. C. Heins, editors. Community and evolutionary ecology of North American stream fishes. University of Oklahoma Press, Norman.

Minshall, G. W., and six coauthors. 1985. Developments in stream ecosystem theory. Canadian Journal of Fisheries and Aquatic Sciences 42:1045–1055.

Mitsch, W. J. 1992. Applications of ecotechnology to the creation and rehabilitation of temperate wetlands. Pages 309–331 *in* M. K. Wali, editor. Ecosystem rehabilitation. Volume 2: ecosystem analysis and synthesis. SBP Academic Publishing, The Hague, Netherlands.

Moore, J. N., S. N. Luoma, and D. Peters. 1991. Downstream effects of mine effluent on an intermontane riparian system. Canadian Journal of Fisheries and Aquatic Sciences 48:222–232.

Moreau, J. K. 1984. Anadromous salmonid enhancement by boulder placement in

Hurdygurdy Creek, California. Pages 97–116 *in* T. J. Hassler, editor. Proceedings, Pacific Northwest stream habitat management workshop. California Cooperative Fishery Research Unit, Humboldt State University, Arcata.

Moyer, E. J., M. W. Hulon, J. J. Sweatman, R. S. Butler, and V. P. Williams. 1995. Fishery responses to habitat restoration in Lake Tohopekaliga, Florida. North American Journal of Fisheries Management 15:591–595.

Moyle, P. B., H. W. Li, and B. A. Barton. 1986. The Frankenstein effect: impact of introduced fishes on native fishes in North America. Pages 415–426 *in* R. H. Stroud, editor. Fish culture in fisheries management. American Fisheries Society, Fish Culture Section and Fisheries Management Section, Bethesda, Maryland.

Moyle, P. B., and P. R. Moyle. 1995. Endangered fishes and economics: intergenerational obligations. Environmental Biology of Fishes 43:29–37.

Moyle, P. B., and G. M. Sato. 1991. On the design of preserves to protect native fishes. Pages 155–169 *in* Minckley and Deacon (1991).

MRC (Mattole Restoration Council). 1988. Distribution of old growth coniferous forests in the Mattole River watershed. Map sheet. MRC, Petrolia, California.

MRC (Mattole Restoration Council). 1989. Elements of recovery: an inventory of upslope sources of sedimentation in the Mattole River watershed, with rehabilitation prescriptions and additional information for erosion control prioritization. Report prepared by the MRC for the California Department of Fish and Game, Sacramento.

MRC (Mattole Restoration Council). 1995. Dynamics of recovery: a plan to enhance the Mattole Estuary. Report prepared by the MRC for the California State Coastal Conservancy, Petrolia.

Mrowka, J. P. 1974. Man's impact on stream regimen and quality. Pages 79–104 *in* I. R. Manners and M. W. Mikesell, editors. Perspectivces on environment. Association of American Geographers, Washington, DC.

MSG (Mattole Salmon Group). 1995. Five-year management plan for salmon stock rescue operations 1995–96 to 1999–2000 seasons. Report prepared by MSG for California Department of Fish and Game, Sacramento.

Muir, J. 1894. The mountains of California. Century, New York.

Muir, J. 1901. Our national parks. Houghton Mifflin, Boston.

MWCOG (Metropolitan Washington Council of Governments). 1991. A commitment to restore our home river: a six-point action plan to restore the Anacostia River. Metropolitan Washington Council of Governments, Washington, DC.

Myers, L. H. 1989. Grazing and riparian management in southwestern Montana. Pages 117–120 *in* R. E. Gresswell, B. A. Barton, and J. L. Kershner, editors. Practical approaches to riparian resource management: an educational workshop. U.S. Bureau of Land Management, Billings, Montana.

Myers, R. L., and J. J. Ewel, editors. 1990. Ecosystems of Florida. University of Central Florida Press, Orlando.

Naiman, R. J., editor. 1992a. Watershed management: balancing sustainability and environmental change. Springer-Verlag, New York.

Naiman, R. J. 1992b. New perspectives for watershed management: balancing long-term sustainability with cumulative environmental change. Pages 3–11 *in* Naiman (1992a).

Naiman, R. J., and eight coauthors. 1992. Fundamental elements of ecologically healthy watersheds in the Pacific Northwest coastal ecoregion. Pages 127–188 *in* Naiman (1992a).

Naiman, R. J., C. A. Johnston, and J. C. Kelley. 1988. Alteration of North American streams by beaver. BioScience 38:753–762.

Naiman, R. J., J. J. Magnuson, D. M. McKnight, and J. A. Stanford, editors. 1995. The freshwater imperative: a research agenda. Island Press, Washington, DC.

Nash, R. F. 1989. The rights of nature. University of Wisconsin Press, Madison.

NDCNR (Nevada Department of Conservation and Natural Resources). 1963. Water and related land resources, Humboldt River basin, Nevada. Nevada Department of Conservation and Natural Resources, Marys River subbasin, Report 4, Reno.

Needham, P., and A. Jones. 1959. Flow, temperature, solar radiation and ice in relation to activities of fishes in Sagehen Creek, California. Ecology 40:465–474.

NEETF (National Environmental Education and Training Foundation). 1996. Report card: environmental attitudes and knowledge in America. The fifth survey of adult Americans. Roper Starch, New York.

Nehlsen, W. 1995. Historical salmon and steelhead runs of the upper Deschutes River and their environments. Portland General Electric Report, Portland, Oregon.

Nehlsen, W., J. E. Williams, and J. A. Lichatowich. 1991. Pacific salmon at the crossroads: stocks at risk from California, Oregon, Idaho, and Washington. Fisheries 16(2):4–21.

Neill, E. D. 1871. Fleet Henry. A brief journal of a voyage made in the bark "Warwick" to Virginia and other parts of the continent of America. As printed in The English Colonization of America during the Seventeenth Century. London.

Neitzel, D. A., M. J. Scott, S. A. Shankle, and J. C. Chatters. 1991. The effect of climate change on stream environments: the salmonid resources of the Columbia River basin. Northwest Environmental Journal 7:271–293.

NEP (New England Power). 1991. Deerfield River project: application for new license for major project—existing dam greater than five megawatts before the U.S. Federal Energy Regulatory Commission (FERC project 2323). NEP, Westborough, Massachusetts.

NEP (New England Power). 1994. Deerfield River project supporting documentation for the offer of settlement before the U.S. Federal Energy Regulatory Commission (FERC project 2323-012). NEP, Westborough, Massachusetts.

Newbury, R. W. 1994. STREAMLAB hydraulic demonstration flume: teaching and construction guide. Newbury Hydraulics, Ltd. Gibsons, British Columbia.

Newbury, R., and M. Gaboury. 1993. Exploration and rehabilitation of hydraulic habitats in streams using principles of fluvial behaviour. Freshwater Biology 29:195–230.

Nibley, H. W. 1978. Nibley on the timely and the timeless. Publishers Press, Salt Lake City, Utah.

Nickelson, T. E., and six coauthors. 1992b. Status of anadromous salmonids in Oregon coastal basins. Oregon Department Fish and Wildlife Report, Portland.

Nickelson, T. E., M. F. Solazzi, S. L. Johnson, and J. D. Rodgers. 1992a. Effectiveness of selected stream improvement techniques to create suitable summer and winter rearing habitat for juvenile coho salmon (*Oncorhynchus kisutch*) in Oregon coastal streams. Canadian Journal of Fisheries and Aquatic Sciences 49:790–794.

Nickum, D., J. McGurrin, and C. Ubert. 1995. Saving a stream. Trout Unlimited, Arlington, Virginia.

Nielsen, L. A. 1995. The practical uses of fishery history. Fisheries 20(8):16–18.

Niemi, G. J., and seven coauthors. 1990. Overview of case studies on recovery of aquatic systems from disturbance. Environmental Management 14:571–587.

Niemi, E., and E. Whitelaw. 1995. The economic consequences of protecting salmon habitat in Idaho (preliminary report). ECONorthwest, Eugene, Oregon. (Available from ECONorthwest, 99 W. Tenth, Suite 400, Eugene, Oregon 97401.)

Nolan, K. M., H. M. Kelsey, and D. C. Marron, editors. 1995. Geomorphic processes and aquatic habitat in the Redwood Creek basin, northwestern California. U.S. Geological Survey Professional Paper 1454, Menlo Park, California.

Nolte, D. A. 1997. Involving local school systems in watershed restoration: Crooked River of Oregon. Pages 198–215 *in* Williams et al. (1997b).

Norton, B. G. 1983. On the inherent danger of undervaluing species. Pages 110–137 *in* B. G. Norton, editor. The preservation of species. Princeton University Press, Princeton, New Jersey.

Norton, B. G. 1987. Why preserve natural variety? Princeton University Press, Princeton, New Jersey.

Noss, R. F. 1990. Indicators for monitoring biodiversity: a hierarchical approach. Conservation Biology 4:355–364.

Noss, R. F., and A. Y. Cooperrider. 1994. Saving nature's legacy: protecting and restoring biodiversity. Island Press, Washington, DC.

NPC (Northwest Policy Center). 1995. An evaluation of the Henry's Fork Watershed Council. Graduate School of Public Affairs, University of Washington, Seattle.

NRC (National Research Council). 1992a. Restoration of aquatic ecosystems: science, technology, and public policy. National Academy Press, Washington, DC.

NRC (National Research Council). 1992b. Water transfers in the West: efficiency, equity and the environment. National Academy Press, Washington, DC.

NRC (National Research Council). 1993a. A biological survey for the nation. National Academy Press, Washington, DC.

NRC (National Research Council). 1993b. Soil and water quality: a new agenda for agriculture. National Academy Press, Washington, DC.

NYSDEC (New York State Department of Environmental Conservation). 1990. Biennial report, rotating intensive basin studies Water Quality Assessment Program 1987–1988. Lake Erie–Niagara River, Lake Champlain, Upper Hudson River, Delaware River, Atlantic Ocean–Long Island Sound. NYSDEC, Bureau of Monitoring and Assessment, Division of Water, Albany, New York.

Obeysekera, J., and M. K. Loftin. 1990. Hydrology of the Kissimmee River basin: influence of man-made and natural changes. Pages 211–222 in Loftin et al. (1990b).

ODEQ (Oregon Department of Environmental Quality). 1992. Oregon's 1992 water quality status assessment report 305(b). ODEQ, Portland.

ODFW (Oregon Department of Fish and Wildlife). 1992. Coquille basin fish management plan. ODFW, Charleston.

ODFW (Oregon Department of Fish and Wildlife). 1994. The natural production program: 1994 biennial report on the status of wild fish in Oregon and the implementation of fish conservation policies. ODFW, Portland.

ODFW (Oregon Department of Fish and Wildlife). 1995. Biennial report of wild fish of Oregon. ODFW, Portland.

Odum, E. P. 1969. The strategy of ecosystem development. Science 164:262–270.

Office of Surface Mining. 1996. America's coal mining reclamation is turning green. Office of Surface Mining 1996 annual report. U.S. Department of the Interior, Washington, DC.

Office of the President. 1995. Budget of the United States government, fiscal year 1996, budget supplement. U.S. Government Printing Office, Washington, DC.

Ogden, P. S. 1950. Peter Skene Ogden's snake country journals 1824–25 and 1825–26. The Hudson's Bay Record Society, London.

Ohnstad, B. 1985. Scissors and comb haircutting: a cut-by-cut guide for home haircutters. You Can Publishing, Minneapolis, Minnesota.

Olson, M. 1987. Will the North rise again? This World Newspaper: 6 December 1987. Coos Bay, Oregon.

Ontario Ministry of Natural Resources. 1994. Natural channel systems: an approach to management and design. Draft. Natural Resources Information Centre, Toronto, Ontario.

Ornstein, R., and P. R. Ehrlich. 1991. New world new mind. Simon and Schuster, New York.

Orr, D. W. 1992. Ecological literacy: education and the transition to a postmodern world. State University of New York Press, Albany.

Orr, D. W. 1994. Earth in mind: on education and the human prospect. Island Press, Washington, DC.

Osborn, G., and B. H. Luckman. 1988. Holocene glacier fluctuations in the Canadian Cordillera (Alberta and British Columbia). Quaternary Science Reviews 7:115–128.

Osborne, L. L., and five coauthors. 1993. Restoration of lowland streams: an introduction. Freshwater Biology 29:187–194.

Osborne, L. L., and D. A. Kovacic. 1993. Riparian vegetated buffer strips in water-quality restoration and stream management. Freshwater Biology 29:243–258.

OSGC (Oregon State Game Commission). 1951. Oregon State Game Commission annual report 1950. OSGC, Fishery Division, Portland.

OTA (Office of Technology Assessment). 1987. Technologies to maintain biological diversity. Congress of the United States, OTA, OTA-F-339, Washington DC.

Overbay, J. C. 1992. Ecosystem management. Pages 3-15 *in* Taking an ecological approach to management. Proceedings of national workshop. U.S. Forest Service, WO-WSA-3, Washington, DC.

Parker, G. C. 1955. Geomorphology. Pages 127-155 *in* G. C. Parker, G. E. Ferguson, and S. K. Love, editors. Water resources of southeastern Florida. U.S. Geological Survey, Water Supply Paper 1255, U.S. Government Printing Office, Washington, DC.

Parr, J. F., R. I. Papendick, S. B. Hornick, and M. E. Meyer. 1992. Soil quality: attributes and relationship to alternative and sustainable agriculture. American Journal of Alternative Agriculture 7:5-11.

Pastor, J., and C. A. Johnston. 1992. Using simulation models and geographical information systems to integrate ecosystem and landscape ecology. Pages 324-346 *in* Naiman (1992a).

Pauley, J. 1993. Minnesota's comprehensive watershed management initiative. Pages 489-491 *in* Watershed '93: a national conference on watershed management. U.S. Environmental Protection Agency, Report 1994-300-781/12415, U.S. Government Printing Office, Washington, DC.

Perkins, S. J. 1994. The shrinking Cedar River: channel changes following flow regulation and bank armoring. American Water Resources Association June 1994:649-659.

Perrin, L. S., and five coauthors. 1982. A report of fish and wildlife studies in the Kissimmee River basin and recommendations for restoration. Florida Game and Fresh Water Fish Commission, Office of Environmental Services, Okeechobee.

Perry, D. A., M. P. Amaranthus, J. G. Borchers, S. L. Borchers, and R. E. Brainerd. 1989. Bootstrapping in ecosystems. BioScience 39:230-237.

Perry, D. A., and J. Maghembe. 1989. Ecosystem concepts and current trends in forest management: time for reappraisal. Forest Ecology and Management 26:123-140.

Peterjohn, W. T., and D. L. Correll. 1984. Nutrient dynamics in an agricultural watershed: observations on the role of a riparian forest. Ecology 65:1466-1475.

Peterman, R. M. 1989. Application of statistical power analysis to the Oregon coho salmon (*Oncorhynchus kisutch*) problem. Canadian Journal of Fisheries and Aquatic Sciences 46:1183-1187.

Peters, D. 1990. Inventory of fishery resources in the Blackfoot River and major tributaries. Montana Department Fish, Wildlife and Parks, Missoula.

Peters, D., and R. Spoon. 1989. Preliminary fisheries inventory of the Big Blackfoot River. Montana Department Fish, Wildlife and Parks, Missoula.

Peterson, G. 1996. Mattole salmon runs show improvement. The Mattole Spawning News. Winter 1996:1-4. Petrolia, California.

Peterson, N. P. 1982. Immigration of juvenile coho salmon (*Oncorhynchus kisutch*) into riverine ponds. Canadian Journal of Fisheries and Aquatic Sciences 39:1308-1310.

PFMC (Pacific Fishery Management Council). 1996. Review of 1995 ocean salmon fisheries. PFMC, Portland, Oregon.

Pickett, S. T. A., V. T. Parker, and P. L. Fiedler. 1992. The new paradigm in ecology: implications for conservation biology above the species level. Pages 65-88 *in* P. L. Fiedler and S. K. Jain, editors. Conservation biology: the theory and practice of nature conservation, preservation, and management. Chapman and Hall, New York.

Pierce, R. 1991. A stream habitat and fisheries analysis for six tributaries to the Big Blackfoot River. Montana Department of Fish, Wildlife and Parks, Missoula.

Pierce, R., D. Peters, and F. Swanberg. 1997. Blackfoot River restoration project progress report. Montana Department of Fish, Wildlife and Parks, Helena. (Available from FWP, Fisheries Division Library, P. O. Box 200701, Helena, Montana 59620-0701.)

Pinchot, G. 1947. Breaking new ground. Harcourt, Brace and Company, New York.

Pister, E. P. 1976. A rationale for the management of nongame fish and wildlife. Fisheries 1(1):11–14.

Pister, E. P. 1985. Desert pupfishes: reflections on reality, desirability, and conscience. Environmental Biology of Fishes 12:3–12.

Pister, E. P. 1987. A pilgrim's progress from group A to group B. Pages 221–232 in J. B. Callicott, editor. Companion to a Sand County almanac. University of Wisconsin Press, Madison.

Pister, E. P. 1991a. Desert Fishes Council: catalyst for change. Pages 55–68 in Minckley and Deacon (1991).

Pister, E. P. 1991b. Environmental water ethics in the eastern Sierra. In Mono Lake symposium. Supplement to the Bulletin of the Southern California Academy of Sciences 90:20–26.

Pister, E. P. 1992. Ethical considerations in conservation of biodiversity. Transactions of the North American Wildlife and Natural Resources Conference 57:355–364.

Pister, E. P. 1993. Species in a bucket. Natural History 102(1):14–18.

Pister, E. P. 1995. The rights of species and ecosystems. Fisheries 20(4):28–29.

Pivnicka, K. 1992. The Klicava Reservoir, Czechoslovakia: a 30 year study of the fish community. Fisheries Research 14:1–20.

Plafkin, J. L., M. T. Barbour, K. D. Porter, S. D. Gross, and R. M. Hughes. 1989. Rapid bioassessment protocols for use in streams and rivers: benthic macroinvertebrates and fish. U.S. Environmental Protection Agency, EPA/444/4-89-001, Washington, DC.

Platt, S., editor. 1992. Respectfully quoted: a dictionary of quotations from the Library of Congress. Congressional Quarterly Inc., Washington, DC.

Platts, W. S., and R. L. Nelson. 1985. Stream habitat and fisheries response to livestock grazing and instream improvement structures, Big Creek, Utah. Journal of Soil and Water Conservation 40:374–379.

Platts, W. S., and R. L. Nelson. 1988. Fluctuations in trout populations and their implications for land-use evaluation. North American Journal of Fisheries Management 8:333–345.

Platts, W. S., and J. N. Rinne. 1985. Riparian and stream enhancement, management, and research in the Rocky Mountains. North American Journal of Fisheries Management 5:115–125.

Platts, W. S., R. J. Torquemada, M. L. McHenry, and C. K. Graham. 1989. Changes in salmon spawning and rearing habitat from increased delivery of fine sediment to the South Fork Salmon River, Idaho. Transactions of the American Fisheries Society 118:274–283.

Poff, N. L., and J. D. Allan. 1995. Functional organization of stream fish assemblages in relation to hydrological variability. Ecology 76:606–627.

Poff, N. L., and J. V. Ward. 1990. Physical habitat template of lotic systems: recovery in the context of historical pattern of spatiotemporal heterogeneity. Environmental Management 14:629–645.

Power, T. M. 1991. Ecosystem preservation and the economy in the Greater Yellowstone area. Conservation Biology 5:395–404.

Power, T. M., editor. 1995a. Economic well-being and environmental protection in the Pacific Northwest. University of Montana, Economic Department, Missoula.

Power, T. M. 1995b. Economic well-being and environmental protection in the Pacific Northwest. Illahee 11(3–4):142–150, Seattle.

Powers, R. F., and eight coauthors. 1990. Sustaining site productivity in North American forests: problems and prospects. Pages 49–79 in S. P. Gessel, D. S. Locate, G. F. Westman, and R. F. Powers, editors. Sustained productivity of forest soils. Proceedings of the 7th North American forest soils conference. University of British Columbia, Faculty of Forest Publication, Vancouver.

Prange, R. 1995. An historical account of the Henrys Lake Foundation. Henrys Lake Foundation Newsletter, Fall 1995. Boise, Idaho.

PRC (Pacific Rivers Council). 1992. The economic imperative of protecting riverine habitat in the Pacific Northwest. PRC, Research Report 5, Eugene, Oregon.

Preister, K. 1989. Our lives are on hold: the prospects for issue management with the Aspen Superfund Site. U.S. Environmental Protection Agency, Region 8, Denver, Colorado.

Preister, K. 1994. Words into action: a community assessment of the Applegate Valley. Prepared for the Applegate Partnership, Rogue River National Forest and others. The Rogue Institute for Ecology and Economy, Ashland, Oregon.

Preister, K., and J. A. Kent. 1984. The issue-centered approach to social impacts: from assessment to management. Clinical Sociology Review 2:120–132.

Preister, K., and J. A. Kent. 1997. Social ecology: a new pathway to watershed restoration. Pages 28–48 *in* Williams et al. (1997b).

Preston, W. C. 1995. Letters to editor. Environmental Solutions (Gelman Sciences) Summer issue:5.

Prichard, D., and eight coauthors. 1993. Process for assessing proper functioning condition. U.S. Bureau of Land Management Technical Reference 1737-9, Denver, Colorado.

Pringle, C. M., C. F. Rabeni, A. C. Benke, and N. G. Aumen. 1993. The role of aquatic science in freshwater conservation: cooperation between the North American Benthological Society and organizations of conservation and resource management. Journal of the North American Benthological Society 12(2):177–184.

Pruitt, B. C., and S. E. Gatewood. 1976. Kissimmee River floodplain vegetation and cattle carrying capacity before and after canalization. Florida Division of State Planning, Tallahassee.

Putnam, R. D. 1995. Bowling alone: America's declining social capital. Journal of Democracy 6 (1):65–78.

Quick, H. 1925. One man's life. Bobbs-Merrill Company, Indianapolis, Indiana.

Rafle, P. A. 1994. The BeaMoc Project. Trout 35(Autumn):28–30.

Rasker, R., N. Tirrell, and D. Kloepfer. 1992. The wealth of nature: new economic realities in the Yellowstone region. The Wilderness Society, Washington, DC.

Raven, P. H., and G. B. Johnson. 1986. Biology. Times Mirror/Mosby College Publication, St. Louis, Missouri.

Ray, G. C., and J. F. Grassle. 1991. Marine biological diversity. BioScience 41:453–457.

Redford, R. 1987. Search for common ground. Harvard Business Review (May–June): 111.

Redwood National Park. 1987. Ninth annual report to Congress on the status of implementation of the Redwood National Park Expansion Act of 27 March 1978. National Park Service, Redwood National Park, Crescent City, California.

Reeves, G. H., L. E. Benda, K. M. Burnett, P. A. Bisson, and J. R. Sedell. 1995. A disturbance-based ecosystem approach to maintaining and restoring freshwater habitats of evolutionarily significant units of anadromous salmonids in the Pacific Northwest. Pages 334–349 *in* J. L. Nielson, editor. Evolution and the aquatic ecosystem: defining unique units in population conservation. American Fisheries Society Symposium 17, Bethesda, Maryland.

Reeves, G. H., F. H. Everest, and T. E. Nickelson. 1989. Identification of physical habitats limiting the production of coho salmon in western Oregon and Washington. U.S. Forest Service General Technical Report PNW-245.

Reeves, G. H., F. H. Everest, and J. R. Sedell. 1991b. Responses of anadromous salmonids to habitat modification: how do we measure them? Pages 62–67 *in* J. Colt and R. J. White, editors. Fisheries bioengineering symposium. American Fisheries Society Symposium 10, Bethesda, Maryland.

Reeves, G. H., F. H. Everest, J. R. Sedell, and D. B. Hohler. 1990. Influence of habitat modifications on habitat composition and anadromous salmonid populations in Fish Creek, Oregon, 1983–1988. Annual Report, Bonneville Power Administration, Division of Fish and Wildlife, Project 84-11, Portland, Oregon.

Reeves, G. H., J. D. Hall, T. D. Roelofs, T. L. Hickman, and C. O. Baker. 1991a. Rehabilitating and modifying stream habitats. Pages 519–557 *in* Meehan (1991).

Reeves, G. H., and J. R. Sedell. 1992. An ecosystem approach to the conservation and management of freshwater habitat for anadromous salmonids in the Pacific Northwest. Transactions of the North American Wildlife and Natural Resources Conference 57:408–415.

Reeves, G. H., and six coauthors. 1997. Fish habitat restoration in the Pacific Northwest: Fish Creek of Oregon. Pages 335–359 *in* Williams et al. (1997b).

Regan, D. H. 1983. Duties of preservation. Pages 195–220 *in* B. G. Norton, editor. The preservation of species. Princeton University Press, Princeton, New Jersey.

Regier, H. A., R. L. Welcomme, R. J. Steedman, and H. F. Henderson. 1989. Rehabilitation of degraded river systems. Pages 86–97 *in* D. P. Dodge, editor. Proceedings of the international large river symposium. Canadian Special Publication of Fisheries and Aquatic Sciences 106.

Regional Interagency Executive Committee. 1995a. Ecosystem analysis at the watershed scale: federal guide for watershed analysis, version 2.2, revised August 1995. U.S. Regional Ecosystem Office, Portland, Oregon.

Regional Interagency Executive Committee. 1995b. Ecosystem analysis at the watershed scale: federal guide for watershed analysis, section II, analysis methods and techniques, version 2.2, November 1995. U.S. Regional Ecosystem Office, Portland, Oregon.

Reich, R. 1992. The work of nations. Vintage Books, New York.

Reid, R. L., and B. Russell. 1995. Analysis of demographic and economic aspects of the Applegate Watershed. A report prepared for the Rogue River National Forest. Southern Oregon State College, Regional Services Institute, Ashland.

Reid, W. V., and K. R. Miller. 1989. Keeping options alive; the scientific basis for conserving biodiversity. World Resources Institute, Washington, DC.

Reimers, P. E. 1971. The length of residence of juvenile fall chinook salmon in Sixes River, Oregon. Doctoral dissertation. Oregon State University, Corvallis.

Reisner, M. 1993. Caddilac Desert. Penguin Books, New York.

Rice, R. M., R. R. Ziemer, and S. C. Hankin. 1982. Slope stability effects of fuel management strategies—inferences from Monte Carlo simulations. Pages 365–371 *in* C. E. Conrad and W. C. Oechel, editors. Proceedings of symposium on dynamics and management of Mediterranean-type ecosystems. U.S. Forest Service General Technical Report PSW-58.

Rieman, B. E., and J. D. McIntyre. 1993. Demographic and habitat requirements for conservation of bull trout. U.S. Forest Service Intermountain Research Station General Technical Report GTR-INT-302.

Rieman, B. E., and J. D. McIntyre. 1995. Occurrence of bull trout in naturally fragmented habitat patches of varied size. Transactions of the American Fisheries Society 124:285–296.

Rifkin, J. 1995. The end of work. Putnam Books, New York.

Rinne, J. N., and P. R. Turner. 1991. Reclamation and alteration as management techniques, and a review of methodology in stream renovation. Pages 219–244 *in* Minckley and Deacon (1991).

Risser, P. G. 1995. Biodiversity and ecosystem function. Conservation Biology. 9:742–746.

River Network. 1995. Watershed 2000: River Network's 1996–2000 strategic plan. River Voices 6(3):1–2, 22.

Rogers, D. V., and M. J. Packer. 1993. Vector-borne diseases, models, and global change. Lancet 342:1282–1285.

Rogers, E. M. 1962. Diffusion of innovations, 3rd edition. The Free Press, New York.

Rohrer, R. L. 1982. Evaluation of Henrys Lake management program. Idaho Department of Fish and Game, Federal Aid in Fish Restoration Project F-73-R-4, Subproject III, Study III, Jobs 1 and 2, Idaho Falls.

Rohrer, R. L. 1983. Henrys Fork fisheries investigations. Idaho Department of Fish and Game, Federal Aid in Fish Restoration Project F-73-R-4, Subproject IV, Study XI, Idaho Falls.

Rolston, H., III. 1981. Values in nature. Environmental Ethics 3(2):113–128.

Rolston, H., III. 1986. Philosophy gone wild. Prometheus Books, Buffalo, New York.

Rolston, H., III. 1988. Environmental ethics. Temple University Press, Philadelphia, Pennsylvania.

Rolston, H., III. 1994. Conserving natural value. Columbia University Press, New York.

Roseboom, D.P., and W. White. 1990. The Court Creek restoration project: erosion control technology in transition. Pages 25–40 *in* Proceedings of XXI conference of the IECA (International Erosion Control Association). IECA, Steamboat Springs, Colorado.

Rosgen, D. L. 1994. A classification of natural rivers. Catena 22:169–199.

Rubinstein, L., editor. 1990. Deerfield River comprehensive management plan. Franklin County Planning Department, Greenfield, Massachusetts.

Sager, M. 1977. Political problems inherent in environmental protection legislation and implementation. Pages 499–521 *in* J. Cairns, Jr., K. L. Dickson, and E. E. Herricks, editors. Recovery and restoration of damaged ecosystems. University Press of Virginia, Charlottesville.

Sampson, N. R. 1991. The politics of the environment. Journal of Soil and Water Conservation 46:398–400.

Sanford, D. K. 1991. Lower Beaverkill fisheries management. New York Department of Environmental Conservation, Stamford.

Sawyer, J. O., D. A. Thornburgh, and J. R. Griffin. 1988. Mixed evergreen forest. Pages 359–381 *in* M. G. Barbour and J. Major, editors. Terrestrial vegetation of California. California Native Plant Society Special Publication 9, second edition. Sacramento, California.

Schlosser, I. J. 1985. Flow regime, juvenile abundance, and the assemblage structure of stream fishes. Ecology 66:1484–1490.

Schlosser, I. J. 1990. Environmental variation, life history attributes, and community structure in stream fishes: implications for environmental management and assessment. Environmental Management 14:621–628.

Schlosser, I. J. 1991. Stream fish ecology: a landscape perspective. BioScience 41:704–712.

Schlosser, I. J. 1995. Dispersal, boundary processes, and trophic-level interactions in streams adjacent to beaver ponds. Ecology 76:908–925.

Schlosser, I. J., and P. L. Angermeier. 1995. Spatial variation in demographic processes of lotic fishes: conceptual models, empirical evidence, and implications for conservation. Pages 392–401 *in* J. L. Nielsen, editor. Evolution and the aquatic ecosystem: defining unique units in population conservation. American Fisheries Society Symposium 17, Bethesda, Maryland.

Schultz, R. C., and five coauthors. 1995. Design and placement of a multispecies riparian buffer strip system. Agroforestry Systems 29:201–226.

Schumm, S. A. 1977. The Fluvial River. Wiley-Interscience, New York.

Schwartz, J. S. 1991. Influence of geomorphology and land use on distribution and abundance of salmonids in a coastal Oregon basin. Master's thesis. Oregon State University, Corvallis.

Schwiebert, E. 1979. The puzzle of the Henrys Fork. Fly Fisherman 10(3):57–60.

Scrivener, J. C., and M. J. Brownlee. 1989. Effects of forest harvesting on spawning gravel and incubation survival of chum (*Oncorhynchus keta*) and coho salmon (*O. kisutch*) in Carnation Creek, British Columbia. Canadian Journal of Fisheries and Aquatic Sciences 46:681–696.

Sear, D. A. 1994. River restoration and geomorphology. Aquatic Conservation, Marine and Freshwater Ecosystems 4:169–177.

Sear, D. A., S. E. Darby, C. R. Thorne, and A. B. Brookes. 1994. Geomorphological approach to stream stabilization and restoration: case study of the Mimmshall Brook, Hertfordshire, U.K. Regulated Rivers Research and Management 9:205–223.

Sedell, J. R., P. A. Bisson, F. J. Swanson, and S. V. Gregory. 1988. What we know about large trees that fall into streams and rivers. U.S. Forest Service General Technical Report PNW-GTR-229:47–81.

Sedell, J. R., and F. H. Everest. 1990. Historic changes in habitat for Columbia River Basin salmon under study for TES listing. U.S. Forest Service, Pacific Northwest Research Station, Corvallis, Oregon.

Sedell, J. R., and K. J. Luchessa. 1982. Using the historical record as an aid to salmonid habitat enhancement. Pages 210–223 *in* N. B. Armantrout, editor. Acquisition and utilization of aquatic habitat inventory information. American Fisheries Society, Western Division, Bethesda, Maryland.

Sedell, J. R., G. H. Reeves, F. R. Hauer, J. A. Stanford, and C. P. Hawkins. 1990. Role of refugia in recovery from disturbances: modern fragmented and disconnected river systems. Environmental Management 14:711–724.

Seideman, D. 1996. Out of the woods. Audubon 98(4):66–75.

Shabecoff, P. 1993. A fierce green fire: the American environmental movement. Hill and Wang, New York.

Shea, R. E. 1991. Monitoring of trumpeter swans in conjunction with trapping efforts at Red Rock Lakes National Wildlife Refuge, Montana, and Harriman State Park, Idaho, during winter 1990–91. U.S. Fish and Wildlife Service, Southeast Idaho Refuge Complex, Pocatello.

Sheldon, A. L. 1988. Conservation of stream fishes: patterns of diversity, rarity and risk. Conservation Biology 2:149–156.

Shepp, D. L., and J. D. Cummins. 1997. Restoration in an urban watershed: Anacostia River of Maryland and the District of Columbia. Pages 297–317 *in* Williams et al. (1997b).

Shields, F. D., Jr., C. M. Cooper, and S. S. Knight. 1995. Experiment in stream restoration. Journal of Hydraulic Engineering 121:494–502.

Siddique, A. K., and six coauthors. 1991. Survival of classic cholera in Bangladesh. Lancet 337:1125–1127.

Simpson, D., G. Peterson, and M. Roche. 1996. The salmon work goes on. Mattole Restoration Newsletter 11:10.

Skinner, B. F. 1938. The behavior of organisms. Appleton-Century Crafts, New York.

Smith, J. E. 1993. Retrospective analysis of changes in stream and riparian habitat characteristics between 1935 and 1990 in two eastern Cascade streams. Master's thesis. University of Washington, Seattle.

Smith, R. W., and J. S. Griffith. 1994. Survival of rainbow trout during their first winter in the Henrys Fork of the Snake River, Idaho. Transactions of the American Fisheries Society 123:747–756.

Sokal, R. R., and F. J. Rohlf. 1981. Biometry. Freeman, San Francisco.

Sparks, R. E. 1995. Need for ecosystem management of large rivers and their floodplains. BioScience 45:168–182.

Sparks, R. E. 1996. Ecosystem effects. Pages 132–162 *in* S. A. Changnon, editor. The great flood of 1993. Westview Press, Boulder, Colorado.

Sparks, R. E., P. B. Bayley, S. L. Kohler, and L. L. Osborne. 1990. Disturbance and recovery of large floodplain rivers. Environmental Management 14:699–709.

Spence, L. E. 1975. Upper Blackfoot River study: a pre-mining inventory of aquatic and wildlife resources. Montana Department Fish and Game, Missoula.

Spotila, J. R., K. M. Terpin, R. R. Koons, and R. L. Benate. 1979. Temperature requirements of fishes from eastern Lake Erie and the upper Niagara River: a review of the literature. Environmental Biology of Fishes 4(3):281–307.

Stanford, J. A., and six coauthors. 1996. A general protocol for restoration of regulated rivers. Regulated Rivers Research and Management 12:391–413.

Stanford, J. A., and J. V. Ward. 1988. The hyporheic habitat of river ecosystems. Nature 335:64–66.

Stanford, J. A., and J. V. Ward. 1992. Management of aquatic resources in large catchments: recognizing interactions between ecosystem connectivity and environmental disturbance. Pages 91–124 *in* Naiman (1992a).

State of Idaho. 1994. House resolution 32. A concurrent resolution stating legislative finds and recognizing the legislative charter of the Henrys Fork Watershed Council. Idaho House of Representatives, Boise.

Stemler, R. 1996. Attraction flows for Mattole Canyon Creek. Mattole Restoration Newsletter 10:3–4, Petrolia, California.

Stephens, W. N. 1907. Biennial report of the State Game Warden, 1905–06. Idaho Department of Fish and Game, Boise.

Stephens, W. N. 1909. Second biennial report of the State Game Warden, 1907–08. Idaho Department of Fish and Game, Boise.

Stevens, W. K. 1995. New rules for old dams can revive rivers. New York Times Science Times (November 28):C1.

Strahler, A. N. 1957. Quantitative analysis of watershed geomorphology. American Geophysical Union EOS Transactions 38:913–920.

Stroud, R. H., editor. 1992. Fisheries management and watershed development. American Fisheries Society Symposium 13, Bethesda, Maryland.

Swanson, F. J., J. L. Clayton, W. F. Megahan, and G. Bush. 1989. Erosional processes and long-term site productivity. Pages 67–81 *in* D. A. Perry, R. Meurisse, B. Thomas, et al., editors. Maintaining the long-term productivity of Pacific Northwest forest ecosystems. Timber Press, Portland, Oregon.

Swanston, D. N. 1991. Natural processes. Pages 139–179 *in* Meehan (1991).

Tangley, L. 1994. The importance of communicating with the public (essay 18B). Pages 535–536 *in* G. K. Meffe and C. R. Carroll, editors. Principles of conservation biology. Sinauer, Sunderland, Massachusetts.

Tarnow, K. 1995. Analysis and recommendations for the federal land Jobs in the Woods program. Pacific Rivers Council, Eugene, Oregon.

Templet, P. H., and S. Farber. 1994. The complementarity between environmental and economic risk: an empirical analysis. Ecological Economics 9:153–165.

The World Commission on Environment and Development. 1987. Our common future. Oxford University Press, Oxford.

Thomas, J. W., technical editor. 1979. Wildlife habitats in managed forests: the Blue Mountains of Oregon and Washington. U.S. Department of Agriculture, Agriculture Handbook 553, Washington, DC.

Thomas, L. M. 1985. The policy perspective: a look to the grass roots. Pages 1–2 *in* M. L. Moore, editor. Perspectives on nonpoint source pollution. Reprinted by Terrene Institute, Washington, DC (1994).

Thomas, R. E. 1925. Tenth biennial report of the State Game Warden, 1923–34. Idaho Department of Fish and Game, Boise.

Thoreau, H. D. 1863. Excursions. Ticknor and Fields, Boston.

Thorp, W. H. 1919. Seventh biennial report of the State Game Warden. Idaho Department of Fish and Game, Boise.

Ticknor, W. D. 1994. Ecosystem management: two obstinate issues. SUNY (State University of New York) College of Environmental Science and Forestry, Faculty of Forestry, Miscellaneous Publication 29, Syracuse.

Tilman, D., and J. A. Downing. 1994. Biodiversity and stability in grasslands. Nature 367:363–365.

Tilt, W. 1993. Public and private cooperation in wetland conservation—from advocacy to investment. Pages 229–234 *in* Waterfowl and wetland conservation in the 1990s, a global perspective. International Waterfowl and Wetlands Research Bureau, Special Publication 26, Slimbridge, UK.

Tilt, W. 1996. Moving beyond the past: a grant-maker's vision for effective environmental education. Wildlife Society Bulletin 24:621–626.

Tilt, W., and C. A. Williams. 1997. Building public and private partnerships. Pages 145–157 *in* Williams et al. (1997b).

Toth, L. A. 1990a. Impacts of channelization on the Kissimmee River ecosystem. Pages 47–56 *in* Loftin et al. (1990b).

Toth, L. A. 1990b. An ecosystem approach to Kissimmee River restoration. Pages 125–133 *in* Loftin et al. (1990b).

Toth, L. A. 1991. Environmental responses to the Kissimmee River demonstration project. South Florida Water Management District, Technical Publication 91-2, West Palm Beach.

Toth, L. A. 1993. The ecological basis of the Kissimmee River restoration plan. Florida Scientist 56:25–51.

Toth, L. A. 1994. Principles and guidelines for restoration of river/floodplain ecosystems—Kissimmee River, Florida. Pages 49–73 *in* J. Cairns, Jr., editor. Rehabilitating damaged ecosystems, 2nd edition. Lewis Publishers/CRC Press, Boca Raton, Florida.

Toth, L. A. 1996. Restoring the hydrogeomorphology of the channelized Kissimmee River. Pages 369–383 *in* A. Brookes and F. D. Shields, Jr., editors. River channel restoration: guiding principles for sustainable projects. Wiley, Chichester, UK.

Toth, L. A., D. A. Arrington, and G. Begue. 1997. Headwater restoration and reestablishment of natural flow regimes: Kissimmee River of Florida. Pages 425–442 *in* Williams et al. (1997b).

Toth, L. A., D. A. Arrington, M. A. Brady, and D. A. Muszick. 1995. Conceptual evaluation of factors potentially affecting restoration of habitat structure within the channelized Kissimmee River ecosystem. Restoration Ecology 3:160–180.

Toth, L. A., and N. G. Aumen. 1994. Integration of multiple issues in environmental restoration and resource enhancement projects in southcentral Florida. Pages 61–78 *in* J. Cairns, Jr., T. V. Crawford, and H. Salwasser, editors. Implementing integrated environmental management. Virginia Polytechnic Institute and State University, Blacksburg.

Toth, L. A., J. T. B. Obeysekera, W. A. Perkins, and M. K. Loftin. 1993. Flow regulation and restoration of Florida's Kissimmee River. Regulated Rivers Research and Management 8:155–166.

Trauger, D. L., W. Tilt, and C. B. Hatcher. 1995. Partnerships: innovative strategies for wildlife conservation. Wildlife Society Bulletin 23(1):114–119.

Trotter, P. C. 1987. Cutthroat, native trout of the West. Colorado Associated University Press, Boulder.

Tullock, G. 1994. The economies of non-human societies. Pallas Press, Tucson, Arizona.

TVA (Tennessee Valley Authority). 1995. Aquatic diversity: a vital natural resource. Page 8. A report on the conditions of the Tennessee River and its tributaries in 1994. RiverPulse, Chattanooga, Tennessee.

Tyrrell, T. J., V. Maharaj, and M. F. Devitt. 1995. Economic impact assessment of the Beaverkill–Willowemoc trout fishery. Impact Research Associates, Wakefield, Rhode Island.

Udall, S. C. 1966. Study of strip and surface mining in the Appalachians. Report to Appalachian Regional commission. U.S. Department of the Interior, Washington, DC.

Udall, S. L. 1991. Foreword. Pages ix–xi *in* Minckley and Deacon (1991).

University of Colorado. 1996. The watershed source book: watershed-based solutions to natural resource problems. Natural Resources Law Center, University of Colorado School of Law, Boulder.

UNPF (United Nations Population Fund). 1991. The state of the world population. United Nations, New York.

USACOE (U.S. Army Corps of Engineers). 1985. Central and southern Florida, Kissimmee

River. Final feasibility report and environmental impact statement. U.S. Army Corps of Engineers, Jacksonville District, Jacksonville, Florida.

USACOE (U.S. Army Corps of Engineers). 1994. Anacostia River and tributaries District of Columbia and Maryland integrated feasibility report and environmental impact statement, Volume I (draft) January 1994.

USBLM (U.S. Bureau of Land Management). 1979. Stream survey reports. Elko District BLM, Elko, Nevada.

USBLM (U.S. Bureau of Land Management). 1987a. Stream survey reports. Elko District BLM, Elko, Nevada.

USBLM (U.S. Bureau of Land Management). 1987b. Marys River riparian/aquatic habitat management plan. Elko District BLM, Elko, Nevada.

USBLM (U.S. Bureau of Land Management). 1988. Wildcat Creek stream survey report. Elko District BLM, Elko, Nevada.

USBLM (U.S. Bureau of Land Management). 1990a. Stream survey reports. Elko District BLM, Elko, Nevada.

USBLM (U.S. Bureau of Land Management). 1990b. Marys River land exchange proposal environmental assessment. Elko District BLM, Elko, Nevada.

USBLM (U.S. Bureau of Land Management). 1991. Wildcat Creek stream survey report. Elko District BLM, Elko, Nevada.

USBLM (U.S. Bureau of Land Management). 1992a. Lower Marys River stream survey report. Elko District BLM, Elko, Nevada.

USBLM (U.S. Bureau of Land Management). 1992b. Marys River master plan. Elko District BLM, Elko, Nevada.

USBLM (U.S. Bureau of Land Management). 1993. Stream survey reports. Elko District BLM, Elko, Nevada.

USBLM (U.S. Bureau of Land Management). 1994a. Ecosystem management in the BLM: from concept to commitment. USBLM, Washington, DC.

USBLM (U.S. Bureau of Land Management). 1994b. Coos Bay district resource management plan. Bureau of Land Management, Coos Bay, Oregon.

USBLM (U.S. Bureau of Land Management). 1996. Partners against weeds: an action plan for the Bureau of Land Management. Bureau of Land Management, Billings, Montana.

USBLM (U.S. Bureau of Land Management) and USFS (U.S. Forest Service). 1994. Applegate Adaptive Management Area, ecosystem health assessment. U.S. Bureau of Land Management, Medford District, et al., Medford, Oregon.

USBOR (U.S. Bureau of Reclamation). 1939. Island Park Dam and Reservoir topography (map). Upper Snake River Project, Burley, Idaho (originally published in Ashton, Idaho).

USBOR (U.S. Bureau of Reclamation). 1992. Prineville Reservoir resource management plan. U.S. Bureau of Reclamation, Northwest Region, Boise, Idaho.

USCFF (U.S. Commission of Fish and Fisheries). 1901. Report of the Commissioner. Washington, DC.

USDOC (U.S. Department of Commerce). 1990. Population Hiwassee River basin. County projections to 2040. Prepared by the U.S. Bureau of the Census, Bureau of Economic Analysis, Regional Economic Analysis Division, Washington, DC.

USDOI (U.S. Department of the Interior) and USDA (U.S. Department of Agriculture). 1996. Watershed restoration/Jobs in the Woods summary report: fiscal year 1995. U.S. Department of the Interior and U.S. Department of Agriculture, Portland, Oregon.

USEPA (U.S. Environmental Protection Agency). 1976. Quality criteria for water. U.S. Environmental Protection Agency, Washington, DC.

USEPA (U.S. Environmental Protection Agency). 1987. National water quality inventory: 1986 report to Congress. U.S. Environmental Protection Agency, Washington, DC.

USEPA (U.S. Environmental Protection Agency). 1991. Near coastal waters national pilot project. "Action plan for Oregon coastal watersheds, estuary, and ocean waters" 1988–1991. Oregon Department of Environmental Quality, Portland.

USEPA (U.S. Environmental Protection Agency). 1993. Region 8 Record of Decision, Smuggler Mountain, Pitkin County, Colorado. U.S. Environmental Protection Agency, Denver, Colorado.

USFS (U.S. Forest Service). 1989. Environmental assessment for the Mary's River area analysis. Humboldt National Forest, Elko, Nevada.

USFS (U.S. Forest Service). 1994. Fish Creek watershed analysis. U.S. Forest Service, Mt. Hood National Forest, Gresham, Oregon.

USFS (U.S. Forest Service). 1996a. Draft environmental impact statement, Targhee National Forest plan revision. St. Anthony, Idaho.

USFS (U.S. Forest Service). 1996b. Scoping document on the proposed Warm River hatchery, Targhee National Forest, Ashton Ranger District, Ashton, Idaho.

USFS (U.S. Forest Service) and USBLM (U.S. Bureau of Land Management). 1994. Record of decision for amendment to Forest Service and Bureau of Land Management planning document within the range of the northern spotted owl. USFS and USBLM, Portland, Oregon.

USFS (U.S. Forest Service) and USBLM (U.S. Bureau of Land Management). 1995. Environmental assessment: interim strategies for managing anadromous fish-producing watersheds on federal lands in eastern Oregon and Washington, Idaho, and portions of California. USFS and USBLM, Washington, DC.

USFWS (U.S. Fish and Wildlife Service). 1959. A detailed report of the fish and wildlife resources in relation to the Corps of Engineers plan of development, Kissimmee River basin, Florida. Appendix A in Central and Southern Florida Project for Flood Control and Other Purposes. Part II. Kissimmee River Basin and Related Areas. Supplement 5. U.S. Army Corps of Engineers, Jacksonville, Florida.

USFWS (U.S. Fish and Wildlife Service). 1986. North American waterfowl management plan, a strategy for cooperation. U.S. Department of the Interior, Washington, DC.

USFWS (U.S. Fish and Wildlife Service). 1994. Fish and Wildlife Coordination Act report: Kissimmee headwater lakes revitalization. U.S. Fish and Wildlife Service, Vero Beach, Florida.

USFWS (U.S. Fish and Wildlife Service). 1995. Lahontan cutthroat trout, *Oncorhynchus clarki henshawi*, recovery plan. U.S. Fish and Wildlife Service, Portland, Oregon.

USGAO (U.S. General Accounting Office). 1988. Public rangelands: some riparian areas restored but widespread improvement will be slow. Report to Congressional requesters. U.S. General Accounting Office, Resources, Community, and Economic Development Division, GA/RCED-88-105, Washington, DC.

USGSA (U.S. General Services Administration). 1990. Total and federally owned land, 1960 to 1990, and by state, 1990. Page 219 *in* U.S. Bureau of the Census, statistical abstracts of the United States, 1994 (114 edition), Washington, DC.

U.S. House. 1949. Comprehensive report on central and southern Florida for flood control and other purposes. 80th Congress, 2nd session, House Document 643, Washington, DC.

Van Kirk, R. 1995. Putting it all together: integration of research results helps explain trout population trends. Henry's Fork Foundation Newsletter, Fall 1995:7–10. Island Park, Idaho.

Van Kirk, R. 1996. A history of fisheries management on the Henry's Fork. Henry's Fork Foundation Newsletter, Winter 1996:1, 5–8. Island Park, Idaho.

Van Kirk, R. W., and C. B. Griffin. 1997. Building a collaborative process for restoration: Henrys Fork of Idaho and Wyoming. Pages 253–276 *in* Williams et al. (1997b).

Van Put, E. 1996. The Beaverkill. Lyons & Burford, New York.

Vanderweth, W. C., editor. 1971. Indian oratory: famous speeches by noted indian chieftains. University of Oklahoma Press, Norman.

VDEC (Vermont Department of Environmental Conservation). 1992. Comprehensive river plan for the Deerfield River watershed. VDEC, Agency of Natural Resources, Waterbury.

Velinski, D. J., and J. D. Cummins. 1996. Distribution of chemical contaminants in

1993–95: wild fish species in the District of Columbia. Interstate Commission on the Potomac River Basin, Report 96-1, Rockville, Maryland.

Vinson, D. K. 1991. Base flow determination for wintering trumpeter swans on the Henrys Fork of the Snake River. U.S. Fish and Wildlife Service, Boise, Idaho.

Vinson, D. K., M. R. Vinson, and T. R. Angradi. 1992. Aquatic macrophytes and instream flow characteristics of a Rocky Mountain river. Rivers 3(4):260–265.

Vitousek, P. M., P. R. Ehrlich, A. H. Ehrlich, and P. A. Matson. 1986. Human appropriation of the products of photosynthesis. BioScience 36:368–373.

Vogl, R. J. 1980. The ecological factors that produce perturbation-dependent ecosystems. Pages 63–94 *in* J. Cairns, Jr., editor. The recovery process in damaged ecosystems. Ann Arbor Science, Ann Arbor, Michigan.

Wahrhaftig, C., and R. Curry. 1967. Geologic implications of sediment discharge records from the northern coast ranges, California. Pages 35–88 *in* Man's effect on California watersheds. A report for the Subcommittee on Forest Practices and Watershed Management, California State Assembly, Sacramento.

Wallace, A. R. 1863. On the physical geography of the Malay archipelago. Journal of the Royal Geographical Society (London) 33:217–234.

Wallace, D. R. 1983. The Klamath knot. Sierra Club Books, San Francisco.

Walters, C. J. 1986. Adaptive management of renewable resources. Macmillan Press, New York.

Walters, C., L. Gunderson, and C. S. Holling. 1992. Experimental policies for water management in the Everglades. Ecological Applications 2:189–202.

Walker, G. T. 1977. Recreational use of the lower Blackfoot River. Montana Department Fish and Game, Missoula.

Ward, J. V. 1989. The four-dimensional nature of lotic ecosystems. Journal of the North American Benthological Society 8:2–8.

Warner, A., D. Shepp, K. Corish, and J. Galli. 1996. An existing source assessment of pollutants to the Anacostia watershed. Metropolitan Washington Council of Governments for the Department of Consumer and Regulatory Affairs, Washington, DC.

Warren, C. E. 1979. Toward classification and rationale for watershed management and stream protection. U.S. Environmental Protection Agency Ecological Research Series EPA-600/3-79-059, Washington, DC.

Warren, C. E., and W. J. Liss. 1980. Adaptation to aquatic environments. Pages 15–40 *in* R. L. Lackey and L. Nielsen, editors. Fisheries management. Blackwell Scientific, Oxford, UK.

Warren, M. L., Jr., and B. M. Burr. 1994. Status of freshwater fishes of the United States: overview of am imperiled fauna. Fisheries 19(1):6–18.

Washington Forest Practices Board. 1993. Standard methodology for conducting watershed analysis under Chapter 222-22 WAC, version 2.0. Washington Department of Natural Resources, Forest Practices Division, Olympia, Washington.

Water Quality 2000. 1994. Evaluation of a watershed approach to clean water. Water Quality 2000, Alexandria, Virginia.

Weatherly, A. H., and S. C. Rogers. 1978. Some aspects of growth. Pages 52–74 *in* S. D. Gerking, editor. Ecology of freshwater fish production. Wiley, New York.

Weaver, W. E., and five coauthors. 1987. An evaluation of experimental rehabilitation work: Redwood National Park. National Park Service, Redwood National Park Technical Report 19, Arcata, California.

Wegener, W. and V. Williams. 1975. Fish population responses to improved lake habitat utilizing an extreme lake drawdown. Proceedings 28th Annual Conference of Southeastern Association of Game and Fish Commissioners 28(1974):144–161.

Wehnes, R. E. 1992. Streams for the future: integrating public involvement in a stream improvement program in Missouri. Pages 229–236 *in* Stroud (1992).

Wernecke, C. 1936. Part II. Cedar River: on flooding in King County. King County Department of Conservation and Development, Seattle.

Wesche, T. A. 1985. Stream channel modifications and reclamation structures to enhance fish habitat. Pages 103–159 *in* J. A. Gore, editor. The restoration of rivers and streams. Butterworth, Boston.

Westman, W. E. 1978. How much are nature's services worth? Science 197:960–964.

White, R. 1991. It's your misfortune and none of mine, a new history of the American West. University of Oklahoma Press, Norman.

Whitehead, R. L. 1978. Water resources of the Henrys Fork basin in eastern Idaho. Idaho Department of Water Resources, Water Information Bulletin 46, Boise.

Whitford, W. G. 1995. Desertification: implications and limitations of the ecosystem health metaphor. Pages 273–293 *in* D. J. Rapport, C. L. Gaudet, and P. Calow, editors. Evaluating and monitoring the health of large-scale ecosystems. Springer-Verlag, Berlin.

Wilkinson, C. F. 1992. Crossing the next meridian: land, water, and the future of the West. Island Press, Washington, DC.

Willard, D. E., and L. D. Kosmond. 1995. A watershed–ecosystem approach. Pages 37–52 *in* Wetlands and watershed management: meeting landowner and resource conservation needs through partnership approaches. Association of State Wetland Managers, Berne, New York.

Williams, G. G. 1996. A watershed approach to assessing brook trout (*Salvelinus fontinalis*) distribution and ecological health in the Hiwassee River watershed. Tennessee Valley Authority, Chattanooga, Tennessee.

Williams, J. D., M. L. Warren, Jr., K. S. Cummings, J. L. Harris, and R. J. Neves. 1993. Conservation status of freshwater mussels of the United States and Canada. Fisheries 18(9):6–22.

Williams, J. E. 1991. Preserves and refuges for native western fishes: history and management. Pages 171–189 *in* Minckley and Deacon (1991).

Williams, J. E., and R. J. Neves. 1992. Opening remarks [special session on biological diversity in aquatic management]. Transactions of the North American Wildlife and Natural Resources Conference 57:343–344.

Williams, J. E., and J. N. Rinne. 1992. Biodiversity management on multiple-use federal lands: an opportunity whose time has come. Fisheries 17(3):4–5.

Williams, J. E., and seven coauthors. 1989. Fishes of North America endangered, threatened, or of special concern: 1989. Fisheries 14(6):2–20.

Williams, J. E., and C. D. Williams. 1995. Who speaks for the Snake River salmon? Fisheries 20(11):24–26.

Williams, J. E., and C. D. Williams. 1997. An ecosystem-based approach to management of salmon and steelhead habitat. Pages 541–556 *in* D. J. Stouder, P. A. Bisson, and R. J. Naiman, editors. Pacific salmon and their ecosystems: status and future options. Chapman and Hall, New York.

Williams, J. E., C. A. Wood, and M. P. Dombeck. 1997a. Understanding watershed-scale restoration. Pages 1–13 *in* Williams et al. (1997b).

Williams, J. E., C. A. Wood, and M. P. Dombeck, editors. 1997b. Watershed restoration: principles and practices. American Fisheries Society, Bethesda, Maryland.

Williams, P. B. 1995. Overcoming technical barriers to integrating watershed, river floodplain, and wetland management. Pages 65–67 *in* Wetlands and watershed management: meeting landowner and resource conservation needs through partnership approaches. Association of State Wetland Managers, Berne, New York.

Williams, V. 1990. Management and mis-management of the upper Kissimmee River basin chain of lakes. Pages 9–29 *in* Loftin et al. (1990b).

Wilson, E. O. 1985. The biological diversity crisis. BioScience 35:700–706.

Wilson, E. O. 1992. The diversity of life. The Belknap Press of Harvard University Press, Cambridge, Massachusetts.

Wilson, P. S. 1992. What Chief Seattle said. Environmental Law 22(4):1451–1468.

Winograd, I. J., and B. J. Szabo. 1991. Time of isolation of *Cyprinodon diabolis* in Devils Hole: geologic evidence. Proceedings of the Desert Fishes Council 20:49–50.

Wise, L. P. 1993. Visions for the future. Pages 69–73 *in* Watershed '93: a national conference on watershed management. U.S. Environmental Protection Agency, Report 1994-300-781/12415, U.S. Government Printing Office, Washington, DC.

Wissmar, R. C. 1990. Recovery of lakes in the 1980 blast zone of Mount St. Helens. Northwest Science 64(5):268–270.

Wissmar, R. C. 1993. The need for long-term monitoring in stream ecosystems of the Pacific Northwest. Environmental Assessment and Monitoring 26:219–234.

Wissmar, R. C. 1997. Historical perspectives. Pages 66–79 *in* Williams et al. (1997b).

Wissmar, R. C., and W. N. Beer. 1994. Distribution of fish and stream habitats and influences of watershed conditions, Beckler River, Washington. University Washington, School of Fisheries, Fisheries Research Institute Technical Report FRI-UW-9418, Seattle.

Wissmar, R. C., and R. Beschta. In press. Restoration and the management of riparian ecosystem. Freshwater Biology.

Wissmar, R. C., A. H. Devol, J. T. Staley, and J. R. Sedell. 1982. Biological responses in lakes of Mount St. Helens blast zone. Science 216:178–181.

Wissmar, R. C., and five coauthors. 1994. A history of resource use and disturbance in riverine basins of eastern Oregon and Washington. Northwest Science (Special Issue) 68:233–267.

Wissmar, R. C., and F. J. Swanson. 1990. Landscape disturbance and lotic ecotones. Pages 65–89 *in* R. J. Naiman and H. Decamps, editors. Ecology and management of aquatic–terrestrial ecotones. Parthenon Press, London.

Woody, T. 1993. Grassroots in action: the Sierra Club's role in the campaign to restore the Kissimmee River. Journal of the North American Benthological Society 12:201–205.

Wullschleger, J. G., S. J. Miller, and L. J. Davis. 1990. An evaluation of the effects of the restoration demonstration project on the Kissimmee River fishes. Pages 67–81 *in* Loftin et al. (1990b).

Young, M. K., editor. 1995. Conservation assessment for inland cutthroat trout. U.S. Forest Service General Technical Report RM-GTR-256, Fort Collins, Colorado.

Yount, J. D., and G. J. Niemi. 1990. Recovery of lotic communities from disturbance—a narrative review of case studies. Environmental Management 14:547–569.

Zedler, J. B. 1988. Salt marsh restoration: lessons from California. Pages 123–138 *in* J. Cairns, Jr., editor. Rehabilitating damaged ecosystems. CRC Press, Boca Raton, Florida.

Zedler, J. B. 1994. Restoring a nation's wetlands: why, where, and how? Pages 417–418 *in* G. K. Meffe and C. R. Carroll, editors. Principles of conservation biology. Sinauer, Sunderland, Massachusetts.

Zedler, J. B., and R. Langis. 1991. Comparisons of constructed and natural salt marshes of San Diego Bay. Restoration and Management Notes 9(1):21–25.

Ziemer, R. R. 1992. Evaluating long-term cumulative hydrologic effects of forest management: a conceptual approach. Pages 47–51 *in* C. E. Conrad and L. A. Newell, editors. Proceedings of the session on tropical forestry for people of the Pacific, XVII Pacific Science Congress. U.S. Forest Service General Technical Report PSW-GTR-129.

Ziemer, R. R. 1997. Temporal and spatial scales. Pages 80–95 *in* Williams et al. (1997b).

Zippen, C. 1958. The removal method of population estimation. Journal of Wildlife Management 22:82–90.

Zuckerman, S. 1990. The next tallest tree: restoration in Redwood National Park. Pacific Discovery 43(3):23–29.

Zuckerman, S. 1997. Thinking like a watershed: Mattole River of California. Pages 216–234 *in* Williams et al. (1997b).

Glossary

Adaptive management: Monitoring or assessing the progress toward meeting objectives and incorporating what is learned into future management plans.

Adfluvial: Life history strategy in which adult fish spawn and juveniles subsequently rear in streams but migrate to lakes for feeding as subadults and adults. Compare *fluvial*.

Aggradation: Geologic process by which streambeds and floodplains are raised in elevation by the deposition of material eroded elsewhere.

Anadromous: Fish that leave freshwater and migrate to the ocean to mature then return to freshwater to spawn.

Anchor ice: Ice that is attached to underwater objects, including the substrate. It is often formed from accumulations of adhesive frazil ice.

Aquifer: Water-bearing rock formation or other subsurface layer.

Base flow: Portion of stream discharge derived from such natural storage sources as groundwater, large lakes, and swamps but does not include direct runoff or flow from stream regulation, water diversion, or other human activities.

Benthic: Bottom dwelling or substrate-oriented; at or in the bottom of a stream or lake.

Bioengineering: Combining structural, biological, and ecological concepts to construct living structures for erosion, sediment, or flood control.

Biological diversity (biodiversity): Variety and variability among living organisms and the ecological complexes in which they occur; encompasses different ecosystems, species, and genes.

Biological integrity: Capability of supporting and maintaining a balanced, integrated, adaptive community of organisms having a species composition, diversity, and functional organization comparable to that of natural habitat of the region; a system's ability to generate and maintain adaptive biotic elements through natural evolutionary processes.

Biological oxygen demand: Amount of dissolved oxygen required by decomposition of organic matter.

Biomass: Summed mass or weight of individuals in one or more species, usually related to a defined area or volume. See also *standing crop*.

Bio-social ecosystem: Existence of social and ecological systems as equal partners.

Boulder weirs: Barrier constructed with boulders (rocks 10 inches and larger [Wentworth Scale]) across a stream. See *weir*.

Braided stream: Stream that forms an interlacing network of branching and recombining channels separated by branch islands or channel bars.

Capping: Physical isolation of contaminated sediments by a medium such as sand.

Carrying capacity: Maximum average number or biomass of organisms that can be sustained in a habitat over the long term. Usually refers to a particular species, but can be applied to more than one.

Channelization: Straightening the meanders of a river; often accompanied by placing riprap or concrete along banks to stabilize the system.

Check dams: Series of small dams placed in gullies or small streams in an effort to control erosion. Commonly built during the 1900s.

Combined sewer overflow: Overflow from sanitary and storm sewers.

Confidence intervals: Probability, based on statistics, that a number will be between an upper and lower limit.

Confluence: Joining.

Debris flow: Movement of materials in intermittent or small, nonfish-bearing, perennial upslope channels.

Debris torrent: Deluge of water charged with soil, rock, and organic debris down a steep stream channel.

Depauperate: Biologically impoverished; having relatively few species.

Disturbance regime: Characteristics (timing, duration, and intensity) of natural (occasionally artificial) disruptions such as floods, wildfires, volcanoes, etc. Natural disturbance regime is the regime that occurred historically.

Diversity: Variation that occurs in plant and animal taxa (i.e., species composition), habitats, or ecosystems. See *species richness*.

Ecological restoration: Involves replacing lost or damaged biological elements (populations, species) and reestablishing ecological processes (dispersal, succession) at historical rates.

Ecosystem: Biological community together with the chemical and physical environment with which it interacts.

Ecosystem management: Management that integrates ecological relationships with sociopolitical values toward the general goal of protecting or returning ecosystem integrity over the long term.

Emergence trap: Trap that captures young fish or insects as they emerge from gravel into the water column.

Escapement: Number of migratory—usually anadromous—fish that reach suitable spawning areas in a given year, typically after passing through a fishery area.

Estuarine: A partly enclosed coastal body of water that has free connection to an open sea, and within which seawater is measurably diluted by fresh river water.

Eutrophic: Water body rich in dissolved nutrients, photosynthetically productive, and often deficient in oxygen during warm periods. Compare *oligotrophic*.

Exotic: Introduced into a habitat or region; not naturally occurring. See *native*.

Extirpate: Make locally extinct.

Fecal coliform: Colon bacteria that indicate fecal contamination; aerobic bacteria found in the colon or feces, often used as indicators of fecal contamination of water supplies.

Fishway (fish ladder): Passageway, often an ascending series of pools, designed to permit fish to pass over dams, waterfalls, or other obstructions.

Floodplain: Lowland areas that are periodically inundated by the lateral overflow of streams or rivers.

Flow regime: Characteristics of stream discharge over time. Natural flow regime is the regime that occurred historically.

Fluvial: Pertaining to streams or rivers; also, organisms that migrate between main rivers and tributaries. Compare *adfluvial*.

Fluvial geomorphology: How running waters shape land formations.

Function of an ecosystem: Ecosystem productivity and functions of hydrology, feeding, and transport.

Functional flexibility: Ability to use a wide range of resources.

Functional groups: Groups composed of different species that perform a similar ecological function within an ecosystem.

Gabion: Wire basket filled with stones, used to stabilize streambanks, control erosion, and divert stream flow.

Generalist species: Organisms having broad environmental tolerance ranges and relatively unspecific resource requirements.

Geomorphology: Study of the form and origins of surface features of the Earth.

Glides: Stream habitat having a slow, relatively shallow run of water with little or no surface turbulence.

Gradient (of a stream): Rate of fall of a stream, typically expressed as so many feet of elevation change per mile.

Hard water: Water with high concentrations of dissolved calcium and magnesium salts; generally found in areas of carbonate rocks such as limestone and dolomite. Compare *soft water*.

Holocene epoch: Most recent geological epoch, comprising the 10,000 or so years since the last major continental glaciation.

Hydrograph: Chart of water levels over time.

Hydrology: Study of the properties, distribution, and effects of water on the Earth's surface, subsurface, and atmosphere.

Hypolimnion: Deepest, coldest, and relatively undisturbed region of a thermally stratified water body.

Hyporheic: Pertaining to a saturated zone beneath a river or stream consisting of sand, gravel, or rock with water-filled interstitial space.

Intermittent stream: Stream that has interrupted flow or does not flow continuously. Compare *perennial stream*.

Intraspecific interactions: Interactions within a species.

Issue management: Ability of an organization to identify three stages of problems—emerging, existing, and disruptive—and to respond to public concerns in such a way that the goals of both the organization and the community can be met.

Limiting factor: Single factor that limits a system or population from reaching its highest potential.

Littoral: Shoreline; shore area.

Macroinvertebrates: Invertebrates large enough to be seen with the naked eye (e.g., most aquatic insects, snails, and amphipods).

Native: Occurring naturally in a habitat or region; not introduced by humans.

Nonpoint source pollution: Polluted runoff from sources that cannot be defined as discrete points, such as areas of timber harvesting, surface mining, agriculture, and livestock grazing.

Oligotrophic: Water body low in dissolved nutrients and organic matter, with dissolved oxygen near saturation, and low chlorophyll levels. Compare *eutrophic*.

Parr: Young trout or salmon actively feeding in freshwater; usually refers to young anadromous salmonids before they migrate to the sea. See *smolt*.

Partnership: Collection of entities (often individuals, agencies, or institutions) where each contributes a combination of time, talent, and treasury—for the benefit of all parties.

Penstock: Gate or sluice used to control the flow of water.

Perennial stream: Stream that flows continually, without interruption. Compare *intermittent stream*.

pH: Measure of the negative logarithm of the hydrogen ion concentration to determine the acidity or alkalinity of water. Water of pH 7 is neutral; lesser values are acidic, higher values (pH 14 maximum) are alkaline.

Photopoints: Photographic record of the project over time at established locations.

Plunge pool: Basin scoured out by vertically falling water.

Pluvial lakes: Wet-climate lakes; relating to a time of wet climate caused by high precipitation (snow and rain) and low evaporation. During pluvial periods, streamflow is strong and lake levels are high.

Point source pollution: Pollution occurring at discrete points, such as sewage or factory effluent. Compare *nonpoint-source pollution*.

Productive harmony: Healthy, balanced state of an environment in which both social and physical resources have high levels of permanence and diversity, enabling their sustainability.

Productivity (biologic): Ability of a given area to produce biological matter (e.g., plants and animals).

Redds: Nests made in gravel (particularly by salmonids); consisting of a depression that is created and then covered.

Refugia: Geographic locations where a species or population has persisted during changed or adverse conditions such as glaciation.

Riffle: Stream habitat having a broken or choppy surface (white water), moderate or swift current, and shallow depth.

Riparian: Type of wetland transition zone between aquatic habitats and upland areas. Typically, lush vegetation along a stream or river.

Riprap: Large rocks, broken concrete, or other structure used to stabilize streambanks and other slopes.

Rootwad: Exposed root system of an uprooted or washed-out tree.

Rotenone: Commonly used fish poison derived from the root of some species of South American plants of the genus *Derris*.

Salmonid: Fish of the family Salmonidae, including salmon, trout, chars, whitefish, ciscoes, and grayling.

Silviculture: Tending, harvesting, and replacing forests, resulting in tree stands of distinctive form.

Sinuosity: Degree to which a stream channel curves or meanders laterally across the land surface.

Smolt: Juvenile salmon migrating seaward; a young anadromous trout, salmon, or char undergoing physiological changes that will allow it to change from life in freshwater to life in the sea. The smolt stage follows the parr stage. See *parr*.

Social ecology: Study of geographic place, local control (through the concept of the municipality), empowerment of citizens, and the meshing of social and environmental goals.

Social ecosystem: Culturally defined geographic area within which people manage their lives and resources.

Soft water: Water with a low concentration of dissolved calcium and magnesium salts; tends to be acidic or circumneutral. Compare *hard water*.

Species composition: Various species inhabiting a habitat or region.

Species richness: Number of species in a given area or habitat. See *diversity*.

Splash dam: Dam built to create a head of water for driving logs.

Standing crop: Summed mass of individuals in a given volume or area; also standing stock. See *biomass*.

Stock: Group of fish that is genetically self-sustaining and isolated geographically or temporally during reproduction. Generally, a local population of fish. More specifically, a local population—especially that of salmon, steelhead (rainbow trout), or other anadromous fish—that originates from specific watersheds as juveniles and generally returns to its birth streams to spawn as adults.

Stream order: Classification system for streams based on the number of tributaries it has. The smallest unbranched tributary in a watershed is designated order 1. A stream formed by the confluence of 2 order 1 streams is designated as order 2. A stream formed by the confluence of 2 order 2 streams is designated order 3, and so on.

Stream reach: Section of a stream between two points.

Structure of an ecosystem: Ecosystem's native species diversity.

Substrate: Material (silt, sand, gravel, cobble, etc.) that forms a stream or lake bed.

Subwatershed: One of the smaller watersheds that combine to form a larger watershed.

Succession (of plants): Sequence of plant communities that replace one another in a given area.

Superfund: Money available under the Comprehensive Response, Compensation, and Liability Act of 1980 for cleaning up the designated worst hazardous waste sites in the United States.

Sustainability: Use of natural resources at a rate that resources can be sustained through time without decline.

Terrace: Floodplain abandoned by a river as flows decreased.

Thalweg: Portion of a stream or river with the deepest water and greatest flow.

Thermal refugia: Geographic places that serve as shelters from adverse instream temperatures.

Trophic: Referring to nourishment; for example, the productivity of a body of water, or aspects of feeding and digestion in species.

Vadose zone: Zone between the land surface and the water table.

Watershed: Entire area that contributes both surface and underground water to a particular lake or river.

Watershed rehabilitation: Used primarily to indicate improvement of watershed condition or certain habitats within the watershed. Compare *watershed restoration*.

Watershed restoration: Reestablishing the structure and function of an ecosystem, including its natural diversity; a comprehensive, long-term program to return watershed health, riparian ecosystems, and fish habitats to a close approximation of their condition prior to human disturbance.

Watershed-scale approach: Consideration of the entire watershed in a project or plan.

Weir: Device across a stream to divert fish into a trap or to raise the water level or divert its flow. Also a notch or depression in a dam or other water barrier through which the flow of water is measured or regulated.

List of Species

Throughout this book, species are cited only by common name. The respective scientific names of these species are listed below. The reference for fishes is *Common and Scientific Names of Fishes from the United States and Canada* (5th edition, 1990), published by the American Fisheries Society. A variety of references were used by authors and editors to verify other animal and plant names.

Fishes

alewife . . . . *Alosa pseudoharengus*
American eel . . . . *Anguilla rostrata*
American shad . . . . *Alosa sapidissima*
Atlantic menhaden . . . . *Brevoortia tyrannus*
Atlantic salmon . . . . *Salmo salar*
Atlantic sturgeon . . . . *Acipenser oxyrhynchus*

bay pipefish . . . . *Sygnathus leptorhynchus*
black crappie . . . . *Pomoxis nigromaculatus*
black rockfish . . . . *Sabastes melanops*
blacknose dace . . . . *Rhinichthys atratulus*
blue catfish . . . . *Ictalurus furcatus*
blueback herring . . . . *Alosa aestivalis*
bluegill . . . . *Lepomis macrochirus*
bocaccio . . . . *Sabastes paucispinis*
brook trout . . . . *Salvelinus fontinalis*
brown bullhead . . . . *Ameiurus nebulosus*
brown Irish lord . . . . *Hemilepidotus spinosus*
brown trout . . . . *Salmo trutta*
buffalo sculpin . . . . *Enophrys bison*
bull trout . . . . *Salvelinus confluentus*

cabezon . . . . *Scorpaenichthys marmoratus*
channel catfish . . . . *Ictalurus punctatus*
chinook salmon . . . . *Oncorhynchus tshawytscha*
chiselmouth . . . . *Acrocheilus alutaceus*
chum salmon . . . . *Oncorhynchus keta*
coastrange sculpin . . . . *Cottus aleuticus*
coho salmon . . . . *Oncorhynchus kisutch*
Columbia River redband trout . . . . *Oncorhynchus mykiss gairdneri*
common carp . . . . *Cyprinus carpio*
common shiner . . . . *Luxilus cornutus*
copper rockfish . . . . *Sebastes caurinus*
creek chub . . . . *Semotilus atromaculatus*

cutlips minnow .... *Exoglossum maxillingua*
cutthroat trout .... *Oncorhynchus clarki*

Devils Hole pupfish .... *Cyprinodon diabolis*

English sole .... *Pleuronectes vetulus*
eulachon .... *Thaleichthys pacificus*

fallfish .... *Semotilus corporalis*

gizzard shad .... *Dorosoma cepedianum*
golden shiner .... *Notemigonus crysoleucas*
green sturgeon .... *Acipenser medirostris*
green sunfish .... *Lepomis cyanellus*

hickory shad .... *Alosa mediocris*

jacksmelt .... *Atherinopsis californiensis*
johnny darter .... *Etheostoma nigrum*

kelp greenling .... *Hexagrammos decagrammus*

Lahontan cutthroat trout .... *Oncorhynchus clarki henshawi*
Lahontan redside .... *Richardsonius egregius*
lake trout .... *Salvelinus namaycush*
largemouth bass .... *Micropterus salmoides*
largescale sucker .... *Catostomus macrocheilus*
lingcod .... *Ophiodon elongatus*
longear sunfish .... *Lepomis megalotis*
longnose dace .... *Rhinichthys cataractae*
longnose sucker .... *Catostomus catostomus*

margined madtom .... *Noturus insignis*
mottled sculpin .... *Cottus bairdi*
mountain sucker .... *Catostomus platyrhynchus*
mountain whitefish .... *Prosopium williamsoni*

northern anchovy .... *Engralis mordax*

Owens pupfish .... *Cyprinodon radiosus*

Pacific herring .... *Clupea pallasi*
Pacific lamprey .... *Lampetra tridentata*
Pacific sand lance .... *Ammodytes hexapterus*
Pacific staghorn sculpin .... *Leptocottus armatus*
Pacific tomcod .... *Microgadus proximus*
Paiute sculpin .... *Cottus beldingi*
pile perch .... *Rhacochilus vacca*
poolfish .... *Empetrichthys* spp.
prickly sculpin .... *Cottus asper*
pumpkinseed .... *Lepomis gibbosus*

quillback rockfish . . . . *Sabastes maliger*

rainbow smelt . . . . *Osmerus mordax*
rainbow trout . . . . *Oncorhynchus mykiss*
razorback sucker . . . . *Xyrauchen texanus*
redbreast sunfish . . . . *Lepomis auritus*
redtail surfperch . . . . *Amphistichus rhodoterus*
rock greenling . . . . *Hexagrammos lagocephalus*

saddleback gunnel . . . . *Pholis ornata*
sand sole . . . . *Psettichthys melanostictus*
sea lamprey . . . . *Petromyzon marinus*
shield darter . . . . *Percina peltata*
shiner perch . . . . *Cymatogaster aggregata*
shortnose sturgeon . . . . *Acipenser brevirostrum*
silver surfperch . . . . *Hyperprosopon ellipticum*
slimy sculpin . . . . *Cottus cognatus*
smallmouth bass . . . . *Micropterus dolomieu*
sockeye salmon . . . . *Oncorhynchus nerka*
speckled dace . . . . *Rhinichthys osculus*
speckled sanddab . . . . *Citharichthys stigmaeus*
starry founder . . . . *Platichthys stellatus*
steelhead . . . . *Oncorhynchus mykiss*
striped bass . . . . *Morone saxatilis*
striped seaperch . . . . *Embiotoca lateralis*
surf smelt . . . . *Hypomesus pretiosus*

Tahoe sucker . . . . *Catostomus tahoensis*
tessellated darter . . . . *Etheostoma olmstedi*
threespine stickleback . . . . *Gasterosteus aculeatus*
topsmelt . . . . *Atherinops affinis*
torrent sculpin . . . . *Cottus rhotheus*
tui chub . . . . *Gila bicolor*

walleye surfperch . . . . *Hyperprosopon argenteum*
western brook lamprey . . . . *Lampetra richardsoni*
western mosquitofish . . . . *Gambusia affinis*
westslope cutthroat trout . . . . *Oncorhynchus clarki lewisi*
white catfish . . . . *Ameiurus catus*
white perch . . . . *Morone americanus*
white seaperch . . . . *Phanerodon furcatus*
white sturgeon . . . . *Acipenser transmontanus*
white sucker . . . . *Catostomus commersoni*
whitespotted greenling . . . . *Hexagrammos stelleri*
wolf-eel . . . . *Anarrhichthys ocellatus*

yellow bullhead . . . . *Ameiurus natalis*
yellow perch . . . . *Perca flavescens*

Yellowstone cutthroat trout .... *Oncorhynchus clarki bouvieri*
yellowtail rockfish .... *Sabastes flavidus*

Other Animals

bald eagle .... *Haliaeetus leucocephalus*
beaver .... *Castor canadensis*
black tern .... *Chlidonias niger*

Canadian lynx .... *Lynx canadensis*
common loon .... *Gavia immer*
Cumberland bean mussel .... *Villosa trabalis*

deer .... *Odocoileus* spp.

elk .... *Cervus elaphus*

ferruginous hawk .... *Buteo regalis*

great egret .... *Casmerodius albus*
green sea turtle .... *Chelonia mydas agassiz*
gray wolf .... *Canis lupus*

harlequin duck .... *Histrionicus histrionicus*

light-footed clapper rail .... *Rallus longirostris levipes*
lion .... *Panthera leo*

mottled duck .... *Anas fulvigula*

northern goshawk .... *Accipiter gentilis*
northern spotted owl .... *Strix occidentalis caurina*
northwestern pond turtle .... *Clemmys marmorata marmorata*

olive-sided flycatcher .... *Contopus borealis*
osprey .... *Pandion haliaetus*

ring-necked duck .... *Aythya collaris*

snail kite .... *Rostihamus sociabilis plumbeus*
snowy egret .... *Egretta thula*

tan riffleshell mussel .... *Epioblasma florentina walkeri*
tiger .... *Panthera tigris*
tsetse fly .... *Glossin* spp.

white-faced ibis .... *Plegadis chihi*
white-tailed deer .... *Odocoileus virginianus*
wolf .... *Canis* spp.
wood duck .... *Aix sponsa*
wood stork .... *Mycteria americana*

zebra mussel .... *Dreissena polymorpha*

Plants

alders . . . . *Alnus* spp.
American beech . . . . *Fagus grandifolia*
American sycamore . . . . *Platanus occidentalis*

Baltic rush . . . . *Juncus balticus*
basin wildrye . . . . *Elymus cinereus*
bebb willow . . . . *Salix bebbiana*
big bluestem . . . . *Andropogon gerardi*
big sagebrush . . . . *Artemisia tridentata*
birches . . . . *Betula* spp.
bitterbrush . . . . *Purshia tridentata*
black cottonwood . . . . *Populus trichocarpa*
black hawthorne . . . . *Crataegus douglasii*
black sagebrush . . . . *Artemisia arbuscula nova*
black walnut . . . . *Juglans nigra*
black-eyed Susan . . . . *Rudbeckia hirta*
bluebunch wheatgrass . . . . *Agropyron spicatum*
also *Pseudoroegneria spicata*
bog birch . . . . *Betula glandulosa*
bottlebrush squirreltail . . . . *Sitanion hystrix*
bulrushes . . . . *Scirpus* spp.
bur oak . . . . *Quercus macrocarpa*

camas . . . . *Camassia quamush*
cattails . . . . *Typha* spp. (including *T. glauca*)
chokecherry . . . . *Prunus virginiana*
coastal redwood . . . . *Sequoia sempervirens*
common chokecherry . . . . *Prunus virginiana*
cottonwoods . . . . *Populus* spp.
currant . . . . *Ribes* spp.

Douglas fir . . . . *Pseudotsuga menziesii*
duckweed . . . . *Lemna* spp.

eastern hemlock . . . . *Tsuga canadensis*
eastern redcedar . . . . *Juniperus virginiana*
Engelmann spruce . . . . *Picea engelmannii*

fescues . . . . *Festuca* spp.

Geyer willow . . . . *Salix geyeriana*
gray dogwood . . . . *Cornus racemosa*
gray-headed coneflower . . . . *Ratibda pinnata*
greasewood . . . . *Sarcobatus vermiculatus*
green ash . . . . *Fraxinus pennsylvanica*

hazel . . . . *Corylus americana*

Howell's gumweed .... *Grindelia howellii*
hydrilla .... *Hydrilla verticillata*

Idaho fescue .... *Festuca idahoensis*
Indian ricegrass .... *Oryzopsis hymenoides*
Indiangrass .... *Sorghastrum nutans*

leafy spurge .... *Euphorbia esula virgata*
limber pine .... *Pinus flexilis*
lodgepole pine .... *Pinus contorta*
low sagebrush .... *Artemisia arbuscula*

maples .... *Acer* spp.
mountain mahogany .... *Cercocarpus ledifolius*

Nanking cherry .... *Prunus tomentosa*
nannyberry .... *Viburnum lentago*
narrowleaf cottonwood .... *Populus augustifolia*
needlegrasses .... *Stipa* spp.
ninebark .... *Physocarpus opulifolius*
northern red oak .... *Quercus rubra*

oaks .... *Quercus* spp.
oatgrasses .... *Danthonia* spp.

ponderosa pine .... *Pinus ponderosa*
poplars .... *Populus* spp.
Port Orford cedar .... *Chamaecyparis lawsoniana*
purple prairie clover .... *Dalea purpurea*

quaking aspen .... *Populus tremuloides*

rabbitbrushes .... *Chrysothamnus* spp.
red osier dogwood .... *Cornus sericea*
Rocky Mountain juniper .... *Juniperus scopulorum*
rough fescue .... *Festuca scabrella*
Ruth's golden aster .... *Pityopsis ruthii*

saltgrass .... *Distichlis stricta*
sandbar willow .... *Salix interior*
Saskatoon serviceberry .... *Amelanchier alnifolia*
sedges .... *Carex* spp.
serviceberries .... *Amelanchier* spp.
silver maple .... *Acer saccharinum*
slender wheatgrass .... *Elymus trachycaulus*
sodar streambank wheatgrass .... *Agropyron riparium*
spikerushes .... *Eleocharis* spp.
spotted knapweed .... *Centaurea maculosa*
spruces .... *Picea* spp.
swamp white oak .... *Quercus bicolor*

switchgrass . . . . *Panicum virgatum*

tanoak . . . . *Lithocarpus densiflorus*
thickspike wheatgrass . . . . *Elymus lanceolatus*
Thurber needlegrass . . . . *Stipa thurberiana*
tubercled orchid . . . . *Platanthera flava*

western hemlock . . . . *Tsuga heterophylla*
western redcedar . . . . *Thuja plicata*
wheatgrasses . . . . *Agropyron* spp.
wild rose *also* woods rose . . . . *Rosa woodsii*
willows . . . . *Salix* spp.

yellow toadflax . . . . *Linaria vulgaris*

Index

About the Editors

Jack E. Williams is a Senior Scientist for the U.S. Bureau of Land Management stationed in Boise, Idaho. Prior to his assignment in Boise, Dr. Williams served as Science Advisor to the Director of the U.S. Bureau of Land Management, and as National Fisheries Program Manager. Previously, he worked for the Endangered Species Program of the U.S. Fish and Wildlife Service in California, and as a Visiting Scholar at the University of California-Davis. In 1992, he was awarded the Conservation Achievement Award by the Pacific Rivers Council and the Conservation Award for Communications from Trout Unlimited for his work to save salmon in the Pacific Northwest. He also was awarded the Trout Unlimited President's Award for Excellence in 1994. Dr. Williams completed undergraduate and graduate work at Arizona State University and the University of Nevada-Las Vegas prior to receiving his doctorate in fisheries science from Oregon State University.

Christopher A. Wood is a Special Assistant to the Chief of the U.S. Forest Service in Washington, D.C. Previously he was a Policy Analyst for the U.S. Bureau of Land Management where he worked on strategic planning and external affairs and was a leading authority for the agency on ecosystem management and watershed health. Previously, Mr. Wood worked for the U.S. Forest Service in Boise, Idaho, and for American Rivers in Washington, D.C. He is an avid fly fisher and has even been known to stalk *Cyprinus carpio* in the Potomac River. Mr. Wood received his undergraduate degree from Middlebury College in Vermont.

Michael P. Dombeck is Chief of the U.S. Forest Service in Washington, D.C. Until recently he served as the Acting Director of the U.S. Bureau of Land Management, and before that Dr. Dombeck served as Chief of Staff for the Assistant Secretary of the Interior, Science Advisor to the Director of the U.S. Bureau of Land Management, and National Fisheries Program Manager for the U.S. Forest Service. Prior to coming to Washington, D.C., he spent 12 years with the U.S. Forest Service primarily in the West and Midwest. He is an accomplished angler and serves on the Board of Governors for the Freshwater Fishing Hall of Fame. Dr. Dombeck completed undergraduate and graduate work at the University of Wisconsin-Stevens Point and the University of Minnesota prior to receiving his doctorate in fisheries biology from Iowa State University.